WHERE MATHEMATICS, COMPUTER SCIENCE,
LINGUISTICS AND BIOLOGY MEET

Where Mathematics, Computer Science, Linguistics and Biology Meet

Essays in honour of Gheorghe Păun

Edited by

Carlos Martín-Vide
Rovira i Virgili University,
Tarragona, Spain

and

Victor Mitrana
University of Bucharest,
Bucharest, Romania

A C.I.P. Catalogue record for this book is available from the Library of Congress.

ISBN 978-90-481-5607-8

Published by Kluwer Academic Publishers,
P.O. Box 17, 3300 AA Dordrecht, The Netherlands.

Sold and distributed in North, Central and South America
by Kluwer Academic Publishers,
101 Philip Drive, Norwell, MA 02061, U.S.A.

In all other countries, sold and distributed
by Kluwer Academic Publishers,
P.O. Box 322, 3300 AH Dordrecht, The Netherlands.

Printed on acid-free paper

Contents

Preface

In the last years, it was observed an increasing interest of computer scientists in the structure of biological molecules and the way how they can be manipulated in vitro in order to define theoretical models of computation based on genetic engineering tools. Along the same lines, a parallel interest is growing regarding the process of evolution of living organisms. Much of the current data for genomes are expressed in the form of maps which are now becoming available and permit the study of the evolution of organisms at the scale of genome for the first time.

On the other hand, there is an active trend nowadays throughout the field of computational biology toward abstracted, hierarchical views of biological sequences, which is very much in the spirit of computational linguistics. In the last decades, results and methods in the field of formal language theory that might be applied to the description of biological sequences were pointed out. Let us briefly provide an example. The structural representation of the syntactic information used by any parsing algorithm is a parse tree, which would appear to any biologist to be a reasonable representation of the hierarchical construction of a typical gene. We can fairly ask, then, to what extent a grammar-based approach could be generalized. Moreover, is such an approach suitable to be used for computing?

It may be assumed that the distinction between the structural and the functional or informational view of biological sequences corresponds to the conventional (and controversial) one drawn between syntax and semantics of natural languages. The functional view will allow us to expand our horizons beyond the local phenomena of syntactic structure to large regions of DNA. It appears crucial, in this respect, to define the semantics of DNA, which is mainly based on evolutionary selection, in order to linguistically reason about the processes of evolution as well as about computational capacity matters.

We may certainly say that a new field has emerged at the crossroads of molecular biology, linguistics and formal language theory. It can be considered as a young brother of another field born in the last decade, that of Bioinformatics. At the same time, such a field is a part of a set of biologically-inspired, formally rigorous new disciplines that shows close connections with linguistics and includes neural networks, genetic algorithms, etc.

Our volume has two goals. One is to present some recent results in active areas of the three domains that converge in the new field. The other one is to celebrate the 50th birthday of Gheorghe Păun, who, from formal language theory, promoted the new research area and made seminal contributions to it; his scientific career is simply exceptional from any point of view we may look at it, and all people having wished to contribute to this collective work are pride of having him as a friend.

The contributions are grouped into four sections, plus an introductory paper written by Solomon Marcus, who has been Gheorghe Păun's mentor for many years. The first section, Grammars and Grammar Systems, is devoted to grammatical formalisms. Classical aspects of formal grammar theory are presented, like regulated rewriting in connection with descriptional complexity or contextual grammars, as well as more recent ones as grammar systems and colonies, with motivations coming from robotic systems design. It includes also papers that use grammar formalisms for modeling aspects of natural languages, concurrency or data knowledge in multi-agent systems.

New trends in a traditional model of computation, that of Automata, are presented in the second section. The papers contained in it show that finite automata, transducers or Turing machines still pose lots of interesting problems. Deterministic iterated finite transducers, probabilistic automata, and distributed automata that cooperate under strategies similar to those met in the area of grammar systems are investigated from the viewpoint of their computational power. Furthermore, relevant issues in database theory are approached by means of automata.

The next section, Languages and Combinatorics, provides recent results about algebraic properties of languages, regulated rewriting in the framework of formal power series, combinatorics on words, sequences, graphs and language representations suggested by techniques widely utilized in the field of signal coding. Combinatorial algorithms are also discussed for problems that appear in some information processing activities.

The last section, Models of Molecular Computing, gives a glimpse of a new and attractive area in computer science that is looking for grammar models of computations based on DNA manipulation. One tries to provide an answer to a hot and quite exciting problem: do molecules compute (and how, if it is the case)? Both enthusiastic and rather pessimistic views are expressed with respect to the possibility of having 'wet' computers using DNA strands. Other molecules that have the potential capacity of computing are presented, too.

Most of the contributions deal with the basic operation of splicing, viewed as a very simple model of DNA recombination.

All the papers are contributed by Gheorghe Păun's collaborators, colleagues, friends and students in the five continents, who wanted to show in this way their recognition to him for his tremendous work. We have collected 38 papers by 75 authors here. (Another set of 38 papers by 65 authors will be published soon in the future.) The title of this volume intends to reflect the temporal sequence of Gheorghe Păun's scientific concerns along the time.

Summing up, this book makes an interdisciplinary journey from classical formal grammars and automata topics, which still constitute the core of mathematical linguistics, to some of their most recent applications in natural language processing and molecular computing.

The editors must stress M. Dolores Jiménez-López's crucial collaboration in the technical preparation of the volume as well as Kluwer's warm receptiveness from the very beginning. We hope this book will be a further step in the reviving of formal language theory as an essentially interdisciplinary field, and will be understood as an act of scientific justice and gratitude towards Gheorghe Păun.

Tarragona, July 2000

Carlos Martín-Vide
Victor Mitrana

Contributing Authors

Arasu, Arvind

Arroyo, Fernando

Asveld, Peter R.J.

Atanasiu, Adrian

Bălănescu, Tudor

Bartzis, Constantinos

Bel-Enguix, Gemma

Bonizzoni, Paola

Bottoni, Paolo

Bozapalidis, Symeon

Calude, Cristian S.

Calude, Elena

Castellanos, Juan

Ceterchi, Rodica

Choffrut, Christian

Ciobanu, Gabriel

Csuhaj-Varjú, Erzsébet

Dassen, Ray

Dassow, Jürgen

Dömösi, Pál

Drewes, Frank

Ehrenfeucht, Andrzej

Fernau, Henning

Ferretti, Claudio

Freund, Rudolf

Gheorghe, Marian

Goode, Elizabeth

Harju, Tero

Head, Tom

Holcombe, Mike

Hoogeboom, Hendrik Jan

Ibarra, Oscar H.

Ilie, Lucian

Ito, Masami

Karhumäki, Juhani

Kari, Jarkko

Kari, Lila

Kelemen, Jozef

Kelemenová, Alica

Kobayashi, Satoshi

Kreowski, Hans-Jörg

Krithivasan, Kamala

Kuich, Werner

Kutrib, Martin

Manca, Vincenzo

Marcus, Solomon

Margenstern, Maurice

Martín-Vide, Carlos

Mateescu, Alexandru

Mauri, Giancarlo

Mingo, Luis Fernando

Mitrana, Victor

Mráz, František

Păun, Andrei

Păun, Mihaela

Petre, Ion

Pixton, Dennis

Plátek, Martin

Polkowski, Lech

Prescott, David M.

Procházka, Martin

Rodrigo, José

Rogozhin, Yurii

Rozenberg, Grzegorz

Sakakibara, Yasubumi

Salomaa, Arto

Skowron, Andrzej

Staiger, Ludwig

Su, Jianwen

Svozil, Karl

Thierrin, Gabriel

Vugt, Nikè van

Yokomori, Takashi

Yu, Sheng

Zandron, Claudio

Chapter 1

THE GAMES OF HIS LIFE

Solomon Marcus
Mathematical Sciences
Romanian Academy
Calea Victoriei 125, 71102 Bucharest, Romania
solomon.marcus@imar.ro

I met first time Gheorghe Păun (G.P.) when he was a student in my class of mathematical linguistics. This happened 1973, at the Faculty of Mathematics, University of Bucharest. Despite the fact that I used to challenge the students with all kinds of questions, G.P. liked to remain silent during classes. He preferred to write down the solution, bringing it to me privately a few days later, instead of answering aloud, in the presence of all his colleagues. Later I realized that this is characteristic of the man; even today he is a little shy and his natural gift is more for research and less for teaching; to be more exact, we should say that his gift for teaching is more visible in his published works than in his spoken presentations. Many scholars have claimed that their way of researching was always via teaching. This seems to be the general pattern. A famous example in this respect are the great French mathematicians (like Cauchy) in the 19th century, most of whose results appear in their textbooks. G.P. does not belong to this pattern, but to its opposite. His lectures, impregnated with suggestive ideas, metaphors and humor, are a result of his research activity. This is one of the reasons G.P. avoided teaching activities, although he spent many years at the University of Bucharest.

His attention to open problems is always alive. I remember the visits he paid to me in my office at the Mathematical Institute of the Romanian Academy, bringing written answers to some problems I raised. The first problem was concerned with the place, in Chomsky's hierarchy, of the language in alphabet $\{a\}$ containing exactly those powers of a whose exponent is a prime number. In my *Grammars and Finite Automata* (in Romanian, 1964) I showed that this language was not context-free and I asked whether it was context-sensitive.

C. Martin-Vide and V. Mitrana (eds.), Where Mathematics, Computer Science, Linguistics and Biology Meet, 1-10.

An affirmative answer was obtained by J. Friant, but his grammar was very complicated. A simplification of this grammar was proposed by B. Brodda. G.P. succeeded in obtaining a further simplification, by reducing both the number of non-terminals and the number of rules; the whole generative process became much simpler. Moreover, G.P. extended his approach by finding relatively simple context-sensitive grammars for other related non-context-free languages, where the property of n to be prime is replaced by $n!$, n^2, or by 2^n, or by n as the nth term in the Fibonacci sequence beginning with 1, 2, 3, 5,... These results became the object of his first published paper (in Romanian, 1974). It was the same year that G.P. graduated in Mathematics, in the branch Mathematical Computer Science.

His second published work (in Romanian, 1974) was concerned with contextual grammars. Besides the simple, the generalized and the selective contextual grammars I introduced in 1969, G.P. proposed two more types: linear contextual grammars and generalized selective contextual grammars. As in his preceding research, here G.P. again shows his ability to push the investigation as far as possible and in as many directions as possible. He looks for the generative capacity of all these devices, he establishes their place in Chomsky's hierarchy, and he proves that any generalized contextual language is context-sensitive. He goes further, by showing that any generalized contextual language is a finite index matrix language. But G.P. does not stop there. This paper, of almost 20 printed pages, follows a scenario typical of many further papers he has published over the years: finding, in any possible mathematical, computational, natural or social framework, some aspects leading to a finite alphabet and to some languages in it, and, in a next step, investigating the respective class of languages under all basic aspects: generative capacity, closure properties in respect to various possible operations, relations to other classes of languages, mainly those in the Chomsky hierarchy, decision problems, complexity problems, and counter examples showing the limits of some theorems. He became very interested in various measures of syntactic complexity; see his fourth and fifth published papers (in Romanian, 1975). Moreover, already in his first articles one can observe a feature which has remained a permanent interest for G.P. He was always interested in the open problems proposed by other authors and, in his turn, he always liked to propose open problems. To many of these problems he was able to give complete answers. The papers discussed above were already elaborated when G.P. was a student. Usually, in such situations, a teacher, acting as a guide, indicates to the student the appropriate bibliography. Most students do not mention this fact, as it is considered to be understood. G.P. did not proceed in this way. His genuine honesty, visible throughout his whole scientific career, always lead him to mention anyone who had proposed a problem for him, who had provided him with some bibliographic references (see, for instance, his fourth published paper) or who had suggested a helpful idea.

Moreover, we can conjecture that in many of his joint works his contribution is the main one.

Already in his first years as a researcher, G.P. extended his interest to the formal linguistic structure of some fields which appeared far from mathematics and computer science. In 1976, he published two articles related to economics, a paper related to the conversation process and another one concerning theatrical plays. In order to understand this variety of directions of interest, we need to look to G.P.'s biography. Despite his obvious gift for scientific research, G.P. was unable to get a position in a research institute or at a university, when he graduated in 1974. Between 1974 and 1978 he had to work in the computing center of a kind of engineering institute aiming to improve the computational background and competence of specialists in various branches of economics. Trying to combine this job with his scientific interests in formal language theory, G.P. began to study the formal language structure of various economic processes systematically. This fact explains why the topic of his PhD thesis in mathematics (1977) was *The Simulation of Some Economic Processes by Means of the Formal Language Theory*. This thesis became his first book (in Romanian) *Generative Mechanisms of Economic Processes* (1980). It is a pity that this work of both economic and mathematical-computational interest is not yet translated in English. In the seventies, a book by Maria Nowakowska, *The Language of Motivations and the Language of Actions*, proposed a praxiological viewpoint in the field of human action, using as a main tool the algebraic analytical approach we developed in our monograph Algebraic Linguistics (1967). Nowakowska's approach may have had some influence on G.P.'s interest in this direction, although G.P. directed his attention to the generative approach. In G.P.'s view, the generative approach is both theoretical and practical, because economic processes are seen as a result of a generative mechanism, so we can distinguish between economic competence and economic performance (in terms of an algorithmic process whose complexity can be measured).

It was at that moment that I became interested in knowing the person hidden behind the mathematician G.P. The son of peasants from the village of Cicăneşti, in the region of Argeş, in the central southern Romania, G.P. was guided during his childhood and his youth by the only things he inherited from his parents, who both died in 1999: a very clever mind, a genuine honesty and the passion for continuous work. To these qualities, G.P. added a deep curiosity to understand the how and why of things; for him, this took the form of a symbiosis between work and play. It is difficult to get tired when your work is almost identical to your play, your way of resting. This is one of the secrets of his prolific research activity: tens of books and hundreds of articles published in international research journals. G.P's slogan was to make research his main activity and this is the reason he avoided being engaged in teaching.

But this picture is not complete. G.P. is not the type of researcher satisfied to practice the definition-theorem-proof ritual. From the beginning he has shown himself to be a complete human being. He wanted to transform Blaise Pascal's opposition *esprit geometrique – esprit de finesse* to a harmonious relationship. At the beginning of his research career, he published a poem in prose about his fascination on entering the world of science in *Contribuţii ştiinţifice studenţeşti*, a students' scientific bulletin of the Faculty of Mathematics, University of Bucharest. His pleasure in making up and telling stories has often taken book form (*The Parallel Sphere*, stories, 1984, *The Generous Circles*, stories, 1988, *Nineteen Ninety-Four*, novel, 1983, a reply to George Orwell's *1984*, *Moving Mirrors*, novel, 1994. He has been a member of the Romanian Union of Writers for a long time. The scholar and the writer in G.P. came together in such books as *The Mathematical Spectacle*, 1983, and *Mathematics? A Spectacle!*, 1988. In 1978, G.P. left the engineering computing center, where he had been working since his graduation, and got a position in the Section of System Studies of the University of Bucharest, where I was the director. Successive administrative changes did not affect his position as a researcher and he was able to devote most of his time to what attracted him. But this happened partially due to his capacity to convert some problems he had to face in his new position into problems of the type he enjoyed. Despite his long stay at the University of Bucharest, he developed no teaching activities there. However, his scientific works became more widely known among the best students and some of them became his disciples.

In 1990, he moved to the Institute of Mathematics, Romanian Academy. But the years at the University had offered him the chance of experimenting in the connection between mathematics, computer science and the social fields. At the end of the seventies, he became a member of the Romanian team representing the University of Bucharest in the international project *Goals, Processes, and Indicators of Development* (GPID) proposed by the United Nations University (Tokyo). As a chief of this team, I had the opportunity to observe G.P.'s contribution in this respect. GPID was a huge project, bringing together representatives from about 30 countries and specialists from virtually all fields of research. The meetings of this project were organized in various countries involved in the project, but G.P. was unable to attend them, due to passport and visa restrictions. It happened that two meetings took place in Bucharest (Romania), which gave G.P. the occasion to lecture in the presence of the representatives of tens of countries. I will try to describe his main contribution in this respect.

The work of G.P. concerning the global problems of humanity revealed his capacity to extract a mathematical problem from a non-mathematical phenomenon, impregnated with all kinds of empirical facts. One of the results he obtained in this field deserves a special attention, because it is of a very singular nature and shows how G.P. can use his imagination to go beyond the standard mathematical approach. It concerns a central problem in the field of social indicators; how can we aggregate several individual indicators into a synthetic one? As a matter of fact, such operations are performed almost everyday, but in an implicit way and by empirical rather than logical means. However, as the number of indicators increases and they become more complex, empirical aggregations are less and less acceptable. Social indicators become more and more heterogeneous. Imagine, for instance, the sheer diversity and the number of indicators accounting for the quality of life; they refer to health, to food, to shelter, to ecological, psychological and social factors, to education etc. In turn, indicators for shelter, for example, are numerous, as are those for food, for health insurance, for education etc. An indicator synthesising some more specific indicators will obviously be an approximation, a compromise; it will not be able to retain all the information contained in the specific indicators; but we can try to diminish the effect of the loss of this information as much as possible.

As a point of departure, G.P. took K. Arrow's famous theorem and showed the difficulty of aggregating several individual choices into a global one. Arrow formulates five natural requirements that such an aggregation should satisfy and shows that any possible aggregation fails to satisfy all of them. For this reason, Arrow's theorem is called the impossibility theorem. (As a matter of fact, is not Gödel's incompleteness theorem such an impossibility theorem, as well? It shows how it is impossible for a sufficiently comprehensive formal system to fulfill both the requirement of consistency and the requirement of completeness.) G.P. proceeds similarly in the case of indicators; but due to their possible variety, the problem approached by G.P. is of a completely different nature and this is the reason why G.P.'s proof has nothing in common with Arrow's proof.

The first requirement formulated by G.P. is a *sensibility* requirement. If the particular indicators improve, then the aggregated indicator should improve, too (for instance, if indicators are numerical, then when the particular indicators increase in value, the aggregated indicator increases in value, too). The second requirement formulated by G.P. is that the aggregation should be *anti-catastrophic*: when small modifications are made to the particular indicators, there should be a correspondingly small modification of the synthetic indicator (this requirement receives a rigorous treatment in G.P's approach; as a matter of fact, one can observe its analogy with the requirement of stability, in the theory of differential equations and with the anti-chaotic requirement, in the theory of non-linear dynamic systems). The third requirement is that aggregation should not be excessively compensatory: for instance, with respect to the indicators of the quality of life, one cannot accept that the deterioration in food could be compensated by some improvements in the work conditions. G.P. proves that any possible aggregation fails to satisfy all these three requirements.

I remember the international GPID session in Bucharest, when G.P. gave a lecture presenting the above theorem. It made such a powerful impact that the chairman of the session, the well-known political scientist Johan Galtung, picked up a flute and played a tune to celebrate the event. Both the statement and the proof of G.P.'s theorem outdid the expectations of the participants, most of them well-known specialists in social sciences. G.P.'s interest in these problems, at the point where the important fields of choice theory and of multicriterial decisions meet today, lead him to write a book (in Romanian): *The Paradoxes of Hierarchies* (1987). This book deserves to be translated in English, too.

You cannot understand G.P. as a person and as a researcher if you do not pay attention to his passion for games of any sort. He likes to make reference to Schiller's remark: 'You are not a complete human being if you don't like to play, from time to time.' In 1982, during a visit in Japan, I was given a beautiful game of GO, the national Japanese strategic game, with the corresponding documentation. My Japanese colleagues invited me to introduce GO into Romania. Back in Bucharest, I felt I could not carry out this task and I tried to transfer it to G.P., taking into account his interest in games. Now I can say that, in doing this, I was very inspired. Combining his cleverness with his passion, G.P. succeeded not only in learning to play GO successfully, but also to become the main animator of GO in Romania, by founding the Romanian GO Association and by organizing the first competitions of GO between various Romanian cities. All these activities became possible and were stimulated by the publication of two books (in Romanian) written by G.P.: *Introduction to GO* in three successive editions (1985, 1986, 1988) and *250 GO Problems* (1986, 1988). If the highest reward for a teacher is to produce a pupil who becomes better than him, then G.P. has achieved this. Relatively recently, one of the Romanian GO players became an European champion.

In other books written in Romanian and dedicated to games (Between Mathematics and Games, 1986, Solutions to 50 Solitary Logical Games, 1987, Competitive Logical Games 1989), G.P. added some original games to various others proposed by other authors. Stressing the strategic and logical aspects of games, these books are excellent tools for developing what is usually called the mathematical way of thinking. It is usually assumed that this style of thinking is incorporated as an obligatory, unavoidable component of the mathematical curriculum and of the mathematical textbooks. Unfortunately, this is not true. To a large extent, the mathematics taught to children and teenagers consists of manipulation of ready-made formulas, equations and algorithms (what the Americans call cook-book mathematics). However, everyday life is full of situations requiring a mathematical way of thinking, but in the absence of any mathematical jargon. G.P.'s books on games are a splendid illustration of this possibility.

Many of his games deserve to be introduced to mathematical textbooks for children aged between 10 and 19 years. The capacity to observe a situation in its totality, including both local and global aspects, to proceed step by step, by distinguishing various possible cases and sub cases; to develop some combinatorial and analogical arguments; to check the deductive consequences of some premises are all involved in most logical and strategic games. Obviously, G.P. is familiar with the work of the great masters of mathematical games, such as M. Gardner, and he uses and develops their approach. But he also adds something specific, related to his passion for words over a given finite alphabet, which lies behind most of his research work. There are many bridges between the games he is proposing and some contemporary mathematical-computational fields, such as combinatorics, graph theory, formal language theory, picture grammars, combinatorial geometry, computational geometry and cryptography.

In this respect, let us refer to the most recent book published by G.P. (December 1999): *The Theory of the Match* (in Romanian: Teoria chibritului). The title is full of humor, but only for those who are introduced in the subtleties of the Romanian language; *teoria chibritului* is an ironic idiomatic expression meaning excessive speculation around trivial things. In contrast with this popular saying, G.P. takes the opposite approach and claims that, far from being a trivial object, a match may become an extremely rich source of intellectual activities. Implicitly, he considers the alphabet of only one element, a match, and various bidimensional grammars on this alphabet: every game with its grammar. G.P.'s aim is to try to obtain as much as possible by means of as little as possible. Problem 1 proposes a square formed with four matches and requires the formation of a construction consisting of three squares by a suitable addition of two more matches; starting from the same square as in Problem 1, Problem 2 asks how we can obtain a construction of three squares, by a suitable addition of four more matches. Step by step, the problems become more and more difficult.

I will now describe what happened to me a few days ago with one of these problems. I was invited to attend a mathematical competition for school students aged between 13 and 18 years, selected from those who had won some local competitions in various cities of Romania. The school teachers I met there brought problems which were typical of the existing textbooks, based mainly on calculations. I wanted to change these habits to some extend and I brought with me three problems of a different type: one of the spot-the- mistake type? (for students aged 13), a second one where students of age 17 were asked to establish the class of real functions of a real variable satisfying the condition of uniform continuity (this terminology was not used there), with the only modification that $|x-y|$ was replaced by $x-y$ and a third problem, for students aged 15, which was in fact Problem 19 from G.P.'s book: Starting from a square formed by three matches on each side, consider the nine small squares where the existing matches are involved as sides, by adding 12 more matches.

We get a total of 24 matches leading to a construction of 14 squares: one whose sides are three long, four whose sides are two long and nine whose sides are one long (in the hypothesis that the length of one match is choosen as the unit of length). The candidates were asked to determine the minimum number of matches that should be deleted in order to get a construction with no squares, irrespective of the length of the sides. The result of this experiment was surprisingly negative; almost no candidate was able to propose a complete solution. This happened because such problems, far from being really difficult, go against the training received by high school students during their education and the existing curricula and textbooks. As a matter of fact, some of the teachers objected that the proposed problems did not refer explicitly to the curriculum of the class to which the candidates belonged. They did not realize that what remains in the intellectual habits of a student, when all the specific formulas are forgotten, is just a way of organizing our thinking. In most cases, life is not a direct application of some ready-made algorithms; it requires habits of a more general relevance, concerning an ordered, step by step procedure, whose details need to be discovered in each separate, specific case. Otherwise, in the absence of the explicit formulas or equations learned in school, we get lost, just as happened to the candidates at the above-mentioned competition.

What is the difference between G.P.'s research work and his writings on games? Is it what is traditionally considered to be the difference between work and play? More than hundred years ago, Lewis Carroll, in the Introduction to his *Symbolic Logic and the Game of Logic*, wrote: "[...] when you have made yourself a first-rate player at any one of these Games, you have nothing real to *show* for it, as a *result*! You enjoyed the Game, and the victory, no doubt, *at the time*; but you have no *result* that you can treasure up and get real *good* out of". In contrast with the gratuitousness of the logical game, Lewis Carroll stresses the utility of symbolic logic as a scientific discipline: "Once [you] master the machinery of Symbolic Logic, [...] you have a mental occupation always at hand, of absorbing interest, and one that will be of real *use* to you in *any* subject you may take up."

Is this observation valid today? The answer seems to be rather negative. On the one hand, logical-strategic games became a field of scientific inquiry in the 20th century, the player being the first who tries to learn the corresponding rational behavior; on the other hand, the game component of any scientific research, once only implicit and marginalized, is today more and more explicit and central. The border between gratuitousness and usefulness is more and more ambiguous, fuzzy, and sometimes requires long historical periods in order to be operable. The more the researcher follows his natural curiosity and his spontaneous need to understand the world, the higher will be the chance that his findings could lead to useful results. So, psychologically, the most advantageous attitude of a researcher is that of a player, even if we want to obtain some practical results, more than some cognitive progress. G.P. practices this work-game interplay with talent and with passion. For him it is a way to rest, when he gets tired of one or other of these activities. The situation is perfectly symmetric, thus different from the vision proposed by Lewis Carroll (idem, ibid.): "Mental recreation is a thing that we all of us need for our mental health; and you may get much healthy enjoyment, no doubt, from Games [...]". Maybe, we could ask G.P. to annotate his games in the way Martin Gardner annotated *Alice in Wonderland*; moreover, we could ask him to annotate his research works, by pointing out their game moments. Maybe, his readers could take up this task on their own.

Until 1989, in absence of the most fundamental human rights in Romania, games were for G.P. the favored intellectual refuge, where his freedom was total. After December 1989, his life changed completely. With his new position at the Institute of Mathematics of the Romanian Academy, consisting exclusively of research duties, he became an international traveller. G.P. began to explain his ideas and results to his colleagues in various universities around the world. The interaction with his listeners and his colleagues, his capacity for intellectual seduction made him very attractive for joint research work. Many authors became absorbed by G.P.'s problems, ideas and results. The high impact of his personality is illustrated by more than a thousand works by other authors, which quote, use and develop G.P.'s works further. This increasing impact is due to the fact that G.P. is not only the author of many concepts and theorems; he is also author or co-author of some new theories, leading to new branches in theoretical computer science: grammar systems, splicing as a theoretical counterpart of DNA computing, computing with membranes, internal contextual grammars are only some of them.

An important event in G.P.'s life was, in the nineties, the meeting with Professor Arto Salomaa, from the University of Turku, Finland. The interaction with the leading figure of the formal language theory extended the scientific horizon of many Romanian researchers in theoretical computer science. Turku became the place where Romanian computer scientists, particularly G.P. and his disciples, found continuous support. Many of these Romanians became, in their turn, successful partners of Professor Salomaa. Particularly, G.P. became the third vertex in a triangle where the two other vertices were already brothers: Grzegorz Rozenberg and Arto Salomaa. The culminating moment of this collaboration was the year 1998, when these three authors published the monograph *DNA Computing; New Computing Paradigms* (Springer), already translated into Japanese (in 1999). Let us also mention the three-volume *Handbook of Formal Languages* (eds. G. Rozenberg, A. Salomaa) published in 1997, where the Romanian contribution and particularly the chapters written by G.P., by Cristian Calude, by Alexandru Mateescu and by the author of this article make an important contribution.

Already in the eighties, the Romanian philosopher Constantin Noica suggested that G.P. apply for a Humboldt scholarship. The application was successful, but travelling to Germany was impossible, due to passport and visa restrictions. Only in 1992 did G.P. have the opportunity to begin to benefit from this prestigious scholarship, due to the scientific support offered by Professor Jürgen Dassow (Magdeburg Univ.). The Dassow-Păun collaboration, mainly known from their joint monograph, remains very important today.

1995 saw the development of a new center of interest for G.P.: Professor Carlos Martín-Vide, University of Tarragona, Spain, began to offer hospitality to G.P., his team and his colleagues and Tarragona became the place where many joint works by G.P., Martín-Vide and others were elaborated. In the same place, G.P. finished his monograph *Marcus Contextual Grammars* (Kluwer, 1997). Other authors, such as E. Csuhaj-Varjú, R. Freund, T. Head, M. Ito, J. Kelemen, A. Kelemenová, P. Bottoni, T. Yokomori have also been involved in important collaborations with G.P. The list could be continued and includes many Romanian co-authors.

In 1997, G.P. was elected as a member of the Romanian Academy.

Already in the early eighties, it was widely felt that formal languages had exhausted their relevance for theoretical computer science, because after the results of Ginsburg and Floyd, related to the place of programming languages in Chomsky's hierarchy, nothing else happened. A similar feeling existed among linguists; the idea that natural languages were not context-free considerably diminished the interest of linguists in formal grammars and languages. This attitude proved to be wrong for both computation and the study of natural languages and G.P. has made a considerable contribution in this change of opinion. The study of economic processes, of L systems and their applications to computer graphics, the importance of different classes of CS languages that are not CF, of the syntax and semantics of programming languages, the role of different types of word complexity, the importance of formal language techniques in the field of cryptography and in the study of splicing as a basic approach to DNA computing, the rediscovery of the relevance of various classes of generative grammars (including contextual grammars, too) in the field of computational linguistics, the role of infinite words in the study of some discrete dynamical systems, in connection with the properties of almost periodicity and quasi-periodicity are only a few indications of the continuing relevance of the field to which G.P. has devoted his research.

So, what of G.P.'s life? He married 25 years ago to a beautiful girl from his own village, who became a teacher of chemistry in a Bucharest high school. They have two sons, the older graduated in mathematical computer science, while the younger will graduate soon in economics. G.P. followed "the rules of the game". Let the unforeseeable games of his life continue, to his joy and to our joy!

I

GRAMMARS AND GRAMMAR SYSTEMS

Chapter 2

DETERMINISTIC STREAM X-MACHINES BASED ON GRAMMAR SYSTEMS

Tudor Bălănescu
Department of Computer Science
Faculty of Sciences
Piteşti University
Str. Targu din Vale, Piteşti, Romania
balanesc@oroles.cs.unibuc.ro

Marian Gheorghe
Department of Computer Science
Faculty of Sciences
Piteşti University
Str. Targu din Vale, Piteşti, Romania
marian@oroles.cs.unibuc.ro

Mike Holcombe
Department of Computer Science
Sheffield University
Regent Court, 211, Portobello Street, Sheffield S1 4DP, England
M.Holcombe@dcs.shef.ac.uk

Abstract This paper investigates the power of the stream X-machines based on cooperating distributed grammar systems which replace relations by functions. The case of regular rules is considered. The deterministic case is introduced in order to allow the application of some already developed testing strategies. Some conditions for getting equivalent stream X-machines in the $= 1$ derivation mode are studied.

C. Martin-Vide and V. Mitrana (eds.), Where Mathematics, Computer Science, Linguistics and Biology Meet, 13-23.

1. INTRODUCTION

The use of formal methods for the specification and verification of software and the development of sophisticated methods of software and system testing are two current areas of emphasis in the development of higher quality software. Testing issues are very seldom addressed by those within the formal methods community and very few of the testing methods are based on a firm scientific approach or theory. In [13], [10] a formal method based on X-machines is provided as a more convincing approach to the problem of detecting *all faults*. This method extends an approach developed by Chow [6] for a very restrictive model, namely the finite state machine. The model based on Eilenberg's X-machine concept [8] considers the control structure, represented by a finite state machine, a data or memory structure, as well as some processing functions at the same time. The model based on X-machines is a convenient tool for specifying real systems. More recently, the communicating stream X-machines as a model for specifying the system having components running in parallel has also been addressed [5], [4].

In this paper we will concentrate on the use of formal cooperating distributed grammar systems (CDGS for short) [7] to define a refinement of the processing relations/functions. The power of these mechanisms has been investigated already [3]. In this context we will go on to study the behaviour of the X-machines which have processing functions expressed as formal grammars under the usual testing requirements [13], namely the deterministic stream X-machines based on CDGS. The power of these mechanisms is studied when relations, provided in [3], are replaced by functions. The deterministic case is investigated for regular rules and $= 1$ derivation mode. Some conditions for getting equivalent deterministic stream X-machines are also studied.

By coupling a stream X-machine with a set of formal grammars we get a new machine called an X-machine based on CDGS (Xmdg for short) which acts as a translator. Two types of input sets are considered, finite and regular languages. Regular rules and the usual derivation strategies studied for CDGS, $\leq k$, $= k$, $\geq k$, $*$ and t [7] are considered.

Our approach combines two models, CDGS [16] and X-machines [8], developed as topics in the frame of formal language theory and applied to artificial intelligence [7], [15], and for the formal specification and verification of complex software systems [9], [10], respectively.

2. BASIC DEFINITIONS

For a set Q we denote by 2^Q the family containing all subsets of Q. Let Σ be a finite alphabet; Σ^* is the set of all strings having symbols from Σ. By λ we denote the *empty string*. The length of a string s is denoted by $\mid s \mid$.

Let P be a set of productions over a set of *nonterminals* N and a *terminal* alphabet A. We define

$$dom(P) = \{A \mid A \in N \text{ and } \exists A \longrightarrow \alpha \in P\}.$$

By $\Longrightarrow_P$ we denote the usual derivation in P. We also define the following derivations $\Longrightarrow_P^{=r}$, $\Longrightarrow_P^{*}$, $\Longrightarrow_P^{\geq r}$, $\Longrightarrow_P^{t}$, $\Longrightarrow_P^{\leq r}$, where $r \geq 1$. Let $\alpha, \beta \in (N \cup A)^*$. Then we write $\alpha \Longrightarrow^{=r} \beta$ if and only if there are some words $x_0, \ldots x_r$ such that $\alpha = x_0$, $\beta = x_r$ and $x_j \Longrightarrow_P x_{j+1}$ for all $0 \leq j \leq r-1$. Moreover, we write $\alpha \Longrightarrow^* \beta$ when $\alpha = \beta$ or $\alpha \Longrightarrow^{=r} \beta$ for some r. We also write $\alpha \Longrightarrow^{\geq r} \beta$ when $\alpha \Longrightarrow^{=k} \beta$ for some $k \geq r$. By $\alpha \Longrightarrow_P^t \beta$ we understand $\alpha \Longrightarrow^* \beta$ and there is no derivation step $\beta = xBy \Longrightarrow_P xzy$ with $B \longrightarrow z \in P$. Note that if $x \in A^*$, then $x \Longrightarrow_P^t x$. We then have $\alpha \Longrightarrow_P^{\leq r} \beta$, if and only if there exists $\alpha_0, \ldots \alpha_m$, such that $\alpha_0 = \alpha, \alpha_m = \beta, 1 \leq m \leq r$ and $\alpha_i \Longrightarrow_P \alpha_{i+1}$ for all $i, 0 \leq i \leq m-1$. By DM we denote the set of all derivation modes, i.e.

$$DM = \{t, *\} \cup \{\leq r, = r, \geq r \mid r \geq 1\}.$$

The *regular rules* have the forms $B \longrightarrow a$ or $B \longrightarrow aC$, where $a \in A$ and $B, C \in N$.

The cooperating/distributed grammar system (CDGS) has been introduced first in [16] with motivations related to two level grammars. An intensive study of CDGS has been started after relating them to artificial intelligence concepts [7] such as the blackboard models in problem solving. A CDGS is a construct consisting of several usual grammars, working together on the same sentential form to generate words. Informally, such systems and their work can be described as follows (see [7]): initially, the axiom is the common sentential form. At any moment, one grammar is active; that means it rewrites the common string, while the others are not active. The language of terminal strings generated in this way is the language generated by the system. As basic stop conditions usually considered, we mention: each component, when active, has to work for exactly k, at least k, at most k, or the maximal number of steps (a step means the application of a rewriting rule). Many other stopping conditions were considered or added to the above mentioned one (see [7], [1], [2]).

CDGS has been also used for modelling the synchronization of various parallel processes sharing some common resources [1], [2].

Definition 2.1 *A cooperating/distributed grammar system is*

$$\Gamma = (N, A, S, P_1, \ldots P_n),$$

where N is the set of nonterminal symbols, A is the set of terminal symbols, S is the nonterminal start symbol, and $P_1, \ldots P_n$ are sets of production rules.

For $d \in DM$, the language generated by Γ using the d derivation mode is $L_d(\Gamma)$.

Definition 2.2 *Let us consider the following elements:*

- *A, a finite alphabet of terminal symbols;*
- *N, a finite set of nonterminal symbols with $A \cap N = \emptyset$;*
- *$P = \{P_1, \ldots P_n\}$ contains n sets of rules with symbols from $A \cup N$.*

A stream X-machine with underlying distributed grammars *(Xmdg, for short) is a system $X = (\Sigma, A, Q, M, P, F, I, T, m_0, N)$ where:*

- *$\Sigma \subseteq N \cup A$ is the input alphabet;*
- *A, the alphabet introduced above, is the output alphabet;*
- *Q is the finite set of states;*
- *$M = (N \cup A)^*$ is the* memory*;*
- *$m_0 \in N \cup A$ is the* initial *memory value;*
- *F is the* next state *partial function $F : Q \times P \to 2^Q$;*
- *I and T are the sets of* initial *and* final *states respectively.*

For a given derivation mode $d \in DM$ we can define a stream X- machine

$$X_d = (\Sigma, A, Q, M, \Phi_d, F, I, T, m_0),$$

where $\Phi_d = \{\phi_1, \ldots \phi_n\}$, and $\phi_i : M \times \Sigma \longrightarrow 2^{A^* \times M}$. Each ϕ_i is associated to P_i and it is defined by

$$(x, m') \in \phi_i(m, \sigma) \text{ if and only if } m\sigma \Longrightarrow^{d}_{P_i} xm'$$

and $\sigma \in \Sigma$, $x \in A^*$, $m, m' \in M$, but m' does not begin with a symbol of A. A computation *in X_d, denoted by $\vdash$, is defined as being*

$$P_i : (q, g, m, \sigma\sigma_1 \ldots \sigma_k) \vdash_{P_i} (r, gx, m', \sigma_1 \ldots \sigma_k)$$

if and only if

$$q, r \in Q, r \in F(q, P_i); g, x \in A^*; \sigma, \sigma_1, \ldots \sigma_k \in \Sigma$$

and

$$m\sigma \Longrightarrow^{d}_{P_i} xm',$$

where $m, m' \in M$ and m' does not begin with a symbol of A.

Definition 2.3 *For a given path $R_1 \ldots R_k : q_1 \to^* q_{k+1}$, we have a* computation closure

$$R_1 \ldots R_k : (q_1, o_1, m_1, \sigma_1 \ldots \sigma_k \alpha) \vdash^* (q_{k+1}, o_{k+1}, m_{k+1}, \alpha)$$

Note 2.1 *By f_d (see [13]) is denoted the relation computed by the stream X-machine X_d under the assumption that the* final memory value is empty. *It follows that $f_d(s)$ is the set of all $x \in A^*$ such that there exist $R_1, \ldots R_k$ with:*

$$R_1 \ldots R_k : (q, \lambda, m_0, s) \vdash^* (r, x, \lambda, \lambda), q \in I, r \in T, k \geq 1.$$

That is, for an input sequence $s \in \Sigma^*$ the set $f_d(s)$ contains all the output strings produced by the machine X_d starting from an initial state, *an empty output string and the* initial memory value m_0 *and leading to a* terminal state, *an* empty input *and an* empty memory value. *For a language L it is defined as $f_d(L) = \bigcup_{s \in L} f_d(s)$.*

Note 2.2 *Two stream X-machines are said to be equivalent if they produce the same output sequences when they receive the same input sequences.*

Note 2.3 *For a derivation mode $d \in DM$, we consider the families $DS_{d,n}REG$ containing all the stream X- machines X_d obtained from regular Xmdg mechanisms with n sets of rules.*

Let us remark that the elements of Φ_d are relations. *The families of stream X-machines X_d with the set Φ_d containing only functions are denoted by $FDS_{d,n}REG$.*

Note 2.4 *By FIN, REG are denoted the families of finite, regular languages respectively. For a given derivation mode $d \in DM$ and a family Y of languages over the input alphabet Σ we define*

$$DS_{d,n}REG(Y) = \{f_d(L) \mid L \in Y \text{ and } X_d \in DS_{d,n}REG\}.$$

When the considered stream X-machines have only functions *in their sets Φ_d, then the prefix F is systematically used. In every case the union of the corresponding families, for all n, is taken. The notation is obtained by replacing the index d, n by d.*

Definition 2.4 *Deterministic stream X-machine (see [13]). A stream X-machine*

$$M = (\Sigma, A, Q, M, \Phi, F, I, T, m_0)$$

is said to be deterministic if the following conditions hold:

1. *F maps each pair $(q, \phi) \in Q \times \Phi$ into at most a single next state, i.e. $F : Q \times \Phi \longrightarrow Q$. A partial function is used because every $\phi \in \Phi$ will not necessarily be defined as the label to an edge in every state.*

2. *I contains only one element, i.e.* $I = \{q_0\}$

3. *If* ϕ *and* ϕ' *are distinct arcs emerging from the same state, then* $dom(\phi) \cap dom(\phi') = \emptyset$

Definition 2.5 *Deterministic Xmdg. An Xmdg*

$$X = (\Sigma, A, Q, M, P, F, I, T, m_0, N)$$

is called deterministic if the following are true:

1. *the condition 1 from Definition 2.4 holds;*

2. *the condition 2 from Definition 2.4 holds;*

3. *If* P_i *and* P_j *are distinct arcs emerging from the same state, then* $dom(P_i) \cap dom(P_j) = \emptyset$.

Remark 2.1 *Any deterministic Xmdg (see definition 2.5) has all the associated stream X-machines* X_d, $d \in DM$, *deterministic according to the definition 2.4. But there are nondeterministic Xmdg having the associated stream X-machine* X_d, *a deterministic machine (see definition 2.4), for some* $d \in DM$. *For example, the following Xmdg:*

$$X_d = (\{a\}, \{a\}, \{1, 2, 3\}, \{a, S\}^*, P, F, \{1\}, \{2, 3\}, S, \{S\}),$$

where

$$P = \{P_1, P_2\},$$

with

$$P_1 = \{S \longrightarrow a\} \text{ and } P_2 = \{S \longrightarrow aS, S \longrightarrow a\};$$

$$F(1, P_1) = \{2\}, F(1, P_2) = \{3\};$$

is not deterministic according to the definition 2.5 (the domains of P_1 *and* P_2 *are not disjoint sets). But for* $d \in \{= k, \geq k \mid k \geq 2\}$, *the stream X-machine* X_d *is deterministic according to the definition 2.4.*

Remark 2.2 *For any stream X-machine (Xmdg) an equivalent one may be found fulfilling the first two conditions provided by definition 2.4 (2.5) by applying the usual method for getting an equivalent deterministic one from a finite state machine [11].*

3. MAIN RESULTS

In [3] the following results are proved:

Theorem 2.1 *The following relations hold:*

1. $DS_dREG(FIN) \subset FIN, d \in \{\leq k, = k \mid k \geq 1\}$;
2. $DS_dREG(FIN) \subset REG, d \in \{*, t\} \cup \{\geq k \mid k \geq 1\}$.

Theorem 2.2 *For* $d \in \{t, *\} \cup \{\leq k, = k, \geq k \mid k \geq 1\}$ *the family* $DS_dREG(REG)$ *contains non context-free languages.*

First let us consider the case of Xmdg mechanisms defining, for a given derivation mode d, stream X-machines with the set Φ_d containing only functions.

Lemma 2.1 $FDS_dREG(FIN) \subset FIN, d \in \{*, t\} \cup \{\geq k \mid k \geq 1\}$

Proof. For a derivation mode $d \in \{*, t\} \cup \{\geq k \mid k \geq 1\}$, a stream X-machine

$$X_d = (\Sigma, A, Q, M, \Phi_d, F, I, T, m_0),$$

and an input sequence $s = \sigma_1 \ldots \sigma_r$, $\sigma_i \in \Sigma$, gives the set of all successful paths, denoted by $Paths(s)$,

$$p : q_0 \longrightarrow^{\phi_{i_1}} q_1 \ldots q_{r-1} \longrightarrow^{\phi_{i_r}} q_r, q_0 \in I; q_r \in T; \phi_{i_j} \in \Phi_d,$$

such that there exists $g \in A^*$, with

$$\phi_{i_r} \ldots \phi_{i_1}(m_0, s) = (g, \lambda)$$

is always finite (see Note 2 in [3]). We also denote by $L_{p,d}(s)$ the set of all output sequences computed from s along the successful path $p \in Paths(s)$, using the derivation mode d. Taking into account that Φ_d contains only functions, it follows that $L_{p,d}(s)$ is finite and hence

$$f_d(s) = \bigcup_{p \in Paths(s)} L_{p,d}(s)$$

is finite. So, it follows that $FDS_dREG(FIN)$ is a finite set. For proving the strict inclusion, Lemma 4 in [3] may be applied. □

Theorem 2.3 $FDS_dREG(FIN) \subset FIN$, $d \in \{*, t\} \cup \{\leq k, = k, \geq k \mid k \geq 1\}$.

Proof. For $d \in \{\leq k, = k \mid k \geq 1\}$, Theorem 2.1, point 1, is used and for $d \in \{*, t\} \cup \{\geq k \mid k \geq 1\}$, Lemma 2.1 is applied. □

Similar to [3], non-context-free languages are obtained when regular languages are supplied to an Xmdg for some derivation strategies.

Theorem 2.4 *The families $FDS_dREG(REG)$, $d \in \{\leq k, = k, \geq k \mid k \geq 1\}$, contain non-context-free languages.*

Proof. Let us consider the Xmdg

$$X = (\Sigma, A, Q, M, P, F, I, T, m_0, N),$$

where

$$\Sigma = \{B, c\}; A = \{a, b, c\}; Q = \{1, 2, 3, 4, 5\};$$
$$M = \{x \mid x \in (A \cup N)^*\};$$
$$P = \{P_i \mid 1 \leq i \leq 6\}, \text{ and}$$
$$P_1 = \{S \longrightarrow a\}, P_2 = \{S \longrightarrow aD\}, P_3 = \{C \longrightarrow bB\},$$
$$P_4 = \{B \longrightarrow b\}, P_5 = \{B \longrightarrow bC\}, P_6 = \{D \longrightarrow aS\};$$
$$F(1, P_1) = \{2\}, F(1, P_2) = \{3\}, F(4, P_3) = \{1\},$$
$$F(2, P_4) = \{5\}, F(3, P_5) = \{5\}, F(5, P_6) = \{4\};$$
$$I = \{1\}; T = \{5\}; N = \{S, B, C, D\}.$$

For $d \in \{= 1, \geq 1\} \cup \{\leq k \mid k \geq 1\}$, it may be shown, for example, that $a^3b^3c^5$ may be obtained in X. Indeed

$$SB \Longrightarrow_{P_2} aDB \Longrightarrow_{P_5} aDbCc \Longrightarrow_{P_6} aaSbCcc$$

$$\Longrightarrow_{P_3} aaSbbBccc \Longrightarrow_{P_1} aaabbBcccc \Longrightarrow_{P_4} aaabbbccccc$$

For $d \in \{= 1, \geq 1\} \cup \{\leq k \mid k \geq 1\}$, the stream X- machine, X_d, computes the relation

$$f_d(\{Bc^{4n+1} \mid n \geq 0\} = \{a^{2n+1}b^{2n+1}c^{4n+1} \mid n \geq 0\},$$

which is not a context-free language. When $d \in \{= k, \geq k \mid k \geq 2\}$ the example above is slightly modified such that the language

$$\{Bc^{4n+1} \mid n \geq 0\} \text{ is translated into } \{a^{(2n+1)k}b^{(2n+1)k}c^{4n+1} \mid n \geq 0\},$$

also a non-context-free language. □

Now, consider the deterministic stream X-machines. First we state some properties of the deterministic Xmdg devices, or the associated stream X-machines, in the case = 1.

Theorem 2.5 *There exists $L \in FDS_{=1}REG(\{C^p\}), p \geq 1$, such that any stream X-machine*

$$X_{=1} = (\{C\}, A, Q, M, \Phi_{=1}, F, I, T, m_0),$$

with

$$\mid I \mid = 1, m_0 \in A, F : Q \times \Phi_{=1} \longrightarrow Q, f_{=1}(\{C^p\}) = L,$$

is not a deterministic stream X-machine.

Proof. In fact we will show that the third condition of Definition 2.4 is not satisfied. Let us consider $L = \{ca^p, cb^p\}$ and

$$X_{=1} = (\{C\}, A, Q, M, \Phi_{=1}, F, I, T, m_0),$$

$$I = q_0, m_0 = c, F : Q \times \Phi_{=1} \longrightarrow Q.$$

There exist $\phi_1, \phi_2 \in \Phi_{=1}$ and two states $q_1, q_2 \in Q$ such that

$$F(q_0, \phi_1) = q_1 \text{ and } F(q_0, \phi_2) = q_2.$$

It follows that

$$\phi_1(c, C) = (ca, \lambda), \phi_2(c, C) = (cb, \lambda),$$

as a consequence of the $= 1$ derivation mode. □

Some necessary conditions are settled in the next theorem for getting a deterministic Xmdg from a more general one in the case $= 1$ and using only regular rules.

Theorem 2.6 *Let $X = (\Sigma, A, Q, M, P, F, I, T, m_0, N)$ be an Xmdg having the following properties:*

1. *the sets of P contain only regular rules*
2. *$\Sigma \subseteq A$, $\mid I \mid = 1$, $F : P \times Q \longrightarrow Q$ and P_4 are sets of rules.*

Then there exists a deterministic Xmdg Y such that $Y_{=1}$ is equivalent with $X_{=1}$.

4. CONCLUSIONS

A mechanism for accepting classes of languages by using a generalized state machine to control a family of distributed grammars is described. The case where relations [3] associated to processing operations are replaced by functions is mainly studied. The deterministic case, in the sense of [10], is also investigated for the derivation mode $= 1$ and for regular rules. Some conditions for getting equivalent deterministic stream X-machines are defined and investigated. A number of problems remain to be studied further:

- the deterministic case for other derivation strategies,
- the context-free case for the rules defining the components,
- the application of the testing strategies [13] for stream X-machines in the case of the Xmdg mechanisms.

In a forthcoming paper (a part) of these problems will be addressed.

References

[1] Bălănescu, T.; H. Georgescu & M. Gheorghe (1997), Grammar Systems with Counting Derivation and Dynamical Priorities, in Gh. Păun & A. Salomaa, eds., *New Trends in Formal Languages:* 150–166. Springer, Berlin.

[2] Bălănescu, T.; H. Georgescu & M. Gheorghe (1999), A New Type of Counting Derivation for Grammar Systems, in Gh. Păun & A. Salomaa, eds., *Grammatical Models of Multi-Agent Systems:* 1–17. Gordon and Breach, London.

[3] Bălănescu, T.; H. Georgescu & M. Gheorghe, Stream X-machines with underlying distributed grammars, submitted to *Informatica.*

[4] Bălănescu, T.; T. Cowling; H. Georgescu; M. Gheorghe; M. Holcombe & C. Vertan (1999), Communicating Stream X-Machines Systems are no more than X-Machines. *Journal of Universal Computer Science* 5.9: 494–507.

[5] Barnard, J.; J. Whitworth & M. Woodward (1996), Communicating X-machines, *Journal of Information and Software Technology,* 38: 401–407.

[6] Chow, T.S. (1978), Testing software design modeled by finite-state machines, *IEEE Transactions on Software Engineering,* 4.3: 178–187.

[7] Csuhaj-Varju, E.; J. Dassow; J. Kelemen & Gh. Păun (1994), *Grammar Systems. A Grammatical Approach to Distribution and Cooperation,* Gordon & Breach, London.

[8] Eilenberg, S. (1974), *Automata, Languages and Machines,* Academic Press, New York.

[9] Holcombe, M. (1988), X-machines as a basis for dynamic system specification, *Software Engineering Journal,* 3.2: 69–76.

[10] Holcombe, M. & F. Ipate (1988), *Correct Systems: Building a Business Process Solution,* Springer, Berlin.

[11] Hopcroft, J.H. & J.D. Ullman (1979), *Introduction to Automata Theory, Languages and Computation.* Addison-Wesley, Reading, Mass.

[12] Ipate, F. & M. Holcombe (1996), Another look at computability, *Informatica,* 20: 359–372.

[13] Ipate, F. & M. Holcombe (1997), An Integration Testing Method That is Proved to Find all Faults, *International Journal of Computer Mathematics*, 63: 159–178.

[14] Ipate, F. & M. Holcombe (1998), A method for refining and testing generalized machine specifications, *International Journal of Computer Mathematics*, 68: 197–219.

[15] Kelemen, J. (1991), Syntactical models of distributed cooperative systems, *Journal of Experimental and Theoretical Artificial Intelligence*, 3: 1–10.

[16] Meersman, R. & G. Rozenberg (1978), Cooperating grammar systems, in *Proceedings of MFCS'78*: 364–374. Springer, Berlin.

Chapter 3

SOME GHOSTS THAT ARISE IN A SPLICED LINGUISTIC STRING: EVIDENCE FROM CATALAN

Gemma Bel-Enguix
Research Group on Mathematical Linguistics
Department of Romance Philologies
Faculty of Arts
Rovira i Virgili University
Pl. Imperial Tàrraco, 43005 Tarragona, Spain
gbe@astor.urv.es

Abstract DNA recombination methods used up to now to generate formal languages can also be applied to the manipulation of natural language because of the biological similarity between the two systems. But if we want to formalize natural language according to parameters of molecular combinations we find that its behaviour has specific features in the linguistic field. One of these distinctive characteristics is that in the result of the recombination of two linguistic sequences some new elements appear which were not involved in any of the initial strings. In this article we study the nature, regularities and behaviour of those elements, named ghosts, when carrying out a splicing.

1. INTRODUCTION

Let us consider a sentence. Although its physical performance refers us to the lineal production of sounds, we are actually dealing with a complex structure with several strata. Thus, when we deal with a sentence we must take into account not only the final string of terminal symbols, but the other levels that compound it. The first level would be the *pattern*. A simple sentence is considered as a pattern composed by three variables: S, V, O. This structure seems universal, but not the order of the variables, as they differ in different languages. In Catalan, the order is SVO.

C. Martín-Vide and V. Mitrana (eds.), Where Mathematics, Computer Science, Linguistics and Biology Meet, 25-35.

At this point, we define the *basic linguistic pattern* as the one composed by the variables S, V, O, noticing the following formation criteria:

- Precedence: SVO is an ordered pattern. For a SVO language such as Catalan, S precedes V and O. V precedes O.

- Simplicity: A basic pattern has only one occurrence for each variable.

Terminal strings are the last level in any linguistic production, obtained by means of the substitution of each one of the variables of the pattern for a string belonging to its domain. Therefore, a basic string is the result of replacing SVO, in this order, by terminal strings from their domains.

$\mathcal{S}$ denotes the domain of S, defined as the set of strings $s_1, s_2, ...s_n$ which can replace S. The domain of V, defined as the set of strings $v_1, v_2, ...v_n$ which can replace V, is denoted by $\mathcal{V}$. $\mathcal{O}$ means the domain of O defined as the set of strings $o_1, o_2, ...o_n$ which can replace O.

s, s' are strings belonging to $\mathcal{S}$. In Catalan $\lambda \in \mathcal{S}$. v, v' are strings belonging to $\mathcal{V}$. o, o' are strings belonging to $\mathcal{O}$. $\lambda \in \mathcal{O}$.

Let us now consider a sentence unfolded in:

- the basic pattern of the sentence with the variables S (subject), V (verb), O (object),

- terminal strings of linear production.

It can be represented as follows:

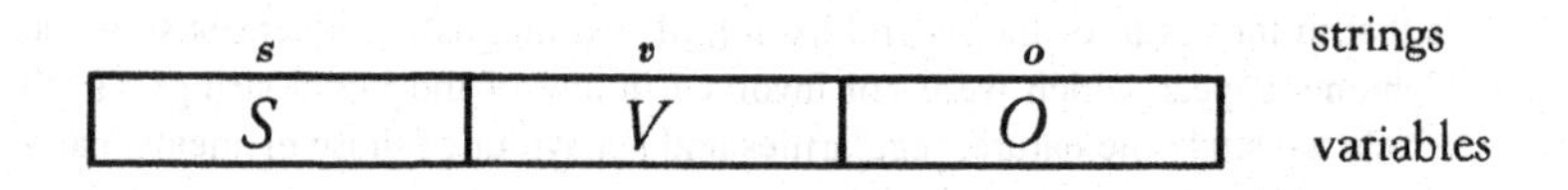

Figure 3.1 ULPS.

This representation with two strata of the same sentence leads to what we call *unfolded linguistic pattern structure* (ULPS). The first letters of the Greek alphabet, α, β, γ... are used to designate both ULPS and terminal strings.

1.1 RELATIONS AMONG SVO

Linguistic strings are different from the other ones because they contain relations among the diverse elements that conform them.

1.1.1 Relations between sv. s and v are subscripted to values of number (N) and person (P). N can be e (singular) or p (plural). P may be 1, 2 or 3. If the subscripts of s and v are the same we consider the structure to be correct. This implies that s and v must exist. We must remember that in Catalan $\exists\, s = \lambda$ because $\lambda \in S$. It is necessary to distingish between languages where $\lambda \in S$ and languages where $\lambda \notin S$. Catalan belongs to the first group, English to the second. We may represent a correct sentence in Catalan as either $s_{PN}v_{PN}o$ or $\lambda_{PN}v_{PN}o$, whereas in English only the first form would be right.

1.1.2 Relations between so. The relationship between these two elements rarely exist. There can only be a binding so if we take into account an element which does not have a very important role in any other agreements: gender. The gender G may be f (feminine) and m (masculine). In this case we find crossed–structures, where svo are bound by N; s and v are bound by P; s, o by G. We call this type of crossed–structures $s_{GPN}v_{PN}o_{GN}$ "subject predicative". We will not try to establish any rule as regards G. Thus, s and o are seen as independent.

1.1.3 Relations between vo. In most languages we cannot find any morphological rule that could show a strong relation between v and o, although v chooses which kind of o it prefers to be combined with. This choice implies an imposition of condition called *level assignment*. v assigns level, and it does so all the time. Thus, each o brings about a level assignment for v. Many levels can be assigned, and they could be marked with subscripts: ∅, q, r.... For instance, it is possible to find svo_q or svo_r.

βll is the set of all level assignments. $\varnothing, q, r$ are elements of this set. βllv is the βll subset for v. $|\beta llv|$ is the number of level assignments βllv. v may have more than one level. When $\beta llv = \varnothing$ then we can talk about a *lonely* v.

An o brings about, in an inherent way, one or more levels that can be represented. Each o may have its own capacity plus a level from v. There is only one level not assigned by v, but free, and that is level L. Every o_L is an o not linked to v so it may appear or disappear without producing any changes to the string. We are dealing with a transpositor element by definition, because of its freedom.

The set of levels that may be represented by $\mathcal{O}$ is $^{-}\beta ll$. $L, q, r \ldots$ are elements of $^{-}\beta ll$. $^{-}\beta llo$ is the $^{-}\beta ll$ subset for an o. $|^{-}\beta llo|$ is the number of assignments of level $^{-}\beta llo$.

We mark with l any level belonging to βllv or to $^{-}\beta llo$.

1.2 STRUCTURE OF STRINGS

We have said above that a basic string is accomplished by replacing the variables of a basic pattern with strings of terminal symbols. But we should give a formal definition taking *svo* relations into account:

The basic string is made up of s, v, o following these criteria:

- Precedence: *svo* is an ordered string. In an SVO language such as Catalan, s precedes v and o, and v precedes o.
- Simplicity: In a basic string there is only one occurrence for each element.
- Agreement:
  - $PN_s = PN_v$
  - $\exists l \mid l \in \beta llv, l \in {}^{-}\beta llo$

We have defined a basic string depending on the structure and the relationship existing among its elements. Following the application rules criterium and not the structural ones we may find the following sorts of strings:

a) **primary string**: it is one in which no rule has been applied.

b) **derivative string**: it is the string emerging from a process.

c) **stage string**: it is one in which a rule has been applied in order to obtain a new string.

d) **ending string**: it is the last string in a derivation process. So it is a derivative string at the same time. Being an ended string, it must have some validation rules. An ending string is valid if:

   (i) $\exists s$

   (ii) $\exists v$

   (iii) agreement criterium is accomplished.

   (iv) precedence criterium is accomplished following the final pattern arrangement.

We think that a primary string is always basic because the generation of language consists of recombining simple sentences in order to produce complex chains. The first string or the first couple of stage strings in a derivation process are also basic.

2. GHOSTS

Ghosts are elements that:

- only arise in derivative linguistic strings,
- are not present in basic strings,

- have not been introduced into a string directly by means of insertion.

Ghosts appear in the last stratum of the sentence. We do not find ghosts among variables, but only when we replace variables by terminal strings, that have strict rules of nearness and combination.

In this article we analyze ghosts produced by the splicing of two strings. They are called *emergent ghosts*. Splicing ([1], [2]) is brought about by means of ULPS in which a blunt cut is applied.

It must be said that although ghosts produced by splicing are emergent, not every emergent ghost is produced by splicing.

2.1 EMERGENT GHOSTS

Emergent ghosts $f\cdot$ can only arise as an outcome of an operation made up of at least two stage strings. This occurs because they take charge of relating repeated elements which never take place in a basic string svo.

Emergent ghosts are produced by friction. By friction we refer to the concatenation of elements or sequences of equal elements (called friction groups) in a string. This contact provokes the appearance of this new symbol between the two groups in a systematic way.

There are six friction contexts, or groups, if we follow the possible combinations with the application of a single rule in basic stage strings:

[1] $ss \rightarrow s \smile s$

[2] $vv' \rightarrow v \smile v$

[3] $oo' \rightarrow o \smile o$

[4] $(sv)(s'v') \rightarrow sv \smile sv$

[5] $(vo)(v'o') \rightarrow vo \smile vo$

[6] $(svo)(s'v'o') \rightarrow svo \smile svo$

Depending on the types of combinations we may establish two kinds of emergent ghosts provoked by splicing: *connectors* and *bounders*. As they are different we will denote connector ghosts with the sign $\cdot$ and the bounder ones by $\star$. An example of $\cdot$ could be "and", an example of $\star$ "because".

2.1.1 *Connector ghosts.* They simply link two equal things. They could be readily assimilated to an addition rule, because they just pile elements up. As a general rule we may say that the appearance of a connector ghost between two friction groups is always possible in a splicing, although the connector ghosts bear readjustment string rules implicitly.

Let us see what happens, then, with the appearance of · among the different friction groups.

2.1.1.1. $ss \to s \cdot s$

"$s \cdot s$" can be brought about by means of a splicing rule such as:
string $\alpha = svo$
string $\beta = s'v'o'$

$$s\#vo \ \ \$ \ \ \#s'v'o' \ \vdash \ s \cdot s'v'o' \tag{3.1}$$

Example 3.1 *En Joan # escrivia una carta $ # Les noies compraven pomes ⊢ En Joan i les noies compraven pomes*
John # wrote a letter $ # The girls bought some apples ⊢ John and the girls bought some apples

But the appearance of the ghost is not always easy. Notice that in applying 3.1 we may find:
*Ells # escrivien una carta $ # La noia comprava pomes ⊢ *Ells i la noia compra pomes*
*They # wrote a letter $ # The girl bought some apples ⊢ *They and the girl buys some apples*

Those examples where $f\cdot$ produces wrong sentences because of the lack of concordance between sv. This make us question which values $s_\alpha s_\beta$ must be accepted by the connector ghost directly, so that this lack of concordance with v is avoided:

This is the table showing the combinations accepted by v in Catalan:

	β						
	$f\cdot$	s_{1e}	s_{2e}	s_{3e}	s_{1p}	s_{2p}	s_{3p}
	s_{1e}	?	–	–	?	–	–
	s_{2e}	–	?	–	yes	?	–
α	s_{3e}	–	–	?	yes	yes	yes
	s_{1p}	–	–	–	?	–	–
	s_{2p}	–	–	–	yes	?	–
	s_{3p}	–	–	–	yes	yes	yes

Table 1

Thus, there are strong restrictions caused by the lack of concordance of $s_\alpha \cdot s_\beta$ with v (-) and by the logical disconnectedness $s_\alpha \cdot s_\beta$ (?):

a. Lack of concordance $s_\alpha \cdot s_\beta$ with v

a.1 $s_\alpha \cdot s_\beta$ only can be carried out if N_{v_β} is plural. The reason is clear: the connector ghost adds and, the result of this addition is, at least, two. That is to say, plural. Except for cases $s_{1e}s_{1e} = s_{1e}$, $s_{2e}s_{2e} = s_{2e}$ and, sometimes, $s_{3e}s_{3e}$, which have special treatment (b.2.)

a.2 $s_\alpha \cdot s_\beta$ only can be brought about if $P_{s_\beta} \leq P_{s_\alpha}$. That is why the combinations $s_{3p}s_{1p}$, $s_{2p}s_{1p}$, $s_{2p}s_{2p}$ are allowed, but neither are $s_{1p}s_{3p}$, $s_{1p}s_{2p}$ nor $s_{2p}s_{3p}$.

a.3 If we join conditions a.1 i a.2, we find that if the subscript of s_β is 1p the ghost arises without needing any readjustment in the resultant string.

b. Logical disconnectedness $s_\alpha \cdot s_\beta$:

b.1 The combinations $s_{1e}s_{1p}$, $s_{2e}s_{2p}$, although possible, are marked "?" on the tale. This is because of $s_{1e} \subset s_{1p}$, $s_{2e} \subset s_{2p}$. We should, them, reformulate the outcome of these combinations: $s_{1e}s_{1p} = s_{1p}$ and $s_{2e}s_{2p} = s_{2p}$.

b.2 $s_{1e}s_{1e}$, $s_{2e}s_{2e}$, $s_{1p}s_{1p}$, $s_{2p}s_{2p}$ and when $s_\alpha = s_\beta$, $s_{3e}s_{3e}$, $s_{3p}s_{3p}$ are replications which the language does not allow.

To solve b.1, b.2 is relatively easy if we introduce some rules called *s reduction*:

1. $s_{1e}s_{1e} = s_{1e}$
2. $s_{2e}s_{2e} = s_{2e}$
3. if $s = s' \mid s_{3e}s_{3e} = s_{3e}$
4. $s_{1p}s_{1p} = s_{1p}$
5. $s_{2p}s_{2p} = s_{2p}$
6. if $s = s' \mid s_{3p}s_{3p} = s_{3p}$
7. $s_{1e}s_{1p} = s_{1p}$; and commutatively $s_{1p}s_{1e} = s_{1p}$
8. $s_{2e}s_{2p} = s_{2p}$; and commutatively $s_{2p}s_{2e} = s_{2p}$

If we apply these rules we can get rid of those question marks in Table 1. But in order to solve a. we have to be able to make sentences with the structure:

- $N_\alpha > N_\beta$
- $P_\beta > P_\alpha$
- $s_e s_e$ if $s_\alpha \neq s_\beta$

If what we want is to $f\cdot$ to join two s there must be a readjustment of PN values of v in the resultant string. That is why we must describe an addition table ss that takes into account all the possible combinations and describes how v is affected:

	β						
	ADD	s_{1e}	s_{2e}	s_{3e}	s_{1p}	s_{2p}	s_{3p}
	s_{1e}	v_{1e}	v_{1p}	v_{1p}	v_{1p}	v_{1p}	v_{1p}
α	s_{2e}	v_{1p}	v_{2e}	v_{2p}	v_{1p}	v_{2p}	v_{2p}
	s_{3e}	v_{1p}	v_{2p}	$v_{3e}, v_{3p}*$	v_{1p}	v_{2p}	v_{3p}
	s_{1p}	v_{1p}	v_{1p}	v_{1p}	v_{1p}	v_{1p}	v_{1p}
	s_{2p}	v_{1p}	v_{2p}	v_{2p}	v_{1p}	v_{2p}	v_{2p}
	s_{3p}	v_{1p}	v_{2p}	v_{3p}	v_{1p}	v_{2p}	v_{3p}

Table 2

According to these results it is possible to formulate the following general rule:

$$s_{PN}\#v_{PN}o \quad \$ \quad \#s'_{PN}v'_{PN}o' \quad \vdash \quad s_{PN} \cdot s'_{PN}v'_{PN_{s \cdot s'}}o' \qquad (3.2)$$

where the value of the subscripts of v is given by the table.

2.1.1.2. $vv' \to v \cdot v$

Where $\alpha = svo, \beta = s'v'o'$

$$sv\#o \quad \$ \quad s'\#v'o' \quad \vdash \quad sv \cdot v'o' \qquad (3.3)$$

There are two conditions for this operation to be carried out. They are:

(i) $PN_{v_\alpha} = PN_{v_\beta}$

(ii) $\varnothing \in \beta ll_{v_\alpha}$

Example 3.2 *En Joan escrivia # una carta $ La Maria # llegia un llibre* ⊢ *En Joan escrivia i llegia un llibre*
John wrote # a letter $ Mary # read a book ⊢ *John wrote and read a book*

2.1.1.3. $oo' \to o \cdot o$

This friction context appears, where $\alpha = svo$, $\beta = s'v'o'$ with rules such as:

$$svo\# \quad \$ \quad s'v'\#o' \quad \vdash \quad svo \cdot o' \qquad (3.4)$$

Example 3.3 *Aquests nois escriuen moltes cartes# $ La Maria llegeix # un llibre* ⊢ *Aquests nois escriuen moltes cartes i un llibre*
These boys are writing many letters # $ Mary is reading # a book ⊢ *These boys are writing many letters and a book*

$v - o$ relations in a string must go through a level assignment, explained in 1.1.3, which is a form of syntactic concordance. In this case vo relations in α do not vary after the splicing. However, some new ones are established between $v_\alpha o_\beta$ which cannot break the level that v_α assigns to o_α. This makes $v_\alpha o_\alpha o_\beta$ necessary in the structure to give an agreement of levels among the three elements: $lv_\alpha = lo_\alpha = lo_\beta$.

2.1.1.4. $(sv)(s'v') \rightarrow sv \cdot sv$

Where $\alpha = svo,\ \beta = s'v'o'$

$$sv\#o \quad \$ \quad \#s'v'o' \ \vdash \ sv \cdot s'v'o' \tag{3.5}$$

Notice that there are two conditions that must be fulfilled, so that ending strings produced by means of this procedure can be considered as valid:

(i) $|\beta llv_\alpha| > 1$

(ii) $\varnothing \in \beta llv_\alpha$

Example 3.4 *En Joan escriu # una carta $ # Les noies juguen a tennis* ⊢ *En Joan escriu i les noies juguen a tennis*
John is writing # a letter $ # The girls are playing tennis ⊢ *John is writing and the girls are playing tennis*

Where $|\beta llwrite| \geq 2,\ \varnothing \in \beta llwrite$

2.1.1.5. $(vo)(v'o') \rightarrow vo \cdot v'o'$

Where $\alpha = svo,\ \beta = s'v'o'$

$$svo\# \quad \$ \quad s'\#v'o' \ \vdash \ svo \cdot v'o' \tag{3.6}$$

Example 3.5 *En Joan escrivia cartes# $ La Maria # jugava a futbol* ⊢ *En Joan escrivia cartes i jugava a futbol*
John wrote some letters# $ Mary # played football ⊢ *John wrote some letters and played football*

The only condition for the result to be acceptable is that $PN_{v_\alpha} = PN_{v_\beta}$

2.1.1.6 $(svo) \quad (s'v'o') \rightarrow svo \cdot svo$

Where $\alpha = svo,\ \beta = s'v'o'$

$$svo\# \quad \$ \quad \#s'v'o' \ \vdash \ svo \cdot s'v'o' \tag{3.7}$$

This is a recursive rule without any restriction beyond our own memory.

Example 3.6 *En Joan escriu cartes# $ # La Maria juga a futbol* $\vdash$ *En Joan escriu cartes i la Maria juga a fubol*
John writes some letters# $ # Mary plays football $\vdash$ *John writes some letters and Mary plays football*

2.2 BOUNDER GHOSTS

There is an only condition under which bounders arise in splicing operations, and that is the existence of sv in the friction groups. They are valid, then, in the following contexts:

[4. $](sv)(s'v') \rightarrow (sv) \star (s'v')$

[6. $](svo)(s'v'o') \rightarrow (svo) \star (s'v'o')$

For [4] the conditions are the same as for the appearance of connector ghosts (2.1.1.4):

1. $|\beta llv_\alpha| > 1$

2. $\exists\, \varnothing \in \beta llv_\alpha$

If v_α observes this condition, the ghost can arise. Then:
where $\alpha = svo,\ \beta = s'v'o'$

$$sv\#o \quad \$ \quad \#s'v'o' \ \vdash \ sv \star s'v'o' \tag{3.8}$$

Example 3.7 *En Joan escriu # una carta $ # La Maria juga a futbol* $\vdash$ *En Joan escriu perquè la Maria juga a futbol*
John is writing # a letter $ # Mary is playing football $\vdash$ *John is writing because Mary is playing football*

However in [6] the appearance of $f\star$ is completely free:
where $\alpha = svo,\ \beta = s'v'o'$

$$svo\# \quad \$ \quad \#s'v'o' \ \vdash \ svo \star s'v'o' \tag{3.9}$$

Example 3.8 *En Joan viu a Barcelona# $ #La Maria juga a futbol* $\vdash$ *En Joan viu a Barcelona perquè la Maria juga a futbol*
John lives in Barcelona# $ # Mary plays football $\vdash$ *John lives in Barcelona because Mary plays football*

3. CONCLUSIONS

With regards to strings in general, ghosts are elements that confer special properties on linguistic strings in terms of their behaviour in molecular operations. In natural language, ghosts are needed for concatenation and string hierarchy or fragments of strings. Therefore, they are of special importance when establishing relations among the diverse elements of a sequence.

The level at which emergent ghosts arise is that of terminal strings, in other words, in the last stratum of a ULPS. This seems to suggest that their appearance is a superficial phenomenon although it carries out a sequence of readjustments in the linguistic string in the level of concordance sv and vo.

Emergent ghosts are the only ones that can be brought about by the application of splicing rules in two linguistic strings. But there are several molecular operations which produce these kinds of ghosts, such as sticker systems. It remains for us to study emergent ghosts produced by other molecular operations, as well as other ghosts that can arise because of both the manipulation of strings produced by staggered cuts and mutations carried out in a string.

References

[1] Head, T. (1987), Formal language theory and DNA: an analysis of the generative capacity of specific recombinant behaviours, *Bulletin of Mathematical Biology*, 49: 737–759.

[2] Păun, Gh.; G. Rozenberg & A. Salomaa (1998) *DNA Computing: New Computing Paradigms*, Springer, Berlin.

Chapter 4

ON SIZE COMPLEXITY OF CONTEXT-FREE RETURNING PARALLEL COMMUNICATING GRAMMAR SYSTEMS

Erzsébet Csuhaj-Varjú

Computer and Automation Research Institute

Hungarian Academy of Sciences

Kende u. 13-17, H-1111 Budapest, Hungary

csuhaj@sztaki.hu

Abstract We show that there exists a natural number k such that every recursively enumerable language can be generated by a context-free returning parallel communicating grammar system where the number of nonterminals is less than or equal to this constant. Moreover, the component grammars of the system have a limited number of productions. The result demonstrates that context-free returning parallel communicating grammar systems are economical tools for language generation.

1. INTRODUCTION

Parallel communicating grammar systems (PC grammar systems, in short) have been introduced for modelling parallel and distributed computation in terms of formal grammars and languages ([6]). These systems consist of several grammars which derive their own sentential forms in parallel being organized in a communicating system to generate a language. The work of the component grammars is synchronized by a universal clock, each component executes one rewriting step in each time unit, and communication is done by request through so-called query symbols, one different symbol referring to each grammar in the system. When a query symbol appears in the sentential form of a component, the rewriting process stops and one or more communication steps are performed by replacing all occurrences of the query symbols in the string with the current

*Research supported by Hungarian Scientific Research Fund "OTKA" Grant no. T 029615.

C. Martin-Vide and V. Mitrana (eds.), Where Mathematics, Computer Science, Linguistics and Biology Meet, 37-49.

sentential forms of the queried component grammars. When no more query symbol is present in any of the sentential forms, the rewriting process starts again. In so-called returning systems after communicating its current sentential form the component returns to its start symbol and begins to generate a new string. In non-returning systems the components continue the rewriting of their current sentential forms. Rewriting steps and communication steps determine a computation. The language defined by the system is the set of terminal words which appear as sentential forms of a dedicated component grammar, the master, at some step of a computation starting from the initial configuration of the system.

Parallel communicating grammar systems have been studied in detail over the last years: see [1] and [4] for a summary of results and open problems. The investigations mainly concentrated on the generative power of PC grammar systems and on examining how this power is influenced by changes in the basic characteristics of these systems.

An important characteristic of PC grammar systems is how economical the descriptions of languages they provide are. For example, how many nonterminals and productions are necesssary for a PC grammar system (for a component grammar) to generate a language in a certain language class. In [3] it was shown that there exists a natural number k such that for any context-free returning parallel communicating grammar system an equivalent system of the same type can be constructed where the total number of symbols used for describing each component grammar is limited by this constant. The result implies that both the number of productions and the number of nonterminals appearing in the productions of a component are limited by constant numbers, too. Since context-free returning PC grammar systems determine the class of recursively enumerable languages ([2]), the statement gives information on the conciseness of the description of languages generated by parallel communicating grammar systems. In this paper we add a further property confirming that PC grammar systems are economical tools for language generation. Namely, we show that there exists a natural number k such that every recursively enumerable language can be generated by a context-free returning parallel communicating grammar system where the number of nonterminals of the system is less than or equal to this number. Moreover, the components of the system are with a limited number of productions, too.

2. BASIC DEFINITIONS

We assume that the reader is familiar with the basics of formal language theory; for further details consult [7].

We denote the set of all nonempty words over an alphabet V by V^+; if the empty word, λ, is included, we use notation V^*. The number of elements of a finite set X is denoted by $card(X)$ and $|w|_X$, $w \in V^*$, $X \subseteq V$, denotes the number of occurrences of symbols from set X in w.

In the following we recall some basic notions concerning parallel communicating grammar systems.

Definition 4.1 *A parallel communicating grammar system of degree* n *(a PC grammar system, in short) is an* $(n+3)$*-tuple* $\Gamma = (N, K, T, G_1, \ldots, G_n)$, $n \geq 1$, *where* N *is a nonterminal alphabet,* T *is a terminal alphabet, and* $K = \{Q_1, Q_2, \ldots, Q_n\}$ *is an alphabet of query symbols.* N, T, *and* K *are pairwise disjoint sets.* $G_i = (N \cup K, T, P_i, S_i)$, $1 \leq i \leq n$, *called a component of* Γ, *is a usual Chomsky grammar with nonterminal alphabet* $N \cup K$, *terminal alphabet* T, *set of rewriting rules* P_i, *and axiom (or start symbol)* S_i. G_1 *is said to be the master grammar (or master) of* Γ.

An n-tuple $(x_1, \ldots, x_n)$, where $x_i \in (N \cup T \cup K)^*$, $1 \leq i \leq n$, is called a *configuration* of Γ. $(S_1, \ldots, S_n)$ is said to be the *initial configuration.*

PC grammar systems change their configurations by direct derivation steps.

Definition 4.2 *Let* $\Gamma = (N, K, T, G_1, \ldots, G_n)$, $n \geq 1$, *be a parallel communicating grammar system and let* $(x_1, \ldots, x_n)$ *and* $(y_1, \ldots, y_n)$ *be two configurations of* Γ.

We say that $(x_1, \ldots, x_n)$ *directly derives* $(y_1, \ldots, y_n)$, *denoted by*

$$(x_1, \ldots, x_n) \Rightarrow (y_1, \ldots, y_n),$$

if one of the following two cases holds:

1. There is no x_i *which contains any occurrence of a query symbol, that is,* $x_i \in (N \cup T)^*$ *for* $1 \leq i \leq n$. *Then for each* i, $1 \leq i \leq n$, $x_i \Rightarrow_{G_i} y_i$ *for* $x_i \notin T^*$ *(*y_i *is obtained from* x_i *by a direct derivation step in* G_i*) and* $x_i = y_i$ *for* $x_i \in T^*$.

2. There is some x_i, $1 \leq i \leq n$, *which contains at least one occurrence of a query symbol.*

Then for each x_i *with* $|x_i|_K \neq 0$, $1 \leq i \leq n$, *we write* $x_i = z_1 Q_{i_1} z_2 Q_{i_2} \ldots z_t Q_{i_t} z_{t+1}$, *where* $z_j \in (N \cup T)^*$, $1 \leq j \leq t+1$, *and* $Q_{i_l} \in K$, $1 \leq l \leq t$. *If* $|x_{i_l}|_K = 0$ *for each* l, $1 \leq l \leq t$, *then* $y_i = z_1 x_{i_1} z_2 x_{i_2} \ldots z_t x_{i_t} z_{t+1}$ *and in returning systems* $y_{i_l} = S_{i_l}$, *in non-returning systems* $y_{i_l} = x_{i_l}$, $1 \leq l \leq t$. *If* $|x_{i_l}|_K \neq 0$ *for some* l, $1 \leq l \leq t$, *then* $y_i = x_i$. *For all* j, $1 \leq j \leq n$, *for which* y_j *is not specified above* $y_j = x_j$.

The transitive and reflexive closure of $\Rightarrow$ *is denoted by* $\Rightarrow^*$.

The first case is the description of a rewriting step. If no query symbol is present in any of the sentential forms, then each grammar has to apply one of its productions except those which have a terminal string as sentential form. The derivation gets blocked if a sentential form is not a terminal string but no rule can be applied to it.

The second case describes communication: if a query symbol Q_j appears in a sentential form x_i, $1 \leq i, j \leq n$, then the rewriting stops and communication must be performed.

Then all the query symbols Q_{i_j}, $1 \leq j \leq t$, which appear in a sentential form x_i must be replaced by the current sentential form x_{i_j} of component G_{i_j} in the same communication step provided that no x_{i_j} has any occurrence of a query symbol. If one of these sentential forms, x_{i_j}, contains a query symbol, then first x_{i_j} must be made free from the queries before changing anything in x_i. The derivation gets blocked if in the obtained configuration none of the sentential forms with an occurrence of a query symbol can be made free from the queries in the above manner.

After communicating its sentential form to another component, the grammar can continue its own work in two ways: In *returning* systems the component must return to its axiom and begin to generate a new string. In *non-returning* systems the components continue the generation of their current strings. Throughout the paper we shall deal with returning systems.

Definition 4.3 *The language generated by a parallel communicating grammar system* $\Gamma = (N, K, T, G_1, \ldots, G_n)$ *with* $G_i = (N \cup K, T, P_i, S_i)$, $1 \leq i \leq n$, *is*

$$L(\Gamma) = \{\alpha_1 \in T^* \mid (S_1, \ldots, S_n) \Rightarrow^* (\alpha_1, \ldots, \alpha_n)\}.$$

A PC grammar system is said to be context-free if its components are context-free grammars.

Size properties of PC grammar systems give information on the conciseness of the description of the grammar system.

Definition 4.4 *Let* $\Gamma = (N, K, T, G_1, \ldots, G_n)$, $n \geq 1$, *be a parallel communicating grammar system. We define*

$$\begin{aligned} GVar(\Gamma) &= card(N), \\ CVar(\Gamma) &= max\{card(N_i) \mid N_i \text{ is the set of nonterminals of } \Gamma \\ &\qquad \text{which appear in } P_i, 1 \leq i \leq n\}, \text{ and} \\ CProd(\Gamma) &= max\{card(P_i) \mid 1 \leq i \leq n\}. \end{aligned}$$

The first measure, $GVar$ expresses how economical the system is presented with respect to the number of nonterminals, while the last two measures give information on the size of the component grammars in the system: $CVAR$ refers to the maximum number of nonterminals which appear in the production set of a component grammar, $CProd$ denotes the maximum of the number of productions of the components.

These measures are extended to languages and language classes in the usual manner.

For a language L, for a class of PC grammar systems $\mathcal{G}$ with $L \in \mathcal{L}(\mathcal{G})$, and for a size complexity measure $M \in \{GVar, CVar, CProd\}$, we define $M_{\mathcal{G}}(L) = min\{M(\Gamma) \mid \Gamma \in \mathcal{G} \text{ and } L(\Gamma) = L\}$. ($\mathcal{L}(\mathcal{G})$ denotes the class of languages generated by elements of $\mathcal{G}$). For a language class $\mathcal{C}$ and a class of PC grammar systems $\mathcal{G}$ with $\mathcal{C} \subseteq \mathcal{L}(\mathcal{G})$, we define $M_{\mathcal{G}}(\mathcal{C}) = sup\{M_{\mathcal{G}}(L) \mid L \in \mathcal{C}\}$.

We say that a size complexity measure M is *bounded* on a class of languages $\mathcal{C}$ with respect to a class of PC grammar systems $\mathcal{G}$ with $\mathcal{C} \subseteq \mathcal{L}(\mathcal{G})$ if there is a natural number k such that for every $L \in \mathcal{C}$ $M_{\mathcal{G}}(L) \leq k$ holds.

3. SIZE COMPLEXITY OF CONTEXT-FREE RETURNING PC GRAMMAR SYSTEMS

In this section we show that size complexity measure $GVar$ is bounded on the class of recursively enumerable languages with respect to the class of returning context-free parallel communicating grammars systems. Moreover, the proof implies the boundedness of complexity measure $CProd$ on this language class, too. (The boundedness of complexity measure $CVar$ obviously follows from the result.) The proof of the statement is based on simulation of Geffert normal form grammars for recursively enumerable languages [5] by context-free returning parallel communicating grammar systems.

By [5] it is known that for every recursively enumerable language L over an alphabet $T = \{a_1, \ldots, a_n\}$ there exist words $z'_{a_i}, v'_j \in \{1', 0'\}^*$, $u_j \in \{1, 0\}^*$, $1 \leq i \leq n$, $1 \leq j \leq m$, such that L can be generated by a grammar $G = (\{S, 1, 0, 1', 0'\}, T, P, S)$, with $P = P_1 \cup P_2$, where $P_1 = \{S \to (z'_{a_i})^R S a_i \mid 1 \leq i \leq n\} \cup \{S \to (v'_j)^R S u_j \mid 1 \leq i \leq m\} \cup \{S \to (v'_j)^R u_j \mid 1 \leq i \leq m\}$ and $P_2 = \{1'1 \to \lambda, 0'0 \to \lambda\}$. (For a word w, we denote its reversal by w^R.)

Thus, a word $w \in T^*$ belongs to L if and only if at some step of the derivation in G the obtained sentential form is of the form $\alpha\beta w$, where $\alpha \in \{0', 1'\}^*$, $\beta \in \{0, 1\}^*$ and $\alpha = (\beta')^R$. (β' denotes the string obtained from β by replacing each letter with its primed version.) Then α and β consist of an equal number of symbols and the value of α as a binary number is equal to the value of the reversal of β as a binary number. Thus, the application of a production p of G to a sentential form $u = \alpha\gamma\beta\delta$, where $\alpha \in \{0', 1'\}^*$, $\beta \in \{0, 1\}^*$, $\gamma \in (\{S\} \cup \{\lambda\})$, $\delta \in T^*$, implies changes in some of the following parameters: in δ, in the number of symbols of α and β, and in the value of α and β as binary numbers, respectively.

The idea of the construction of a context-free returning PC grammar system Γ simulating G for L is based on this observation. The property that the simulating parallel communicating grammar system Γ has a limited number of nonterminals follows from the fact that any Geffert normal form grammar above has the same number of nonterminals, five. Before presenting the formal details, we give some informal explanations to the proof.

For each production of G in P_1, the PC grammar system Γ contains a separate collection of grammars which are able to simulate both the application of the production or skipping of its execution. In such a collection there is a component grammar which is able to simulate the change in the maximal terminal substring of the sentential form in generation, there is a component responsible for realizing the change in the number of symbols from $\{0, 1\}$ in the sentential form when applying the production, another component responsible for the change in the number of symbols from $\{0', 1'\}$, a component for realizing the change in the value of the maximal substring over $\{0, 1\}$, and, finally, there is a grammar responsible for realizing the change in the value of the maximal substring over $\{0', 1'\}$. (Each number is represented at the corresponding component as a sentential form having the same number of occurrences of a dedicated symbol, A, as the considered number.) The grammar collection contains other components for simulating the skipping the execution of the production and for assisting the work of the others. The simulation of productions of P_2 is done in other collections of grammars, which establish whether the corresponding components (which are for counting the number of symbols from $\{0, 1\}$ and $\{0', 1'\}$ and which contain the values of the corresponding substrings in the simulated sentential form) have sentential forms of the same length. These grammar collections are able to simulate the skipping of the corresponding checking phase as well.

Derivations in G are simulated by derivations in Γ as follows: suppose that Γ consists of m collections of grammars and at some step of a derivation in Γ the r-th collection of grammars obtains from the $(r-1)$-th collection a 5-tuple of sentential forms, described above, which corrresponds to a sentential form of G. Then, the grammar collection decides whether the application of the r-th production or skipping its execution will be simulated and performs the corresponding activity. After finishing the procedure, the resulted 5-tuple of sentential forms is forwarded to the next, $(r+1)$-th collection of grammars. Thus, the 5-tuples of strings representing sentential forms of G move from every grammar collection r to the next one, $r+1$, and return from the last collection, m, to the first one. Since a move from a collection to the next one or a series of such moves corresponds to either execution or skipping the application of some production of G, derivations in G can be simulated in Γ in this manner.

Now we present the result.

Theorem 4.1 *Size complexity measure $GVar$ is bounded on the class of recursively enumerable languages with respect to the class of context-free returning PC grammar systems.*

Proof. To prove the statement we show that there exists a natural number k such that for any recursively enumerable language L there is a generating context-free returning PC grammar system Γ with at most k nonterminals. Let L be a recursively enumerable language over an alphabet $T = \{a_1, \dots, a_n\}$, and let $G = (\{S_0, S_1, S_2, 1, 0, 1', 0'\}, T, P, S_0)$ with $P = P_1 \cup P_2 \cup P_3 \cup P_4 \cup P_5 \cup P_6$, where $P_1 = \{S_0 \to (z'_{a_i})^R S_0 a_i \mid 1 \le i \le n\}$, $P_2 = \{S_0 \to S_1\}$, $P_3 = \{S_1 \to (v'_j)^R S_1 u_j \mid 1 \le j \le m\}$, $P_4 = \{S_1 \to S_2\}$, $P_5 = \{S_2 \to (v'_j)^R u_j \mid 1 \le j \le m\}$, and $P_6 = \{1'1 \to \lambda, 0'0 \to \lambda\}$, be a grammar equivalent to a Geffert normal form grammar generating L.

Let $h = n+2m+2$ and let us denote the productions of G by p_k, $1 \le k \le h$, where $p_i \in P_1$ for $1 \le i \le n$, $p_{n+1} \in P_2$, $p_j \in P_3$ for $n+2 \le j \le n+1+m$, $p_{n+1+m+1} \in P_4$, and $p_l \in P_5$ for $n+1+m+2 \le l \le h$.

We fix some further notations.

For a production $p = A \to \alpha$ in G, where $A \in \{S_0, S_1, S_2\}$, $\alpha \in (T \cup \{S_0, S_1, S_2, 1, 0, 1', 0'\})^*$, let $dig(p) = |\alpha|_{\{0,1\}}$ and let $dig'(p) = |\alpha|_{\{0',1'\}}$.

Let $\alpha = \beta\gamma\delta\rho$, where $\beta \in \{0', 1'\}^*$, $\delta \in \{0, 1\}^*$, $\gamma \in (\{S_0, S_1, S_2\} \cup \{\lambda\})$, and $\rho \in T^*$.

Let us denote by $val'(p)$ the value of β as a binary number and by $val_R(p)$ the value of the reversal of δ as a binary number.

Now we construct a context-free returning PC grammar system Γ simulating G. Γ will consist of $h+9$ collections of grammars, U_j, $0 \le j \le h+8$, $h = n+2m+2$, defined above and a master grammar. Grammar collection U_0 is for initialization, grammar collections U_k, $1 \le k \le h$, simulate the effect of applying/skipping production p_k of G and the remaining 8 collections are responsible for simulating the effect of productions $1'1 \to \lambda$ and $0'0 \to \lambda$ of G. Each grammar collection U_j, $0 \le j \le h+8$, consists of grammars $G^{st}_{j,i}$, G^{sel}_j, G^{ini}_j, $G^{prep}_{j,i}$, $G^{exe}_{j,i}$, and $G^{fw}_{j,i}$, $1 \le i \le 5$, where $G^{st}_{j,i}$, $G^{prep}_{j,i}$, $G^{exe}_{j,i}$, and $G^{fw}_{j,i}$ are for storing, modifying and forwarding the 5-tuples of sentential forms representing the generated string in G, G^{sel}_j is for deciding the chosen activity, and G^{ini}_j is for storing this information.

Let

$$\begin{aligned}\Gamma = {} & (N, K, T, G^M, \\ & G^{st}_{0,1}, \dots, G^{st}_{0,5}, G^{sel}_0, G^{ini}_0, G^{prep}_{0,1}, \dots, G^{prep}_{0,5}, \\ & G^{exe}_{0,1}, \dots, G^{exe}_{0,5}, G^{fw}_{0,1}, \dots, G^{fw}_{0,5}, \dots, \\ & G^{st}_{h+8,1}, \dots, G^{st}_{h+8,5}, G^{sel}_{h+8}, G^{ini}_{h+8}, G^{prep}_{h+8,1}, \dots, G^{prep}_{h+8,5}, \\ & G^{exe}_{h+8,1}, \dots, G^{exe}_{h+8,5}, G^{fw}_{h+8,1}, \dots, G^{fw}_{h+8,5}).\end{aligned}$$

Let

$$\begin{aligned} N &= \{S, S_0, S_1, S_2, S_3, S_4, S_5, S_6, S_7, S_8, S_9, A\} \cup \\ &\quad \{E, F, E', F', E", F", \tilde{E}, \tilde{F}\} \cup \\ &\quad \{D, D', S^{(1)}, S'^{(1)}, S', Z, Z'\} \end{aligned}$$

and let S be the start symbol of each component grammar of Γ.

Let us define the production sets of the components as follows:

Let $P^M = \{S \rightarrow S_1, S_1 \rightarrow S_2, S_2 \rightarrow S_3, S_3 \rightarrow S, S_3 \rightarrow Q^{exe}_{h+8,1}\}$.

(The master component selects the words of the language.)

Let for $0 \leq k \leq h+8$ and $1 \leq i \leq 5$

$$\begin{aligned} P^{st}_{k,i} &= \{D \rightarrow D', D' \rightarrow \lambda, S \rightarrow S', S' \rightarrow S^{(1)}\} \cup \\ &\quad \{S \rightarrow Q^{ini}_j, \tilde{F} \rightarrow Q^{fw}_{j,i} D, \tilde{E} \rightarrow Q^{exe}_{j,i} D\} \end{aligned}$$

and $j = k-1$ for $1 \leq k \leq h+8$ and $j = h+8$ for $k = 0$.

(These components store the 5-tuples of sentential forms representing the string in generation in G before starting the simulation/skipping of the execution of the corresponding production (or activity). The strings are obtained from the preceding grammar collection, U_j.)

For each k, $0 \leq k \leq h+8$, let

$$\begin{aligned} P^{sel}_k &= \{S \rightarrow E, S \rightarrow F, E \rightarrow E', F \rightarrow F', E' \rightarrow S, F' \rightarrow S\} \text{ and} \\ P^{ini}_k &= \{S \rightarrow Q^{sel}_k, E \rightarrow E', F \rightarrow F', E' \rightarrow \tilde{E}, F' \rightarrow \tilde{F}\}. \end{aligned}$$

(P^{sel}_k decides whether the application of some production -execution of some activity- or skipping its execution will be simulated. P^{ini}_k stores this information.)

The following components simulate the effect of the application of productions in G.

Let for $0 \leq k \leq h+8$ and $1 \leq i \leq 5$

$$\begin{aligned} P^{prep}_{k,i} &= \bar{P}^{prep}_{k,i} \cup \{S \rightarrow Q^{sel}_k, F \rightarrow F', F' \rightarrow F", F" \rightarrow S\}, \\ P^{exe}_{k,i} &= \bar{P}^{exe}_{k,i} \cup \{S \rightarrow Q^{sel}_k, E \rightarrow E', E' \rightarrow Q^{prep}_{k,i}, F \rightarrow F', \\ &\quad F' \rightarrow F", F" \rightarrow S\}, \end{aligned}$$

where

$\bar{P}^{prep}_{0,1} = \{E \to Q^{st}_{0,1}, S^{(1)} \to S'^{(1)}\},$ $\qquad \bar{P}^{exe}_{0,1} = \{S'^{(1)} \to S_0\},$
$\bar{P}^{prep}_{k,1} = \{E \to Q^{st}_{k,1}, S_0 \to S_0\},$
$\bar{P}^{exe}_{k,1} = \{S_0 \to S_0 a_{i_k} \mid p_k = S_0 \to (z'_{a_{i_k}})^R S_0 a_{i_k}\},$
$\quad for\ 1 \le k \le n,$
$\bar{P}^{prep}_{n+1,1} = \{E \to Q^{st}_{n+1,1}, S_0 \to S_0\},$ $\qquad \bar{P}^{exe}_{n+1,1} = \{S_0 \to S_1\},$
$\bar{P}^{prep}_{k,1} = \{E \to Q^{st}_{k,1}, S_1 \to S_1\},$ $\qquad \bar{P}^{exe}_{k,1} = \{S_1 \to S_1\},$
$\quad for\ n+2 \le k \le n+1+m,$
$\bar{P}^{prep}_{n+1+m+1,1} = \{E \to Q^{st}_{n+1+m+1,1}, S_1 \to S_1\},$
$\bar{P}^{exe}_{n+1+m+1,1} = \{S_1 \to S_2\},$
$\bar{P}^{prep}_{k,1} = \{E \to Q^{st}_{k,1}, S_2 \to S_2\},$ $\qquad \bar{P}^{exe}_{k,1} = \{S_2 \to S_3\},$
$\quad for\ n+1+m+2 \le k \le h.$

(These components simulate the change in the maximal terminal substring in the string in generation in G by applying the corresponding production from P_1, P_2, P_3, P_4 or P_5. Components $P^{prep}_{0,l}$ and $P^{exe}_{0,l}$, for $l = 1, 2, 3, 4, 5$ are for initialization.)

Let for $j = 2, 3, 4, 5$

$$\bar{P}^{prep}_{0,j} = \{E \to Q^{st}_{0,j}, S^{(1)} \to S'^{(1)}\}, \quad \bar{P}^{exe}_{0,j} = \{S'^{(1)} \to A\},$$

and for $1 \le k \le h$

$$\begin{aligned}
\bar{P}^{prep}_{k,2} &= \{E \to (Q^{st}_{k,2})^{2^{dig(p_k)}} Z, Z \to Z'\}, & \bar{P}^{exe}_{k,2} &= \{Z' \to A^{val_R(p_k)}\},\\
\bar{P}^{prep}_{k,3} &= \{E \to (Q^{st}_{k,3})^{2^{dig'(p_k)}} Z, Z \to Z'\}, & \bar{P}^{exe}_{k,3} &= \{Z' \to A^{val'(p_k)}\},\\
\bar{P}^{prep}_{k,4} &= \{E \to Q^{st}_{k,4} Z, Z \to Z'\}, & \bar{P}^{exe}_{k,4} &= \{Z' \to A^{dig(p_k)}\},\\
\bar{P}^{prep}_{k,5} &= \{E \to Q^{st}_{k,5} Z, Z \to Z'\}, & \bar{P}^{exe}_{k,5} &= \{Z' \to A^{dig'(p_k)}\},
\end{aligned}$$

(These components are for simulating the change in the value of the maximal substrings over $\{0,1\}$ and $\{0',1'\}$ in the string in generation in G and for counting the lengths of these subwords.)

The following components simulate the execution of productions $1'1 \to \lambda$ and $0'0 \to \lambda$ in G. Let $j = 2, 3, 4, 5$, and let $i = 2, 3$ and $l = 4, 5$.

Let

$$\begin{aligned}
\bar{P}^{prep}_{h+1,1} &= \{E \to Q^{st}_{h+1,1}, S_3 \to S_4\}, & \bar{P}^{exe}_{h+1,1} &= \{S_4 \to S_4\},\\
\bar{P}^{prep}_{h+1,j} &= \{E \to Q^{st}_{h+1,j} Z, Z \to \lambda\}, & \bar{P}^{exe}_{h+1,j} &= \{A \to A\}.
\end{aligned}$$

(These components indicate that checking the equality of the corresponding strings over $\{0,1\}$ and $\{0',1'\}$ started.)

Let

$$\begin{aligned}
\bar{P}^{prep}_{h+2,1} &= \{E \to Q^{st}_{h+2,1}, S_4 \to S_4\}, & \bar{P}^{exe}_{h+2,1} &= \{S_4 \to S_4\},\\
\bar{P}^{prep}_{h+2,i} &= \{E \to Q^{st}_{h+2,i} Z, Z \to \lambda\}, & \bar{P}^{exe}_{h+2,i} &= \{A \to \lambda\},\\
\bar{P}^{prep}_{h+2,l} &= \{E \to Q^{st}_{h+2,l} Z, Z \to \lambda\}, & \bar{P}^{exe}_{h+2,l} &= \{A \to A\}.
\end{aligned}$$

(These components decrease the values of the binary numbers defined by the maximal substrings over $\{0,1\}$ and $\{0',1'\}$ in the generated string in G by one.)

Let

$$\bar{P}^{prep}_{h+3,1} = \{E \to Q^{st}_{h+3,1}, S_4 \to S_5\}, \quad \bar{P}^{exe}_{h+3,1} = \{S_5 \to S_5\},$$
$$\bar{P}^{prep}_{h+3,j} = \{E \to Q^{st}_{h+3,j}Z, Z \to \lambda\}, \quad \bar{P}^{exe}_{h+3,j} = \{A \to A\}.$$

(Checking the equality of the lengths of the maximal substrings over $\{0,1\}$ and $\{0',1'\}$ in the generated string in G started.)

Let

$$\bar{P}^{prep}_{h+4,1} = \{E \to Q^{st}_{h+4,1}, S_5 \to S_5\}, \quad \bar{P}^{exe}_{h+4,1} = \{S_5 \to S_5\},$$
$$\bar{P}^{prep}_{h+4,i} = \{E \to Q^{st}_{h+4,i}Z, Z \to \lambda\}, \quad \bar{P}^{exe}_{h+4,i} = \{A \to A\},$$
$$\bar{P}^{prep}_{h+4,l} = \{E \to Q^{st}_{h+4,l}Z, Z \to \lambda\}, \quad \bar{P}^{exe}_{h+4,l} = \{A \to \lambda\}.$$

(These components check the equality of the lengths of the maximal substrings over $\{0,1\}$ and $\{0',1'\}$ in the generated string in G by decreasing the number of A-s by one from time to time.)

Let

$$\bar{P}^{prep}_{h+5,1} = \{E \to Q^{st}_{h+5,1}, S_5 \to S_6\}, \quad \bar{P}^{exe}_{h+5,1} = \{S_6 \to S_6\},$$
$$\bar{P}^{prep}_{h+5,j} = \{E \to Q^{st}_{h+5,j}Z, Z \to Z\}, \quad \bar{P}^{exe}_{h+5,j} = \{A \to \lambda\}.$$

(These components can work successfully only if there is at least one A in the sentential forms representing the value and the length of the maximal substrings over $\{0,1\}$ and $\{0',1'\}$ in the generated string in G, respectively.)

Let

$$\bar{P}^{prep}_{h+6,1} = \{E \to Q^{st}_{h+6,1}, S_6 \to S_7\}, \quad \bar{P}^{exe}_{h+6,1} = \{S_7 \to S_7\},$$
$$\bar{P}^{prep}_{h+6,j} = \{E \to Q^{st}_{h+6,j}Z, Z \to Z\}, \quad \bar{P}^{exe}_{h+6,j} = \{Z \to \lambda\},$$
$$\bar{P}^{prep}_{h+7,1} = \{E \to Q^{st}_{h+7,1}, S_7 \to S_8\}, \quad \bar{P}^{exe}_{h+7,1} = \{S_8 \to S_8\},$$
$$\bar{P}^{prep}_{h+7,j} = \{E \to Q^{st}_{h+7,j}\}, \quad \bar{P}^{exe}_{h+7,j} = \emptyset.$$

(These components can work successfully only if no A is present at the corresponding components, that is, both the lengths of the maximal substrings over $\{0,1\}$ and $\{0',1'\}$ in the generated string in G and their values as binary numbers are equal.)

Then, let

$$\bar{P}^{prep}_{h+8,1} = \{E \to Q^{st}_{h+8,1}, S_8 \to S_9\}, \quad \bar{P}^{exe}_{h+8,1} = \{S_9 \to \lambda\},$$
$$\bar{P}^{prep}_{h+8,j} = \{E \to Q^{st}_{h+8,j}\}, \quad \bar{P}^{exe}_{h+8,j} = \emptyset.$$

(These components produce the generated word at component $\bar{P}^{exe}_{h+8,1}$, which can be forwarded to the master component.)

Finally, let for $0 \leq k \leq h+8$ and $1 \leq i \leq 5$

$$P_{k,i}^{fw} = \{S \to Q_k^{sel}, E \to E', E' \to E", E" \to S, F \to Q_{k,i}^{st}D, D \to D', D' \to \lambda\}.$$

(These components are for storing the 5-tuples of sentential forms representing the string in generation in G before forwarding them to the next grammar collection.)

We shall prove that $L(G) = L(\Gamma)$ holds. We first show that grammar collections U_j, $0 \leq j \leq h+8$, are able to simulate both the application and the skipping the execution of a production of G (an activity simulating the application of productions $0'0 \to \lambda$ and $1'1 \to \lambda$). (U_0 is for initialization.) Suppose that at some step of the derivation, $G_{k,i}^{st}$ has $\alpha_{k,i}D$ as its sentential form, where $\alpha_{k,i} \in ((N \setminus \{D\}) \cup T)^*$, $0 \leq k \leq h+8$, $1 \leq i \leq 5$, and all the other components have S as their sentential form (*). Then, by the next step G_k^{sel} decides whether the activity represented by U_k will be executed (E) or skipped (F). This information will be given by request to components G_k^{ini}, $G_{k,i}^{prep}$, $G_{k,i}^{exe}$ and $G_{k,i}^{fw}$ and after a rewriting step and a communication step $\alpha_{k,i}$ or some power of it, depending on k and i, will be forwarded to $G_{k,i}^{prep}$ (if E was selected) or $\alpha_{k,i}$ will be forwarded to $G_{k,i}^{fw}$ (if the choice was F). Meantime the other components of U_k perform synchronization steps. After these derivation steps $G_{k,i}^{st}$ will have S as its sentential form which in the next rewriting step can enter either S' or Q_j^{ini}, where $j = k-1$ for $1 \leq k \leq h+8$, $j = h+8$ for $k = 0$. The derivation can continue only if Q_j^{ini} is chosen, otherwise it gets blocked by the following rewriting step. Then, if G_k^{sel} selected E at the beginning, after a rewriting step, the following communication step and the next rewriting step $G_{k,i}^{exe}$ will either have a sentential form $\beta_{k,i}$ which corresponds to the string obtained from $\alpha_{k,i}$ by performing the activity represented by U_k or the derivation will get blocked if the activity cannot be performed. At the same step $G_{k,i}^{st}$ will ask for the result of the work of U_j, $j = k-1$ for $1 \leq k \leq h+8$ and $j = h+8$ for $k = 0$. The other components execute steps maintaining synchronization. Then, after the induced communication, $G_{k,i}^{st}$ will have string $\gamma_{j,i}D$, $\gamma_{j,i} \in ((N \setminus \{D\}) \cup T)^*$, as a result of the work of U_j, j above, and $G_{r,i}^{st}$ will have $\beta_{k,i}D$, where $r = k+1$, $0 \leq r \leq h+7$, and $r = 0$ for $k = h+8$. All the other components in U_k will have S as their sentential form. Thus, we obtained a configuration of the form (*), and the above procedure can be repeated. If G_k^{sel} selected F at the beginning, then $\alpha_{k,i}$ is stored and preserved in a series of derivation steps, and it is forwarded through $G_{k,i}^{fw}$ to $G_{r,i}^{st}$, r defined above. Similarly to the previous considerations, it can be shown that in this case we shall obtain a configuration of the form (*), too. Together with the informal explanations added to the productions of the components of Γ and the short explanations at the beginning of the chapter, we can see the following. Since any move from a grammar collection to the next one, described above, corresponds to either execution or skipping the application of some production of G (or some activity simulating productions $1'1 \to \lambda$ and $0'0 \to \lambda$), if the 5-tuple $\alpha_{k,i}$, for $i = 1,2,3,4,5$, at $G^{st}_{k,i}$ in a configuration of the form (*) represents a sentential form u of G, then by choosing appropriate activities for the grammar collections passed by the 5-tuples of strings arising from $\alpha_{k,i}$ moving in the system, any derivation in G starting from u can be simulated in Γ. We can observe that in a series of derivation steps from a configuration in the form (*) to another one in the same form, $h + 8$ 5-tuples of strings make one move from a grammar collection to the next one; thus, several derivations in G can be simulated in Γ at the same time. The derivation in Γ aborts if the grammar collections are not able to perform the chosen activity.

Now we show that starting from the initial configuration the first few steps of any derivation in Γ lead to a configuration of the form (*) above. At the initial configuration each component has S as its sentential form. Then G^{sel}_k, $0 \le k \le h + 8$, selects either E or F and this information is forwarded by request to G^{ini}_k, $G^{prep}_{k,i}$, $G^{exe}_{k,i}$ and $G^{fw}_{k,i}$, $1 \le i \le 5$. Meantime S at $G^{st}_{k,i}$ can enter either S' or Q^{ini}_j, $j = k-1$ for $1 \le k \le h+8$ and $j = h+8$ for $k = 0$, but only S' can be chosen, otherwise the derivation gets blocked. Suppose now that the choice of G^{sel}_k at the beginning was E. Then by the rewriting step following the previous derivation step, S' at $G^{st}_{k,i}$ enters $S^{(1)}$ which is forwarded to $G^{prep}_{k,i}$ by request in several copies, depending on k and i. The other components meantime perform synchronization steps. The derivation can continue if E was selected only for $k = 0$ and in the other cases F was chosen, because components $G^{prep}_{k,1}$ cannot continue the derivation for the lack of production for $S^{(1)}$. Suppose that this is the case. Meantime, after communication $G^{st}_{k,i}$ returns to the axiom, S. Again, from the two possibilities to continue, S' and Q^{ini}_j, only the second one can be chosen, otherwise the derivation gets blocked by the next rewriting steps. Then, similarly to the general case, above, we can show that in few steps $G^{st}_{1,1}$ will have S_0D, and $G^{st}_{1,l}$ will have AD, for $l = 2,3,4,5$. By similar considerations we can prove that if F is selected by G^{sel}_k at the beginning, then the first few steps of the derivations lead to a configuration where $G^{st}_{r,i}$ will have $S^{(1)}D$, for $i = 1,2,3,4,5$ and $r = k + 1$ for $0 \le k \le h + 7$ and $r = 0$ for $k = h + 8$. Thus, after some initial steps Γ enters a configuration of the form (*). If at the initial step G^{sel}_0 selects E, then the simulation of a derivation in G starts, otherwise 5-tuples of $S^{(1)}D$s move among the grammar collections. It can be seen that in the course of derivation the moving 5-tuples of sentential forms either correspond to a sentential form in G or they are 5-tuples of $S^{(1)}D$. Thus, Γ simulates derivations of G. It also can be seen that the words of the language of Γ, which are selected by the master grammar asking $G^{exe}_{h+8,1}$ at a right moment, are the same as the terminal words generated by G. No more terminal word can be obtained by Γ. This implies that $L(\Gamma) = L(G)$. Since the number of nonterminals of Γ is 27, the theorem is proven. □

By Theorem 4.1 and its proof we immediately obtain the following result.

Theorem 4.2 *Size complexity measure $CVar$ and $CProd$ are bounded on the class of recursively enumerable languages with respect to the class of context-free returning PC grammar systems.*

Acknowledgments

Thanks are due to Gy. Vaszil for his useful comments to the previous version of the paper.

References

[1] Csuhaj-Varjú, E.; J. Dassow; J. Kelemen & Gh. Păun (1994), *Grammar Systems: A Grammatical Approach to Distribution and Cooperation*, Gordon and Breach, London.

[2] Csuhaj-Varjú, E. & Gy. Vaszil (1999), On the computational completeness of context-free parallel communicating grammar systems, *Theoretical Computer Science*, 215: 349–358.

[3] Csuhaj-Varjú, E. & Gy. Vaszil (1999), Parallel communicating grammar systems with bounded resources, submitted.

[4] Dassow, J.; Gh. Păun & G. Rozenberg (1997), Grammar Systems, in G. Rozenberg & A. Salomaa, eds., *Handbook of Formal Languages*, II: 155–213. Springer, Berlin.

[5] Geffert, V. (1988), Context-free-like forms for phrase-structure grammars, in *Proceedings of MFCS'88*: 309–317. Springer, Berlin.

[6] Păun, Gh. & L. Sântean, L. (1989), Parallel communicating grammar systems: the regular case, *Ann. Univ. Bucharest, Ser. Mat.-Inf.*, 38.1: 55–63.

[7] Salomaa, A. (1973), *Formal Languages*, Academic Press, New York.

Chapter 5

SUBREGULARLY CONTROLLED DERIVATIONS: RESTRICTIONS BY SYNTACTIC PARAMETERS

Jürgen Dassow
Department of Computer Science
Faculty of Informatics
Otto-von-Guericke-University Magdeburg
PSF 4120, D–39016 Magdeburg, Germany
dassow@iws.cs.uni-magdeburg.de

Abstract We consider regularly controlled context-free grammars and restrict the number of nonterminals or productions which are necessary to generate the regular control language. We investigate the hierarchies with respect to these parameters. Moreover, we compare the succinctness of the description of regular languages by regular grammars (without control) and regular grammars with regular control.

1. INTRODUCTION AND NOTATION

It is a well-known fact that context-free grammars are not able to cover all phenomena of programming languages and natural languages. Therefore a lot of regulating mechanisms have been introduced in the last three decades in order to cnlarge the generative power and to preserve the context-freeness of the rules (see e.g. [2] and [3]). The use of a control language is one of these mechanisms, i.e. only such derivations are allowed where the sequence of applied productions – considered as a word – belongs to a given language. If one considers context-free grammars with a control by context-free languages one already obtains the family of all recursively enumerable languages. If one considers context-free grammars controlled by regular languages one gets the same family of languages as in the case of control by matrices or programs. In this paper we consider

*The first paper written jointly by Gheorghe and me contributed to the syntactic complexity of Lindenmayer systems. Our first joint monograph and most joint papers concerned grammars with controlled derivations. This paper discusses one possiblity to combine both subjects.

C. Martin-Vide and V. Mitrana (eds.), Where Mathematics, Computer Science, Linguistics and Biology Meet, 51-61.

context-free grammars controlled by languages belonging to a subfamily of the family of all regular languages.

Investigations of this type were started in [1], where families of regular languages given by special algebraic or combinatorial properties were used for control. In this paper we restrict the regular languages by syntactic parameters. We require that the language can be generated by a (regular) grammar with a restricted number of nonterminals or productions.

We now recall some notions of the theory of formal (regulated) grammars and languages, mostly, in order to specify our notation. For undefined concepts we refer to [5].

The empty word is designated by λ. By $\#_M(w)$ we denote the number of occurrences of letters from $M \subseteq V$ in the word $w \in V^*$. The set of all positive integers is designated by $\mathbf{N}$. $\#(M)$ denotes the cardinality of the set M.

A *context-free* grammar $G = (N, T, P, S)$ is specified by
– the disjoint alphabets N and T of nonterminals and terminals, respectively,
– the finite set $P \subseteq N \times (N \cup T)^*$ of context-free productions, and
– the axiom or start element $S \in N$.
As usual we use the notation $A \to w$ for $(A, w) \in P$.

A grammar $G = (N, T, P, S)$ is called *regular* if all rules of P are of the form $A \to wB$ or $A \to w$ with $A, B \in N$ and $w \in T^*$.

A regular grammar is in *normal form*, if all rules are of the form $A \to aB$ or $A \to a$ with $A, b \in N$ and $a \in T \cup \{\lambda\}$.

It is well-known that any regular language can be generated by a regular grammar in normal form.

By CF and REG we denote the families of context-free and regular grammars, respectively. By $\mathcal{L}(CF)$ and $\mathcal{L}(REG)$ we designate the families of languages generated by context-free and regular grammars, respectively.

For a regular grammar $G = (N, T, P, S)$ and a regular language L, we define

$$v(G) = \#(N) \quad \text{and} \quad p(G) = \#(P)$$

and

$$\begin{aligned}
v(L) &= \min\{v(G) \mid L = L(G), G\ is\ regular\}, \\
p(L) &= \min\{p(G) \mid L = L(G), G\ is\ regular\}, \\
nv(L) &= \min\{v(G) \mid L = L(G), G\ is\ a\ regular\ grammar \\
&\qquad in\ normal\ form\}, \\
np(L) &= \min\{p(G) \mid L = L(G), G\ is\ a\ regular\ grammar \\
&\qquad in\ normal\ form\}.
\end{aligned}$$

A *grammar with regular control* is a pair (G, R) where
– $G = (N, T, P, S)$ is a grammar, called the underlying grammar, and
– R is a regular language over P, called the control language.

The language $L(G,R)$ generated by (G,R) consists of all words $z \in T^*$ such that there is a derivation

$$S \underset{p_1}{\Longrightarrow} w_1 \underset{p_2}{\Longrightarrow} w_2 \underset{p_3}{\Longrightarrow} \dots \underset{p_n}{\Longrightarrow} w_n = z$$

with $p_1 p_2 \dots p_n \in R$.

For a context-free grammar $G = (N,T,P,S)$ and a word $p \in P^*$, the language $L(G,\{p\})$ is empty or consists of a single word, which we denote by $w(G,p)$. Moreover, let $o(G)$ be the set of all words $p \in P^*$ such that $L(G,\{p\})$ is non-empty. Then, for an arbitrary language $R \subseteq P^*$, we get

$$L(G,R) = \{w(G,p) \mid p \in R \cap o(G)\} = L(G, R \cap o(G)) . \tag{5.1}$$

By $\mathcal{L}(CF, REG)$ and $\mathcal{L}(REG, REG)$ we denote the families of all languages which can be generated by grammars (G,R) with regular control where the underlying grammar G is a context-free and regular grammar, respectively.

For $X \in \{CF, REG\}$, $c \in \{v, p, nv, np\}$ and a positive integer i, by $\mathcal{L}(X,c,i)$ we denote the family of languages which can be generated by grammars (G,R) with regular control where $G \in X$ and $c(R) \leq i$.

Through these definitions we immediately get the following statement.

Lemma 5.1 *For any $c \in \{v, p, nv, np\}$, $X \in \{CF, REG\}$ and any integers i and j with $i \leq j$,*

$$\mathcal{L}(X,c,i) \subseteq \mathcal{L}(X,c,j) \subseteq \mathcal{L}(X, REG) .$$

For any $d \in \{v, p\}$, $X \in \{CF, REG\}$ and $i \in \mathbf{N}$,

$$\mathcal{L}(X, nd, i) \subseteq \mathcal{L}(X, d, i) .$$

The aim of this paper is the investigation of the hierarchies of the families $\mathcal{L}(X,c,i)$, $c \in \{v, p, nv, np\}$, $X \in \{CF, REG\}$ and $i \in \mathbf{N}$. We shall prove that, for underlying context-free grammars,
- for the measure v, the hierarchy is finite and consists of a single set,
- for the measures p and np, the hierarchies are infinite, and
- for the measure nv, the first two elements of the hierarchy are different.

Moreover, we show that,
- for $i \in \mathbf{N}$, $\mathcal{L}(REG, n, i) = \mathcal{L}(REG, nv, i) = \mathcal{L}(REG)$ and
- for the measures p and np and underlying regular grammars, the hierarchies are infinite.

Furthermore, we compare the succinctness of description of regular languages by regular grammars with regular control and by regular grammars (without control).

2. UNDERLYING CONTEXT-FREE GRAMMARS

We start with the consideration of the number of nonterminals.

Theorem 5.1 *For $i \in \mathbf{N}$, $\mathcal{L}(CF, v, i) = \mathcal{L}(CF, REG)$*

Proof. We first recall the notion of a matrix grammar.

A context-free matrix grammar is a pair (G, M) where $G = (N, T, P, S)$ is a context-free grammar and M is a finite set of sequences of rules from P. We say that $x \in (N \cup T)^*$ derives $y \in (N \cup T)^*$ by an application of the matrix $m = (p_1, p_2, \ldots, p_n) \in M$, denoted as $x \underset{m}{\Longrightarrow} y$ iff

$$x \underset{p_1}{\Longrightarrow} w_1 \underset{p_2}{\Longrightarrow} w_2 \underset{p_3}{\Longrightarrow} \ldots \underset{p_n}{\Longrightarrow} w_n = y.$$

The language $L(G, M)$ generated by the matrix grammar (G, M) consists of all words $z \in T^*$ such that there are an integer k and matrices $m_1, m_2, \ldots, m_k$ in M with

$$S \underset{m_1}{\Longrightarrow} v_1 \underset{m_2}{\Longrightarrow} v_2 \underset{m_3}{\Longrightarrow} \ldots \underset{m_k}{\Longrightarrow} v_k = z.$$

With a matrix $m = (p_1, p_2, \ldots, p_n) \in M$ we now associate a word $m' = p_1 p_2 \ldots p_n$ over P. From the definitions we obtain

$$S \underset{m_1}{\Longrightarrow} v_1 \underset{m_2}{\Longrightarrow} v_2 \underset{m_3}{\Longrightarrow} \ldots \underset{m_k}{\Longrightarrow} v_k = z$$

if and only if

$$S \underset{q_1}{\Longrightarrow} u_1 \underset{q_2}{\Longrightarrow} u_2 \underset{m_3}{\Longrightarrow} \ldots \underset{q_t}{\Longrightarrow} v_t = z$$

where $q_1 q_2 \ldots q_t = m'_1 m'_2 \ldots m'_k$. Hence

$$z \in L(G, M) \quad \text{iff} \quad z \in L(G, \{m' \mid m \in M\}^+).$$

With the notation $R = \{m' \mid m \in M\}^*$ we get $L(G, M) = L(G, R)$ for the (context-free) grammar (G, R) with the regular control language R. It is well-known that, for any language $L \in \mathcal{L}(CF, REG)$ there is a matrix grammar (G, M) such that $L = L(G, M)$. From the above relations we obtain, $L = L(G, R)$.

In order to prove the statement, we have to show that $n(R) = 1$. This follows from the fact that the regular grammar

$$G' = (\{S'\}, P, \{S' \to m'S' \mid m \in M\} \cup \{S' \to m' \mid m \in M\}, S')$$

generates R. □

Obviously, the grammar G' in the proof of Theorem 5.1 is not in normal form. Thus this proof is not valid for the measure nv. We now show that the statement does not hold for nv either.

Theorem 5.2 $\mathcal{L}(CF) = \mathcal{L}(CF, nv, 1) \subset \mathcal{L}(CF, nv, 2)$.

Proof. i) $\mathcal{L}(CF) \subseteq \mathcal{L}(CF, nv, 1)$.

Let $L \in \mathcal{L}(CF)$. Then $L = L(G)$ for some context-free grammar $G = (N, T, P, S)$. Obviously, $L = L(G, P^+)$ since the control language allows any derivation. Because P^+ is generated by the regular grammar

$$H = (\{U\}, P, \{U \to pU \mid p \in P\} \cup \{U \to p \mid p \in P\}, U),$$

which is in normal form and has only one nonterminal, we get $L \in \mathcal{L}(CF, nv, 1)$.

ii) $\mathcal{L}(CF, nv, 1) \subseteq \mathcal{L}(CF)$.

Let $H = (\{U\}, P, Q, U\}$ be a regular grammar in normal form with exactly one nonterminal U. Without loss of generality – we can assume that Q does not contain the rule $U \to \lambda$. We set

$$A = \{p \mid p \in P, U \to pU \in Q\} \text{ and } B = \{p \mid p \in P, U \to p \in Q\}.$$

It is easy to see that $L(H) = A^*B$.

Let $L \in \mathcal{L}(CF, nv, 1)$. Then $L = L(G, R)$ where $G = (N, T, P, S)$ is a context-free grammar and $R \subseteq P^*$ is a regular language generated by a regular grammar in normal form with one nonterminal. By the above considerations, $L = L(G, A^*B)$ for some sets $A \subseteq P$ and $B \subseteq P$.

Assume that B contains a rule $p = A \to \alpha$ with $\#_N(\alpha) \geq 1$. Then, for any word $z = z'p$, $L(G, \{z\})$ is empty because the sequence of rules is not applicable or the generated word contains a nonterminal. Thus – without loss of generality (see (5.1)) - we can assume that all rules in B are of the form $A \to \alpha$ with $\alpha \in T \cup \{\lambda\}$. Let N' be the set of left-hand sides of rules in B. We define the finite substitution $\tau : (N' \cup T)^* \to T^*$ by

$$\tau(x) = \begin{cases} \{\alpha \mid A \to \alpha \in B\} & \text{for } x = A \in N' \\ \{x\} & \text{for } x \in T \end{cases}.$$

We now consider the context-free grammar $G' = (N, T, A, S)$. Obviously, any word – not necessarily terminal word – generated by (G', A^*) is a sentential form of G' and conversely. Because the set of sentential forms of a context-free grammar is a context-free language and $\mathcal{L}(CF)$ is closed under intersection with regular sets, the set V of sentential forms of G' of the form x_1Ax_2 with $n \geq 1$, $A \in N'$ and $x_1, x_2 \in T^*$ is a context-free language. Since $\mathcal{L}(CF)$ is closed under finite subsitutions and $L(G, R) = \tau(V)$, we obtain that $L(G, R)$ is context-free.

iii) $\mathcal{L}(CF, nv, 1) \subset \mathcal{L}(CF, nv, 2)$.
The inclusion follows by Lemma 5.1. By the equality $\mathcal{L}(CF) = \mathcal{L}(CF, nv, 1)$ shown in i) and ii) it is sufficient to give a non-context-free language which is in $\mathcal{L}(CF, nv, 2)$.

We consider the context-free grammar

$$G = (\{S, A, B\}, \{a, b, c\}, \{p_1, p_2, p_3, p_4, p_5\}, S)$$

with the rules

$$p_1 = S \to AB,\ p_2 = A \to aAb,\ p_3 = B \to cB,\ p_4 = A \to ab,\ p_5 = B \to c$$

and the control language

$$R = (\{p_1, p_3\}\{p_2, p_4\})^* \{p_5\}.$$

R is generated by the regular grammar

$$H = (\{T, U\}, \{p_1, p_2, p_3, p_4, p_5\}, P, T)$$

with

$$P = \{T \to p_1U, T \to p_3U, T \to p_5, U \to p_2T, U \to p_4T\}.$$

Moreover, H is in normal form and has only two nonterminals. Thus

$$L(G, R) \in \mathcal{L}(CF, nv, 2). \tag{5.2}$$

We now determine $L(G, R)$. A word $z \in R \cup o(G)$ has to satisfy the following conditions:

- p_1 is the first letter of z (since p_3 cannot be applied to the axiom S),
- $\#_{p_1}(z) = 1$ (the symbol S does not occur in any sentential form different from the axiom; thus p_1 is only applicable to the axiom),
- $\#_{p_4}(z) \geq 1$ (otherwise, A is in any sentential form besides the axiom),
- $\#_{p_4}(z) \leq 1$ (the first application of p_4 deletes A; hence A is not present in any sentential form generated after the first application of p_4),
- p_4p_5 is the suffix of z of length 2 (otherwise p_2 has to be applied sometimes after p_4 which is impossible for reasons analogous to those given in the preceding condition).

Thus $z = p_1(p_2p_3)^n p_4p_5$ where $n \geq 0$. Moreover,

$$w(G, p_1(p_2p_3)^n p_4p_5) = a^{n+1}b^{n+1}c^{n+1}.$$

Thus, by (5.1) we obtain

$$L(G,R) = L(G, \{p_1(p_2p_3)^n p_4p_5 \mid n \geq 0\} = \{a^{n+1}b^{n+1}c^{n+1} \mid n \geq 0\}.$$

By (5.2), $\{a^n b^n c^n \mid n \geq 1\} \in \mathcal{L}(CF, nv, 2)$. On the other hand $\{a^n b^n c^n \mid n \geq 1\} \notin \mathcal{L}(CF)$ is well known. □

We now study the parameter number of productions.

Theorem 5.3 *Let $c \in \{p, np\}$ and $i \in \mathbf{N}$. Then there is an integer $j > i$ such that*

$$\mathcal{L}(CF, c, i) \subset \mathcal{L}(CF, c, j).$$

Proof. We only give the proof for p. The proof for np follows from the same arguments.

For $n \in \mathbf{N}$, we consider the language

$$L_n = a_1^+ \cup a_2^+ \cup \ldots \cup a_n^+.$$

Let $G_n = (N, \{a_1, a_2, \ldots, a_n\}, P, S)$ be a context-free grammar and R_n a regular language such that $L = L(G_n, R_n)$, and let $H_n = (N', P, Q, T)$ be a regular grammar such that $L(H_n) = R_n$. Without loss of generality we can assume that

$$P = P_0 \cup P_1 \cup P_2 \cup \ldots \cup P_n$$

with

$$\begin{aligned} P_0 &= \{A \to w \mid w \in N^*\}, \\ P_i &= \{A \to w \mid w \in (N \cup \{a_i\})^*, \#_{a_i}(w) \geq 1\} \text{ for } 1 \leq i \leq n. \end{aligned}$$

From the structure of the words in L_n, any control word $z \in o(G_n)$ has to be in $(P_i \cup P_0)^* P_i (P_i \cup P_0)^*$ for some i, $1 \leq i \leq n$. Hence in order to generate z we need a rule $q_i \in Q$ with $q_i = B \to v$, $\#_{P_i}(v) \geq 1$. Moreover, $\#_{P_j}(v) = 0$ has to hold for $j \neq i$, $1 \leq j \leq n$ because the existence of an element of P_j in v leads to a word $w(G_n, z)$ which contains a_i and a_j in contrast to the structure of L_n. Thus $q_i \neq q_j$ for $1 \leq i, j \leq n$, $i \neq j$. Therefore H_n has at least n productions and

$$L \notin \mathcal{L}(CF, p, i) \text{ for } i < n. \tag{5.3}$$

On the other hand, it is easy to see that $L = L(G_n, R_n)$ and $R_n = L(H_n)$ for

$$\begin{aligned} &G_n = (\{S\}, \{a_1, a_2, \ldots, a_n\}, \{q, p_1, p_2, \ldots, p_n\}, S), \\ &q = S \to \lambda \quad \text{and} \quad p_i = S \to a_i S \text{ for } 1 \leq i \leq n, \end{aligned}$$

$$R_n = p_1^+ q \cup p_2^+ q \cup \ldots \cup p_n^+ q,$$
$$H_n = (\{T, A_1, A_2, \ldots, A_n\}, \{q, p_1, p_2, \ldots, p_n\},$$
$$\{p_1', q_1', r_1', p_2', q_2', r_2', \ldots, p_n', q_n', r_n'\}, T),$$
$$p_i' = T \to p_i A_i,\ q_i' = A_i \to p_i A_i,\ r_i' = A_i \to q \text{ for } 1 \le i \le n.$$

Therefore $L_n \in \mathcal{L}(CF, p, 3n)$. Now $\mathcal{L}(CF, p, i) \subset \mathcal{L}(CF, p, 3n)$ for $i < n$ follows from Lemma 5.1 and (5.3). Hence it is sufficient to choose $j = 3(i+1)$ in order to prove the statement. □

3. UNDERLYING REGULAR GRAMMARS

In the case of regular grammars with control all hierarchies with respect to the number of nonterminals are finite.

Theorem 5.4 *For $c \in \{v, nv\}$ and $i \in \mathbf{N}$, $\mathcal{L}(REG, c, i) = \mathcal{L}(REG)$.*

Proof. As in part i) of the proof of Theorem 5.2 $\mathcal{L}(REG) \subseteq \mathcal{L}(REG, nv, 1)$ can be shown. If we take into consideration that $\mathcal{L}(REG, REG) = \mathcal{L}(REG)$ (see [2], [3]) we obtain from Lemma 5.1

$$\begin{aligned} \mathcal{L}(REG) &\subseteq \mathcal{L}(REG, nv, 1) \subseteq \mathcal{L}(REG, nv, i) \\ &\subseteq \mathcal{L}(REG, v, i) \subseteq \mathcal{L}(REG, REG) \\ &= \mathcal{L}(REG). \end{aligned}$$

Hence

$$\mathcal{L}(REG, nv, i) = \mathcal{L}(REG, v, i) = \mathcal{L}(REG).$$

□

Theorem 5.4 can be interpreted as follows: any regular language can be generated by a regular grammar with regular control such that the control language has minimal complexity. We now prove that an analogous result holds for the underlying grammar.

Theorem 5.5 *For any regular language L there are a regular grammar G (in normal form) with $v(G) = 1$ and a regular language R such that $L = L(G, R)$.*

Proof. Let $L = L(G')$ for some regular grammar $G' = (N, T, P, S)$. With any production $p = A \to \alpha B \in P$ and $q = A \to \beta \in P$ we associate the productions

$$p' = U \to \alpha U,\ p'' = A \to p'B \quad \text{and} \quad q' = U \to \beta,\ q'' = A \to q'.$$

We construct the regular grammars

$$\begin{aligned} G &= (\{U\}, T, \{p' \mid p \in P\}, U), \\ H &= (N, \{p' \mid p \in P\}, \{p'' \mid p \in P\}, S). \end{aligned}$$

By construction,
– $v(G) = 1$ (and G is in normal form if G' is in normal form),
– for $z = p_1''p_2''\dots p_k'' \in L(H)$, either $z \notin o(G)$ or $w(G,z) = \alpha_1\alpha_2\dots\alpha_k \in L(G')$.
Therefore, by (5.1), $L(G, L(H)) = L(G')$ which proves the statement. □

In the proofs of Theorems 5.4 and 5.5 the underlying grammar or the regular control language have minimal complexity whereas the other component has the same complexity as the given regular language itself. However, the following result shows that both components can be essentially less complex than the generated language considered as a language generated by a regular grammar.

Theorem 5.6 *For any $n \in \mathbf{N}$, there is a regular language L_n such that $v(L_n) = n^2+1$ and $L_n = L(G_n, R_n)$ holds for a regular grammar G with $v(G_n) = n+1$ and $v(R_n) \le n+1$.*

Proof. For $n = 1$ the statement follows from $n(a^+b^+) = 2$ and part i) of the proof of Theorem 5.2.

For $n \ge 2$, we consider the language

$$L_n = (ab)^+ \cup (a^2b)^+ \cup (a^3b)^+ \cup \dots \cup (a^{n^2}b)^+ .$$

By [4],

$$v(L_n) = n^2 + 1 .$$

We set

$$\begin{aligned}
G_n &= (\{S, A_0, A_1, \dots, A_{n-1}\}, \{a, b\}, P_n, S),\\
P_n &= \{p_i \mid 0 \le i \le n-1\} \cup \{q_{in+j} \mid 0 \le i \le n-1, 1 \le j \le n\}\\
&\qquad \cup\{r_{in+j} \mid 0 \le i \le n-1, 1 \le j \le n\},\\
p_i &= S \to A_i,\ q_{in+j} = A_i \to a^{in+j}bA_i,\ r_{in+j} = A_i \to a^{in+j}b\\
&\qquad \text{for } 0 \le i \le n-1,\\
R_n &= \bigcup_{j=1}^{n} \{p_i \mid 0 \le i \le n-1\}\{q_{in+j} \mid 0 \le i \le n-1\}\{r_{in+j} \mid\\
&\qquad 0 \le i \le n-1\},\\
H_n &= (\{T, B_1, B_2, \dots, B_n\}, P_n, Q_n, T),\\
Q_n &= \{T \to p_iB_j \mid 0 \le i \le n-1\} \cup \{B_j \to q_{in+j}B_j \mid 0 \le i \le n-1,\\
&\qquad 1 \le j \le n\} \cup \{B_j \to r_{in+j} \mid 0 \le i \le n-1, 1 \le j \le n\}.
\end{aligned}$$

We first note that $R_n = L(H_n)$. Since $v(H_n) = n+1$, we obtain $v(R_n) \le n+1$. Because $v(G_n) = n+1$, it is sufficient to show that $L_n = L(G_n, R_n)$.

For the first step of a series of derivations in G_n we choose a nonterminal A_i and for the first step of a series of derivations in H_n we choose a nonterminal

B_j. Both chosen symbols cannot be changed during the derivation. Therefore any word $z \in R_n \cap o(G_n)$ has the form $p_i q_{in+j}^k r_{in+j}$ with $k \geq 0$. Furthermore, $w(G_n, p_i q_{in+j}^k r_{in+j}) = (a^{in+j}b)^k$ holds. Now it is easy to show that $L_n = L(G_n, R_n)$. □

We note that the statement of Theorem 5.6 also holds for regular grammars in normal form. In order to prove this one has to consider the languages L_n, $n \in \mathbf{N}$, of the proof of Theorem 5.3. However, the proof of Theorem 5.6 works with languages over a fixed alphabet of two letters whereas the size of the alphabets of the languages of the proof of Theorem 5.3 grows with the parameter n.

Since the proof of Theorem 5.3 has concerned regular languages we can analogously prove the following statement.

Theorem 5.7 *Let $c \in \{p, np\}$ and $i \in \mathbf{N}$. Then there is an integer $j > i$ such that*

$$\mathcal{L}(REG, c, i) \subset \mathcal{L}(REG, c, j)\,.$$

4. CONCLUDING REMARKS

In this paper we started the investigation of grammars with regular control languages which are limited by their complexity measured by the number of nonterminals and number of productions. We have left *open* the following *problems*:

- Is the hierarchy given by the families $\mathcal{L}(CF, nv, i)$, $i \geq 1$, infinite?

- Which of the inclusions $\mathcal{L}(X, c, i) \subseteq \mathcal{L}(X, c, i+1)$, $X \in \{CF, REG\}$, $c \in \{p, np\}$ and $i \in \mathbf{N}$, is proper?

- Do Theorems 5.3 and 5.7 hold, if (the size of) the alphabet is fixed?

The family of regular languages can also be characterized by finite (deterministic or nondeterministic) automata. The number of states is a very natural measure of the complexity of automata and can be extended to languages by taking the minimum over all automata accepting this language. We recall that the number of nonterminals of the grammar and the number of states of the automaton differ at most by 1 if we use the standard constructions to transform regular grammars in normal form into nondeterministic automata accepting the same language and vice versa. Hence the consideration of nondeterministic automata does not essentially differ from that of regular grammars in normal forms. However, the investigation of finite deterministic automata and their state complexity as a measure for regular control remains open.

References

[1] Dassow, J. (1988), Subregularly controlled derivations: The context-free case, *Rostock Math. Kolloq.*, 34: 61–70.

[2] Dassow, J. & Gh. Păun (1989), *Regulated Rewriting in Formal Language Theory*, Springer, Berlin.

[3] Dassow, J.; Gh. Păun & A. Salomaa (1997), Grammars with controlled derivations, in Rozenberg, G. & A. Salomaa, eds., *Handbook of Formal Languages*, II: 101–154. Springer, Berlin.

[4] Gruska, J. (1967), On a classification of context-free languages, *Kybernetika*, 13: 22–29.

[5] Rozenberg, G. & A. Salomaa, eds. (1997), *Handbook of Formal Languages*, Springer, Berlin.

Chapter 6

NEO-MODULARITY AND COLONIES

Jozef Kelemen
Institut of Computer Science
Silesian University
Bezručovo nám. 13, 746 01 Opava, Czech Republic
kelemen@fpf.slu.cz

Alica Kelemenová
Institut of Computer Science
Silesian University
Bezručovo nám. 13, 746 01 Opava, Czech Republic
kelemenova@fpf.slu.cz

Victor Mitrana
Department of Mathematics
Faculty of Mathematics
University of Bucharest
Str. Academiei 14, 70109 Bucharest, Romania
mitrana@funinf.math.unibuc.ro

Abstract The contribution connects the paradigm arising from so called *neo-modularity*, used, e.g., in robotic systems design, with the formal framework of *colonies* – a part of the theory of grammar systems. Some problems studied in the field of colonies are informally connected with some problems of system design according to the neo-modularity concept. Along these lines, a dynamic descriptional complexity measure, namely the degree of cooperation, is introduced and studied.

*Research supported by the Grant Agency of Czech Republic Grant No. 201/99/1086 and the Dirección General de Enseñanza Superior e Investigación Científica, SB 97–00110508.

C. Martin-Vide and V. Mitrana (eds.), Where Mathematics, Computer Science, Linguistics and Biology Meet, 63-74.

1. ON THE MODULARITY CONCEPT

An almost generally accepted conviction in systems analysis and design is that systems can be modularized into *functional modules.* In fact, decomposition of systems into functional modules is such a basic principle that we have tended to take it for granted. Traditional computer science reflects this tendency in the form of the so called *procedural abstraction* according to which computational units (e.g. procedures) are defined as functions which generate the corresponding outputs from given inputs.

The following example illustrates the situation.

Example 6.1 *Consider a behavior given as an infinite set of strings*

$$L = (\{a, a^2, \dots, a^p\} \cdot \{b, b^2, \dots, b^q\})^+$$

over an alphabet $T = \{a, b\}$. Suppose that L is generated by some generative device and we try to specify this device. It seems to be very "natural" to imagine this device as working in two "basic steps" performed by two specialized "modules":

(1) Construction *of appropriate segments formed by the symbols a and b in the required order, namely $a^i b^j$, $1 \leq i \leq p, 1 \leq j \leq q$, in a certain way.*

(2) Iteration *of the results of (1), received as inputs, leading to outputs of the form $a^{i_1} b^{j_1} a^{i_2} b^{j_2} \dots a^{i_s} b^{j_s}$, for some $s \geq 1$ and $1 \leq i_r \leq p, 1 \leq j_r \leq q$ for all $1 \leq r \leq s$. The two "modules" can be described by simple formal grammars as follows.*

Construction of the appropriate segments can be realized by starting from the symbol S and using the rules

$$
\begin{aligned}
& S \to aA_1, \\
& A_1 \to aA_2, A_1 \to bB_1, A_1 \to B_1, \\
& \dots \\
& A_{p-1} \to aA_p, A_{p-1} \to bB_1, A_{p-1} \to B_1, \\
& A_p \to B_1, \\
& B_1 \to bB_2, B_1 \to b, \\
& B_2 \to bB_3, B_2 \to b, \\
& \dots \\
& B_{q-1} \to bB_q, B_{q-1} \to b,
\end{aligned}
$$

operating on the set $N = \{S, A_1, ...A_p, B_1, ...B_q\}$ of nonterminal symbols.

Iteration generating strings in the required form starting from the strings prepared before can be described by the rules:

$$B_1 \to S, B_2 \to S, \dots B_q \to S.$$

This style of thinking and the resulting system decomposition are examples of systems decomposition into functionally specified modules. Actually, all variants of classical *formal grammars* are in a certain sense examples of functionally modularized systems for generating (infinite) sets of symbol strings – *formal languages*.

But while functional decomposition is certainly a legitimate architectural principle, it is not the only one applied in the present time.

Stein states in [12] that many systems are neither analyzable nor synthesizable as in the previously sketched situation. This follows mainly from the interactive nature of most *prima face* non-computational systems (like behavior-based robots or biological living systems) but also of many computational systems (like software agents or computer networks). The common characteristics of such systems are their massive mutual interaction and/or their massive interaction with dynamically changing environments in which they are (physically or informationally) situated.

Generally, an *interactive system* is – according to [12] – one in which information is continually being sensed and behavior is continually being generated. Interactive systems exploit properties of their environment to generate future behaviors. They are always operating concurrently with their environment. While traditional computer procedures – while executing – are blissfully ignorant of their environment, an interactive system, in contrast, is always responding to its environment. It means also, that interactive systems are opportunistic in the sense that they depend on simpler systems without necessarily preserving modular-functional encapsulation with all of the redundancies, interdependences, etc., that this implies.

Fortunately, this does not mean that we must give up all our convictions about principled analysis or synthesis. Stein formulates three principles – imagination, representational alignment through shared grounding, and incremental adaptation – for the analysis and synthesis of the so called *neo-modular systems*. However, the basic principle distilled from experience in, for example, behavior-based robotics, software agent technology or computer networking and open information system design is that the neo-modular systems are set up *opportunistically* from parts working in the massive interaction with their environment(s) where interactive systems exploit their environments to generate future behaviors, and that the behavior of the whole neo-modular system *emerges* from this interaction.

Example 6.2 *Consider the language L from Example 6.1 as a required behavior of a system again. However, suppose that for some reason which was obvious before the goal of generating the behavior L is stated, we have at hand a device G_1 for generating a finite behavior*

$$L_1 = \{a^i \mid 1 \leq i \leq p\} \cup \{a^i B a^j \mid 1 \leq i, j \leq p\}$$

over an alphabet $T_1 = \{a, B\}$ *starting from an initial symbol* A.

Now, are we able to proceed opportunistically by constructing some further generative device which will cooperate with the existing one in order to have a device generating the behavior L *using the device* G_1, *previously developed for generating the behavior* L_1, *instead of defining a completely new device? In other words, let us consider the current string generated by* G_1 *as a specific kind of symbolic environment in which* G_1 *is situated. Are we able to add some new generative device, say* G_2, *operating in the same environment and exploiting the results of the work of* G_1 *such that the common activity of both devices will result in the desired behavior* L?

The answer is affirmative. We consider a generative device G_2 *generating the behavior*

$$L_2 = \{b^j \mid 1 \leq j \leq q\} \cup \{b^j A b^k \mid 1 \leq j, k \leq q\}$$

starting from the initial symbol B.

If the initial state of the environment is AB, *the devices* G_1 *and* G_2 *are able to generate all states of the form required by* L.

2. SHARED GROUNDING AND ADAPTATION

The modules G_1 and G_2 operate without any direct communication, and perform only individual changes in the shared environment. This style of work is an example of the *representational alignment through shared grounding* supposed in [12] as one among the main architectural principles of neo-modularity.

The interactive systems placed in the same environment will act independently of each other, according to their own rules, and their acts will cause some changes in the environment states. As Example 6.2 illustrates, we can capitalize on the "bootstrap" of systems sharing the environment, which may result. What G_1 produced, G_2 is able to change, and the result can be the input for G_1 again. In such a way, individual activities of the two devices (each of them with finite behavior) can be combined – without any direct communication between them, purely through the shared environment – into a more complex infinite behavior L. Example 6.2 illustrates how – in contrast to the principle of funcional modularization illustrated by Example 6.1 for engineering systems – neo-modularity does not require interfaces for direct communication between modules. Instead, the required behavior is achieved in such a way that each device placed in an environment is designed to work independently only by providing interactions with the environment (note that the environment does not mean necessarily some physical outer reality of some physically embodied agent; it may be understood also as some kind of "data reality" or "virtual reality", which is, in fact, the case in the aforementioned examples).

Another principle of neo-modularity which can be illustrated by Example 6.2 is the principle of *adaptation*. In [12] *incremental adaptation* is mentioned. However, we will concentrate on the adaptation only, because in our context the adjective "incremental" – bearing in mind its meaning in literature on learning – is not very appropriate. What is often needed in order to build a system with a more complex behavior starting from components (modules) behaving as simply as possible? This problem can be solved in two main ways: through changing components or through changing internal behaviors of components.

Example 6.2 illustrates how by adding a new component with a very simple behavior one can change the global behavior and make it considerably more complex than before. In [8], the possibility of obtaining a kind of a low level rational behavior from non-rational ones in such a way is discussed.

Another possibility (more similar to L. Steins' concept of incremental adaptation) consists in associating some kind of weights (e.g. probabilities). Such a model, very closely related to the devices used in Example 6.2, is studied in [5, 6] with rules appearing in devices mentioned in Example 6.2. These weights may change with respect to the success of the devices' interactions; thus the behavior of devices as well as the one emerging from their activities in the shared environment can change in an appropriate way.

3. COLONIES

What we have defined in Example 6.2 is actually an example of a well-known structure in the theory of formal grammars and languages – that of a *grammar system* or, more precisely, an example of a version of grammar systems called *colonies* introduced in [10]. For a monographic presentation of the theory of grammar systems see [1]. Basic information is also presented also in [4]. More about colonies can be found in [9].

We consider a colony as a collection of *regular grammars* generating *finite languages* only – operating on a common string of symbols (a *sentential form*) in a sequential manner.

A colony behaves in a symbolic environment – a finite string – and changes its state only through acts performed by the *components* of the colony. The states may change only through acts performed by components. The set of all "final" states of the environment (these states cannot be changed any more) will be considered as the language generated by the colony. The environment itself is quite passive – it does not change its state autonomously, but only through the *sequential* activities of the components of a colony.

Because of the lack of any predefined *strategy of cooperation* of components, each component participates in the rewriting of the current string whenever it can. Conflicts are solved nondeterministically, as usual in the classical theory of formal grammars. For the definition and for some basic properties of *parallel* colonies see [2].

A *behavior* of a colony is defined as a set of all terminal strings which can be generated by the colony from a given *initial string*.

More formally, a *colony* C is a 4-tuple $C = (\mathcal{R}, V, T, w)$, where

– $\mathcal{R} = \{R_i : 1 \leq i \leq n\}$ is a finite set of regular grammars $R_i = (N_i, T_i, P_i, S_i)$ producing the finite languages $L(R_i) = F_i$. R_i are called the *components* of C.

– $V = \bigcup_{i=1}^{n} (T_i \cup N_i)$ is the *alphabet* of the colony,

– $T \subseteq V$ is the *terminal alphabet* of the colony, and

– $w \in V^*$ is the *initial string*.

We note that a terminal symbol of one grammar can occur as a nonterminal symbol of another grammar.

The activity of components in a colony is done by transformations performed on the shared string. Elementary changes of strings are determined by a *basic derivation step*:

For $x, y \in V^*$ and $1 \leq i \leq n$ we define $x \Longrightarrow_i y$ iff

$$x = x_1 S_i x_2,\ y = x_1 z x_2, \text{ where } z \in F_i.$$

The *language* (or the *behavior*) determined by a colony C starting with a word is given by

$$\begin{aligned} L(C) \quad &= \quad \{v \in T^* \mid w \Longrightarrow_{i_1} w_1 \Longrightarrow_{i_2} w_2 \Longrightarrow \ldots \Longrightarrow_{i_m} w_m = v, \\ &\qquad 1 \leq i_1, i_2, \ldots, i_m \leq n\}. \end{aligned}$$

Colonies highlight the decomposition of behavior generating devices into modules – components – which interact on rewriting symbols appearing in the sentential form they share. Individually each component R_i of a colony C changes the state of its environment rewriting some symbols in it if the conditions for rewriting them by R_i are (or come to be) satisfied (as results of activities of other component). In such a case each R_i acts by performing several rewriting steps (according to the previously defined mode of derivation in C). So, the strategy of component cooperation in a colony C is determined by

– the history of the whole activity of C in the past (the current structure of the sentential form generated by it),

– its own disposition to be active.

A colony generates a behavior which really *emerges* from the individual more or less coordinated behaviors of its components. We may say that the languages generated by colonies characterize common structural properties of sequences of acts which can be performed by their components. From such a perspective colonies seem to represent a suitable theoretical framework for expressing some fundamental features of systems based on the neo-modularity idea.

As one can easily see, it is sufficient to simply identify the components by pairs (S_i, F_i), where strings in F_i do not contain the symbol S_i. For technical reasons we shall adopt this definition in what follows.

4. SOME PROBLEMS

The connection of the idea of neo-modularity with the theory of colonies leads to some interesting new questions for the study of colonies. The aim of this section is to formulate and solve some of them.

As proved in [10], for each context-free language L there exists a colony generating this language. This means that if a required behavior of a system is context-free then there exists a colony which generates such a behavior. Moreover, it is proved that for any k component colony there exists a context-free language which cannot be generated by this colony. In other words, the class $\mathcal{L}_c$ of all languages which can be generated by colonies cannot be defined as a union of a finite number of classes $\mathcal{L}_{c_i}$, where $\mathcal{L}_{c_i}$ stands for the class of languages which can be generated by colonies with no more than i components.

We will now define a dynamic measure of descriptional complexity of a colony which tries to express how much the components of a colony cooperate. This measure has been already investigated for context-free grammars [3]. Although colonies cannot generate more than context-free languages, we shall prove that there exists an infinite set of context-free languages which can be generated by colonies with a smaller degree of communication than that of any context-free grammar generating them.

Obviously, it is not compulsory that all components of a colony C participate in generating some strings in $L(C)$. So, we may ask what is the minimal number of components of C which take part in the generating process of a given string in $L(C)$. We define a function, denoted by *coop*, which associates the minimal number of components $coop(x, C)$ that are necessary for generating x with each string x in the language generated by a colony C, and investigate the computability of this function.

Formally, for every word $x \in L(C)$ and every derivation D of x in C, of the following form

$$S \Longrightarrow_{i_1} x_1 \Longrightarrow_{i_2} x_2 \ldots \Longrightarrow_{i_m} x_m = x$$

we define

$$coop(x, D) = \text{card}(\{i_j \mid 1 \leq j \leq m\}),$$

where $\text{card}(A)$ denotes the cardinality of the finite set A,

$$coop(x, C) = min\{coop(x, D) \mid D \text{ is a derivation of } x \text{ in } C\}$$

and

$$coop(C) = max\{coop(x, C) \mid x \in L(C)\}.$$

If $L(C) = \emptyset$, then $coop(C)$ delivers 0. Finally, for every language L generated by a colony we write

$$coop(L) = min\{coop(C) \mid L = L(C)\}.$$

In a very similar way the degree of cooperation in a context-free grammar has been introduced in [3]. For a context-free language L we denote by $coop_{CF}(L)$ and $coop_{COL}(L)$ the degree of cooperation in context-free grammars and colonies, respectively, for generating L.

The next result states that an infinite number of context-free languages require less cooperation in colonies than in context-free grammars.

Theorem 6.1 *There exists a sequence of context-free languages L_n, $n \geq 2$, such that*

$$coop_{CF}(L_n) = n + 1 \quad \textit{and} \quad coop_{COL}(L_n) = 2.$$

Proof. Let us consider the Dyck language D_n over $T_n = \{a_1, b_1, \ldots, a_n, b_n\}$, $n \geq 1$ i.e. the context-free language generated by the grammar

$$G = (\{S\}, T_n, S, \{S \longrightarrow \lambda, S \longrightarrow SS\} \cup \{S \longrightarrow a_i S b_i | 1 \leq i \leq n\})$$

By induction on n, one can prove that $coop_{CF}(D_n) = n + 1$, for all $n \geq 1$, see, e.g., [3]. On the other hand the following colony has the behavior D_n.

$$C = (T_n \cup \{S_1, S_2\}, T_n, (S_1, F_1), (S_2, F_2), S_1),$$

where

$$\begin{aligned} F_1 &= \{\lambda, S_2 S_2\} \cup \{a_i S_2 b_i \mid 1 \leq i \leq n\}, \\ F_2 &= \{S_1\}. \end{aligned}$$

As one can easily see, $coop(C) \leq 2$, hence $coop_{COL}(D_n) = 2$ for all $n \geq 1$. □

We now investigate the possibility of algorithmically computing this measure for a given string as well as for a given colony.

Theorem 6.2 *One can compute the degree of cooperation for every word in the language generated by a given colony.*

Proof. Let $C = (V, T, (S_1, F_1), (S_2, F_2), \ldots, (S_n, F_n), S_1)$ be an arbitrary colony. For every $1 \leq q \leq n$, we define

$$L_{\leq q}(C) = \{x \in L(C) | coop(x, C) \leq q\}$$

Let q be a given integer between 1 and n. Let $R_1, R_2, \ldots, R_{\binom{n}{q}}$ be all subsets of q integers of $\{1, 2, \ldots, n\}$. The language $L_{\leq q}(\mathcal{C})$ can be expressed now as a finite union of languages as follows:

$$L_{\leq q}(\mathcal{C}) = \bigcup_{j=1}^{\binom{n}{q}} L(\mathcal{C}_j)$$

where $\mathcal{C}_j = (V, T, (S_r, F_r)_{r \in R_j}, S_1), 1 \leq j \leq \binom{n}{q}$. Since all colonies generate context-free languages ([10]) and the family of context-free languages is closed under union ([11]) it follows that all languages $L_{\leq q}(\mathcal{C})$ are also context-free.

Now, for every word $x \in L(\mathcal{C})$, $coop(x, \mathcal{C})$ is the minimal q such that $x \in L_{\leq q}(\mathcal{C})$. The decidability of the membership problem for context-free languages concludes the proof. □

On the other hand,

Theorem 6.3 *There is no algorithm which computes* $coop(\mathcal{C})$ *for any given colony* $\mathcal{C}$.

Proof. Let $\mathcal{C} = (V, \{a, b, c\}, (S_1, F_1), (S_2, F_2), \ldots, (S_n, F_n), S_1)$ be a colony generating

$$L = \{wcv | w, v \in \{a, b\}^+, v \neq w^R\}.$$

Here w^R denotes the mirror image of w. Moreover, let

$$I = \{(u_1, v_1), (u_2, v_2), \ldots, (u_m, v_m)\}$$

be an instance of the Post Correspondence Problem ([7]) over $\{a, b\}$, $m \geq n$.

We say that the instance I has a maximal solution if there are the integers $1 \leq i_1, i_2, \ldots, i_k \leq m$, for some $k \geq m$, such that the following hold:

(i) $x_{i_1} x_{i_2} \ldots x_{i_k} = y_{i_1} y_{i_2} \cdots y_{i_k}$

(ii) for any $1 \leq q \leq m$ there exists $1 \leq j \leq k$ such that $i_j = q$

As any algorithm that decides whether the instance I has a maximal solution can be used in order to decide whether the instance I has an arbitrary solution, it follows that the existence of a maximal solution is an undecidable question.

Now we construct the colony

$$\begin{aligned}\mathcal{C}_I = {} & (V \cup \{S_0, S\} \cup \{X_i, Y_i \mid 1 \leq i \leq m\}, \{a, b, c\}, (S_0, F_0), (S_1, F_1), \\ & (S_2, F_2), \ldots, (S_n, F_n), (X_1, F_{n+1}), (X_2, F_{n+2}), \ldots, (X_m, F_{n+m}), \\ & (Y_1, F_{n+m+1}), (Y_2, F_{n+m+2}), \ldots, (Y_m, F_{n+2m}), S_0)\end{aligned}$$

where S_0 and X_i, Y_i, $1 \leq i \leq m$, are additional symbols not contained in V, and

$$\begin{aligned} F_0 &= \{S, X_1, X_2, \ldots, X_m, Y_1, Y_2, \ldots, Y_m\} \\ F_{n+i} &= \{u_i Y_j v_i^R \mid 1 \leq j \leq m\} \cup \{u_i c v_i^R\}, 1 \leq i \leq m, \\ F_{n+m+i} &= \{X_j \mid 1 \leq j \leq m\}, 1 \leq i \leq m. \end{aligned}$$

Then any derivation of a word in $L(C_I)$ is either of the form

$$D_1 : S_0 \Longrightarrow_0 S_1 \Longrightarrow^* z_1,$$

where $S \Longrightarrow^* z_1$ is a derivation according to C, $z_1 \in L$ and $coop(z_1, D_1) \leq n + 1$, or of the form

$$\begin{aligned} D_2 : \quad & S_0 \Longrightarrow_0 X_{i_1} \Longrightarrow_{n+i_1} u_{i_1} Y_{j_1} v_{i_1}^R \Longrightarrow_{n+m+j_1} u_{i_1} X_{i_2} v_{i_1}^R \Longrightarrow_{n+i_2} \\ & u_{i_1} u_{i_2} Y_{j_2} v_{i_2}^R v_{i_1}^R \Longrightarrow \ldots \Longrightarrow_{n+i_r} u_{i_1} u_{i_2} \ldots u_{i_r} c v_{i_r}^R \ldots v_{i_2}^R v_{i_1}^R = z_2. \end{aligned}$$

We consider a word $z = wcv \in L(C_I)$. If $v \neq w^R$, then there is a derivation of the form D_1 generating z and therefore $coop(z, C) \leq n + 1$.

On the other hand, if $v = w^R$, then z is generated by a derivation of the form D_2 and the sequence $i_1 i_2 \ldots i_r$ is a solution of the instance I. Furthermore, $coop(z, C) = m + 2$ if and only if $i_1, i_2, \ldots, i_r$ is a maximal solution of the instance I.

Hence $coop(C) = 2m + 1$ holds if and only if the instance I has a maximal solution. Thus, the computability of $coop(C)$ would imply the decidability of the Post Correspondence Problem. □

From the standpoint of the neo-modularity concept it is interesting to investigate the effect of the modification of an existing colony by adding new components, from a given set of available components, on the behavior of the colony.

More formally, given a colony which contains the components $G_1, \ldots G_n$ and generates the language $L(C)$, and given a (possibly infinite) set of new components $\mathcal{S} = \{G_{n+1}, G_{n+2}, \ldots\}$, we are allowed to add components from $\mathcal{S}$ to C. In this way, we construct a set of colonies denoted by $COL(C, \mathcal{S})$. Now, many problems naturally arise: Does $L(C_1) = L(C_2)$ hold for two arbitrary colonies from $COL(C, \mathcal{S})$? In this case we say that the original colony is *stabilized* with respect to the set $\mathcal{S}$; that is, one cannot modify its behavior by adding new components from $\mathcal{S}$ to it. If $L(C_1)$ and $L(C_2)$ are incomparable for arbitrary many pairs of colonies from $COL(C, \mathcal{S})$, one may say that the original colony has a *divergent behavior* with respect to $\mathcal{S}$ by continuously changing its behavior.

Another important situation can be observed when the above condition holds for a finite number of pairs of colonies; we say that the original colony has a *convergent behavior* with respect to $\mathcal{S}$.

The universal problem (does a colony generate all strings over a given alphabet) is known to be undecidable due to the undecidability of the same problem for context-free grammars. It follows directly that

Theorem 6.4 *One cannot decide algorithmically whether or not a given colony has a stabilized, divergent or convergent behavior with respect to a given set of components.*

Usually (in practical situations) we are interested in systems generating only finite - maybe very large - languages. In such a case, how can we approximate the desired language in the process of modifying some components of the "starting" colony? And then, what kind of components should we modify?

Such types of problem can be formulated with theoretical rigour within the framework of the theory of colonies, and the eventual results might be interesting from the standpoint of a better understanding of the neo-modular concept of systems, and may be useful for those developing systems who accept the methodology of neo-modularity.

References

[1] Csuhaj-Varjú, E.; J. Dassow; J. Kelemen & Gh. Păun (1994), *Grammar Systems – A Grammatical Approach to Distribution and Cooperation*. Gordon and Breach, London.

[2] Dassow, J.; J. Kelemen & Gh. Păun (1993), On parallelism in colonies, *Cybernetics and Systems*, 24: 37–49.

[3] Dassow, J. & V. Mitrana (1997), Cooperation in context-free grammar, *Theoretical Computer Science*, 180: 353–361.

[4] Dassow, J.; Gh. Păun & G. Rozenberg (1997), Grammar systems, in G. Rozenberg & A. Salomaa, eds, *Handbook of Formal Languages*, II: 155–214. Springer, Berlin.

[5] Gašo, J. (1998), Unreliable colonies as systems of stochastic grammars, in Kelemenova, A., ed., *Grammar Systems – Proceedings of the MFCS'98 Satellite Workshop*: 53–54, Institute of Computer Science, Silesian University, Opava.

[6] Gašo, J. (2000), Unreliable colonies – the sequentional case, *Journal of Automata, Languages and Combinatorics*, accepted.

[7] Harju, T. & J. Karhumäki (1997), Morphisms, in G. Rozenberg & A. Salomaa, eds., *Handbook of Formal Languages*, I: 439–510. Springer, Berlin.

[8] Kelemen, J. (1996), A note on achieving low-level rationality from pure reactivity, *Journal of Experimental & Theoretical Artificial Intelligence*, 8: 121–127.

[9] Kelemen, J. (1997), Colonies as models of reactive systems, in Gh. Păun & A. Salomaa, eds, *New Trends in Formal Languages*: 220–235. Springer, Berlin.

[10] Kelemen, J. & A. Kelemenová (1992), A grammar-theoretic treatment of multiagent systems, *Cybernetics and Systems*, 23: 621–633.

[11] Rozenberg, G. & Salomaa, A., eds. (1997), *Handbook of Formal Languages*, Springer, Berlin.

[12] Stein, L. A. (1997), Post-modular systems – architectural principles for cognitive robotics, *Cybernetics and Systems*, 28: 471–487.

Chapter 7

SEWING CONTEXTS AND MILDLY CONTEXT-SENSITIVE LANGUAGES

Carlos Martín-Vide
Research Group on Mathematical Linguistics
Department of Romance Philologies
Faculty of Arts
Rovira i Virgili University
Pl. Imperial Tàrraco, 1, 43005 Tarragona, Spain
cmv@astor.urv.es

Alexandru Mateescu
Turku Centre for Computer Science (TUCS)
Lemminkäisenkatu 14 A, 20520 Turku, Finland
and
Faculty of Mathematics
University of Bucharest
Str. Academiei 14, 70109 Bucharest, Romania
alexmate@pcnet.pcnet.ro

Arto Salomaa
Turku Centre for Computer Science (TUCS)
Lemminkäisenkatu 14 A, 20520 Turku, Finland
asalomaa@utu.fi

*This work has been partially supported by Project 137358 of the Academy of Finland. All correspondence to Carlos Martín-Vide.

C. Martín-Vide and V. Mitrana (eds.), Where Mathematics, Computer Science, Linguistics and Biology Meet, 75-84.

Abstract Sewing grammars introduced below are very simple grammars, still able to define families of mildly context-sensitive languages. These grammars are inspired by Marcus contextual grammars and simple matrix grammars.

1. INTRODUCTION

The need for *mildly context-sensitive families of languages* has been emphasized in connection with linguistics, see [3]. The aim of our paper is to introduce very simple grammars that can define families of mildly context-sensitive families of languages.

A mildly context-sensitive family of languages should contain the most significant languages that occur in the study of natural languages. Languages in such a family must be semilinear languages, and, moreover, they should be computationally feasible, i.e., the membership problem for languages in such a family must be solvable in deterministic polynomial time complexity.

Remark 7.1 *By a mildly context-sensitive family of languages we mean a family $\mathcal{L}$ of languages such that the following conditions are fulfilled:*

(i) each language in $\mathcal{L}$ is semilinear,

(ii) for each language in $\mathcal{L}$ the membership problem is solvable in deterministic polynomial time, and

(iii) $\mathcal{L}$ contains the following three non-context-free languages:

- multiple agreements: $L_1 = \{a^i b^i c^i \mid i \geq 0\}$,

- crossed agreements: $L_2 = \{a^i b^j c^i d^j \mid i, j \geq 0\}$, *and*

- duplication: $L_3 = \{ww \mid w \in \{a, b\}^*\}$.

The paper is organized as follows. Firstly, we introduce the basic type of sewing grammar and we show that the corresponding family of languages is a mildly context-sensitive family of languages. This type of grammars is extended to a more general type of grammars. We investigate pumping lemmata for the languages defined by these grammars as well as closure properties of these families of languages.

Next we define some families of languages that are almost mildly context-sensitive, and, show that, for languages in these families the problems of equivalence, of inclusion, etc., are decidable problems.

Finally, we compare these grammars with other grammars.

Now we recall some terminology and definitions that we will use in this paper.

Let Σ be an alphabet and let Σ^* be the free monoid generated by Σ with an identity denoted by λ. The free semigroup generated by Σ is $\Sigma^+ = \Sigma^* - \{\lambda\}$.

Elements in Σ^* (Σ^+) are referred to as *words* (*nonempty words*). λ is the *empty word*. A *context* is a pair of words, i.e., (u, v), where $u, v \in \Sigma^*$.

The families of regular, linear, context-free, context-sensitive and recursively enumerable languages are denoted by REG, LIN, CF, CS and RE, respectively.

In this paper the so called regular, resp. linear, simple matrix grammars are of special interest, see [2].

Contextual grammars were firstly considered in [4] with the aim of modelling some natural aspects from descriptive linguistics as the acceptance of a word (construction) only in certain contexts. For a detailed presentation of this topic, the reader is referred to the recent monograph [7].
The family of all Marcus simple contextual languages is denoted by $\mathcal{SM}$.

The reader is referred to [8] for the basic notions of formal languages we use in the sequel and to [1] and [5] for interrelations between linguistics and formal languages.

2. SEWING CONTEXTS

We will now introduce the basic model of *sewing grammars*, with some examples and we will show some properties of these grammars.

Definition 7.1 *A sewing grammar is a construct $G = (\Sigma, B, C, n, f)$, where Σ is an alphabet, $n \geq 1$ is an integer called the degree of G, $B \subset (\Sigma^*)^n$, B finite, is the base of G, $C \subset (\Sigma^*)^n$, C finite, is the set of contexts (or rules) of G and f is a recursive function, $f : (\Sigma^*)^n \longrightarrow \Sigma^*$, called the zipper function of G.*

Using the above notations, a sewing grammar G defines a relation of *direct derivation*, denoted $\Longrightarrow_G$ or $\Longrightarrow$, between elements in $(\Sigma^*)^n$. By definition

$$(x_1, x_2, \ldots, x_n) \Longrightarrow_G (y_1, y_2, \ldots y_n)$$

iff there exists $(z_1, z_2, \ldots, z_n) \in C$, such that $y_i = x_i z_i$, $1 \leq i \leq n$.

The reflexive and transitive closure of $\Longrightarrow_G$ or $\Longrightarrow$ is denoted by $\Longrightarrow_G^*$ or $\Longrightarrow^*$ and called the relation of *derivation* defined by G.

The n*-ary language* defined by G, denoted by $nL(G)$, is by definition:

$$nL(G) = \{(x_1, x_2, \ldots, x_n) \in (\Sigma^*)^n \mid (u_1, u_2, \ldots, u_n) \Longrightarrow_G^* (x_1, x_2, \ldots, x_n),$$

$$\text{for some } (u_1, u_2, \ldots, u_n) \in B\}.$$

Definition 7.2 *Let $G = (\Sigma, B, C, n, f)$ be a sewing grammar. The language defined by G is:*

$$L(G) = \{f(x_1, x_2, \ldots, x_n) \mid (x_1, x_2, \ldots, x_n) \in nL(G)\}.$$

Therefore, the language defined by a sewing grammar $G = (\Sigma, B, C, n, f)$ is the set of all words obtained by applying the zipper function f to the n-tuples from the language $nL(G)$.
Notation. By $SW_n(f)$ we denote the family of all languages generated by sewing grammars of degree n and with the zipper function f.

Remark 7.2 *In this paper we mainly consider that the zipper function f is the catenation function of arity n, denoted cat_n, i.e., the function, $cat_n : (\Sigma^*)^n \longrightarrow \Sigma^*$, $cat_n(u_1, u_2, \ldots, u_n) = u_1 u_2 \ldots u_n$.*
We will drop the indice f whenever this is the case. For instance SW_n denotes the family $SW_n(f)$, where f is the catenation function.

Theorem 7.1 *Let Σ be an alphabet.*

(i) The languages $\emptyset$, Σ^*, $F \subset \Sigma^*$, F finite are in SW_n for every $n \geq 1$.

(ii) The language:

- multiple agreements, $L_1 = \{a^i b^i c^i \mid i \geq 0\}$, is in SW_n for every $n \geq 3$.

(iii) The language:

- crossed agreements, $L_2 = \{a^i b^j c^i d^j \mid i, j \geq 0\}$ is in SW_n for every $n \geq 4$.

(iv) The language:

- duplication, $L_3 = \{ww \mid w \in \{a, b\}^*\}$ is in SW_n for every $n \geq 2$.

Proof. (i) The language $\emptyset$ is generated by a sewing grammar of degree 1 with $B = \emptyset$. The language Σ^* is generated by a sewing grammar of degree 1 with $B = \{(\lambda)\}$ and the contexts $C = \{(a) \mid a \in \Sigma\}$. Finally, a finite language F is generated by a sewing grammar of degree 1 with $B = \{(u) \mid u \in F\}$ and the contexts $C = \{(\lambda)\}$.
Therefore all these languages are in SW_1 and it follows that they are in all families SW_n, with $n \geq 1$.
(ii) The language L_1 is generated by the sewing grammar

$$G_1 = (\{a, b, c\}, B, C, 3, cat_3),$$

where $B = \{(\lambda, \lambda, \lambda)\}$ and $C = \{(a, b, c)\}$.
(iii) The language L_2 is generated by the sewing grammar

$$G_2 = (\{a, b, c, d\}, B, C, 4, cat_4),$$

where $B = \{(\lambda, \lambda, \lambda, \lambda)\}$ and $C = \{[(a, \lambda, c, \lambda)], [(\lambda, b, \lambda, d)]\}$.

(iv) Finally, the language L_3 is generated by the sewing grammar

$$G_3 = (\{a, b, \}, B, C, 2, cat_2),$$

where $B = \{(\lambda, \lambda)\}$ and $C = \{(a, a), (b, b)\}$. □

Theorem 7.2 *Each language in $\mathcal{SW}_n$, where $n \geq 1$, is a semilinear language.*

The *membership problem* consists of the following: given a language $L \subseteq \Sigma^*$ (defined by a certain type of grammar, automaton, etc.) and a word $w \in \Sigma^*$, how can we decide by an algorithm whether or not w is in L? The existence of such an algorithm as well as its complexity are very important from the practical point of view. The next theorem shows that the membership problem for languages in $\mathcal{SW}_p$, $p \geq 1$ is in P, the class of all deterministic polynomial time complexity problems.

Theorem 7.3 *For every $p \geq 1$ and for every $L \in \mathcal{SW}_p$ the membership problem is solvable in deterministic polynomial time.*

Proof. Let $L \subseteq \Sigma^*$ be a language in $\mathcal{SW}_p$, where $p \geq 1$, p fixed. Moreover, let $G = (\Sigma, B, C, p, cat_p)$ be a sewing grammar of degree p such that $L(G) = L$. Let w be a word from Σ^*. We describe an algorithm that decides whether or not $w \in L(G)$. Assume that $w = a_1 a_2 \ldots a_n$, where $a_i \in \Sigma$, $1 \leq i \leq n$. If i, j are integers such that $1 \leq i \leq j \leq n$, then denote by $w_{i,j}$ the word $a_i a_{i+1} \ldots a_j$. Moreover, if $j < i$, then $w_{i,j} = \lambda$. We define a Turing machine M with two tapes. The first tape contains w and its content is not changed during the computations. The second tape is used by M to store $2p$ counters. Each of the values of these counters will be between 0 and $n + 1$. Therefore, the space used by M on the second tape in order to store these counters is $\mathcal{O}(2p\log n)$. The $2p$ counters are denoted by:

$$i_1, j_1, i_2, j_2, \ldots, i_p, j_p.$$

Initially, $i_1 = 1$ and $i_p = n + 1$.

M guesses the counters i_k, $2 \leq k \leq p - 1$ such that:

$$i_1 \leq i_2 \leq i_3 \leq \ldots \leq i_p.$$

and M assigns to each j_k the value of i_{k+1}, where $1 \leq k \leq p - 1$.

Now M performs the following test:

(∗) if $(w_{i_1,j_1-1}, w_{i_2,j_2-1}, \ldots, w_{i_p,j_p-1}) \in B$, then M accepts w and *Stop*.

Otherwise, M guesses values for j_k such that $i_k \leq j_k \leq i_{k+1}$ for all $1 \leq k \leq p$. If for these values of the counters

$$(w_{i_1,j_1}, w_{i_2,j_2}, \ldots, w_{i_p,j_p}) \in C,$$

then M replaces the values of i_k with j_k, $1 \leq k \leq p$ and M repeats the above test $(*)$.

If there is no guessing for the counters

$$i_1, j_1, i_2, j_2, \ldots, i_p, j_p.$$

such that the test $(*)$ leads to acceptance, then M rejects the input w.

Clearly, M accepts w if and only if $w \in L(G)$.

Note that M is a nondeterministic Turing machine that works in space $\mathcal{O}(2p\log n)$. Hence $L(M) \in NSPACE(\log n)$ and by a well-known theorem of Turing complexity, it follows that $L(M) \in \mathcal{P}$, i.e., $L(M)$ is of deterministic polynomial time complexity. Therefore, the membership problem for languages in $\mathcal{SW}_p$ is solvable in deterministic polynomial time. □

From Theorem 7.1, Theorem 7.2 and Theorem 7.3 we obtain the following:

Theorem 7.4 *For every integer $n \geq 4$, the family $\mathcal{SW}_n$ is a mildly context-sensitive family of languages.*

Now we list some pumping lemmata for languages in $\mathcal{SW}_n$, where $n \geq 1$.

Let $G = (\Sigma, B, C, n, cat_n)$ be a sewing grammar of degree n, $n \geq 1$. Let $x = (x_1, x_2, \ldots, x_n)$ be a vector from B. The *length* of x, denoted $|x|$, is by definition, $|x| = |x_1| + |x_2| + \ldots + |x_n|$. Similarly, the *length of a context* $c \in C, c = (u_1, u_2, \ldots, u_n)$ is by definition $|c| = |u_1| + |u_2| + \ldots + |u_n|$. Note that for every $c \in C$, $|c| > 0$, since we can assume that G does not contain a completely empty context.

Theorem 7.5 *Let $L \subseteq \Sigma^*$ be a language in $\mathcal{SW}_n$, $n \geq 1$. There exist two integers $m \geq 1$ and $k \geq 1$ such that:*

(i) (pumping an arbitrary context) If $w \in L$ such that $|w| > m$, then w has a decomposition $w = x_1u_1x_2u_2 \ldots x_nu_nx_{n+1}$, with $0 < |u_1| + |u_2| + \ldots + |u_n| \leq k$, such that for all $i \geq 0$, the following words are in L:

$$w_i = x_1u_1^ix_2u_2^ix_3 \ldots x_nu_n^ix_{n+1}.$$

(ii) (pumping an innermost context) If $w \in L$ such that $|w| > m$, then w has a decomposition $w = x_1u_1y_1x_2u_2y_2x_3 \ldots x_nu_ny_nx_{n+1}$, with $0 < |u_1| + |u_2| + \ldots + |u_n| \leq k$, and $|y_1| + |y_2| + \ldots + |y_n| \leq m$, such that for all $i \geq 0$, the following words are in L:

$$w_i = x_1u_1^iy_1x_2u_2^iy_2x_3 \ldots x_nu_n^iy_nx_{n+1}.$$

(iii) (pumping an outermost context) If $w \in L$ such that $|w| > m$, then w has a decomposition $w = u_1y_1u_2y_2 \ldots u_ny_n$, with $0 < |u_1| + |u_2| + \ldots + |u_n| \leq k$, such that for all $i \geq 0$, the following words are in L:

$$w_i = u_1^iy_1u_2^iy_2 \ldots u_n^iy_n.$$

(iv) *(pumping all occurring contexts) If* $w \in L$ *such that* $|w| > m$, *then* w *has a decomposition* $w = u_1y_1u_2y_2 \ldots u_ny_n$, *with* $0 < |y_1| + |y_2| + \ldots + |y_n| \leq m$, *such that for all* $i \geq 0$, *the following words are in* L*:*

$$w_i = u_1^i y_1 u_2^i y_2 \ldots u_n^i y_n.$$

(v) *(interchanging contexts) If* $w, w' \in L$ *such that* $|w| > m$ *and* $|w'| > m$, *then* w *and* w' *have decompositions:* $w = x_1u_1x_2u_2 \ldots x_nu_nx_{n+1}$, *and* $w' = x'_1u'_1x'_2u'_2 \ldots x'_nu'_nx'_{n+1}$, *with* $0 < |u_1| + |u_2| + \ldots + |u_n| \leq k$, *and with* $0 < |u'_1| + |u'_2| + \ldots + |u'_n| \leq k$, *such that the following two words,* z *and* z', *are also in* L*:*

$$z = x_1u'_1x_2u'_2 \ldots x_nu'_nx_{n+1}, \text{ and } z' = x'_1u_1x'_2u_2 \ldots x'_nu_nx'_{n+1}.$$

Theorem 7.5 can be used to show that certain languages are not in a family $\mathcal{SW}_n, n \geq 1$.

Theorem 7.6 *If* $m, n \geq 1$ *such that* $m < n$, *then* $\mathcal{SW}_m \subset \mathcal{SW}_n$ *and the inclusion is strict.*

Combining Theorem 7.4 and Theorem 7.6 we obtain:

Theorem 7.7 *The families* $(\mathcal{SW}_n)_{n\geq 4}$ *define an infinite hierarchy of mildly context-sensitive languages.*

3. DECIDABLE PROBLEMS FOR SEWING LANGUAGES

In this section we introduce some special families of sewing languages. Each such family is almost a mildly context-sensitive family of languages, and, moreover, each such family has good decidability properties.

We start by introducing a special type of sewing grammar. If n is an integer, $n \geq 1$, then $[n]$ denotes the set $\{1, 2, \ldots, n\}$. An n-function is a function $g : [n] \longrightarrow N$, where N denotes the set of all positive integers. The length of an n-function g, denoted $|g|$, is defined as $|g| = g(1) + g(2) + \ldots + g(n)$.

Definition 7.3 *Let* $n \geq 1$ *be a fixed integer, let* g *be an* n*-function and let* $k \geq 1$ *be a fixed integer. A sewing grammar* $G = (\Sigma, B, C, n, cat_n)$ *is of type* (g, k) *iff for all* $c = (c_1, c_2, \ldots, c_n) \in C$, $|c_i| = g(i)$, $i = 1, 2, \ldots, n$ *and for all* $b = (b_1, b_2, \ldots, b_n) \in B$, $b_i = \lambda$ *for all* $i \neq k$ *and* $|b_k| \leq |g|$.

Notation. Let $n \geq 1$ be an integer, let g be an n-function and let $1 \leq k \leq n$ be an integer. We denote by $\mathcal{SW}_{n,g,k}$ the following family of languages:

$$SW_{n,g,k} = \{L \mid \text{ there exists a sewing grammar } G = (\Sigma, B, C, n, cat_n) \text{ of type } (g,k) \text{ such that } L(G) = L\}.$$

Remark 7.3 *Assume that $n \geq 1$ is a fixed integer. Let g be an n-function and let $1 \leq k \leq n$ be an integer. Consider an alphabet Σ and let Σ_1 and Σ_2 be the following two alphabets:*

$$\Sigma_1 = \{[\alpha] \mid |\alpha| = |g|, \alpha \in \Sigma^*\}$$

and

$$\Sigma_2 = \{[\beta] \mid |\beta| < |g|, \beta \in \Sigma^*\}$$

Note that for each $w \in \Sigma^*$ there exist two unique integers $p, r \geq 0$ such that

$$|w| = p|g| + r \text{ and } 0 \leq r < |g|.$$

Moreover, notice that a unique decomposition exists for w:

$$w = w_1 w_2 \dots w_k \beta w_{k+1} \dots w_n,$$

such that for all $i = 1, 2, \dots, n$, $|w_i| = pg(i)$ and $|\beta| = r$. Hence, each w_i, $1 \leq i \leq n$ is the catenation of p words from Σ^*, $w_i = w_i^{(1)} w_i^{(2)} \dots w_i^{(p)}$ with $w_i^{(j)} \in \Sigma^{g(i)}$, for $1 \leq j \leq p$.

Using the above notations we define a function:

$$\varphi_g^{n.k} : \Sigma^* \longrightarrow \Sigma_1^* \Sigma_2,$$

such that

$$\varphi_g^{n,k}(w) = [w_1^{(1)} w_2^{(1)} \dots w_n^{(1)}][w_1^{(2)} w_2^{(2)} \dots w_n^{(2)}] \dots [w_1^{(p)} w_2^{(p)} \dots w_n^{(p)}][\beta].$$

One can easily prove the following:

Proposition 7.1 *The function $\varphi_g^{n,k}$ is a bijective function.*

The next two results show the importance of the function $\varphi_g^{n,k}$.

Proposition 7.2 *If $G = (\Sigma, B, C, n, cat_n)$ is a sewing grammar of type (g,k), then the language $\varphi_g^{n,k}(L(G))$ is a regular language.*

Proof. We define a regular grammar $G' = (\Sigma, \{S\}, S, P)$ where S is a new symbol. The set of productions P is defined as follows:

$$P = \{S \longrightarrow [u_1 u_2 \dots u_n] S \mid (u_1, u_2, \dots, u_n) \in C\} \cup$$

$$\cup\{S \longrightarrow [v_1 v_2 \dots v_n] \mid (v_1, v_2, \dots, v_n) \in B\}.$$

One can easily prove that $L(G') = \varphi_g^{n,k}(L(G))$. □

As a consequence of this result we show that each family $\mathcal{SW}_{n.g,k}$ has good decidability properties.

Theorem 7.8 *For each family $\mathcal{SW}_{n,g,k}$ the following problems are decidable:*
(i) the equivalence problem ($L_1 = L_2$?).
(ii) the inclusion problem ($L_1 \subseteq L_2$?).
(iii) the completeness problem ($L = \Sigma^*$?).

4. COMPARISON WITH OTHER FAMILIES OF LANGUAGES

In this section we investigate the interrelations between the families $\mathcal{SW}_n$, $n \geq 1$ and the families of languages in the Chomsky hierarchy as well as with families of simple matrix languages.

Notation. Let $\psi_g^{n,k}$ be the function

$$\psi_g^{n,k} : \Sigma_1^* \Sigma_2 \longrightarrow \Sigma^*$$

such that $\psi_g^{n,k}$ is the inverse of the function $\varphi_g^{n,k}$.

Proposition 7.3 *Let $L \subseteq \Sigma_1^* \Sigma_2$ be a regular language that can be generated by a regular grammar with only one nonterminal and with only one terminal production. Then the language $\psi_g^{n,k}(L)$ is a sewing language of degree n and of type (g, k).*

Proposition 7.4 *A language L is in $\mathcal{SW}_n$ if and only if L can be generated by a regular matrix grammar of degree n having only one nonterminal.*

Proposition 7.5 *Let f and g be the functions $f(u, v) = umi(v)$ and $g(u, v) = mi(u)v$, $u, v \in \Sigma^*$. The following equalities are true:*

$$\mathcal{SW}_2(f) = \mathcal{SW}_2(g) = \mathcal{SM}.$$

Remark 7.4 *The families $\mathcal{SW}_n$, $n > 1$ are not comparable with any of the families REG, LIN and CF. The reason is that the language $a^* \cup b^*$ is not contained in any of the families $\mathcal{SW}_n$, whereas each of the families $\mathcal{SW}_n$, where $n > 1$, contains non-context-free languages.*

References

[1] Delany, P. & G.P Landow, eds. (1991), *Hypermedia and Literary Studies*, MIT Press, Cambridge.

[2] Ibarra, O. (1970), Simple matrix languages, *Information and Control*, 17: 359–394.

[3] Joshi, A.K.; K. Vijay-Shanker & D. Weir (1991), The convergence of mildly context-sensitive grammatical formalisms, in P. Sells; S. Shieber & T. Wasow, eds., *Foundations Issues in Natural Language Processing*, MIT Press, Cambridge.

[4] Marcus, S. (1969), Contextual grammars, *Revue Roumaine des Mathématiques Pures et Appliquées*, 14.10: 1525–1534.

[5] Marcus S. (1997), Contextual grammars and natural languages, in G. Rozenberg & A. Salomaa, eds., *Handbook of Formal Languages*, II: 215–236. Springer, Berlin.

[6] Martín-Vide, C. & A. Mateescu (1999), Contextual grammars with trajectories, in W. Thomas, ed., *Preproceedings of Developments in Language Theory Fourth International Conference (DLT 99)*: 176–185. Rheinisch-Westfälische Technische Hochschule, Aachen.

[7] Păun, Gh. (1997), *Marcus Contextual Grammars*, Kluwer, Dordrecht.

[8] Rozenberg, G. & A. Salomaa, eds. (1997), *Handbook of Formal Languages*, Springer, Berlin.

Chapter 8

TOWARDS GRAMMARS OF DECISION ALGORITHMS

Lech Polkowski
Polish-Japanese Institute of Information Technology
Koszykowa 86, 02008 Warsaw, Poland
& Institute of Mathematics
Warsaw University of Technology
Banacha 2, 02097 Warsaw, Poland
polkow@pjwstk.waw.pl

Andrzej Skowron
Institute of Mathematics
Warsaw University of Technology
Banacha 2, 02097 Warsaw, Poland
skowron@mimuw.edu.pl

Abstract Rough mereology is a paradigm allowing for approximate reasoning in data oriented approach. It is rooted in Rough Set Theory [4]. Rough sets and rough mereology have been applied in particular to problems of formal language and grammar theories [2], [3] by Păun, Polkowski and Skowron and to problems of synthesis of grammars from data about multi-agent systems [8]. Here, we continue the topic of [8] on a meta-level: we introduce rough mereological distances (similarity measures) not only on granules of knowledge but also on pairs of granules (i.e. on decision rules) as well as on collections of pairs (i.e. on decision algorithms). This will provide a tool for extending the construction of synthesis grammars proposed in [8] in terms of granules to synthesis grammars defined in terms of decision algorithms.

1. INTRODUCTION

We begin with basic notions of rough set theory [4], [6], [7], [8], [9]. Knowledge is represented here by means of an information system which is a triple

C. Martin-Vide and V. Mitrana (eds.), Where Mathematics, Computer Science, Linguistics and Biology Meet, 85-95.

$\mathbf{A} = (U, A)$ where U is a (current) set of objects, and A is a (current) set of (conditional) attributes. Each attribute $a \in A$ is a mapping on the set U i.e. $a : U \to V_a$ where V_a is the set of values of a.

The basic assumption about objects described by means of the information system $\mathbf{A}$ is the following: objects with identical descriptions are not discernible to the user of the system. This assumption leads on a formal level to quotient structures on which one operates to generate decision rules. We introduce a way of object denotation: for each object $u \in U$ and a set $B \subseteq A$ of attributes, the *information set* $Inf_B(u)$ of the object u over the set B of attributes is defined by

(INF) $Inf_B(u) = \{(a, a(u)) : a \in B\}$.

The symbol INF_B will denote the set $\{Inf_B(u) : u \in U\}$.

We denote by the symbol $PARTINF_B$ the set $\{Inf_C(u) : u \in U \text{ and } C \subseteq B\}$.

The indiscernibility of objects can be expressed now as the identity of their information sets i.e.

(IND) $IND_B(u, w)$ is $TRUE$ iff $Inf_B(u) = Inf_B(w)$ for any pair u, w of objects in U.

The indiscernibility relation IND_A partitions the set U into classes $[u]_{IND_A}$ of its objects; we let $U/IND_A = \{[u]_A : u \in U\}$. It is obvious that we could have a relativized version as well: given a set $B \subseteq A$ of attributes, we could define a restriction $\mathbf{A}_B$ of the information system $\mathbf{A}$ over the set B by letting $\mathbf{A}_B = (U, B)$. The restriction operator induces a projection operator on the space INF_A of information sets: $proj(A, B) : INF_A \to INF_B$ acting coordinate-wise according to the formula: $proj(A, B)(Inf_A(u)) = Inf_B(u)$. As $proj(A, B)$ considered as a function from U into itself preserves the relation IND_A, it does induce a projection operator $proj(A, B)$ from the equivalence space U/IND_A onto the equivalence space U/IND_B where $IND_B = proj(A, B)(IND_A)$; clearly, $[u]_{IND_B} = proj(A, B)([u]_{IND_A})$ for each $u \in U$.

We have therefore a variety of indiscernibility spaces $\{U/IND_B : B \subseteq A\}$; the Boolean algebra generated over the set of atoms U/IND_B by means of set-theoretical operations of union, intersection and complement is said to be the *B-algebra* $CG(B)$ *of B-pre-granules.* Any member of $CG(B)$ is called a *B-pre-granule.*

We have an alternative logical language in which we can formalize the notion of an *information pre-granule*; for a set of attributes $B \subseteq A$, we recall [2] the definition of the B-logic L_B: elementary formulae of L_B are of the form (a, v) where $a \in B$ and $v \in V_a$. Formulae of L_B are built from elementary formulae by means of logical connectives $\vee, \wedge$; thus, each formula can be represented in DNF as $\bigvee_{j \in J} \bigwedge_{i \in I_j} (a_i, v_i)$. The formulae of L_B, called *information pre-granules*, are interpreted in the set of objects U: the denotation $[(a, v)]$ of an elementary formula (a, v) is the set of objects satisfying the equation $a(x) = v$ i.e. $[(a, v)] = \{u \in U : a(u) = v\}$ and this is extended by structural induction viz. $[\alpha \vee \beta] = [\alpha] \cup [\beta]$, $[\alpha \wedge \beta] = [\alpha] \cap [\beta]$ for $\alpha, \beta \in L_B$. In what follows, we will interpret formulae of L_B in the quotient universe U/IND_B preserving the same symbol [.].

Clearly, given a B -pre-granule $G \in CG(B)$, there exists an information pre-granule α_G of L_B such that $[\alpha_G] = G$;

An atom of the Boolean algebra $CG(B)$ will be called an *elementary B-pre-granule;* clearly, for any atom G of $CG(B)$ there exists an *elementary information pre-granule* α_G of the form $\wedge_{a \in B}(a, v_a)$ such that $[\alpha_G] = G$.

For given non-empty sets $B, C \subseteq A$, a pair (G_B, G_C) where $G_B \in CG(B)$ and $G_C \in CG(C)$ is called a *(B,C)-granule of knowledge.* There exists therefore *an information granule* $(\alpha_{G_B}, \alpha_{G_C})$ such that $\alpha_{G_B} \in L_B, \alpha_{G_C} \in L_C$, $[\alpha_{G_B}]$=G_B and $[\alpha_{G_C}]$=G_C. If G_B, G_C are atoms then the pair (G_B, G_C) is called an *elementary (B,C)-granule.*

In the language of granules, we may express partial dependencies between sets B, C of attributes by relating classes of IND_B to classes of IND_C. There are two characteristics of the granule (G, G') important in applications to decision algorithms viz. the characteristic whose values measure what part of G' is in G (the $'confidence'$ of the decision rule $\alpha_G \Longrightarrow \alpha_{G'}$) and the characteristic whose values measure what part of G is in G' (the *'strength of the support'* for the rule $\alpha_G \Longrightarrow \alpha_{G'}$) (implying clearly the existence of a common domain e.g. U for G, G'.

A standard choice of an appropriate measure may be based on frequency count; the formal rendering is the *standard rough inclusion function* [5] defined for two sets $X, Y \subseteq U$ by the formula $\mu(X, Y) = \frac{card(X \cap Y)}{card(X)}$ when X is non-empty and $\mu(X, Y) = 1$, otherwise.

To select sufficiently strong rules, we would set a threshold ρ_{cr}. We define then, in analogy with machine learning techniques, two characteristics:

(ρ) $\rho(G, G') = \mu([G], [G'])$; (η) $\eta(G, G') = \mu([G'], [G])$

and we call an (η, ρ) *granule of knowledge* any granule (G, G') such that

(i) $\rho(G, G') \geq \rho_{cr}$; (ii) $\eta(G, G') \geq \rho_{cr}$

where ρ_{cr}, η_{cr} may depend on additional parameters i.e. on information vectors from sets $PARTINF_B$, sets B determined by prime implicants of the corresponding formulae α_G, α'_G.

This logical model of granulation may not fit practical demands: the relation IND may be too rigid and ways of its relaxation are among most intensively studied topics [9]. Here, we propose to introduce a rough mereological approach to the granulation problem in which IND-classes are replaced with mereological classes i.e. similarity classes.

2. ROUGH MEREOLOGY

Rough mereology [6], [7], [10] has been proposed and studied as a means of clustering in a relational way. Its primitive notion is that of a *rough inclusion* i.e. a family of functors $\mu(r)$ being a part, in degree at least, of r, where $r \in [0, 1]$.

2.1 MEREOLOGY

It will be convenient to refer here to the mereological approach proposed by Leśniewski [1]. It offers a formal treatment of the predicate of being a part. We begin with the notion of a *part* functor (*pt*, for short; $XptY$ reads: *X is a part of Y*). We rephrase basic requirements about *pt*.

(ME1) $XptY \wedge YptZ \Longrightarrow XptZ$;

(ME2) $non(XptX)$.

The concept of an improper part is reflected in the notion of an *element* (or, *ingredient* cf. [1], [6]) *el* defined as follows: $XelY \Longleftrightarrow XptY \vee X = Y$.

An important theoretical feature of mereology is the presence of the class functor Kl making collections of objects into single objects and defined as follows. Important usage of the class operator is described in the sequel as a tool for granule formation. For a given *collection* Y of objects,

X is $Kl(Y) \Longleftrightarrow$

$\forall Z.(Z \text{ in } Y \Longrightarrow Zel(X)) \wedge$

$$\forall Z.(Zel(X) \Longrightarrow \exists U, W.U \text{ is in } Y \wedge Wel(U) \wedge Wel(Z)).$$

Thus, the class of objects in Y, $Kl(Y)$, consists of all objects which have an element in common with an object in Y.

The notion of a class is subjected to the following restrictions:

(ME3) X is $Kl(Y) \wedge Z$ is $Kl(Y) \Longrightarrow Z = X$ ($Kl(Y)$ is an individual);

(ME4) $\exists Z.Z \text{ in } Y \Longleftrightarrow \exists Z.Z \text{ in } Kl(Y)$ (the class exists for each non-empty Y).

Thus, $Kl(Y)$ is defined for any non-empty collection Y and $Kl(Y)$ is a single object (meaning that two objects satisfying requirements for $Kl(Y)$ are identical cf. [1]).

2.2 ROUGH MEREOLOGY: FIRST NOTIONS

Rough Mereology has been proposed and studied in [6], [7], [10] as a vehicle for reasoning under uncertainty.

The following is a list of basic postulates about Rough Mereology. We introduce a graded family μ_r, where $r \in [0, 1]$ is a real number from the unit interval, of functors which would satisfy ($X \mu_r Y$ is read as the statement *X is a part of Y in degree at least r*) the following:

(RM1) $X\mu_1 Y \iff XelY$;
(RM2) $X\mu_1 Y \Longrightarrow \forall Z.(Z\mu_r X \Longrightarrow Z\mu_r Y)$;
(RM3) $X = Y \wedge X\mu_r Z \Longrightarrow Y\mu_r Z$;
(RM4) $X\mu_r Y \wedge s \leq r \Longrightarrow X\mu_s Y$;

One may have as an archetypical rough mereological predicate the rough membership function [5] defined in an extended form as:
$X\mu_r Y \iff \frac{card(X\cap Y)}{card(X)} \geq r$ in case X non–empty, 1 else
where X, Y are (either exact or rough) subsets in the universe U of an information/decision system (U, A).

2.3 ROUGH MEREOLOGICAL COMPONENT OF GRANULATION

The functors μ_r may enter our discussion of a granule in each of the following ways:

1. Concerning the definitions $(\eta), (\rho)$ of functions η, ρ , we may replace in them the rough membership function μ with a function μ_r possibly better suited to a given context:
(ρ) $G\mu_{\rho_{cr}}(G')$; (η) $G'\mu_{\rho_{cr}}(G)$.

2. The process of clustering may be described in terms of the class functor of mereology:
(ρ) $Gel(Kl_{\rho_{cr}}(G'))$; (η) $G'el(Kl_{\rho_{cr}}(G))$
where $Kl_r(X)$ is the class of objects Z satisfying $Z\mu_r X$. We will adhere to this means of clustering and we denote in the sequel the class $Kl_r(X)$ with the symbol $gr(X, r)$ read as *the granule of size r about X*.

An important advantage of this approach consists in the fact that rough inclusions may be determined directly from data tables/information systems (cf. [7]) which allows for more context-sensitive clustering of data.

3. ADAPTIVE CALCULUS OF GRANULES IN DISTRIBUTED SYSTEMS

We construct a mechanism for transferring granules of knowledge among agents by means of transfer functions induced by rough mereological connectives extracted from their respective information systems [7].

We now recall the basic ingredients of our scheme of agents [7], [10].

3.1 DISTRIBUTED SYSTEMS OF AGENTS

We assume that a pair (Inv, Ag) is given where Inv is an *inventory of elementary objects* and Ag is a set of intelligent computing units called *agents*. We consider an agent $ag \in Ag$. The agent ag is endowed with tools for reasoning and communicating about objects; these tools are defined by components of the agent label.

The *label of the agent ag* is the tuple

$$lab(ag) = (\mathbf{A}(ag), \mu(r)(ag), L(ag), Link(ag), O(ag), St(ag)),$$

$$Unc - rel(ag), Unc - rule(ag), Dec - rule(ag))$$

where

1. $\mathbf{A}(ag)$ is an information system of the agent ag.

2. $\mu_r(ag)$ is a functor of part in a degree at ag.

3. $L(ag)$ is a set of unary predicates (properties of objects) in a predicate calculus interpreted in the set $U(ag)$ (e.g. in logics L_B, cf. Sect. 1).

4. $St(ag) = \{st(ag)_1, ..., st(ag)_n\} \subset U(ag)$ is the set of *standard objects* at ag.

5. $Link(ag)$ is a collection of strings of the form $ag_1 ag_2 ... ag_k ag$ which are elementary teams of agents; we denote by the symbol $Link$ the union of the family $\{Link(ag) : ag \in Ag\}$.

6. $O(ag)$ is the set of operations at ag; any $o \in O(ag)$ is a mapping of the Cartesian product $U(ag_1) \times U(ag_2) \times ... \times U(ag_k)$ into the universe $U(ag)$ where $ag_1 ag_2 ... ag_k ag \in Link(ag)$.

7. $Unc - rel(ag)$ is the set of uncertainty relations

$$\rho_i = \rho_i(o_i(ag), st(ag_1)_i, st(ag_2)_i, ..., st(ag_k)_i, st(ag))$$

depending on parameters $st(ag_1)_i, st(ag_2)_i, ..., st(ag_k)_i, st(ag)$ where $ag_1 ag_2 \, ... \, ag_k ag \in Link(ag)$, $o_i(ag) \in O(ag)$ are such that

$$\rho_i((x_1, \varepsilon_1), (x_2, \varepsilon_2), ., (x_k, \varepsilon_k), (x, \varepsilon))$$

holds for $x_1 \in U(ag_1), x_2 \in U(ag_2), .., x_k \in U(ag_k)$ and $\varepsilon, \varepsilon_1, \varepsilon_2, .., \varepsilon_k \in [0, 1]$ iff $x_j \mu(ag_j)_{\varepsilon_j} st(ag_j)$ for $j = 1, 2, .., k$ and $x\mu(ag)_\varepsilon st(ag)$ where $o_i(st(ag_1), st(ag_2), .., st(ag_k)) = st(ag)$ and $o_i(x_1, x_2, .., x_k) = x$.

Uncertainty relations express the agents knowledge about relationships among uncertainty coefficients of the agent ag and uncertainty coefficients of its children.

8. $Unc - rule(ag)$ is the set of uncertainty rules f_j where $f_j : [0, 1]^k \longrightarrow [0, 1]$ is a function which has the property that **if** $x_1 \in U(ag_1), x_2 \in U(ag_2), ..,$ $x_k \in U(ag_k)$ satisfy the conditions $x_i \mu(ag_i)_{\varepsilon(ag_i)}(st(ag_i))$ for $i = 1, 2, .., k,$ **then** $o_j(x_1, x_2, ..., x_k)\mu(ag)_{f_j(\varepsilon(ag_1), \varepsilon(ag_2), .., \varepsilon(ag_k))} st(ag)$ where all parameters are as in 7.

9. $Dec - rule(ag)$ is a set of decomposition rules $dec - rule_j$ and $(\Phi(ag_1), \Phi(ag_2), ..., \Phi(ag_k), \Phi(ag)) \in dec - rule_j$

(where $\Phi(ag_1) \in L(ag_1), \Phi(ag_2) \in L(ag_2), .., \Phi(ag_k) \in L(ag_k), \Phi(ag) \in L(ag)$ and $ag_1 ag_2 ... ag_k ag \in Link(ag)$) iff there exists a collection of standards $st(ag_1)$, $st(ag_2), .., st(ag_k)$, $st(ag)$ with the properties that $o_j(st(ag_1), st(ag_2), .., st(ag_k)) = st(ag)$, $st(ag_i)$ satisfies $\Phi(ag_i)$ for $i = 1, 2, .., k$ and $st(ag)$ satisfies $\Phi(ag)$.

Decomposition rules are decomposition schemes in the sense that they describe the standard $st(ag)$ and the standards $st(ag_1), ..., st(ag_k)$ from which the standard $st(ag)$ is assembled under o_j in terms of predicates which these standards satisfy.

3.2 APPROXIMATE SYNTHESIS OF COMPLEX OBJECTS

The process of synthesis of a complex object (e.g. signal, action) by the above defined scheme of agents has, in our approach, two communication stages viz. the top-down communication/negotiation process and the bottom-up communication/assembling process. We outline the two stages here in the language of approximate formulae.

3.2.1 Approximate logic of synthesis. We assume for simplicity of notation that the relation $ag' \leq ag$, which holds for agents $ag', ag \in Ag$ iff there exists a string $ag_1 ag_2 ... ag_k ag \in Link(ag)$ with $ag' = ag_i$ for some $i \leq k$, orders the set Ag into a tree. We also assume that $O(ag) = \{o(ag)\}$ for $ag \in Ag$ i.e. each agent has a unique assembling operation.

We recall a logic $L(Ag)$ [7], [10] in which we can express global properties of the synthesis process.

Elementary formulae of $L(Ag)$ are of the form $< st(ag), \Phi(ag), \varepsilon(ag) >$ where $st(ag) \in St(ag), \Phi(ag) \in L(ag), \varepsilon(ag) \in [0, 1]$ for any $ag \in Ag$. Formulae of $L(ag)$ form the smallest extension of the set of elementary formulae closed under propositional connectives $\vee$, $\wedge$, $\neg$ and under the modal operators $\Box$, $\Diamond$.

The meaning of a formula $\Phi(ag)$ is defined classically as the set $[\Phi(ag)] = \{u \in U(ag) : u$ has the property $\Phi(ag)\}$; we denote satisfaction by $u \vdash \Phi(ag)$. For $x \in U(ag)$, we say that x *satifies* $< st(ag), \Phi(ag), \varepsilon(ag) >$, in symbols:

$$x \vdash < st(ag), \Phi(ag), \varepsilon(ag) >,$$

iff

(i) $st(ag) \vdash \Phi(ag)$;

(ii) $x \varepsilon \mu(ag)_{\varepsilon(ag)}(st(ag))$.

We extend satisfaction over formulae by recursion as usual.

By a *selection* over Ag we mean a function sel which assigns to each agent ag an object $sel(ag) \in U(ag)$. For two selections sel, sel' we say that sel *induces* sel', in symbols $sel \to_{Ag} sel'$ when $sel(ag) = sel'(ag)$ for any $ag \in Leaf(Ag)$ and $sel'(ag) = o(ag)(sel'(ag_1), sel'(ag_2), ..., sel'(ag_k))$ for any $ag_1 ag_2 ... ag_k ag \in Link$.

We extend the satisfiability predicate $\vdash$ to selections: for an elementary formula $< st(ag), \Phi(ag), \varepsilon(ag) >$, we let $sel \vdash < st(ag), \Phi(ag), \varepsilon(ag) >$ iff $sel(ag) \vdash < st(ag), \Phi(ag), \varepsilon(ag) >$.

We now let $sel \vdash \Diamond < st(ag), \Phi(ag), \varepsilon(ag) >$ when there exists a selection sel' satisfying the conditions: $sel \to_{Ag} sel'; sel' \vdash < st(ag), \Phi(ag), \varepsilon(ag) >$.

In terms of $L(Ag)$ it is possible to express the problem of synthesis of an approximate solution to the problem posed to Ag. We denote by $head(Ag)$ the root of the tree $(Ag, \leq)$ and by $Leaf(Ag)$ the set of leaf-agents in Ag. In the process of top-down communication, a requirement Ψ received by the scheme from an external source (which may be called a *customer*) is decomposed into approximate specifications of the form $< st(ag), \Phi(ag), \varepsilon(ag) >$ for any agent ag of the scheme. The decomposition process is initiated at the agent $head(Ag)$ and propagated down the tree. We are able now to formulate the synthesis problem.

Synthesis problem

Given a formula $\alpha :< st(head(Ag)), \Phi(head(Ag)), \varepsilon(head(Ag)) >$ *find a selection* sel *over the tree* $(Ag, \leq)$ *with the property* $sel \vdash \alpha$.

A solution to the synthesis problem with a given formula α is found by negotiations among the agents based on uncertainty rules and their successful result can be expressed by a top-down recursion in the tree $(Ag, \leq)$ as follows: given a local team $ag_1 ag_2 ... ag_k ag$ with the formula $< st(ag), \Phi(ag), \varepsilon(ag) >$ already chosen, it is sufficient that each agent ag_i choose a standard $st(ag_i) \in U(ag_i)$, a formula $\Phi(ag_i) \in L(ag_i)$ and a coefficient $\varepsilon(ag_i) \in [0, 1]$ such that

(iii) $(\Phi(ag_1), \Phi(ag_2), ..., \Phi(ag_k), \Phi(ag)) \in Dec - -rule(ag)$ with standards $st(ag), st(ag_1), ..., st(ag_k)$;

(iv) $f(\varepsilon(ag_1), .., \varepsilon(ag_k)) \geq \varepsilon(ag)$ where f satisfies $unc - rule(ag)$ with $st(ag), st(ag_1), ..., st(ag_k)$ and $\varepsilon(ag_1), ..., \varepsilon(ag_k), \varepsilon(ag)$.

For a formula α, we call an α - *scheme* an assignment of a formula $\alpha(ag) : < st(ag), \Phi(ag), \varepsilon(ag) >$ to each $ag \in Ag$ in such manner that (iii), (iv) are satisfied and $\alpha(head(Ag))$ is $< st(head(Ag)), \Phi(head(Ag)), \varepsilon(head(Ag)) >$; we denote this scheme with the symbol $sch(< st(head(Ag)), \Phi(head(Ag)), \varepsilon(head(Ag)) >)$. We say that a selection sel is *compatible* with a scheme $sch(<$

$st(head(Ag))$, $\Phi(head(Ag))$, $\varepsilon(head(Ag)) >)$ in case $sel(ag)\mu(ag)_{\varepsilon(ag)}$ $(st(ag))$ for each leaf agent $ag \in Ag$.

The goal of negotiations can be summarized now as follows.

Proposition 8.1 *Given a formula* $< st(head(Ag))$, $\Phi(head(Ag))$, $\varepsilon(head(Ag)) >$: **if** *a selection sel is compatible with a scheme* $sch(< st(head(Ag)), \Phi(head(Ag)), \varepsilon(head(Ag)) >)$ **then** $sel \vdash \Diamond < st(head(Ag)), \Phi(head(Ag)), \varepsilon(head(Ag)) >$.

3.3 SYNTHESIS IN TERMS OF GRANULES

For a standard $st(ag)$ and a value $\varepsilon(ag)$, we denote by the symbol $gr(st(ag), \varepsilon(ag))$ the pre-granule $Kl_{\varepsilon(ag)}(st(ag))$ i.e. the granule of size $\varepsilon(ag)$ about $st(ag)$; then, *a granule selector* sel_g is a map which for each $ag \in Ag$ chooses a granule $sel_g(ag)=gr(st(ag), \varepsilon(ag))$.
We say that $gr(st(ag), \varepsilon'(ag))$ satisfies a formula $\alpha :< st(ag), \Phi(ag), \varepsilon(ag) >$ $(gr(st(ag), \varepsilon(ag)) \vdash \alpha)$ if and only if $st(ag) \vdash \Phi(ag)$ and $\varepsilon'(ag) \geq \varepsilon(ag)$. Given $ag_1 ag_2 ... ag_k ag \in Link$ and a formula $< st(ag), \Phi(ag), \varepsilon(ag) >$ along with f satisfying $unc-rule(ag)$ with $st(ag)$, $st(ag_1)$, ..., $st(ag_k)$ and $\varepsilon(ag_1)$, ..., $\varepsilon(ag_k), \varepsilon(ag), o(ag)$ maps the product $\times_i gr(st(ag_i), \varepsilon(ag_i))$ into $gr(st(ag), \varepsilon(ag))$.

Composing these mappings along the tree $(Ag, \leq)$, we define a mapping $prod_{Ag}$ which maps any set $\{gr(st(ag), \varepsilon(ag)) : ag \in Leaf(Ag)\}$ into the granule $gr(st(head(Ag), \varepsilon(head(Ag))$. We say that a selection sel_g is *compatible* with a scheme $sch(< st(head(Ag)), \Phi(head(Ag)), \varepsilon(head(Ag)) >)$ if

$$sel_g(ag_i) = gr(st(ag_i), \varepsilon'(ag_i)) \Longrightarrow \varepsilon'(ag_i) \geq \varepsilon(ag_i)$$

for each leaf agent ag_i.

As

$$prod_{Ag}(sel_g) \vdash < st(head(Ag)), \Phi(head(Ag)), \varepsilon(head(Ag)) >$$

we have the pre-granule counterpart of Proposition 1.

Proposition 8.2 *Given a formula* $< st(head(Ag)), \Phi(head(Ag))$, $\varepsilon(head(Ag)) >$: **if** *a selection* sel_g *is compatible with a scheme* $sch(< st(head(Ag)), \Phi(head(Ag)), \varepsilon(head(Ag)) >)$ **then** $sel_g \vdash \Diamond < st(head(Ag)), \Phi(head(Ag)), \varepsilon(head(Ag)) >$.

4. GRANULES AND GRAMMARS OF DECISION ALGORITHMS

We now extend our measures of similarity defined by means of functors μ_r to pairs of granules (i.e. decision rules) and to finite collections of pairs (i.e. decision algorithms).

We recall the function $\top$ which assigns to real numbers x, y from the unit interval [0,1] the value $\top(x, y)$ called a *t-norm* [7].

Consider pairs $(A, B), (C, D)$ of granules with $A\mu_r C$, $B\mu_s D$; we propose in this case that $(A, B)\mu_{\top(r,s)}(C, D)$. This defines the functor μ on pairs of granules which represent decision rules as pointed to above. We agree that the notion of an element be extended to pairs by the following: $(A, B)el(C, D)$ iff $AelC$ and $BelD$. Then we have a proposition whose proof is straightforward.

Proposition 8.3 *Letting* $(A, B)\mu_{\top(r,s)}(C, D)$ *in case* $A\mu_r C$, $B\mu_s D$ *defines a functor* μ *on decision rules which satisfies (RM1)-(RM4).*

The next step is to extend μ over finite collections of decision rules i.e. over decision algorithms. In this case, rough mereology will provide us with an essential tool.

Consider decision algorithms $Alg_1 = \{(A_1, B_1), ..., (A_k, B_k)\}$ and $Alg_2 = \{(C_1, D_1), ..., (C_m, D_m)\}$; we let: A is $Kl\{A_1, .., A_k\}$, B is $Kl\{B_1, ..., B_k\}$, C is $Kl\{C_1, ..., C_m\}$, D is $Kl\{D_1, ..., D_m\}$. We say that $Alg_1 el Alg_2$ iff $AelC$ and $BelD$. We extend μ by letting $Alg_1\mu_{\top(r,s)}Alg_2$ in case $A\mu_r C$, $B\mu_s D$. Again

Proposition 8.4 *Letting* $Alg_1\mu_{\top(r,s)}Alg_2$ *in case* $A\mu_r C$, $B\mu_s D$ *defines a functor* μ *on decision algorithms which satisfies (RM1)-(RM4).*

We may now return to Section 3 and introduce granules of decision algorithms. Then we may define synthesis grammars based on decision granules repeating the construction given in [8].

Acknowledgments

This work has been prepared under the Research Grant No 8T11C 024 17 from the State Committee for Scientific Research (KBN) of the Republic of Poland and under the European ESPRIT Program in CRIT 2 Research Project No 20288.

References

[1] Leśniewski, S. (1927), O podstawach matematyki, *Przegląd Filozoficzny*, 30: 164–206; 31: 261–291; 32: 60–101; 33: 77–105; 34: 142–170.

[2] Păun, Gh.; L. Polkowski & A. Skowron (1996), Parallel communicating grammar systems with negotiations, *Fundamenta Informaticae*, 28: 315–330.

[3] Păun, Gh.; L. Polkowski & A. Skowron (1997), Rough set approximations of languages, *Fundamenta Informaticae*, 32: 149–162.

[4] Pawlak, Z. (1992), *Rough Sets: Theoretical Aspects of Reasoning about Data*, Kluwer, Dordrecht.

[5] Pawlak, Z. & A. Skowron (1994), Rough membership functions, in R.R. Yaeger; M. Fedrizzi & J. Kacprzyk, eds., *Advances in the Dempster Shafer Theory of Evidence:* 251–27. John Wiley, New York.

[6] Polkowski, L. & A. Skowron (1994), Rough mereology, in *Lecture Notes in Artificial Intelligence*, 869: 85–94. Springer, Berlin.

[7] Polkowski, L. & A. Skowron (1996), Rough mereology: a new paradigm for approximate reasoning, *International Journal of Approximate Reasoning*, 15: 333–365.

[8] Polkowski, L. & A. Skowron (1998), Grammar systems for distributed synthesis of approximate solutions extracted from experience, in Gh. Păun & A. Salomaa, eds., *Grammatical Models of Multi-Agent Systems*: 316–333, Gordon and Breach, London.

[9] Polkowski, L. & A. Skowron, eds. (1998), *Rough Sets in Knowledge Discovery*, Studies in Fuzziness and Soft Computing, 18–19, Springer, Berlin.

[10] Skowron, A. & L. Polkowski (1998), Rough mereological foundations for design, analysis, synthesis and control in distributed systems, *Information Sciences. An International Journal*, 104: 129–156.

[11] Zadeh, L.A. (1996), Fuzzy logic = computing with words, *IEEE Transactions on Fuzzy Systems*, 4: 103–111.

[12] Zadeh, L.A. (1997), Toward a theory of fuzzy information granulation and its certainty in human reasoning and fuzzy logic, *Fuzzy Sets and Systems*, 90: 111–127.

II

AUTOMATA

Chapter 9

COMPUTATIONAL COMPLEMENTARITY FOR PROBABILISTIC AUTOMATA

Cristian S. Calude
Institute of Theoretical Physics
Technical University of Wien
Wiedner Hauptstraße 8-10/136, A-1040 Vienna, Austria
and
Computer Science Department
The University of Auckland
Private Bag 92109, Auckland, New Zealand
cristian@cs.auckland.ac.nz

Elena Calude
Institute of Information Sciences
Massey University at Albany
Private Bag 102-904, North Shore MSC, Auckland, New Zealand
E.Calude@massey.ac.nz

Karl Svozil
Institute of Theoretical Physics
Technical University of Wien
Wiedner Hauptstraße 8-10/136, A-1040 Vienna, Austria
svozil@tph.tuwien.ac.at

Abstract Motivated by Mermin's analysis of *Einstein-Podolsky-Rosen* correlations [24] and [7], we study two computational complementarity principles introduced in [8] for a class of probabilistic automata. We prove the existence of probabilistic automata featuring both types of computational complementarity and we present a method to reduce, under certain conditions, the study of computational comple-

C. Martin-Vide and V. Mitrana (eds.), Where Mathematics, Computer Science, Linguistics and Biology Meet, 99-113.

mentarity of probabilistic automata to the study of computational complementarity of deterministic automata.

1. INTRODUCTION

Quantum entanglement [32] and nonlocal correlations [1, 15] are essential features of quantized systems used for quantum computation [12]. Building on Moore's "Gedanken" experiments, in [33, 30] complementarity was modeled by means of finite automata. Two new types of computational complementarity principles have been introduced and studied in [8, 10, 11, 9, 35, 34] using Moore automata *without initial states*. Motivated by Mermin's simple device [24] designed to explain *Einstein-Podolsky-Rosen (EPR)* correlations and the analysis in [7], we study the above mentioned computational complementarity properties for a class of probabilistic automata. We prove the existence of probabilistic automata featuring both types of computational complementarity and we reduce the study of computational complementarity of probabilistic automata to the study of computational complementarity of deterministic automata.

2. MERMIN'S DEVICE

Mermin [24] imagined a simple device to illustrate the EPR conundrum without using the classical quantum mechanical notions of wave functions, superposition, wave-particle duality, uncertainty principle, etc. The device has three "completely unconnected"[1] parts, two detectors (D1) and (D2) and a source (S) emitting particles. The source is placed between the detectors: whenever a button is pushed on (S), shortly thereafter two particles emerge, moving off toward detectors (D1) and (D2). Each detector has a switch that can be set in one of three possible positions–labeled 1,2,3–and a bulb that can flash a red (R) or a green (G) light. The purpose of lights is to "communicate" information to the observer. Each detector flashes either red or green whenever a particle reaches it. Because of the lack of any relevant connections between any parts of the device, the link between the emission of particles by (S), i.e., as a result of pressing a button, and the subsequent flashing of detectors can only be provided by the passage of particles from (S) to (D1) and (D2). Additional tools can be used to check and confirm the lack of any communication, cf. [24], p. 941.

The device is repeatedly operated as follows:

1. the switch of either detector (D1) and (D2) is set randomly to 1 or 2 or 3, i.e., the settings or states 11, 12, 13, 21, 22, 23, 31, 32, 33 are equally likely,

[1]There are no relevant causal connections, neither mechanical nor electromagnetic or any other.

2. pushing a button on (S) determines the emission toward both (D1) and (D2),
3. sometime later, (D1) and (D2) flash one of their lights, G or R,
4. every run is recorded in the form $ijXY$, meaning that D1 was set to state i and flashed X and (D2) was set to j and flashed Y.

For example, the record $31GR$ means "(D1) was set to 3 and flashed G and (D2) was set to 1 and flashed R".

Long recorded runs show the following pattern:

a) In records starting with ii, i.e., 11, 22, 33, both (D1) and (D2) flash the same colours, RR, GG, with equal frequency; RG and GR are never flashed.

b) In records starting with $ij, i \neq j$, i.e., 12, 13, 21, 23, 31, 32, both (D1) and (D2) flash the same colour only 1/4 of the time (RR and GG come with equal frequencies); the other 3/4 of the time, they flash different colours (RG, GR), occurring again with equal frequencies.

The above patterns are statistical, that is they are subject to usual fluctuations expected in every statistical prediction: patterns are more and more "visible" as the number of runs becomes larger and larger.

The conundrum posed by the existence of Mermin's device reveals that the seemingly simplest physical explanation of the pattern a) is incompatible with pattern b). Indeed, as (D1) and (D2) are unconnected there is no way for one detector to "know", at any time, the state of the other detector or which colour the other is flashing. Consequently, it seems plausible to assume that the colour flashed by detectors is determined only by some property, or group of properties, of particles, say speed, size, shape, etc. What properties determine the colour does not really matter; only the fact that each particle carries a "program" which determines which colour a detector will flash in some state is important. So, we are led to the following two hypotheses:

H1 *Particles are classified into eight categories:*

$$GGG, GGR, GRG, GRR, RGG, RGR, RRG, RRR.$$ [2]

H2 *Two particles produced in a given run carry identical programs.*

[2] A particle of type XYZ will cause a detector in state 1 to flash X; a detector in state 2 will flash Y and a detector in state 3 will flash Z.

According to H1–H2, if particles produced in a run are of type RGR, then both detectors will flash R in states 1 and 3; they will flash G if both are in state 2. Detectors flash the same colours when being in the same states because *particles carry the same programs.*

It is clear that from H1–H2 it follows that *programs carried by particles do not depend in any way on the specific states of detectors*: they are properties of particles, not of detectors. Consequently, both particles carry the same program whether or not detectors (D1) and (D2) are in the same states.[3]

One can easily argue that

[L] For each type of particle, *in runs of type* b) *both detectors flash the same colour at least one third of the time.*

The conundrum reveals as a significant difference appears between the data dictated by particle programs (colours agree at least one third of the time) and the quantum mechanical prediction (colours agree only one fourth of the time):

under H1–H2, *the observed pattern b) is incompatible with* [L].

3. MERMIN'S PROBABILISTIC AUTOMATA

Consider now a probabilistic automaton simulating Mermin's device. The states of the automaton are all combinations of states of detectors (D1) and (D2), $Q = \{11, 12, 13, 21, 22, 23, 31, 32, 33\}$, the input alphabet models the lights red and green, $\Sigma = \{G, R\}$, the output alphabet captures all combinations of lights flashed by (D1) and (D2), $O = \{GG, GR, RG, RR\}$, and the output function $f : Q \to O$, modeling all combinations of green/red lights flashed by (D1) and (D2) in all their possible states, is probabilistically defined by:

$$
\begin{aligned}
f(ii) &= XX, \text{ with probability } 1/2, \text{ for } i = 1,2,3, X \in \{G,R\}, \\
f(ii) &= XY, \text{ with probability } 0, \text{ for } i = 1,2,3, X,Y \in \{G,R\}, X \neq Y, \\
f(ij) &= XX, \text{ with probability } 1/8, \text{ for } i,j = 1,2,3, i \neq j, \\
&\quad\ X \in \{G,R\}, \\
f(ij) &= XY, \text{ with probability } 3/8, \text{ for } i,j = 1,2,3, i \neq j, \\
&\quad\ X,Y \in \{G,R\}, X \neq Y.
\end{aligned}
$$

For example, $f(11) = RR$ with probability 1/2, $f(11) = GR$ with probability 0, $f(11) = RG$ with probability 0, $f(11) = RR$ with probability

[3]The emitting source (S) has no knowledge about the states of (D1) and (D2) and there is no communication among any parts of the device.

1/2, $f(12) = GG$ with probability 1/8, $f(12) = GR$ with probability 3/8, $f(12) = RG$ with probability 3/8, $f(12) = RR$ with probability 1/8, etc.

The automaton transition $\delta : Q \times \Sigma \to Q$ is *not specified.* In fact, varying all transition functions δ we get a class of Mermin automata:

$$\mathcal{M} = (Q, \Sigma, O, \delta, (p_{ij}^{XY}, i, j = 1, 2, 3, X, Y \in \{G, R\})),$$

where p_{ij}^{XY} is the probability that the automaton on state ij outputs XY: $p_{ii}^{XX} = 1/2$, $p_{ii}^{XY} = 0$, $X \neq Y$, $p_{ij}^{XX} = 1/8$, $p_{ij}^{XY} = 3/8$, $X \neq Y$.

4. COMPUTATIONAL COMPLEMENTARITY FOR DETERMINISTIC AUTOMATA

Moore [25] has studied some experiments on finite deterministic automata[4] trying to understand what kind of conclusions about the internal conditions of a finite machine could possibly be drawn from input-output experiments. A (simple) Moore experiment can be described as follows: a copy of the machine will be experimentally observed, i.e., the experimenter will input a finite sequence of input symbols to the machine and will observe the sequence of output symbols. The correspondence between input and output symbols depends on the particular chosen machine and on its initial state. The experimenter will study the sequences of input and output symbols and will try to conclude that "the machine being experimented on was in state q at the beginning of the experiment".

A state p is "indistinguishable" from a state q (with respect to a given automaton) if every experiment performed on the automaton starting in state p produces the same outcome as it would starting in state q.

Moore [25] has proven the following important result: *There exists an automaton such that any pair of its distinct states are distinguishable, but there is no experiment which can determine what state the machine was in at the beginning of the experiment.* In Calude, Calude, Svozil and Yu [8] two nonequivalent concepts of computational complementarity were introduced and studied for finite automata. Informally, they can be expressed as follows. Consider the class of all elements of reality (observables) and the following properties:

A Any two distinct elements of reality can be mutually distinguished by a suitably chosen measurement procedure, see Bridgman [4].

B For any element of reality, there exists a measurement which distinguishes between this element and all the others. That is, a distinction between any one of them and all the others is operational.

[4] See Brauer [3], Hopcroft and Ullman [21] for good introductions into automata theory.

C There exists a measurement which distinguishes between any two elements of reality. That is, a single predefined experiment operationally exists to distinguish between an arbitrary pair of elements of reality.

Clearly, **C** implies **B** and **B** implies **A**, but both converse implications fail to be true; consequently, two *principles of complementarity* emerge:

CI Property **A** *but not property* **B** *(and therefore not* **C***):* The elements of reality can be mutually distinguished by experiments, but one of these elements cannot be distinguished from all the other ones by any single experiment.

CII Property **B** *but not property* **C***:* Any element of reality can be distinguished from all the other ones by a single experiment, but there does not exist a single experiment which distinguishes between any pair of distinct elements.

We may regard *CI* as an "uncertainty principle" (cf. Conway [13, p. 21]), later termed "computational complementarity" by Finkelstein and Finkelstein [17].[5]

In *CII* each experiment "generates" a pair of distinct states which exercise a mutual influence, namely, they cannot be separated into proper independent parts by the experiment; this influence mimics, in a sense, the state of *quantum entanglement.*[6] We may be conceive *CII* as a *toy model* for the *EPR effect* (see [16, 27, 28]), as well as for the *Zou-Wang-Mandel effect* [39, 37, 18]. Under *CII*, for each experiment w we have at least two states q, q' (as distant as we like in terms of the emitting outputs) which interact via the experiment w: any measurement of q is affecting q' and, conversely, any measurement of q' is affecting q.

5. COMPUTATIONAL COMPLEMENTARITY FOR PROBABILISTIC AUTOMATA

Motivated by Mermin's automaton probabilistic automaton analysis in [7] we introduce and study computational complementarity for a class of probabilistic automata. In opposition with a more popular model, in which the transition is stochastic, but the output is deterministic (see [26]), here we will work with automata having deterministic transitions but stochastic outputs. So, our probabilistic finite automata consist of a finite set of input symbols (the alphabet), a

[5]These types of models have been intensively studied from the point of view of their experimental logical structure by Grib and Zapatrin [19, 20], Svozil [33], Schaller and Svozil [29, 30, 31], Dvurečenskij, Pulmanová and Svozil [14], Calude and Lipponen [9], Calude, Calude and Ştefănescu [6], Jurvanen, and Lipponen [23]. See Svozil and Zapatrin [36] for a comparison of models.

[6]In particular, this influence cannot be used to send an actual message from a state to the other.

finite set of states, a finite set of outputs, a set of transitions from state to state that occur on input symbols chosen from the alphabet and an *output probabilistic function.* For each symbol there is exactly one transition out of each state, possible back to the state itself. The output function emits an output with some probability.

Formally, a *finite probabilistic automaton* $\mathcal{A} = (Q, \Sigma, O, \delta, (a_{p,o})_{p\in Q, o\in O})$ consists of an input alphabet Σ, a finite set Q of states, an output finite set O, a transition function $\delta : Q\times\Sigma \to Q$ and an output probabilistic function $f : Q \to O$ given by the matrix $(a_{p,o})_{p\in Q,o\in O}$ satisfying the condition $\sum_{o\in O} a_{p,o} = 1$, for every $p \in Q$. The output emitted by $\mathcal{A}$ on p is $f(p) = o$ with probability $a_{p,o}$. In case of a deterministic finite automaton, for every $p \in Q$, there exist one (unique) output $o \in O$ such that $a_{p,o} = 1$ and all other probabilities are 0; that is, $f(p) = o$. As in case of deterministic automata the transition function extends naturally to $Q \times \Sigma^* \to Q$ satisfying the equation $\delta(p, uv) = \delta(\delta(p, u), v)$, for all $p \in Q, u, v \in \Sigma^*$.[7]

Consider, for example, the automaton consisting of $Q = \{1, 2, 3, 4\}$, $\Sigma = \{0, 1\}$, $O = \{G, R\}$, the transition given by the following tables

q	σ	$\delta(q,\sigma)$
1	0	4
1	1	3
2	0	1
2	1	3

q	σ	$\delta(q,\sigma)$
3	0	4
3	1	4
4	0	2
4	1	2

and the output function defined by

$$\begin{array}{ll} f(1) = G, \text{ with probability } 2/3, & f(3) = G, \text{ with probability } 1/3, \\ f(1) = R, \text{ with probability } 1/3, & f(3) = R, \text{ with probability } 2/3, \\ f(2) = G, \text{ with probability } 1/3, & f(4) = G, \text{ with probability } 1/3, \\ f(2) = R, \text{ with probability } 2/3, & f(4) = R, \text{ with probability } 2/3. \end{array}$$

The following graphical representation will be consistently used in what follows:

[7] In what follows the extension will also be denoted by δ.

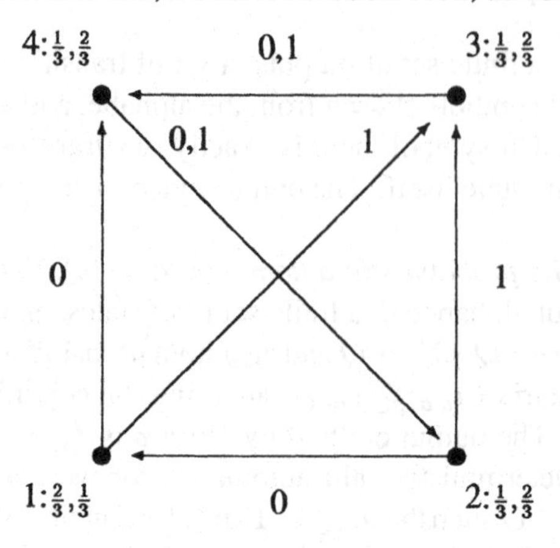

Figure 9.1.

The *response* of the automaton on state p to the "experiment" $x = x_1x_2\cdots x_n \in \Sigma^*$ is defined by a concatenation of random variables:

$$Res_{\mathcal{A}}(p, x_1x_2\cdots x_n) = f(p)f(\delta(p,x_1))\cdots f(\delta(p,x_1x_2\cdots x_n)).$$

For example, considering the automaton in Figure 9.1, the experiment starting in state 1 with input sequence 000100010 leads to the response:

$$Res_{\mathcal{A}}(1,000100010) = f(1)f(4)f(2)f(1)f(3)f(4)f(2)f(1)f(3)f(4).$$

The response is $Res_{\mathcal{A}}(1,000100010) = GRGGGRGGGR$ with probability $\cdot\frac{2}{3} = 2^6 \cdot 3^{-10}$; $Res_{\mathcal{A}}(1,000100010) = GGGGGGGGGG$ with probability $8 \cdot 3^{-10}$, a.s.o.

Let $\alpha \in [1/2,1]$. We say that a state p is *distinguishable* from the state q with *confidence* α if there exists an experiment $x = x_1x_2\cdots x_n \in \Sigma^*$ such that at least one probability that $f(p) \neq f(q)$, $f(\delta(p,x_1x_2\cdots x_i)) \neq f(\delta(q,x_1x_2\cdots x_i))$, $1 leq i leq n$ is greater or equal to α.

This means that at least on one point, during the "measurement" process of the responses of the automaton to the experiment x, the probability that the response of $\mathcal{A}$ on p and x is different (within the fixed level of confidence) to the response of $\mathcal{A}$ on q and x.[8]

[8]The motivation is the following. For a deterministic automaton $\mathcal{B}$, two states p,q are distinguishable if $Res_{\mathcal{B}}(p,x_1\ldots x_n) \neq Res_{\mathcal{B}}(q,x_1\ldots x_n)$, for some experiment $x_1\ldots x_n$ (see [8]), that is, for the corresponding transition and output functions, $f(p) \neq f(q)$ or $f(\delta(p,x_1x_2\cdots x_i)) \neq f(\delta(q,x_1x_2\cdots x_i))$, $1 leq i leq n$. In the probabilistic case we just replace the above conditions with the corresponding probabilistic ones.

For the automaton in Figure 9.1 we have:

$$\begin{aligned}
Res_{\mathcal{A}}(1,001) &= f(1)f(4)f(2)f(3),\\
Res_{\mathcal{A}}(2,001) &= f(2)f(1)f(4)f(2),\\
Res_{\mathcal{A}}(3,001) &= f(3)f(4)f(2)f(3),\\
Res_{\mathcal{A}}(4,001) &= f(4)f(2)f(1)f(3),
\end{aligned}$$

and the probability that $f(1) \neq f(i)$ is 5/9 for $i = 2,3,4$. So, with confidence 5/9 the experiment 001 distinguishes between every pair of distinct states.

We are now in a position to define properties **A, B, C** for a probabilistic automaton and a level of confidence α ($\alpha \in [1/2, 1]$).

- A probabilistic automaton has property **A** with level of confidence α if every pair of different states is distinguishable with *confidence* α.
- A probabilistic automaton has **B** with level of confidence α if every state is distinguishable with *confidence* α from any other distinct state.
- A probabilistic automaton has **C** with level of confidence α if there exists an experiment distinguishing with *confidence* α between each different states.

For example, the automaton in Figure 9.1 has **C**. Here are two examples of probabilistic automata having respectively **A** but not **B** and **B** but not **C**, i.e, *CI*, *CII*.

For *CI* we keep all components of the automaton in Figure 9.1 but modify the output function (see Figure 9.2):

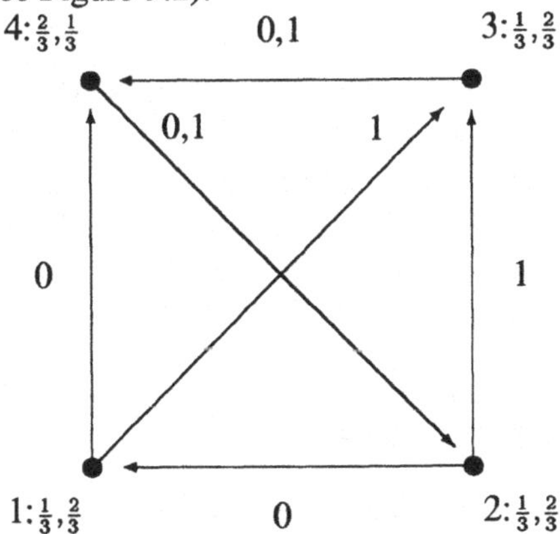

Figure 9.2.

With confidence 5/9 the probabilistic automaton in Figure 9.2 has *CI*. Indeed, using the experiments 1, 10 and 010 we can distinguish with confidence

5/9 between every two distinct states (0 distinguishes between (1,2), (1,4), (2,3), 1 distinguishes between (1,3), 10 distinguishes between (2,4) and 010 distinguishes between (3,4)), but no experiment starting with 1 distinguishes with confidence 5/9 between states 1 and 2, and no experiment starting with 0 distinguishes between the states 1 and 3 ($\delta(1,1x) = \delta(2,1x) = \delta(3,x)$, $\delta(1,0y) = \delta(3,0y) = \delta(4,y)$).

The probabilistic automaton in Figure 9.3 has *CII* with confidence 5/9:

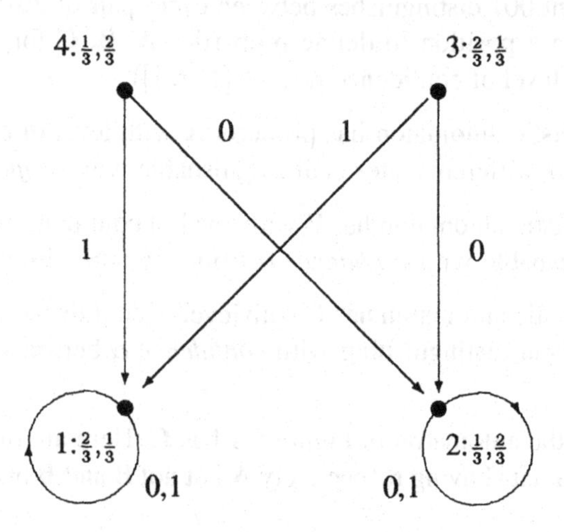

Figure 9.3.

Indeed, the following pairs of states are distinguishable with confidence 5/9 by every experiment: $(1,2), (1,4), (2,3), (3,4)$. Accordingly, with confidence 5/9, 1 is distinguishable from the other states by 0, 2 is distinguishable by 1, 3 is distinguishable by 0, and 4 is distinguishable by 1, so the probabilistic automaton has property **B** with confidence 5/9. It does not have property **C** with confidence 5/9 because:

- any experiment which starts with 1, i.e. $1x,\ x \in \Sigma^*$, does not distinguish with confidence 5/9 between 1 and 3;
- any experiment which starts with 0, i.e. $0y,\ y \in \Sigma^*$, does not distinguish with confidence 5/9 between 2 and 4.

6. MORE ABOUT MERMIN'S PROBABILISTIC AUTOMATA

We now argue that *with confidence 1/2, for every transition function δ, the corresponding Mermin probabilistic automaton has* **C**. Indeed,

- for every $i \neq j$, the probability that $f(ii) \neq f(jj) = 1/2$,
- for every $j \neq k$, the probability that $f(ii) \neq f(jk) = 7/8$,
- for every $i \neq j, k \neq l, ij \neq kl$, the probability that $f(ij) \neq f(kl) = 11/16$.

As all probabilities calculated above are greater than 1/2, it follows that with confidence 1/2, every Mermin automaton has **C**.

7. DECIDABILITY QUESTIONS

With techniques similar to those in Calude, Calude, Svozil, Yu [8] one can show that properties **A**, **B**, **C**, *CI*, *CII* are decidable for every probabilistic automaton. Complementarity properties *CI*, *CII* cannot appear for probabilistic automata with less than four states.

One way to check these properties, with some approximation, is by "simulating" a probabilistic automaton with a deterministic automaton (with some level of confidence). Let $\mathcal{A} = (Q, \Sigma, O, \delta, (a_{p,o})_{p \in Q, o \in O})$ be a probabilistic automaton, and $\alpha \in [1/2, 1]$ a confidence level. We construct a deterministic automaton $\mathcal{A}' = (Q, \Sigma, O', \delta, f')$, where $f : Q \to O'$ is the output function satisfying the following constraints: for every pair of distinct states p, q, if the probability that $f(p) \neq f(q)$ is greater or equal to α, then $f'(p) \neq f'(q)$; otherwise, $f'(p) = f'(q)$.

Note that the above construction cannot be carried on in all cases. For example, consider the probabilistic automaton $\mathcal{A} = (Q, \Sigma, O, \delta, (a_{p,o})_{p \in Q, o \in O})$ where $Q = \{p, q, r\}$, $O = \{G, R\}$, and the output probabilities are $a_{p,G} = 1, a_{p,R} = 0, a_{q,G} = 0, a_{q,R} = 1, a_{r,G} = a_{r,R} = 1/4$ (Σ, δ are arbitrary). It is easy to see that no deterministic automaton $\mathcal{A}'$ simulates $\mathcal{A}$, for every $\alpha \geq 1/2$. Reason: $f'(p) \neq f'(q)$, but $f'(p) = f'(q) = f'(r)$. However, this phenomenon cannot appear for all $\alpha \in [1/2, 1]$.

If the simulation is possible at the level of confidence α, then *every pair of distinct states $p, q \in Q$ are distinguishable by an experiment applied to $\mathcal{A}$ if and only if they are distinguishable by the same experiment applied to $\mathcal{A}'$.* Consequently, the probabilistic automaton $\mathcal{A}$ has **A** (**B**,**C**) if and only if $\mathcal{A}'$ has **A** (**B**, **C**, respectively).

For example, every Mermin probabilistic automaton is simulated with confidence $\alpha \leq 11/16 \approx 68.7\%$ by the automaton having the same components as every Mermin automaton and the output function $f'(11) = f'(22) = f'(33) = 0, f'(12) = f'(13) = f'(21) = f'(23) = f'(31) = f'(32) = 1$. Modifying the output function to $f'(11) = f'(22) = f'(33) = 0, f'(12) = 1, f'(13) = 2, f'(21) = 3, f'(23) = 4, f'(31) = 5, f'(32) = 6$ we get a simulation with confidence $\alpha \leq 7/8 \approx 87.5\%$. An analysis of properties **C**, *CI* and *CII* for these automata can be found in [5].[9]

If $\mathcal{A}$ has only two output states, say $\{G, R\}$, and $\alpha > 1/2$, then O' needs no more than two states. Indeed, O' needs more than two states if there are three distinct states $p, q, s \in Q$ such that

$$\text{the probability that } f(x) \neq f(y) \text{ is greater or equal to } \alpha, \qquad (9.1)$$

for all distinct $x, y \in \{p, q, s\}$. Assume that the probability that $f(x) = G$ is $a_{x,G}$, $x \in \{p, q, s\}$. Then, condition (9.1) can be written as

$$a_{x,G}(1 - a_{y,G}) + a_{y,G}(1 - a_{x,G}) \geq \alpha,$$

or

$$(a_{x,G} - \alpha)(\alpha - a_{y,G}) \geq 2\alpha - 1.$$

We arrive at a system of three inequalities which has no solution for $\alpha > 1/2$. The system has an infinity of solutions for $\alpha = 1/2$. For example, to simulate the probabilistic automaton $\mathcal{A} = (Q, \Sigma, O, \delta, (a_{p,o})_{p \in Q, o \in O})$ where $Q = \{1, 2, 3\}$, Σ, δ are arbitrary, $O = \{G, R\}$ and $f(1) = G$ with probability 1, and $f(2) = f(3) = G$ with probability 1/2, we need a deterministic automaton with three outputs.

References

[1] Bell, J.S. (1964), On the Einstein Podolsky Rosen paradox, *Physics*, 1:195–200. Reprinted in [38]: 403–408, and in [2]: 14–21.

[2] Bell, J.S. (1987), *Speakable and Unspeakable in Quantum Mechanics*, Cambridge University Press, Cambridge.

[3] Brauer, W. (1984), *Automatentheorie*, Teubner, Stuttgart.

[4] Bridgman, P.W. (1934), A physicist's second reaction to Mengenlehre, *Scripta Mathematica*, 2: 101–117, 224–234. Cfr. [22].

[5] Calude, C.S.; E. Calude; T. Chiu; M. Dumitrescu & R. Nicolescu (1999), Testing Computational Complementarity for Mermin Automata, CDMTCS Research Report, 109.

[6] Calude, C.S.; E. Calude & C. Ştefănescu (1998), Computational complementarity for Mealy automata, *Bulletin of the European Association for Theoretical Computer Science*, 66: 139–149.

[9]Due to the large number of transitions, i.e., $9^{18} \approx 150 \cdot 10^{15}$, an exhaustive search was computationally not feasible; instead, simulation techniques were used.

[7] Calude, C.S.; E. Calude & K. Svozil (1999), Quantum correlations conundrum: An automaton-theoretic approach, in *Proceedings of WIA'99*, Potsdam, Germany, in press.

[8] Calude, C.S.; E. Calude; K. Svozil & S. Yu (1997), Physical versus computational complementarity I, *International Journal of Theoretical Physics*, 36: 1495–1523.

[9] Calude, C.S. & M. Lipponen (1998), Computational complementarity and sofic shifts, in X. Lin, ed., *Theory of Computing 98, Proceedings of the 4th Australasian Theory Symposium, CATS'98*: 277–290. Springer, Singapore.

[10] Calude, E. & M. Lipponen (1998), Deterministic incomplete automata: simulation, universality and complementarity, in C.S. Calude; J. Casti & M. Dinneen, eds., *Proceedings of the First International Conference on Unconventional Models of Computation*: 131–149. Springer, Singapore.

[11] Calude, E. & M. Lipponen (1997), Minimal deterministic incomplete automata, *Journal of Universal Computer Science*, 11: 1180–1193.

[12] Calude, C.S. & Gh. Păun (2000), *Computing with Cells and Atoms*. Taylor & Francis Publishers, London, in progress.

[13] Conway, J.H. (1971), *Regular Algebra and Finite Machines*, Chapman and Hall, London.

[14] Dvurečenskij, A.; S. Pulmannová & K. Svozil (1995), Partition logics, orthoalgebras and automata, *Helvetica Physica Acta*, 68: 407–428.

[15] Greenberger, D. M.; M.A. Horne & A. Zeilinger (1989), Going beyond Bell's theorem, in M. Kafatos, ed., *Bell's Theorem, Quantum Theory, and Conceptions of the Universe*: 73–76. Kluwer, Dordrecht.

[16] Einstein, A.; B. Podolsky & N. Rosen (1935), Can quantum-mechanical description of physical reality be considered complete?, *Physical Review*, 47: 777–780. Reprinted in [38, pp. 138-141].

[17] Finkelstein, D. & S. R. Finkelstein (1983), Computational complementarity, *International Journal of Theoretical Physics*, 22.8: 753–779.

[18] Greenberger, D. B.; M. Horne & A. Zeilinger (1993), Multiparticle interferometry and the superposition principle, *Physics Today*, 46: 22–29.

[19] Grib, A. A. & R. R. Zapatrin (1990), Automata simulating quantum logics, *International Journal of Theoretical Physics*, 29.2: 113–123.

[20] Grib, A. A. & R. R. Zapatrin (1992), Macroscopic realization of quantum logics, *International Journal of Theoretical Physics*, 31.9: 1669–1687.

[21] Hopcroft, J. E. & J. D. Ullman (1979), *Introduction to Automata Theory, Languages, and Computation*, Addison-Wesley, Reading, Mass.

[22] Landauer, R. (1994), Advertisement for a paper I like, in J.L. Casti & J.F. Traub, eds., *On Limits*: 39. Santa Fe Institute Report 94-10-056, Santa Fe, NM.

[23] Jurvanen, E. & M. Lipponen (1999), Distinguishability, simulation and universality of Moore tree automata, *Fundamenta Informaticae*, 34: 1–13.

[24] Mermin, N. D. (1981), Bringing home the atomic world: Quantum mysteries for anybody, *American Journal of Physics*, 49: 940–943.

[25] Moore, E. F. (1956), Gedanken-experiments on sequential machines, in C.E. Shannon & J. McCarthy, eds., *Automata Studies*: 128–153. Princeton University Press, Princeton.

[26] Paz, A. (1971), *Introduction to Probabilistic Automata*, Academic Press, New York.

[27] Penrose, R. (1990), *The Emperor's New Mind: Concerning Computers, Minds, and the Laws of Physics*, Oxford University Press, Oxford.

[28] Penrose, R. (1994), *Shadows of the Minds, A Search for the Missing Science of Consciousness*, Oxford University Press, Oxford.

[29] Schaller, M. & K. Svozil (1994), Partition logics of automata, *Il Nuovo Cimento*, 109B: 167–176.

[30] Schaller, M. & K. Svozil (1995), Automaton partition logic versus quantum logic, *International Journal of Theoretical Physics*, 34.8: 1741–1750.

[31] Schaller, M. & K. Svozil (1996) Automaton logic, *International Journal of Theoretical Physics*, 35.5: 911–940.

[32] Schrödinger, E. (1935), Die gegenwärtige Situation in der Quantenmechanik, *Naturwissenschaften*, 23: 807-812, 823-828, 844- 849. English translation in J.A. Wheeler & W. H. Zurek (1983), *Quantum Theory and Measurement*: 152–167. Princeton University Press, Princeton.

[33] Svozil, K. (1993), *Randomness & Undecidability in Physics*, World Scientific, Singapore.

[34] Svozil, K. (1998), *Quantum Logic*, Springer, Singapore.

[35] Svozil, K. (1998), The Church-Turing Thesis as a guiding principle for physics, in C.S. Calude; J. Casti & M. Dinneen, eds., *Proceedings of the First International Conference on Unconventional Models of Computation*: 371–385. Springer, Singapore.

[36] Svozil, K. & R. R. Zapatrin (1996), Empirical logic of finite automata: microstatements versus macrostatements, *International Journal of Theoretical Physics*, 35.7: 1541–1548.

[37] Wang, L. J.; X. Y. Zou & L. Mandel (1991), Induced coherence without induced emission, *Physical Review*, A44: 4614–4622.

[38] Wheeler, J. A. & W. H. Zurek (1983), *Quantum Theory and Measurement*, Princeton University Press, Princeton.

[39] Zou, X. Y.; L. J. Wang & L. Mandel (1991), Induced coherence and indistinguishability in optical interference, *Physical Review Letters*, 67.3: 318–321.

Chapter 10

ACCEPTANCE OF ω-LANGUAGES BY COMMUNICATING DETERMINISTIC TURING MACHINES

Rudolf Freund
Institute of Computer Languages
Technical University of Wien
Karlsplatz 13, A-1040 Wien, Austria
rudi@logic.at

Ludwig Staiger
Institute of Informatics
Martin-Luther-University at Halle-Wittenberg
Kurt-Mothes-Str. 1, D-06120 Halle (Saale), Germany
staiger@informatik.uni-halle.de

Abstract Using a specific model of communicating deterministic Turing machines we prove that the class of ω-languages accepted by deterministic Turing machines via complete non-oscillating (complete oscillating) runs on the input coincides with the class of Π_3-definable (Σ_3-definable, respectively) ω-languages.

1. INTRODUCTION

In the models found in most papers in the literature (e.g., see the recent surveys [4], [10] or [13, 14]), the acceptance of ω-languages by Turing machines is determined by the behaviour of the Turing machines on the input tape as well as by specific final state conditions well-known from the acceptance of ω-languages by finite automata.

Hence, in this paper we consider systems of communicating (deterministic) Turing machines consisting of three components – the input component (where we observe the behaviour on the input tape), the transition component, and the output component (which is checked for the final states condition). The

C. Martin-Vide and V. Mitrana (eds.), Where Mathematics, Computer Science, Linguistics and Biology Meet, 115-125.

purpose of this decomposition of a deterministic Turing machine in three components is to facilitate the analysis and synthesis of the acceptance behaviour of deterministic Turing machines on infinite words.

The following types of behaviour of deterministic Turing machines on the input tape are found in literature:

Type 1 The approach of [15, 12] (cf. also [9, 10]) does not take into consideration the behaviour of the Turing machine on its input tape. Acceptance is based solely on the infinite sequence of internal states the machine runs through during its infinite computation.

Type 2 For X-automata Engelfriet and Hoogeboom [4] require that, in addition to the fulfillment of certain conditions on the infinite sequence of internal states in order to accept an input, the machine has to read the whole infinite input tape. Cohen and Gold [1] considered this same type for pushdown automata, too.

Type 3 The most complicated type of acceptance for Turing machines was introduced by Cohen and Gold [2, 3]. They require – in addition to type 2 – that the machine scans every cell of the input tape only finitely many times. This behaviour is termed as having a complete non-oscillating run.

Type 3' The complementary type of acceptance with respect to type 3 is the acceptance by complete oscillating runs, i.e., almost every cell of the input tape has to be scanned an infinite number of times.

The investigations carried out in [11] for type 1 and type 2 show that taking into account the different possible behaviours of deterministic Turing machines indeed results in different classes of accepted ω-languages. In this paper we focus on type 3 and its complementary type 3'. The class of ω-languages accepted by deterministic Turing machines working with type 3 turns out to coincide with the class Π_3 of the arithmetical hierarchy, regardless of which of the usual final states conditions we take. We also show that deterministic Turing machines working with the complementary type 3' really accept the ω-languages in the complementary class Σ_3 exactly.

2. PRELIMINARY DEFINITIONS

We start with some basic notations. By $\mathbb{N} = \{0, 1, 2, \ldots\}$ we denote the set of natural numbers. We consider the space X^ω of infinite strings (sequences, ω-words) on a finite alphabet of cardinality ≥ 2. By X^* we denote the set (monoid) of finite strings (words) on X, including the *empty* word e. For $w \in X^*$ and $b \in X^* \cup X^\omega$ let $w \cdot b$ be their *concatenation*. This concatenation product extends in an obvious way to subsets $W \subseteq X^*$ and $B \subseteq X^* \cup X^\omega$. As usual, we denote subsets of X^* as languages and subsets of X^ω as ω-languages. Furthermore, $|w|$ is the *length* of the word $w \in X^*$. For an ω-word ξ and every $n \in \mathbb{N}$, ξ/n denotes the prefix of ξ of length n.

As usual, we define Σ_1*-definable ω-languages* $E \subseteq X^\omega$ as

$$E = \{\xi \in X^\omega : \exists n \in \mathbb{N} \, (\xi/n \in W_E)\} \tag{10.1}$$

where $W_E \subseteq X^*$ is a recursive language, and we define Π_2*-definable ω-languages* $F \subseteq X^\omega$ as

$$F = \{\xi \in X^\omega : \forall i \in \mathbb{N} \, \exists n \in \mathbb{N} \, ((i, \xi/n) \in M_F)\} \tag{10.2}$$

where M_F is a recursive subset of $\mathbb{N} \times X^*$. Σ_2-definable and Π_1-definable ω-languages are defined accordingly.

In the sequel we will consider the following class $\mathfrak{P}$ of ω-languages defined as in condition 1 of lemma 3 in [11]:

$$\mathfrak{P} := \{F \subseteq X^\omega : \forall n \in \mathbb{N} \, \exists i \in \mathbb{N} \, ((i, \xi/n) \in M_F)\}, \tag{10.3}$$

where M_F is a recursive subset of $\mathbb{N} \times X^*$. Observe the difference of the ω-languages in $\mathfrak{P}$ to the definition of Π_2-definable ω-languages: For an ω-languages in $\mathfrak{P}$, the first quantifier is related to ξ and the second one to the other variable. The class $\mathfrak{P}$ is a subclass of $\mathbf{\Pi}_2$, the class of all Π_2-definable ω-languages, and it is closed under union and intersection (cf. [9, 11]).

Moreover, we define Π_3-definable ω-languages $E \subseteq X^\omega$ as

$$E = \{\xi \in X^\omega : \forall k \in \mathbb{N} \, \exists i \in \mathbb{N} \, \forall n \in \mathbb{N} \, ((k, i, \xi/n) \in M_E)\} \tag{10.4}$$

where M_E is a recursive subset of $\mathbb{N} \times \mathbb{N} \times X^*$, and we call an ω-language Σ_3-definable, if for some recursive subset $M_F \subseteq \mathbb{N} \times \mathbb{N} \times X^*$ we have

$$F = \{\xi \in X^\omega : \exists k \in \mathbb{N} \, \forall i \in \mathbb{N} \, \exists n \in \mathbb{N} \, ((k, i, \xi/n) \in M_F)\}. \tag{10.5}$$

The higher levels of the arithmetical hierarchy of ω-languages are defined accordingly, see e.g. [8], [9] or [10].

We only mention the following well-known closure properties of the classes of the arithmetical hierarchy with respect to boolean operations.

Lemma 10.1 *Each one of the classes* $\mathbf{\Sigma}_i$ *and* $\mathbf{\Pi}_i$ *of* Σ_i*-definable or* Π_i*-definable subsets of* X^ω*, respectively, is closed under union and intersection, and we have* $F \in \mathbf{\Sigma}_i$ *if and only if* $X^\omega \setminus F \in \mathbf{\Pi}_i$.

3. THE MODEL OF COMMUNICATING DETERMINISTIC TURING MACHINES

For a detailed description of configurations (instantaneous descriptions) and the behaviour (computing process) of Turing machines the reader is referred to the literature (e.g. [8]); for acceptance of ω-languages by Turing machines see [10] or the papers mentioned in the introduction.

We will describe the construction of (the system of communicating) Turing machines and their behaviour only in an informal manner leaving details to the reader. As indicated in Figure 10.1 each of the machines of the system has a finite control, a finite number of working tapes and a tape as input which their head can only read. For the Transition and the Output machines the input tapes are called communication tapes, because they serve also as output tapes for the Input and the Transition machine, respectively.

In addition to the communication from the upper machines to the lower ones via the communication tapes 1 and 2 the system has two feedback lines allowing some information flow from the lower to the upper machines. Formally, we define this as an input to the upper machine providing the current state of the lower machine.

Initially, the input is written on the input tape and the tape-cells of the other tapes are filled with blank symbols. Thus we may require the machines reading on communication tapes 1 and 2 not to read tape-cells where the preceding writing machines have not yet written.

Furthermore, one can easily verify that it is no loss of generality to require that the writing machines do not rewrite the contents of the tape-cells of the communication tapes.

It is obvious that the accepting power of a system of communicating Turing machines cannot exceed the power of a single Turing machine. We introduced this splitting of one machine into three components of a system in order to understand the different items of the accepting process of a Turing machine better:

1. the behaviour on the input tape,
2. the computation process, and
3. the accepting condition.

For machines accepting according to type 1 as defined above, a similar splitting was considered in [12]. In case of type 1 the input machine is not necessary, and it turned out that the accepting process can easily be described by a composition of a transition machine and a finite automaton as output machine without using any feed-back.

The read-write behaviour on the communication tapes as described above makes it possible to use as machines so-called deterministic Turing transducers (tt-transducers) or strictly deterministic Turing machines as they were called in [7], this type of machine having a read-only input tape, work-tapes and a write-only output tape on which the machine, for every input ω-word, produces (in a continuous or sequential way) an infinite output. These Turing transducer

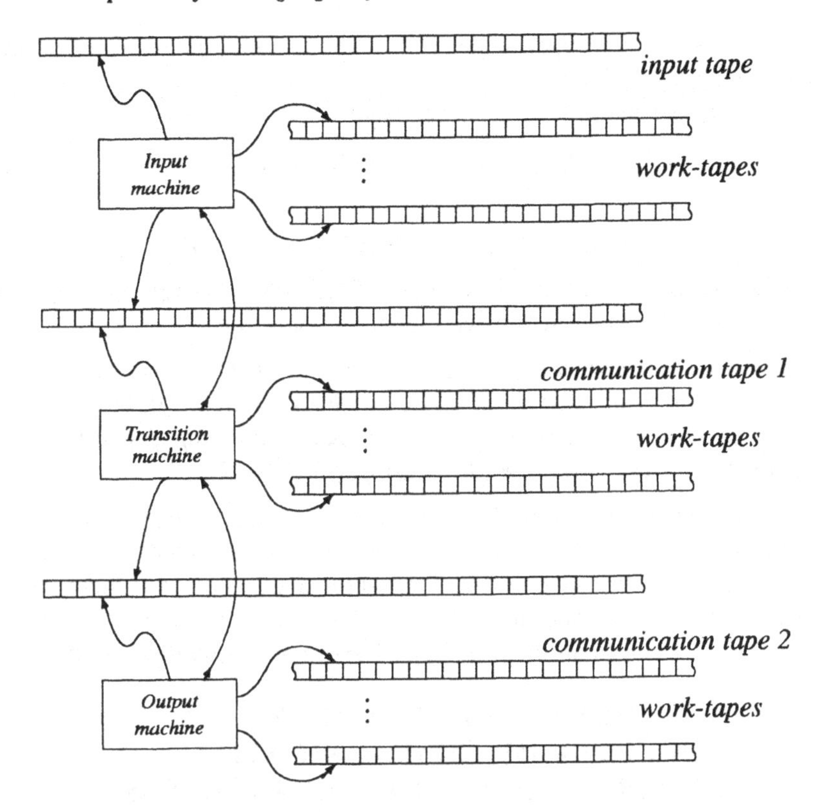

Figure 10.1 System of communicating Turing machines.

mappings or recursive operators on ω-words are described in more detail in [7], or [9] (see also [5] or [10]).

We recall here only the following properties useful with respect to Π_i-definable and Σ_i-definable ω-languages.

A first property is the following one.

Lemma 10.2 *Let $\Psi : X^\omega \to X^\omega$ be a tt-mapping, and let $F \subseteq X^\omega$ be Σ_i-definable and $E \subseteq X^\omega$ be Π_i-definable. Then $\Psi^{-1}(F)$ and $\Psi^{-1}(E)$ are also Σ_i-definable and Π_i-definable, respectively.*

In order to derive the next one we introduce a Cantor triple function, that is, a computable bijection $\langle\cdot,\cdot,\cdot\rangle : \mathbb{N} \times \mathbb{N} \times \mathbb{N} \to \mathbb{N}$. Let $\boldsymbol{l} : \mathbb{N} \to \mathbb{N}$, $\boldsymbol{m} : \mathbb{N} \to \mathbb{N}$ and $\boldsymbol{r} : \mathbb{N} \to \mathbb{N}$ be the left, middle and right inverse functions of $\langle\cdot,\cdot,\cdot\rangle$, respectively, defined via the equations

$$\boldsymbol{l}(\langle n,m,k\rangle) := n, \quad \boldsymbol{m}(\langle n,m,k\rangle) := m \quad \text{and} \quad \boldsymbol{r}(\langle n,m,k\rangle) := k\,.$$

For the sake of convenience, let $\langle \cdot, \cdot, \cdot \rangle$ be monotone in all of its arguments. Then $\boldsymbol{l}(l), \boldsymbol{m}(l), \boldsymbol{r}(l) \leq l$.

We consider the following ω-languages:

$$P_3 = \{\xi \in \{0,1\}^\omega : \forall n \exists m \forall k \, \xi(\langle n, m, k \rangle) = 1\} \text{ , and} \tag{10.6}$$

$$S_3 = \{\xi \in \{0,1\}^\omega : \exists n \forall m \exists k \, \xi(\langle n, m, k \rangle) = 1\} \text{ .} \tag{10.7}$$

In [12] it was shown that P_3 is a Π_3-complete ω-language and S_3 is a Σ_3-complete ω-language, that is, the following holds true.

Lemma 10.3 *For every Π_3-definable ω-language $F \subseteq X^\omega$ there is a tt-mapping $\Phi : X^\omega \to X^\omega$ such that $\Phi(L) \subseteq P_3$ and $\Phi(X^\omega \setminus L) \subseteq \{0,1\}^\omega \setminus P_3$, and for every Σ_3-definable ω-language $E \subseteq X^\omega$ there is a tt-mapping $\Psi : X^\omega \to X^\omega$ such that $\Psi(L) \subseteq S_3$ and $\Psi(X^\omega \setminus L) \subseteq \{0,1\}^\omega \setminus S_3$.*

For more details on Π_i- and Σ_i-complete ω-languages see [12] and also [9].

In Theorem 3.3 of [4] the existence of a similar splitting of X-automata into ω-preserving X-transducers and finite automata was proved. Here the notion of ω-preserving X-transducer differs a little bit from the above tt-mapping, because it takes into account the Type 2-acceptance generally assumed in [4].

4. CLASSES OF ω-LANGUAGES ACCEPTED BY COMMUNICATING DETERMINISTIC TURING MACHINES

In this section we investigate the classes of ω-languages accepted by communicating deterministic Turing machines via complete non-oscillating runs and via complete oscillating runs, respectively.

4.1 ACCEPTANCE WITHOUT FINAL STATES CONDITIONS

In this subsection we are interested in the question of which ω-languages can be accepted by (communicating) deterministic Turing machines when we disregard internal behaviour (states) and consider only the reading behaviour of the head on the input tape.

For a (communicating) deterministic Turing machine $\mathfrak{M}$, the ω-language accepted by $\mathfrak{M}$ via complete non-oscillating runs and complete oscillating runs, respectively, is denoted by $L_{cno}(\mathfrak{M})$ and $L_{co}(\mathfrak{M})$, respectively. The corresponding classes of ω-languages accepted by deterministic Turing machines are denoted by DT_{cno} and DT_{co}.

For any $\mathfrak{M}$, $L_{cno}(\mathfrak{M}) \cup L_{co}(\mathfrak{M})$ is the type 2-accepted ω-language of $\mathfrak{M}$ and therefore in $\mathfrak{P}$ (cf. [11]). The complement of $L_{cno}(\mathfrak{M}) \cup L_{co}(\mathfrak{M})$ consists of those ω-words where $\mathfrak{M}$ does not read the whole input.

In fact, it is easy to see that in the case of communicating deterministic Turing machines accepting an ω-word without taking into account the sequence of internal states of the output component, this component can be omitted in our model without changing the power of the model of communicating deterministic Turing machines for accepting ω-languages. As we shall show at the end of this section, the classes of ω-languages accepted by (communicating) deterministic Turing machines via complete non-oscillating (complete oscillating, respectively) runs do not change when we add one of the well-known final states conditions.

We now are going to show that $DT_{cno} = \mathbf{\Pi}_3$.

Lemma 10.4 $DT_{cno} \subseteq \mathbf{\Pi}_3$.

Proof. Let L be an ω-language accepted by a communicating deterministic Turing machine via type 3; this means that $\xi \in L$ if and only if for every position j on the input tape there is a (smallest) time instance t such that the cell at position j is scanned at time t and for all time instances t' with $t' > t$ the cell at position j is never scanned again. Obviously, this condition is Π_3-definable. □

Lemma 10.5 $P_3 \in DT_{cno}$.

Proof. The actions of a deterministic Turing machine $\mathfrak{M}$ accepting P_3 via complete non-oscillating runs can be sketched as follows:

$\mathfrak{M}$ successively for $l = 0, 1, 2, ...$ computes n, m, k such that $\langle n, m, k\rangle = l$ and reads the contents $c(l)$ of the cell at position l on the input tape.

1. In case we have $k = m = 0$ and $c(l) = 1$, $\mathfrak{M}$ adds the pair $(n, 0)$ to the list l_H of hypotheses.
2. If $(m, k) \neq (0, 0)$ and $c(l) = 1$, $\mathfrak{M}$ proceeds to $l + 1$.
3. In case we have $k = m = 0$ and $c(l) = 0$, $\mathfrak{M}$ adds the pair $(n, 1)$ to the list l_H of hypotheses, because $(n, 1)$ is the first possible hypothesis for fulfilling the required condition. Observe that for the monotonicity of the Cantor numbering we use, $\langle n, 1, 0\rangle > l$.
4. If $(m, k) \neq (0, 0)$ and $c(l) = 0$, $\mathfrak{M}$ has to check whether the pair (n, m) occurs in the list l_H or not.
    (a) (n, m) does not occur in the list l_H, i.e., $c(l) = 0$ is consistent with the current hypothesis for n, therefore $\mathfrak{M}$ proceeds to $l + 1$.
    (b) (n, m) does occur in the list l_H, i.e., $c(l) = 0$ contradicts the current hypothesis for n, therefore $\mathfrak{M}$ has to replace (n, m) in the list l_H by

a new hypothesis (n, m') for n, where m' is computed as follows: For $m' = m+1, m+2, \ldots$ $\mathfrak{M}$ successively checks the condition $c(\langle n, m', k' \rangle) = 1$ for all k' such that $\langle n, m', k' \rangle < l$. If such a minimal number m' with $\langle n, m', 0 \rangle < l$ can be found, we replace the pair (n, m) in the list l_H by the pair (n, m'), otherwise we compute the minimal number m' such that $\langle n, m', 0 \rangle > l$ and take this number m' for our new hypothesis for n. Observe that due to the monotonicity of the Cantor numbering, during this procedure $\mathfrak{M}$ never has to read a cell at a position $< \langle n, 0, 0 \rangle$.

Finally, $\mathfrak{M}$ moves its input head back to the position $\langle n, 0, 0 \rangle$ and continues with reading the contents of $\langle n, 0, 0 \rangle$ again. This oscillation guarantees that a complete non-oscillating run is possible on the infinite input sequence if and only if it fulfills the defining condition of P_3.

□

Lemma 10.6 $DT_{cno} \supseteq \mathbf{\Pi}_3$.

Proof. According to lemma 10.3, for any ω-language $L \in \mathbf{\Pi}_3$ there exists a deterministic Turing machine $\mathfrak{M}_L$ mapping each ω-word ξ on its input tape to an ω-word $\Psi_L(\xi)$ on its output tape in such a way that no cell on the output tape is rewritten any more. Moreover, let $\mathfrak{M}_{P_3}$ be a deterministic Turing machine accepting P_3 via complete non-oscillating runs as constructed in lemma 10.5. Hence, we can construct a communicating deterministic Turing machine $\mathfrak{M}$ accepting L via complete non-oscillating runs in the following way: In essence, $\mathfrak{M}_L$ forms the main part of the input component of $\mathfrak{M}$, whereas $\mathfrak{M}_{P_3}$ builds up the main part of the transition component of $\mathfrak{M}_{P_3}$. $\mathfrak{M}$ simulates $\mathfrak{M}_{P_3}$ and $\mathfrak{M}_L$ step by step in parallel with the constraint that $\mathfrak{M}_{P_3}$ can only read cells which $\mathfrak{M}_L$ has already written. Via the communication channel to the input component $\mathfrak{M}_L$ has to perform the following additional actions after each computation step simulating one step of $\mathfrak{M}_{P_3}$: $\mathfrak{M}$ moves the head on the input tape to the position currently under consideration on the communication tape between $\mathfrak{M}_L$ and $\mathfrak{M}_{P_3}$ and then moves the head of the input tape back to the cell currently to be read by $\mathfrak{M}_L$. To fulfill this intermediate task, $\mathfrak{M}$ has to store (for example, on additional working tapes) the actual positions of $\mathfrak{M}$ on the input tape as well as on the communication tape. This intermediate procedure guarantees that $\mathfrak{M}_{P_3}$ has a complete non-oscillating run on the input ξ if and only if $\mathfrak{M}_{P_3}$ has a complete non-oscillating run on $\Psi_L(\xi)$. □

Combining lemmas 10.4 and 10.6, we obtain

Theorem 10.1 $DT_{cno} = \mathbf{\Pi}_3$.

Lemma 10.7 *For any ω-language L with $L \in DT_{cno}$ there exists a deterministic Turing machine $\mathfrak{M}'$ such that every run of $\mathfrak{M}'$ is complete and $L = L_{cno}(\mathfrak{M}')$.*

Proof. Let $\mathfrak{M}$ be a (communicating) deterministic Turing machine accepting L. We now construct a (communicating) deterministic Turing machine $\mathfrak{M}'$ such that every run of $\mathfrak{M}'$ is complete and $L_{cno}(\mathfrak{M}) = L_{cno}(\mathfrak{M}')$. The (communicating) deterministic Turing machine $\mathfrak{M}'$ simulates the (communicating) deterministic Turing machine $\mathfrak{M}$ step by step, yet between two such simulation steps $\mathfrak{M}'$ takes the following intermediate actions: $\mathfrak{M}'$ scans the next cell on the input tape not scanned so far by $\mathfrak{M}'$ itself and then moves the head of the input tape back to the cell currently under consideration by $\mathfrak{M}$. In order to fulfill this intermediate task, $\mathfrak{M}'$ has to store (for example, on additional working tapes) the actual position of $\mathfrak{M}$ on the input tape as well as the last position scanned until then by $\mathfrak{M}'$ itself. In total, this algorithm guarantees that each run of $\mathfrak{M}'$ is complete, but still $L_{cno}(\mathfrak{M}) = L_{cno}(\mathfrak{M}')$. □

From the preceding lemma 10.7, theorem 10.1 and lemma 10.1, we immediately infer the following results:

Corollary 10.1 *For any ω-language L with $L \subseteq X^\omega$ we have:*

1. $L \in DT_{cno}$ *implies* $X^\omega \setminus L \in DT_{co}$.
2. $\mathbf{\Sigma}_3 \subseteq DT_{co}$.

Theorem 10.2 $DT_{co} = \mathbf{\Sigma}_3$.

Proof. Due to Corollary 10.1, we only have to prove that $\mathbf{\Pi}_3 \supseteq DT_{co}$. As already mentioned on page 120, for every deterministic Turing machine $\mathfrak{M}$, $L := L_{cno}(\mathfrak{M}) \cup L_{co}(\mathfrak{M}) \in \mathfrak{P} \subseteq \mathbf{\Pi}_2 \subseteq \mathbf{\Sigma}_3$. By definition, $L_{cno}(\mathfrak{M}) \cap L_{co}(\mathfrak{M}) = \emptyset$. As $X^\omega \setminus L_{cno}(\mathfrak{M}) \in \mathbf{\Sigma}_3$ and $\mathbf{\Sigma}_3$ is closed under intersection, $L_{co}(\mathfrak{M}) = L \cap (X^\omega \setminus L_{cno}(\mathfrak{M})) \in \mathbf{\Sigma}_3$. □

4.2 STATE-ACCEPTANCE CONDITIONS FOR (COMMUNICATING) DETERMINISTIC TURING MACHINES

In this part we informally consider the possible influence of the six conditions usually found in literature (cf. [4, 10, 12]) as additional conditions for determining the ω-language accepted by a (communicating) deterministic Turing machine. The most powerful final state condition is Muller's condition, requiring that the set of states occurring infinitely often in a (complete) run coincides with one of the given final states sets.

In terms of the arithmetical hierarchy, Muller's condition is a Boolean combination of $\forall\exists$-conditions which have to be combined in a conjunctive way with the input condition, which itself is a $\forall\exists\forall$-condition. According to the Tarski-Kuratowski-algorithm, in total we obtain not more than a $\forall\exists\forall$-condition. Hence we can formulate the following corollary:

Corollary 10.2 *The class of ω-languages accepted by (communicating) deterministic Turing machines via complete non-oscillating (oscillating, respectively) runs fulfilling Muller's condition for the corresponding state sequences is the class of Π_3-definable (Σ_3-definable) ω-languages.*

5. CONCLUSION

The results shown in [11] and in this paper reveal the importance of considering the behaviour of a deterministic Turing machine when reading its input tape for determining the accepted ω-language. As we have proved in this paper, deterministic Turing machines with type 3-acceptance (as well as with the complementary type 3'-acceptance) are strictly more powerful than deterministic Turing machines using type 2-acceptance, which in turn are more powerful than deterministic Turing machines with type 1-acceptance as was shown in [11]. In contrast to types 1 and 2, in the case of type 3- and type 3'-acceptance, the resulting classes of ω-languages do not depend on final states conditions. Moreover, separating the behaviour on the input tape from the final state condition in our model of communicating deterministic Turing machines allowed for a transparent analysis of the accepting power of complete non-oscillating runs of deterministic Turing machines.

References

[1] Cohen, R.S. & A. Y. Gold (1977), Theory of ω-languages I: Characterizations of ω-context-free languages, and II: A study of various models of ω-type generation and recognition, *Journal of Computer and System Sciences*, 15.2: 169–184, 185–208.

[2] Cohen, R.S. & A. Y. Gold (1978), ω-computations on Turing machines, *Theoretical Computer Science*, 6: 1–23.

[3] Cohen, R.S. & A. Y. Gold (1980), On the complexity of ω-type Turing acceptors, *Theoretical Computer Science*, 10: 249–272.

[4] Engelfriet, J. & H.J. Hoogeboom (1993), X-automata on ω-words, *Theoretical Computer Science*, 110.1: 1–51.

[5] Freund, R. & L. Staiger (1996), Numbers defined by Turing machines, *Collegium Logicum (Annals of the Kurt-Gödel-Society)*, II: 118–137. Springer, Wien.

[6] Landweber, L.H. (1969), Decision problems for ω-automata, *Mathematical Systems Theory*, 3.4: 376–384.

[7] Minsky, M.L. (1971), *Berechnung: Endliche und unendliche Maschinen*, Berliner Union, Stuttgart.

[8] Rogers, H. (1967), *Theory of Recursive Functions and Effective Computability*, McGraw Hill, New York.

[9] Staiger, L. (1986), Hierarchies of recursive ω-languages, *Journal of Information Processing and Cybernetics EIK*, 22.5/6: 219–241.

[10] Staiger, L. (1997), ω-languages, in G. Rozenberg & A. Salomaa, eds., *Handbook of Formal Languages*, III: 339–387. Springer, Berlin.

[11] Staiger, L. (1999), On the power of reading the whole infinite input tape, *Grammars*, 2.3: 247–257.

[12] Staiger, L. & K. Wagner (1978), Rekursive Folgenmengen I, *Zeitschrift für Mathematische Logik und Grundlagen der Mathematik*, 24.6: 523–538.

[13] Thomas, W. (1990), Automata on infinite objects, in J. Van Leeuwen, ed., *Handbook of Theoretical Computer Science*, B: 133–191, Elsevier, Amsterdam.

[14] Thomas, W. (1997), Languages, automata, and logic, in G. Rozenberg & A. Salomaa, eds., *Handbook of Formal Languages*, III: 389–455. Springer, Berlin.

[15] Wagner, K. & L. Staiger (1977), Recursive ω-languages, in *Fundamentals of Computation Theory*: 532–537. Springer, Berlin.

Chapter 11

COUNTER MACHINES AND THE SAFETY AND DISJOINTNESS PROBLEMS FOR DATABASE QUERIES WITH LINEAR CONSTRAINTS

Oscar H. Ibarra
Department of Computer Science
University of California
Santa Barbara, CA 93106-5110, USA
ibarra@cs.ucsb.edu

Jianwen Su
Department of Computer Science
University of California
Santa Barbara, CA 93106-5110, USA
su@cs.ucsb.edu

Constantinos Bartzis
Department of Computer Science
University of California
Santa Barbara, CA 93106-5110, USA
bar@cs.ucsb.edu

Abstract We give a brief review of some known decidability results concerning multi-counter machines. Using these results we show that safety and disjointness of database queries with linear integer constraints are decidable.

C. Martin-Vide and V. Mitrana (eds.), Where Mathematics, Computer Science, Linguistics and Biology Meet, 127-137.

1. INTRODUCTION

There is a large class of machines more powerful than finite automata with decidable decision problems similar to those for finite automata. The class consists of finite automata augmented with reversal-bounded counters (i.e., the number of times a counter can change mode from nondecreasing to nonincreasing or vice-versa is bounded by a constant). It is known that the emptiness, infiniteness, disjointness, containment, and equivalence problems for these machines are decidable [7]. The decidability of these problems has been used to show the existence of algorithms for some decision questions in automata theory, formal languages, and recently in database query languages [8].

In this paper we exhibit another application in studying properties of conjunctive queries with linear integer constraints: "safety" and "disjointness". We consider relational databases [3] that consist of finite relations. A conjunctive query is a logic formula involving only relations in the databases and may include equality, order, and linear arithmetic operations. The answer to a conjunctive query may become infinite; the query is called "unsafe" if such a case happens. The decision problem of safety is to determine if a (conjunctive) query always returns a finite relation as its answer regardless of the input database. The disjointness problem, on the other hand, is to determine if two conjunctive queries always return disjoint answers for all databases. In this paper, we use counter machines to show that both problems are decidable. The results extend the decidability results of query containment and equivalence problems studied in [8].

2. REVERSAL-BOUNDED MULTICOUNTER MACHINES

We recall the definition of a reversal-bounded multicounter machine first introduced in [7, 6]. For $k, r \in \mathbb{N}$ (= the set of nonnegative integers), define a nondeterministic (deterministic) r-reversal k-counter machine as a one-way nondeterministic finite automaton augmented with k counters such that on any input, each counter makes no more than r reversals. Note that each counter can be tested whether it is 0 or non-0 and incremented or decreased by 1. We count each alternation from nondecreasing mode to nonincreasing mode or vice-versa as a *reversal.* So, for example, a counter with the following computation pattern:

$$0000011111122222234444 4$$

corresponds to 0-reversal. On the other hand

$$000001111112222223444\underline{43}33222$$

corresponds to 1-reversal (the reversal is underlined).

When k, r are not specified, we refer to the machine as a reversal-bounded multicounter machine. We note that the input head needs not always move right from the symbol it is currently scanning at every time step (i.e., the head may *sit* on a symbol for a while before moving right; in fact, the machine can go into an infinite loop without leaving the symbol). Although the machine is one-way, we assume for convenience that the input has left and right delimiters. We assume, without loss of generality, that the machine accepts (by entering an accepting state) with all counters zero and the input head falling off the right input delimiter.

Reversal-bounded multicounter machines are quite powerful. As an example, let L be the language consisting of all strings $x\%y\%z$, such that x, y, z are pair-wise distinct binary strings (% is a separator). A nondeterministic 1-reversal 3-counter machine M can accept L. M uses one counter to check that x is different from y, a second counter to compare x and z, and a third counter to check that y is different from z. To verify that x is different from y, M "guesses" a position of discrepancy (within the string x). It does this by incrementing the first counter by 1 for every symbol it encounters while moving right on x, and nondeterministically terminating the counting at some point, guessing that a position of discrepancy has been reached. M records in its finite-control the symbol in that position. M uses the value in the counter to arrive at the same location within y where a discrepancy was guessed to occur. The second and third counters are used in a similar way to compare x with z and y with z.

A fundamental result concerning reversal-bounded multicounter machines [7] is the decidability of the emptiness problem: Given a machine M, is the language $L(M)$ accepted by M empty? For completeness, we will sketch the proof of this result briefly.

Let k be a positive integer. A subset S of $\mathbb{N}^k$ is a *linear set* if there exist vectors $v_0, v_1, \ldots, v_t$ in $\mathbb{N}^k$ such that $S = \{v \mid v = v_0 + a_1 v_1 + \cdots + a_t v_t, a_i \in \mathbb{N}\}$. The constant vector v_0 and the periods $v_1, \ldots, v_t$ are called the *generators*. S is *semi-linear* if it is a finite union of linear sets.

An empty set is a trivial (semi)-linear set, where the set of generators is empty. Any finite subset of $\mathbb{N}^k$ is semi-linear (it is a finite union of linear sets whose generators are constant vectors).

Let A be an alphabet consisting of k symbols $a_1, \ldots, a_k$. For each string (word) w in A^*, define the *Parikh map* of w, denoted by $f(w)$, as follows:

$$f(w) = (i_1, \ldots, i_k), \text{ where } i_j \text{ is the number of occurrences of } a_j \text{ in } w.$$

If L is a subset of A^*, define $f(L) = \{f(w) \mid w \in L\}$. $f(L)$ is called the Parikh map of L. It is known [9] that if M is a one-way nondeterministic finite automaton (NFA) then $f(L(M))$ is a semi-linear set effectively computable from M , i.e., the Parikh map of a regular set is semi-linear. This generalizes to reversal-bounded multicounter machines [7]:

Theorem 11.1 *Let M be a nondeterministic reversal-bounded multicounter machine. Then $f(L(M))$ is a semi-linear set effectively computable from M.*

Proof. Without loss of generality, we may assume that the counters of M make exactly one reversal (since a counter which makes r reversals can be replaced by $(r+1)/2$ counters, each making one reversal). Note that M accepts with all counters zero and the input head falling off the right delimiter.

We sketch the proof for the case when the input alphabet is $\{a, b\}$ and M has only one counter which is 1-reversal bounded. The generalization to any alphabet and any number of counters is straightforward.

We construct an NFA M' with input alphabet $\{\#, a, b, +, -\}$. An input $\#w\#$ to M (the #'s are the input delimiters) will correspond to some input $\#z\#$ to M', where z is a string composed of symbols $a, b, +, -$. The string z is such that if the $+$'s and $-$'s are deleted from z, we get w. All occurrences of $+$'s in z preceed all occurrences of $-$'s. Thus if we delete all occurences of symbols a and b from z, we get a string of $+$'s followed by a string of $-$'s. The idea is for "$+$" to represent the increment "$+1$" to the counter and for "$-$" to represent the decrement "-1". Intuitively the NFA M' simulates the computation of M where instead of using the counter, M' reads a "$+$" whenever the counter increments and reads a "$-$" whenever the counter decrements. Note that a valid accepting computation of M requires that the number of $+$'s is equal to the number of $-$'s in z. Of course, M' has no way of detecting if this is the case (since M' is a finite automaton). So M' nondeterministically guesses at some point during the computation that a "last" occurence of "$-$" on the input corresponds to zero counter and no more $-$'s will be encountered on the input (it aborts if this is not the case). M' continues the simulation of M and accepts if M enters an accepting state.

Since M' is an NFA, $f(L(M'))$ is a semi-linear set S_1 effectively computable from M'. Note that S_1 is a set of 5-tuples (i, j, k, l, m) of nonnegative integers corresponding to the number of occurences of symbols a, b, #, $+, -$ (in this order) in the strings in $L(M')$. Note that $k = 2$ since there are exactly two occurences of #'s in the strings.

Let $S_2 = \{(i, j, 2, m, m) \mid i, j, m \in \mathbb{N}\}$. Clearly S_2 is a semi-linear set. Then $S_3 = S_1 \cap S_2$ is a semi-linear set effectively computable from S_1 and S_2, since semi-linear sets are effectively closed under intersection [4]. Finally, the set S_4 obtained from S_3 by deleting the last three coordinates of the 5-tuples is an effectively computable semi-linear set. Clearly, S_4 is $f(L(M))$. □

Since the emptiness problem for semi-linear sets is decidable (it is empty if it has no generators), we have:

Theorem 11.2 *The emptiness problem is decidable for nondeterministic reversal-bounded multicounter machines.*

Theorem 11.3 *The infiniteness and disjointness problems are decidable for nondeterministic reversal-bounded multicounter machines.*

Proof. **Infiniteness:** Given a machine M, $f(L(M))$ is a semi-linear set S effectively computable from M. $L(M)$ is finite if and only if all linear sets comprising S have only constant generators.
Disjointness: Given two machines M_1 and M_2 with k_1 and k_2 counters, respectively, construct a machine M with $k_1 + k_2$ counters which simulates M_1 and M_2 in parallel on the same input. M accepts if and only if both M_1 and M_2 accept. Then $L(M_1)$ and $L(M_2)$ are disjoint if and only if $L(M)$ is empty, which is decidable by Theorem 11.2. □

Containment and equivalence are undecidable for nondeterministic machines. In fact, it is undecidable to determine, given a nondeterministic machine with one 1-reversal counter, whether it accepts all strings [2]. However, for deterministic machines, we have:

Theorem 11.4 *The containment and equivalence problems are decidable for deterministic reversal-bounded multicounter machines.*

Proof. It is sufficient to show that containment is decidable. Suppose M_1 and M_2 are deterministic reversal-bounded multicounter machines. We show below that we can construct a deterministic reversal-bounded multicounter machine M_3 accepting the complement of $L(M_2)$. Then $L(M_1) \subseteq L(M_2)$ if and only if $L(M_1) \cap L(M_3)$ is empty, which is decidable by Theorem 11.2.

Assume, as in the proof of Theorem 11.1, that the counters of M_2 make exactly one reversal and that M_2 accepts with the input head falling off the right delimiter. We construct a machine M_3 which computes like M_2. If M_2 accepts, M_3 rejects; if M_2 rejects and halts, M_3 accepts. However, M_2 can go into an infinite loop without leaving an input symbol. This can only happen if the input head remains on the same input position for more than s steps (where s is the number of states of M_2) without any counter decreasing its value. Clearly, M_3 can detect this situation and moves the head to the right, falling off the right delimiter in an accepting state. □

It can be shown that the complexity of the algorithms for the emptiness, and infiniteness problems in Theorems 11.2 and 11.3 is deterministic $n^{c \cdot r \cdot k}$ time, where n is the length of the binary representation of the r-reversal k-counter machine M, and c is some constant. A similar complexity holds for disjointness, containment, and equivalence, except that now $n = n_1 + n_2$, $r_1 + r_2$ and $k = k_1 + k_2$ for r_1-reversal k_1-counter machine M_1 and r_2-reversal k_2-counter machine M_2.

3. DECIDABLE PROBLEMS CONCERNING CONJUNCTIVE QUERIES

We study queries over relational databases [3] that may involve linear arithmetic operations on integers. A database may consist of a finite set of finite relations over a universal domain. Although in general the universal domain can be some countably infinite set, we consider in this paper the set $\mathbb{Z}$ of integers as the universal domain since our queries may involve arithmetic operations.

We assume the existence of a countably infinite set $\boldsymbol{P}$ of *relation names*, each having an associated *arity*. A *schema* is a finite subset of $\boldsymbol{P}$. A *database* of a schema σ is a total mapping I from σ such that for each k-ary $p \in \sigma$, $I(p)$ is a finite subset of $\mathbb{Z}^k$. Let $inst(\sigma)$ be the set of all databases of σ.

We focus on the class of "conjunctive" queries extended with linear integer arithmetic operations ($<, =, +, -$, and $\equiv_k$ for each integer $k > 1$). The query language is defined in terms of rules (cf [1]). An *atom* is a formula of form $p(y_1, \ldots, y_k)$ where p is a k-ary relation name in $\boldsymbol{P}$ and $y_1, \ldots, y_k$ are variables or integers in $\mathbb{Z}$. Let σ be a schema. A *rule over* σ is an expression of the following form:

$$ans(\bar{x}) \leftarrow p_1(\bar{y}_1), \ldots, p_n(\bar{y}_n), \varphi(\bar{z}) \qquad (\dagger)$$

where

1. n is a non-negative integer,
2. $\bar{x}$, $\bar{y}_i$'s, are $\bar{z}$ are sequences of variables or integers in $\mathbb{Z}$,
3. $p_1, \ldots, p_n$ are relation names in σ and $ans(\bar{x}), p_1(\bar{y}_1), ..., p_n(\bar{y}_n)$ are atoms, and
4. φ is a quantifier-free first-order formula involving linear arithmetic operations and sentential connectives over variables $\bar{z}$.

In a rule of the form ($\dagger$), $\bar{x}, \bar{z}$ and $\bar{y}_i$'s may contain common variables.

Suppose Q is a rule over a schema σ of form ($\dagger$) and I is a database of σ. Then each $I(p_i)$ ($1 \leqslant i \leqslant n$) is a finite relation. The *result* of the rule Q on the database I, denoted by $Q[I]$, is the relation

$$\{(\bar{a}) \mid \text{the formula } \psi(\bar{a}) \text{ is true over } \mathbb{Z}\} ,$$

where the formula $\psi(\bar{x}) = \exists \bar{v} \left(\bigwedge_{1 \leqslant i \leqslant n} I(p_i)(\bar{y}_i)\right) \wedge \varphi(\bar{z})$ and $\bar{v}$ are all variables occurring in Q but not in $\bar{x}$. For each database, the result of a rule on the database can be effectively computed (see [5] for a formal discussion).

Let σ be a schema and k a nonnegative integer. A *k-ary conjunctive query* over σ is a rule Q over σ of form ($\dagger$) where $\bar{x}$ is a vector of k variables or integers in $\mathbb{Z}$. Suppose Q is a conjunctive query and I a database of σ. The *answer* to Q on I, denoted by $Q(I)$, is the result $Q[I]$ if it is finite, and is undefined otherwise.

For a conjunctive query, it is clear that it may contain in the answer elements that not in the input database. Therefore the query result may sometimes be infinite even though the input databases are always finite, i.e., the query may not be closed since the answer does not exist.

Definition 11.1 *A conjunctive query is* safe *if for each database, its answer exists (i.e., the result is finite).*

When arithmetic operations are absent, safety can be syntactically characterized by "range restriction" [1]. Specifically, a conjunctive query Q of form (†) over a schema σ is *range restricted* if each variable in the $\bar{x}$ occurs in at least one atom on the right hand side of Q. When arithmetic operations are present, range restriction reduces the expressive power.

Proposition 11.1 *There exists a safe conjunctive query which is not equivalent to any range restricted conjunctive query.*

Proof. Consider the conjunctive query $Q : ans(x) \leftarrow r(y), x = y + 1$ over a schema with a single unary relation r. Clearly Q is safe. However, Q is not equivalent to any range restricted conjunctive query. This is because the answer always contains an integer that is greater than all integers in the input database. □

For the sake of simplicity, we assume for the remainder of the section that the schema σ contains only one relation name r of arity k. It will become clear that the results can be easily generalized to arbitrary schemas.

Our first result states that safety is decidable for conjunctive queries. To prove the result, we introduce the following notion. Let n be a positive integer. A database I of σ is called *n-bounded* if I (i.e. $I(r)$) contains no more than n tuples. Let $inst_n(\sigma)$ denote the family of all n-bounded databases over σ.

Lemma 11.1 *Let Q be a conjunctive query over the schema σ with n atoms in the left hand side of the rule. Then, Q is safe if and only if for each n-bounded database I, $Q(I)$ is defined (i.e. $Q[I]$ is finite).*

Proof. The only if direction holds obviously. For the if direction, let I be an arbitrary database. If I is n-bounded, $Q(I)$ exists. Otherwise, let $I_1, \ldots, I_m$ be all possible subsets of I with at most n tuples. Since Q is defined for all n-bounded databases, $Q[I_i]$ is finite for each $1 \leqslant i \leqslant m$. It can be verified that $Q[I] = \cup_i Q[I_i]$. Therefore $Q[I]$ is finite and $Q(I)$ is defined. □

The above lemma allows us to focus only on bounded databases in order to determine safety of a conjunctive query. In other words, if for each instantiation of atoms in the conjunctive query Q, there are only a finite number of answers generated, then Q is safe. The key is to examine if the arithmetic formula in Q always admits finite answers. The following lemma gives characterization for unsafe formulas.

Lemma 11.2 *Let $\phi(y_1, \ldots, y_\ell, x)$ be a quantifier-free formula involving linear arithmetic operations. Then, the following are equivalent:*

1. *For all $b_1, \ldots, b_\ell$, the set $\{x \mid \phi(b_1, \ldots, b_\ell, x) \text{ is true }\}$ is finite.*
2. *There do not exist integers $a, b_1, \ldots, b_\ell$ such that $|a| > c_1 + \sum_i c_2|b_i|$ and $\phi(b_1, \ldots, b_\ell, a)$ is true, where c_1 is the sum of the absolute values of all constants in ϕ and c_2 is the largest absolute value of coefficients in ϕ.*

Proof. Note that item (2) states equivalently that for all $b_1, \ldots, b_\ell$, there is a lower and an upper bound on values for x that satisfy the formula $\phi(b_1, \ldots, b_\ell, x)$. Therefore, the direction (2) $\rightarrow$ (1) is obvious. To prove the other direction, suppose there exist integers $a, b_1, \ldots, b_\ell$ that satisfy the stated condition. Without loss of generality, we assume that ϕ contains no negation (which can be easily eliminated by changing "=" to a disjunction of "<" and ">", etc). Let $\psi \equiv (ex+c+\sum_i e_i y_i)\,\theta\,0$ be an atomic formula in ϕ such that $\psi(b_1, \ldots, b_\ell, a)$ is true. It follows that θ cannot be "=" for $b_1, \ldots, b_\ell, a$ since

$$|ea| \geqslant |a| > c_1 + \textstyle\sum_i c_2|y_i| \geqslant |c| + \sum_i |e_i| \cdot |y_i| \geqslant |c + \sum_i e_i y_i|.$$

Therefore, it follows that for all $x \geqslant a$ if a is positive or $x \leqslant a$ if a is negative, $\psi(b_1, \ldots, b_\ell, x)$ remains true. Consequently, $\phi(b_1, \ldots, b_\ell, x)$ is true for infinitely many x values, which is a contradiction. □

The above lemma can be extended to cases with several variables and with existentially quantified variables. In these cases, the bound will need to be increased by a factor determined by the coefficients of these variables and the number of such variables.

To prove the decidability of safety, we reduce the safety problem over bounded databases to the emptiness problem of languages accepted by reversal bounded counter machines. Since the latter is decidable by Theorem 11.2, it follows that safety is decidable by Lemmas 11.1 and 11.2.

The reduction uses a standard "unary" encoding of databases and tuples. Let the alphabet consist of symbols $c_1, \ldots, c_k, +, -$, and $\#$, where k is the (maximal) arity of relations in σ. A tuple $t = (a_1, \ldots, a_k) \in \mathbb{Z}^k$ is encoded as $ENC(t) =$ "$s_1 c_1^{v_1}, \ldots, s_k c_k^{v_k}$", where s_i is the sign of a_i and v_i the absolute value of a_i. If $t_1, \ldots, t_\ell$ is an enumeration of a relation in a database I, the encoding of the relation is "$ENC(t_1) \cdots ENC(t_\ell)$". The encoding of the database I, $ENC(I)$, is the concatenation of encodings of its relations (in some fixed order).

Let Q be a conjunctive query over σ and m a non-negative integer. Suppose k_I is the largest bound on the absolute value of the values in the answer for a given database I, derived from Lemma 11.2. For a tuple t, we denote by $|t| > k_I$ the condition that the absolute value of each component in t is greater than k_I. We define the language $L_m^k(Q)$ as follows:

$$L_m(Q) = \{ENC(t)\#ENC(I) \mid I \in inst_m(\sigma), t \in Q[I], |t| > k_I\}.$$

Lemma 11.3 *For each conjunctive query Q of size n and each non-negative integer m, $L_m(Q)$ is accepted by a nondeterministic one-way 1-reversal $p(n)$-counter machine M for some polynomial p. The size of M is also polynomial in n.*

Proof. Given a conjunctive query Q of form (†), we construct a nondeterministic counter machine M which, on an input "$ENC(t)\#ENC(I)$", first "guesses" an assignment α of values to the variables occurring in the rule Q. M then verifies if $|t| > k_I$ and if $t \in Q(I)$ under the assignment α. Verification of $|t| > k_I$ is obvious (since k_I is linear in terms of values occurring in the database). The verification of $t \in Q(I)$ under the assignment α involves (i) checking if the tuple for each atom is in the corresponding relation of I, which can be done in the straightforward manner, and (ii) checking if the formula ϕ is satisfied, which can be done by a one-way, polynomial-size, polynomial-reversal counter machine M' constructed as follows. The states (finite control) of M' remember all coefficients including the signs and ϕ. For each variable x, if a is the largest absolute value of its coefficients, M' has the following counters: $c_{x,0}, c_{x,1}, \ldots, c_{x,\lceil \log a \rceil}, c'_x$. The counter $c_{x,i}$ will store the value $2^i \times x$. Initially $c_{x,0}$ gets the value from the (one-way) input. For each $0 \leqslant i \leqslant \lceil \log a \rceil - 1$, the value of $c_{x,i+1}$ is computed from $c_{x,i}$ using c'_x as an auxiliary counter. The value of $c_{x,i}$ is restored after computing the value of $c_{x,i+1}$. For each occurrence of a term t, M' includes a counter c_t that will hold the absolute value for the term (the sign is remembered in the control). The values of the counters for terms "ax" are computed from the counters $c_{x,i}$, $0 \leqslant i \leqslant \lceil \log a \rceil$, possibly using the auxiliary space c'_x. The values of the counters $c_{x,i}$'s are again restored after their use. The values of the counters for the other terms are then computed based on the syntax of the terms. Finally, M' performs comparison tests for $=, \leqslant$, and $<$, and then computes the truth value of the formula ϕ with Boolean operations. Clearly M' has polynomially many counters and a polynomial size in the size of ϕ. As for the number of reversals, we note that each "doubling" in computing $c_{x,i}$ and each addition/substraction operation may need a couple of reversals. It follows that the total number of reversals is polynomial in the size of ϕ. □

It follows from the above lemma and Theorem 11.2 that:

Theorem 11.5 *Safety of conjunctive queries with linear arithmetic operations is decidable in single exponential time on the length of the query.*

Next, we introduce the notion of "query disjointness". We only consider safe queries.

Definition 11.2 *Let Q, Q' be two safe conjuntive queries over a schema σ. The query Q is* disjoint from Q', *if for each $I \in \text{inst}(\sigma)$, $Q(I) \cap Q'(I) = \emptyset$.*

Let I, J be two databases over the same schema σ. We define $I \subseteq J$ if for each $p \in \sigma$, $I(p) \subseteq J(p)$. A safe conjunctive query Q over σ is *monotonic* if for all databases $I, J \in inst(\sigma)$, $I \subseteq J$ implies $Q(I) \subseteq Q(J)$.

Lemma 11.4 Every safe conjunctive query is monotonic.

We now prove that disjointness of conjunctive queries with linear arithmetic operations is decidable in exponential time on the length of the queries. The proof is again based on the decidability results for finite-reversal counter machines.

Lemma 11.5 *LetQ, Q' be two conjunctive queries over the schema σ. Then there is an non-negative integer m (depending on Q and Q') such that Q is disjoint from Q' iff for each database $I \in \text{inst}_m(\sigma)$, $Q(I) \cap Q'(I) = \emptyset$.*

Proof. Let m be the sum of the number of atoms in the rules in Q and Q'. The only if direction is trivial since $inst_m(\sigma)$ is a subset of $inst(\sigma)$. For the if direction, suppose Q and Q' are not disjoint. Then for some I, $Q(I) \cap Q'(I) \neq \emptyset$, i.e., there is a tuple t such that $t \in Q(I)$ and $t \in Q'(I)$. Let the rules in Q and Q' respectively be:

$$ans(\bar{x}) \leftarrow p_1(\bar{y}_1), \ldots, p_{m_1}(\bar{y}_{m_1}), \phi(\bar{z})$$

and

$$ans(\bar{x}) \leftarrow q_1(\bar{u}_1), \ldots, q_{m_2}(\bar{u}_{m_2}), \phi'(\bar{v})$$

Since $t \in Q(I)$ and $t \in Q'(I)$, there are assignments α and α' mapping variables to integers such that the rule bodies are true and $\alpha(\bar{x}) = \alpha'(\bar{x}) = t$. Let I' be the database containing the $m = m_1 + m_2$ tuples in the images of α, α'. Then clearly $t \in Q(I')$ and $t \in Q'(I')$. Therefore $Q(I') \cap Q(I') \neq \emptyset$ for some m-bounded database I'. □

Let Q be a conjunctive query and m a nonnegative integer. We define the language $L'_m(Q)$ as follows.

$$L'_m(Q) = \{ENC(t)\#ENC(I) \mid I \in inst_m(\sigma), t \in Q[I]\}.$$

Lemma 11.6 *Let Q_1 and Q_2 be two safe conjunctive queries and m a non-negative integer such that Q and Q' are disjoint iff $\forall I \in \text{inst}_m(\sigma)$, $Q(I) \cap Q(I) \neq \emptyset$ (such m exists by Lemma 11.5). Then $L'_m(Q_1) \cap L'_m(Q_2) = \emptyset$ iff Q_1 is disjoint from Q_2.*

Lemma 11.7 *For each safe conjunctive query Q of size n and each non-negative integer m, $L'_m(Q)$ is accepted by a nondeterministic one-way 1-reversal $p(n)$-counter machine M, where p is a polynomial. The size of M is also polynomial in n.*

The proof of the above lemma is similar to the proof of Lemma 11.3. From Lemmas 11.6 and 11.7 and Theorem 11.3 we can easily conclude:

Theorem 11.6 *Disjointness of safe conjunctive queries with linear arithmetic operations is decidable in single exponential time on the sum of the lengths of the queries.*

Acknowledgments

This work is partially supported by NSF grants IRI-9700370 and IIS-9817432.

References

[1] Abiteboul, S.; R. Hull & V. Vianu (1995), *Foundations of Databases*. Addison-Wesley, Reading, Mass.

[2] Baker, B. & R. Book (1974), Reversal-bounded Multipushdown Machines, *Journal of Computer and System Sciences*, 8: 315–332.

[3] Codd, E.F. (1970), A Relational Model of Data for Large Shared Data Banks, *Communications of the ACM*, 13.6: 377–387.

[4] Ginsburg, S. & E. Spanier (1964), Bounded Algol-like Language, *Transactions of the American Mathematical Society*, 113: 333–368.

[5] Grumbach, S. & J. Su (1997), Finitely representable databases, *Journal of Computer and System Sciences*, 55.2: 273–298.

[6] Gurari, E.M. & O. H. Ibarra (1981), The Complexity of Decision Problems for Finite-turn Multicounter Machines, *Journal of Computer and System Sciences*, 22: 220–229.

[7] Ibarra, O.H. (1978), Reversal-bounded Multicounter Machines and Their Decision Problems, *Journal of the ACM*, 25: 116–133.

[8] Ibarra, O.H. & J. Su (1999), A Technique for Proving Decidability of Containment and Equivalence of Linear Constraint Queries, *Journal of Computer and System Sciences*, 59.1: 1–28.

[9] Parikh, R. (1966), On Context-free Languages, *Journal of the ACM*, 13: 570–581.

Chapter 12

AUTOMATA ARRAYS AND CONTEXT-FREE LANGUAGES

Martin Kutrib
Institute of Informatics
University of Giessen
Arndtstr. 2, D-35392 Giessen, Germany
kutrib@informatik.uni-giessen.de

Abstract From a biological point of view automata arrays have been employed by John von Neumann in order to solve the logical problem of nontrivial self-reproduction. From a computer science point of view they are a model for massively parallel computing systems. Here we are dealing with automata arrays as acceptors for formal languages. Our focus of investigation concerns their capabilities to accept the classical linguistic languages. While there are simple relations to the regular and context-sensitive ones, here we shed some light on the relations to the context-free languages and some of their important subfamilies.

1. INTRODUCTION

One of the cornerstones in the theory of automata arrays is the early result of John von Neumann who solved the logical problem of nontrivial self-reproduction. From the biological point of view he employed a mathematical device which is a multitude of interconnected automata operating in parallel to form a larger automaton, a macroautomaton built by microautomata. He established that it was logically possible for such a nontrivial computing device to replicate itself ad infinitum [18].

Nowadays the so-called polyautomata theory has several branches. One important field is the modelling of natural phenomena in, for example, physics, biology or chemistry. Another one deals with certain types of polyautomata as computational models, in particular as acceptors for formal languages. Those automata arrays are the subject of the present article.

A lot of investigation has been done in order to explore the computing capabilities of such devices but there are still some basic, unanswered questions.

C. Martin-Vide and V. Mitrana (eds.), Where Mathematics, Computer Science, Linguistics and Biology Meet, 139-148.

Usually this investigation is done in terms of and with the methods of automata and complexity theory. We are particularly interested in the relation between some of the most important automata array language families and the classical linguistic language families. It has turned out that the interesting families include the regular languages and are contained in the deterministic context-sensitive languages in such a way that this interest is synonymous with interest in the relations to the context-free languages and their subfamilies.

2. AUTOMATA ARRAYS

We denote the rational numbers by $\mathbb{Q}$, the integers by $\mathbb{Z}$, the positive integers $\{1, 2, ...\}$ by $\mathbb{N}$, the set $\mathbb{N} \cup \{0\}$ by $\mathbb{N}_0$ and the powerset of a set S by 2^S. The empty word is denoted by ε and the reversal of a word w by w^R. For the length of w we write $|w|$. We use $\subseteq$ for inclusions and $\subset$ if the inclusion is strict. For a function f we denote its i-fold composition by $f^{[i]}$, $i \in \mathbb{N}$.

The specification of an automata array includes the type and specification of the microautomata (cells), their interconnection scheme (which can imply a dimension to the system), a local transition function and the input and output modes. One-dimensional synchronous devices with nearest neighbor connections whose cells are (nondeterministic) finite automata, commonly called *cellular arrays* resp. *iterative arrays* in the case of parallel resp. sequential input mode.

For convenience we identify the cells of one-dimensional arrays by positive integers. The state transition depends on the current state of a cell and the current states of its neighbors. The transition function is applied to all cells synchronously at discrete time steps. With an eye towards formal language processing we define formally:

Definition 12.1 *A* nondeterministic (two-way) cellular array *(*NCA*) is a system* $\langle S, \delta_{nd}, \#, A, F\rangle$, *where*

1. *S is the finite, nonempty set of* cell states,
2. *$A \subseteq S$ is the nonempty set of* input symbols,
3. *$F \subseteq S$ is the nonempty set of* accepting states,
4. *$\# \notin S$ is the* boundary state,
5. *$\delta_{nd} : (S \cup \{\#\})^3 \to 2^S \setminus \{\emptyset\}$ is the (nondeterministic)* local transition function.

Let $\mathcal{M} = \langle S, \delta_{nd}, \#, A, F\rangle$ be an NCA. A *configuration* of $\mathcal{M}$ at some time $t \geqslant 0$ is a description of its global state which is actually a mapping $c_t : [1, \dots, n] \to S$ for $n \in \mathbb{N}$. The configuration at time 0 is defined by the initial sequence of states. For a given input word $w = a_1 \cdots a_n \in A^+$ we set $c_{0,w}(i) = a_i$, $1 \leqslant i \leqslant n$. During its course of computation an NCA steps nondeterministically through a sequence of configurations whereby successor configurations are chosen according to the global transition function Δ_{nd}: Let c_t, $t \geqslant 0$, be a configuration, then its successor configurations are as follows:

$$
\begin{aligned}
c_{t+1} \in \Delta_{nd}(c_t) \iff & \\
& c_{t+1}(1) \in \delta_{nd}(\#, c_t(1), c_t(2)) \\
& c_{t+1}(i) \in \delta_{nd}(c_t(i-1), c_t(i), c_t(i+1)), i \in \{2, \ldots, n-1\} \\
& c_{t+1}(n) \in \delta_{nd}(c_t(n-1), c_t(n), \#)
\end{aligned}
$$

Thus, Δ_{nd} is induced by δ_{nd}.

A restricted connection to the nearest neighbor to the right only (i.e. the next state of each cell depends on the state of the cell itself and the state of its immediate neighbor to the right) yields one-way information flow through the array. The corresponding devices are denoted by NOCAs.

An NCA (NOCA) is *deterministic* if $\delta_{nd}(s_1, s_2, s_3)$ ($\delta_{nd}(s_1, s_2)$) is a singleton for all states $s_1, s_2, s_3 \in S \cup \{\#\}$. Deterministic cellular arrays are denoted by CA resp. OCA.

There is a natural way of restricting the nondeterminism of the arrays. One can limit the number of allowed nondeterministic state transitions. For this purpose a *deterministic local transition* $\delta_d : (S \cup \{\#\})^3 \to S$ is provided and the global transition function induced by δ_d is denoted by Δ_d.

Let $g : \mathbb{N} \to \mathbb{N}_0$ be a mapping that gives the number of allowed nondeterministic transitions dependent on the length of the input. The resulting system $\langle S, \delta_{nd}, \delta_d, s_0, \#, A, F \rangle$ is a gG-CA (gG-OCA) if starting with the initial configuration $c_{0,w}$ (for some $w \in A^+$) the possible configurations at some time i are computed by the global transition function as follows: $\{c_{0,w}\}$ if $i = 0$, $\Delta_{nd}^{[i]}(c_{0,w})$ if $i \leqslant g(|w|)$ and $\bigcup_{c' \in \Delta_{nd}^{[g(|w|)]}(c_{0,w})} \Delta_d^{[t-g(|w|)]}(c')$ otherwise.

In cellular arrays the input mode is called parallel. One can suppose that all cells fetch their input symbol during a pre-initial step. Another natural way to supply the input is the so-called sequential input mode. Now the distinguished cell at the origin, the so-called *communication cell*, fetches the input sequence symbol by symbol. In order to define such iterative arrays formally we have to provide an initial (quiescent) state for the cells. Moreover, for historical reasons the space (number of cells) of iterative arrays is not bounded. We assume that once the whole input is consumed an end-of-input symbol is supplied permanently.

Definition 12.2 *An iterative array* (IA) *is a system* $\langle S, \delta, \delta_0, s_0, \#, A, F \rangle$, *where*

1. *S is the finite, nonempty set of* cell states,

2. *A is the finite, nonempty set of* input symbols,
3. $F \subseteq S$ *is the set of* accepting states,
4. $s_0 \in S$ *is the* initial (quiescent) state,
5. $\# \notin A$ *is the* end-of-input symbol,
6. $\delta : S^3 \to S$ *is the* local transition function for non-communication cells *satisfying* $\delta(s_0, s_0, s_0) = s_0$,
7. $\delta_0 : S^3 \times (A \cup \{\#\}) \to S$ *is the* local transition function for the communication cell.

Here iterative arrays are defined as deterministic devices. The reason is that real -resp. linear-time- nondeterministic IAs have exactly the same accepting power as real -resp. linear-time- nondeterministic CAs.

A configuration of an IA at some time $t \geqslant 0$ is a pair (w_t, c_t) where $w_t \in A^*$ is the remaining input sequence and $c_t : \mathbb{Z} \to S$ is a function that maps the single cells to their current states. The configuration (w_0, c_0) at time 0 is defined by the input word w_0 and the mapping $c_0(i) = s_0$, $i \in \mathbb{Z}$. The global transition function Δ is induced by δ and δ_0 as follows: Let (w_t, c_t), $t \geqslant 0$, be a configuration.

$$(w_{t+1}, c_{t+1}) = \Delta((w_t, c_t)) \iff$$
$$c_{t+1}(i) = \delta(c_t(i-1), c_t(i), c_t(i+1)), i \in \mathbb{Z} \setminus \{0\},$$
$$c_{t+1}(0) = \delta_0(c_t(-1), c_t(0), c_t(1), a)$$

where $a = \#$, $w_{t+1} = \varepsilon$ if $w_t = \varepsilon$, and $a = a_1$, $w_{t+1} = a_2 \cdots a_n$ if $w_t = a_1 \cdots a_n$.

An input word w is accepted by an N(O)CA (IA) if at some time i during its course of computation the leftmost cell (communication cell) becomes accepting.

Definition 12.3 *Let* $\mathcal{M} = \langle S, \delta_{nd}, \#, A, F\rangle$ *be an* N(O)CA *(*$\mathcal{M} = \langle S, \delta, \delta_0, s_0, \#, A, F\rangle$ *be an* IA*).*

1. *A word* $w \in A^*$ *is* accepted *by* $\mathcal{M}$ *if it is the empty word or if there exists a time step* $i \in \mathbb{N}$ *such that* $c_i(1) \in F$ *holds for a configuration* $c_i \in \Delta_{nd}^{[i]}(c_{0,w})$ *(*$c_i(0) \in F$ *holds for* $(w_i, c_i) = \Delta^{[i]}((w, c_0))$*).*
2. $L(\mathcal{M}) = \{w \in A^* \mid w$ *is accepted by* $\mathcal{M}\}$ *is the* language accepted *by* $\mathcal{M}$.
3. *Let* $t : \mathbb{N} \to \mathbb{N}$, $t(n) \geqslant n$ *(*$t(n) \geqslant n+1$*), be a mapping and* i_w *be the minimal time step at which* $\mathcal{M}$ *accepts a* $w \in L(\mathcal{M})$. *If all* $w \in L(\mathcal{M})$

are accepted within $i_w \leqslant t(|w|)$ time steps, then L is said to be of time complexity t.

The family of all languages acceptable by NCAs (IAs) with time complexity t is denoted by $\mathcal{L}_{\sqcup}(\text{NCA})$ ($\mathcal{L}_{\sqcup}(\text{IA})$). If t equals the identity function $id(n) = n$ (the function $n + 1$), acceptance is said to be in *real-time* and we write $\mathcal{L}_{\nabla\sqcup}(\text{NCA})$ ($\mathcal{L}_{\nabla\sqcup}(\text{IA})$). Since for nontrivial computations an IA needs to read at least one end-of-input symbol, real-time has to be defined as $(n + 1)$-time. The *linear-time* languages $\mathcal{L}_{\ddagger\sqcup}(\text{NCA})$ ($\mathcal{L}_{\ddagger\sqcup}(\text{IA})$) are defined according to $\mathcal{L}_{\ddagger\sqcup}(\text{NCA}) = \bigcup_{\|\in\mathbb{Q},\, \|\geqslant\infty} \mathcal{L}_{\|\cdot\rangle\lceil}(\text{NCA})$.

In the sequel we will use a corresponding notion for other types of acceptors. Throughout the article the regular, deterministic context-free, context-free resp. deterministic context-sensitive languages are denoted by $\mathcal{L}_{\ni}$, $\mathcal{L}_{\in,\sqcap\sqcup}$, $\mathcal{L}_{\in}$ resp. $\mathcal{L}_{\infty,\sqcap\sqcup}$.

3. RELATIONS BETWEEN AUTOMATA ARRAYS AND THE POSITION OF L_2

In order to compare the context-free languages and some of their important subclasses to the language families definable by time-bounded automata arrays it is convenient to recall some of the known results. Figure 12.1 summarizes inclusions between real-time and unlimited-time (O)CAs, 1G-OCAs, IAs and regular languages.

$$\mathscr{L}_3 \subset \mathscr{L}_{rt}(\text{IA}) \subset \mathscr{L}_{rt}(\text{CA}), \quad \mathscr{L}_3 \subset \mathscr{L}_{rt}(\text{OCA}) \subset \mathscr{L}_{rt}(\text{CA})$$

$$\mathscr{L}_{rt}(\text{CA}) \subseteq \mathscr{L}_{rt}(\text{1G-OCA}) \subseteq \mathscr{L}(\text{OCA}) \subseteq \mathscr{L}(\text{CA})$$

Figure 12.1 Hierarchy of language families.

At the left end of the hierarchy there are the regular languages, which are strictly included in all of the considered array language families. It is known that the families $\mathcal{L}_{\nabla\sqcup}(\text{OCA})$ and $\mathcal{L}_{\nabla\sqcup}(\text{IA})$ are incomparable but both are strictly included in the real-time CA languages. Unfortunately, these are the only inclusions known to be strict. The right hand end of the (shown part of the) hierarchy is marked by the unlimited-time CA languages. These are in turn identical to the deterministic context-sensitive languages. Thus, we have a hierarchy between the linguistic families $\mathcal{L}_{\ni}$ and $\mathcal{L}_{\infty,\sqcap\sqcup}$ and, naturally, the question of the position of $\mathcal{L}_{\in}$ arises.

Since $\mathcal{L}_{\nabla\sqcup}(\text{IA})$ as well as $\mathcal{L}_{\nabla\sqcup}(\text{OCA})$ are containing noncontext-free languages, the strict inclusion $\mathcal{L}_{\in} \subset \mathcal{L}(\text{CA})$ follows immediately.

In 1988 the position could be shifted downwards to the unlimited-time one-way languages. The inclusion $\mathcal{L}_{\in} \subset \mathcal{L}(\text{OCA})$ has been shown in [4]. Due to the hierarchical result $\mathcal{L}_{\nabla\sqcup}(\infty\text{G-OCA}) \subseteq \mathcal{L}(\text{OCA})$ [1] the following theorem is a further improvement.

Theorem 12.1 $\mathcal{L}_{\in} \subset \mathcal{L}_{\nabla\sqcup}(\infty\text{G-OCA})$

Proof. From [1] the equality $\mathcal{L}_{\nabla\sqcup}(\infty\text{G-OCA}) = \mathcal{L}_{\nabla\sqcup}(\infty\text{G-CA})$ is known. Therefore, we may prove the theorem by showing how a 1G-CA can simulate an ε-free nondeterministic pushdown automaton (pda) in real-time.

The leftmost cell of the 1G-CA can identify itself (by the # on its left). It simulates the state transitions of the pda.

During the first time step all the other cells are guessing a state transition of the pda nondeterministically. These guesses as well as the input symbols are successively shifted to the left on some additional tracks. Thereby, at every time step, the leftmost cell receives an input symbol and a previously guessed state transition. Thus, simply by applying the transition to the current situation it simulates a nondeterministic transition of the pda deterministically. It remains to be shown how to simulate the stack. Details of that construction can be found in [2]. □

The next step of positioning the family $\mathcal{L}_{\in}$ would be to prove or disprove the inclusion $\mathcal{L}_{\in} \subset \mathcal{L}_{\nabla\sqcup}(\text{CA})$. The problem whether or not the real-time CA languages contain the context-free ones was raised in [15] and is still open.

4. ITERATIVE ARRAYS

Now we turn to the comparison of $\mathcal{L}_{\in}$ and the language families at the lower end of the hierarchy. The present section is devoted to real-time iterative arrays. (In case of linear-time the accepting capabilities of IAs and CAs are identical.) In [10] it has been proved that iterative arrays can accept the context-free languages in square-time ($\mathcal{L}_{\in} \subset \mathcal{L}_{\setminus\in}(\text{IA})$) and that the time complexity can be reduced to linear-time if the cells are arranged in two dimensions ($\mathcal{L}_{\in} \subset \mathcal{L}_{\ddagger\sqcup}(\text{2D-IA})$). About seventeen years later the second result could be improved to one-way information flow ($\mathcal{L}_{\in} \subset \mathcal{L}_{\ddagger\sqcup}(\text{2D-OIA})$) [3]. Moreover, the corresponding 2D-OIAs are not only acceptors but parsers for the context-free languages.

With the help of a specific witness language the incomparability of $\mathcal{L}_{\in}$ and $\mathcal{L}_{\nabla\sqcup}(\text{IA})$ has been established in [6]. In the following we are going to explore the boundary between acceptable and nonacceptable context-free languages in real-time IAs.

Lemma 12.1 *Every ε-free (real-time) deterministic context-free language belongs to $\mathcal{L}_{\nabla\sqcup}(\text{IA})$.*

Proof. Using the method referred to in the proof of Theorem 12.1 one can construct an iterative array that simulates a deterministic pushdown automaton.

The communication cell controls the stack, fetches the input and simulates the state transition. Since per assumption the pda performs no ε-transitions, the IA operates in real-time. □

Once the stack simulation principle was known the previous lemma became straightforward. But nevertheless it seems to mark a sharp boundary which becomes clear with the next theorem.

Theorem 12.2 *There exists a deterministic, linear context-free language that does not belong to* $\mathcal{L}_{\nabla\sqcup}(\text{IA})$.

Proof. Consider the configuration of an IA $\langle S, \delta, \delta_0, s_0, \#, A, F\rangle$ at time $n - m$, $m < n$, where n denotes the length of the input. The remaining computation of the communication cell depends on the last m input symbols and the states of the cells $-m-1, \dots, 0, \dots, m+1$ only. Due to their distance from the communication cell all the other cells cannot influence the overall computation result. So we are concerned with at most $|S|^{2\cdot(m+1)+1}$ different situations.

Now let $L = \{\$x_k\$\cdots\$x_1@y_1\$\cdots\$y_k\$ \mid x_i^R = y_i z_i, x_i, y_i, z_i \in \{0,1\}^*\}$. Obviously L is a deterministic, linear context-free language. For every two different prefix-words w and w' of the form $\$x_k\$\cdots\$x_1@$ with $|x_i| = k$ there exists at least one $1 \leqslant j \leqslant k$ such that $x_j \neq x'_j$. It follows $w\$^{j-1}x_j^R\$^{k-j+1} \in L$ and $w'\$^{j-1}x_j^R\$^{k-j+1} \notin L$. There exist 2^{k^2} different prefix-words of the given form.

On the other hand, at time $n-2k$ the IA can distinguish at most $|S|^{2\cdot(2k+1)+1}$ $= |S|^{4k+3}$ different situations. For large k we can conclude $|S|^{4k+3} < 2^{k^2}$ and, thus, after processing two different words w and w' the situation must be identical. So we obtain $w\$^{j-1}x_j^R\$^{k-j+1} \in L \iff w'\$^{j-1}x_j^R\$^{k-j+1} \in L$, a contradiction. □

By the previous theorem neither the deterministic nor the linear context-free languages are subfamilies of $\mathcal{L}_{\nabla\sqcup}(\text{IA})$.

5. ONE-WAY CELLULAR ARRAYS

Now we are going to investigate the relations between real-time OCAs and some important subfamilies of $\mathcal{L}_\in$. The results presented will prove the incomparability of $\mathcal{L}_{\nabla\sqcup}(\text{OCA})$ and $\mathcal{L}_{\nabla\sqcup}(\text{IA})$ in terms of context-free languages.

The reason we restrict our considerations to real-time OCAs is due to the known relation $\mathcal{L}_{\updownarrow\sqcup}(\text{OCA}) = \mathcal{L}^{\mathcal{R}}_{\nabla\sqcup}(\text{CA})$ [5, 17]. Besides, the real-time CAs is the subject of the next section.

Some subfamilies of $\mathcal{L}_\in$ are known to be strictly included in $\mathcal{L}_{\nabla\sqcup}(\text{OCA})$. E.g. the Dyck languages [13] and the bracketed context-free languages [7]. In the following we deal with linear languages. In [14] it has been shown that they are acceptable by OCAs in real-time.

Theorem 12.3 *Every linear context-free language belongs to* $\mathcal{L}_{\nabla\sqcup}(\text{OCA})$.

Proof. Let $G = (N, T, X_0, P)$ be a linear context-free grammar ([12]). W.l.o.g. we may assume that all productions are of the form $X \longrightarrow a$, $X \longrightarrow Ya$ or $X \longrightarrow aY$, where $X, Y \in N$ are nonterminals and $a \in T$ is a terminal.

Let $w = a_1 a_2 \cdots a_n$ be a word in $L(G)$. If $X \Rightarrow^* a_i \cdots a_j$ then there must exist a Y such that either $Y \Rightarrow^* a_i \cdots a_{j-1} \wedge X \Rightarrow Ya_j$ or $Y \Rightarrow^* a_{i+1} \cdots a_j \wedge X \Rightarrow a_i Y$. Based on this fact, sets of nonterminals for a given word w are defined as follows:

$N(i,i) = \{X \in N \mid (X \to a_i) \in P\}, 1 \leqslant i \leqslant n$

$N(i,j) = N_1(i,j) \cup N_2(i,j), 1 \leqslant i < j \leqslant n$, where

$N_1(i,j) = \{X \in N \mid (X \to Ya_j) \in P \wedge Y \in N(i, j-1)\}$

$N_2(i,j) = \{X \in N \mid (X \to a_i Y) \in P \wedge Y \in N(i+1, j)\}$.

It holds $w \in L(G) \iff X_0 \in N(1,n)$.

A corresponding OCA behaves as follows. During the first transition a cell i computes $N(i,i)$ and (a_i, a_i). In subsequent transitions, say at time $k+1$, it computes $N(i, i+k)$ and (a_i, a_{i+k}) if $k+1 \geqslant i$ and keeps its state otherwise. The new values can be determined by the state of the cell itself and the state of its neighbor to the right. Thus, the leftmost cell computes $N(1,n)$ at time step n. □

Again, the question arises whether we can relax the condition of Theorem 12.3 for larger subfamilies. An answer follows from the results given in [16] where by counting arguments the closure of $\mathcal{L}_{\nabla\sqcup}(\text{OCA})$ under concatenation has been shown to be negative. In the proof the language $L = \{0^i 1^i \mid i \in \mathbb{N}\} \cup \{0^i 1 x 0 1^i \mid x \in \{0,1\}^*, i \in \mathbb{N}\}$ has been used as a witness. Clearly, L is a linear context-free language and, thus, belongs to $\mathcal{L}_{\nabla\sqcup}(\text{OCA})$. But for the concatenation $L_1 = L \cdot L$ it holds $L_1 \notin \mathcal{L}_{\nabla\sqcup}(\text{OCA})$. Since L_1 is the concatenation of two linear languages it is a 2-linear language.

Corollary 12.1 *There exists a 2-linear context-free language that does not belong to* $\mathcal{L}_{\nabla\sqcup}(\text{OCA})$.

On the other hand, one can show $L_1 \in \mathcal{L}_{\nabla\sqcup}(\text{IA})$ and, hence, with the previous results the incomparability of $\mathcal{L}_{\nabla\sqcup}(\text{IA})$ and $\mathcal{L}_{\nabla\sqcup}(\text{OCA})$ follows in terms of context-free languages.

Although $\mathcal{L}_{\nabla\sqcup}(\text{OCA})$ does not contain the 2-linear languages, real-time OCAs are powerful devices. For example, the non-Parikh-linear language $\{(0^i 1)^* \mid i \in \mathbb{N}_0\}$ [13] and the inherently ambiguous language $\{0^i 1^j 2^k \mid i = j \vee j = k,\ i, j, k \in \mathbb{N}\}$ [15] belong to $\mathcal{L}_{\nabla\sqcup}(\text{OCA})$.

6. TWO-WAY CELLULAR ARRAYS

For structural reasons $\mathcal{L}_{\nabla\sqcup}(\text{IA}) \subseteq \mathcal{L}_{\nabla\sqcup}(\text{CA})$ and $\mathcal{L}_{\nabla\sqcup}(\text{OCA}) \subseteq \mathcal{L}_{\nabla\sqcup}(\text{CA})$ holds. Since $\mathcal{L}_{\nabla\sqcup}(\text{IA})$ and $\mathcal{L}_{\nabla\sqcup}(\text{OCA})$ are incomparable, both inclusions must be strict. Unfortunately, this proof of the strictness gives us no context-free subfamily contained in $\mathcal{L}_{\nabla\sqcup}(\mathrm{CA})$.

Theorem 12.4 $\mathcal{L}_{\in,\sqcap\sqcup} \subset \mathcal{L}_{\nabla\sqcup}(\mathrm{CA})$

Proof. Here we cannot use a stack simulation as in the proof of Lemma 12.1 because we are concerned with deterministic pushdown automata that are allowed to perform ε-transitions. But w.l.o.g. we may assume that a given deterministic pda $\mathcal{M}$ pushes at most $k \in \mathbb{N}$ symbols onto the stack at every non-ε-transition and erases exactly one symbol from the stack at every ε-transition [8].

Due to the equality $\mathcal{L}_{\ddagger\sqcup}(\mathrm{OCA}) = \mathcal{L}^{\mathcal{R}}_{\nabla\sqcup}(\mathrm{CA})$ it suffices to construct a $(k \cdot n)$-time OCA that accepts the language $L^R(\mathcal{M})$. The details of the proof can be found in [11]. □

Combined with Theorem 12.2 the previous result gives us the strictness of the inclusion $\mathcal{L}_{\nabla\sqcup}(\mathrm{IA}) \subset \mathcal{L}_{\nabla\sqcup}(\mathrm{CA})$ *and* a subfamily of $\mathcal{L}_{\in}$ strictly contained in $\mathcal{L}_{\nabla\sqcup}(\mathrm{CA})$. Regarding $\mathcal{L}_{\nabla\sqcup}(\mathrm{OCA})$ the following theorem works in a sense similar sense.

Theorem 12.5 *Every metalinear language belongs to* $\mathcal{L}_{\nabla\sqcup}(\mathrm{CA})$.

Proof. Let L be a metalinear language. Then there exists a $k \in \mathbb{N}$ such that L is k-linear. Therefore, we can represent L as the concatenation $L_1 \cdot L_2 \cdot \ldots \cdot L_k$ of k linear languages.

The family $\mathcal{L}_{\nabla\sqcup}(\mathrm{OCA})$ is closed under reversal [13]. Combined with Theorem 12.3 we conclude that there exist real-time OCAs for each of the languages $L_1^R, \ldots, L_k^R$. Since the concatenation of a real-time and a linear-time OCA language is again a linear-time OCA language [9] we obtain $L_k^R \cdot \ldots \cdot L_1^R \in \mathcal{L}_{\ddagger\sqcup}(\mathrm{OCA})$. Due to the equality $\mathcal{L}_{\ddagger\sqcup}(\mathrm{OCA}) = \mathcal{L}^{\mathcal{R}}_{\nabla\sqcup}(\mathrm{CA})$ it follows $L_1 \cdot \ldots \cdot L_k = L \in \mathcal{L}_{\nabla\sqcup}(\mathrm{CA})$. □

References

[1] Buchholz, Th.; A. Klein & M. Kutrib (1998), On interacting automata with limited nondeterminism, *Fundamenta Informaticae*, to appear. Also, IFIG Research Report 9806, Institute of Informatics, University of Giessen, Giessen.

[2] Buchholz, Th. & M. Kutrib (1997), Some relations between massively parallel arrays, *Parallel Comput.*, 23: 1643–1662.

[3] Chang, J. H.; O.H. Ibarra & M.A. Palis (1987), Parallel parsing on a one-way array of finite-state machines, *IEEE Trans. Comput.*, C-36: 64–75.

[4] Chang, J. H.; O.H. Ibarra & A. Vergis (1988), On the power of one-way communication, *Journal of the Association for Computing Machinery*, 35: 697–726.

[5] Choffrut, C. & K. Čulik II (1984), On real-time cellular automata and trellis automata, *Acta Informatica*, 21: 393–407.

[6] Cole, S. N. (1969), Real-time computation by n-dimensional iterative arrays of finite-state machines, *IEEE Trans. Comput.*, C-18: 349–365.

[7] Dyer, C. R. (1980), One-way bounded cellular automata, *Information and Control*, 44: 261–281.

[8] Ginsburg, S. (1966), *The Mathematical Theory of Context-Free Languages*, McGraw Hill, New York.

[9] Ibarra, O. H. & T. Jiang (1987), On one-way cellular arrays, *SIAM Journal of Computing*, 16: 1135–1154.

[10] Kosaraju, S. R. (1975), Speed of recognition of context-free languages by array automata, *SIAM Journal of Computing*, 4: 331–340.

[11] Kutrib, M. (1999), Array automata and context-free languages, IFIG Research Report 9907, Institute of Informatics, University of Giessen, Giessen.

[12] Salomaa, A. (1973), *Formal Languages*, Academic Press, New York.

[13] Seidel, S. R. (1979), Language recognition and the synchronization of cellular automata, Technical Report 79-02, Department of Computer Science, University of Iowa, Iowa City.

[14] Smith III, A. R. (1970), Cellular automata and formal languages, in *IEEE Symposium on Switching and Automata Theory*: 216–224.

[15] Smith III, A. R. (1972), Real-time language recognition by one-dimensional cellular automata, *Journal of Computer and System Sciences*, 6: 233–253.

[16] Terrier, V. (1995), On real time one-way cellular array, *Theoretical Computer Science*, 141: 331–335.

[17] Umeo, H.; K. Morita & K. Sugata (1982), Deterministic one-way simulation of two-way real-time cellular automata and its related problems, *Information Processing Letters*, 14: 158–161.

[18] von Neumann, J. (1966), *Theory of Self-Reproducing Automata*. Edited and completed by A. W. Burks. University of Illinois Press, Urbana.

Chapter 13

ON SPECIAL FORMS OF RESTARTING AUTOMATA

František Mráz
Department of Computer Science
Charles University
Malostranské nám. 25, 118 00 Praha 1, Czech Republic
mraz@ksvi.ms.mff.cuni.cz

Martin Plátek
Department of Computer Science
Charles University
Malostranské nám. 25, 118 00 Praha 1, Czech Republic
platek@ksi.ms.mff.cuni.cz

Martin Procházka
Department of Computer Science
Charles University
Malostranské nám. 25, 118 00 Praha 1, Czech Republic

Abstract We study transformations of automata from some (sub)classes of restarting automata ($RRWW$-automata) into two types of special forms. We stress particularly the transformations into the linguistically motivated weak cyclic form. Special forms of the second type express a certain degree of determinism of such automata.

*Supported by the Grant Agency of the Czech Republic, Grant-No. 201/99/0236, and by the Grant Agency of Charles University, Grant-No. 157/1999/A INF/MFF.

C. Martin-Vide and V. Mitrana (eds.), Where Mathematics, Computer Science, Linguistics and Biology Meet, 149-160.

1. INTRODUCTION

This paper is a continuation of the paper [5], which studied transformations into weak cyclic forms of automata from some subclasses of *RWW*-automata. These (sub)classes constitute a part of the taxonomy of restarting automata ($RRWW$-automata) presented in [1].

The restarting automata serve as a formal tool to model analysis by reduction in natural languages. This means that restarting automata serve similarly as Marcus contextual grammars [3, 4] to cover (and to understand) the part of syntax which defines well- and ill-formedness of natural language sentences, but which is not (yet) burdened with the explicit effort of describing the relation between form and meaning (semantics). That is relevant in practical terms, for example, in the development of grammar-checkers [2], and in splitting the complex task to develop a parser of a natural language into several well defined (described) simpler tasks.

In this paper we concentrate on the property of a *weak cyclic form* which naturally corresponds to analysis by reduction for natural languages.

Let us take a closer look at the linguistic roots of our motivations. Analysis by reduction consists in stepwise simplification of an extended sentence so that the (in)correctness of the sentence is not affected. Thus, after a certain number of steps a simple sentence is arrived at or an error is found. Let us illustrate it on the reduction of the sentence

'Martin, Peter and Jane work very slowly.'

The sentence can be simplified for example in the following way:

'Martin, Peter and Jane work slowly.'

'Martin, Peter and Jane work.'

'Martin and Peter work.' or *'Peter and Jane work.'* or *'Martin and Jane work.'*

'Martin works.' or *'Peter works.'* or *'Jane works.'*

The last three sentences are all the simple sentences which can be reduced from the original sentence. We can see that each of them consist of two words only. The property of weak cyclic form for restarting automata corresponds to the general belief that for any natural language there exists an upper bound of the length of its simple sentences.

A restarting automaton (of the type of $RRWW$-automaton) can be roughly described as follows. It has a finite control unit, a head with a lookahead window attached to a list, and it works in certain phases. In each phase, it moves the head from left to right along the words on the list; according to its instructions, it can at some point rewrite the contents of its lookahead by a shorter string once and (possibly) move to the right-hand side again. At a certain place, according to its instructions, it can 'restart' – i.e., reset the control unit to the initial state and place its head on the left end of the shortened word. The phase ending by restart is called a cycle. The computation ends in a phase by halting in an accepting or a rejecting state – such a phase is called a tail. A restarting automaton M is in weak cyclic form if the length of any word accepted by M using a tail only (without restart) is not greater than the size of its lookahead.

Several classes of restarting automata were considered in [1], the most general is that of $RRWW$-automata. For such automata, the property of *monotonicity of computations* has been introduced, which allows us to derive three different definitions of *monotonicity of automata*; we call them monotonicity, a-monotonicity and g-monotonicity. The properties of monotonicity allow us to characterize the class of context-free languages, and different types of their analysis by reduction.

Section 2 defines the $RRWW$-automata, their subclasses (*R*-, *RR*-, *RW*-, *RRW*-, RWW-automata), the above mentioned three types of monotonicity properties as well as the weak cyclic form and three degrees of (non)determinism. Some basic properties of restarting automata are mentioned there, too.

The results are presented in Section 3. The transformation into weak cyclic form of RWW-automata and their subclasses (together with their monotonic, a-monotonic and g-monotonic versions) was studied in [5]. Here we show that the weak cyclic form can be constructed for any *RR*-, RW- or $RRWW$-automaton and for any of its monotonic, a-monotonic and g-monotonic versions. We show further that the situation can be slightly more complex if we combine the transformation into weak cyclic form with a transformation into some other special forms of $RRWW$-automata, based on a certain degree of determinism.

2. DEFINITIONS AND BASIC PROPERTIES

We present the definitions informally; the formal technical details can be added in a standard automata theory way. λ denotes the empty word, and $\emptyset$ denotes the empty set.

A *restarting automaton* $M = (Q, \Sigma, \Gamma, k, I, q_0, Q_A, Q_R)$ is a device with a finite state control unit, with the (finite) set of states Q, where $(Q_A \cup Q_R) \subseteq Q$, and with one head moving on a finite linear (doubly linked) list of items (cells). The first item always contains a special symbol ¢, the last one another special symbol \$, and each other item contains a symbol from the union of two disjoint finite alphabets Σ and Γ (not containing ¢, \$). Σ is called *the input alphabet*, Γ *the working alphabet*. The head has a lookahead 'window' of length k ($k > 0$) – besides the current item, M also scans the next $k-1$ right neighbour items (or simply the end of the word when the distance to \$ is less than $k-1$). M contains a fixed *initial* state q_0, ($q_0 \in Q$).

A configuration of the automaton M is (u, q, v), where $u \in \{\lambda\} \cup \{¢\}(\Sigma \cup \Gamma)^*$ is the contents of the list from the left sentinel ¢ to the position of the head,

$q \in Q$ is the current state and $v \in (\Sigma \cup \Gamma)^*\{\$\}$ is the contents of the list from the scanned item to the right sentinel.

In the *restarting configuration* on a *word* $w \in (\Sigma \cup \Gamma)^*$, the word $¢w\$$ is stored in the items of the list, the control unit is in the initial state q_0, and the head is attached to that item which contains the left sentinel (scanning ¢, looking also at the first $k-1$ symbols of the word w). An initial computation of M starts in an *initial configuration*, that is a restarting configuration on *an input word* ($w \in \Sigma^*$). We suppose that the set of states Q is divided into two classes – the set of *nonhalting states* $Q - (Q_A \cup Q_R)$ (there is at least one instruction that is applicable when the unit is in such a state) and the set of *halting states* $Q_A \cup Q_R$ (any computation finishes by entering such a state), where Q_A is the set of *accepting states* and Q_R is the set of *rejecting states*.

The *computation* of M is controlled by a finite set of *instructions* I. Instructions are of the following three types ($q, q' \in Q, a \in \Sigma \cup \Gamma \cup \{¢, \$\}$, $u, v \in (\Sigma \cup \Gamma)^* \cup (\Sigma \cup \Gamma)^* \cdot \{\$\}$):

(1) $(q, au) \to (q', MVR)$

(2) $(q, au) \to (q', REWRITE(v))$

(3) $(q, au) \to RESTART$

The left-hand side of an instruction determines when it is applicable – q means the current state (of the control unit), a the symbol being scanned by the head, and u means the contents of the lookahead window (u being a string of length k or less if it ends with \$). The right-hand side describes the activity to be performed.

In case (1), M changes the current state to q' and moves the head to the right neighbour item of the item containing a.

In case (2), the activity consists of deleting (removing) some items (at least one) of the just scanned part of the list (containing au), and of rewriting some (possibly none) of the nondeleted scanned items (in other words au is replaced with v, where v must be shorter than au). After that, the head of M is moved to the right to the item containing the first symbol after the lookahead and the current state of M is changed to q'. There are two exceptions: if au ends by \$then v also ends by \$ (the right sentinel cannot be deleted or rewritten) and after the rewriting the head is moved to the item containing \$; similarly the left sentinel ¢ cannot be deleted or rewritten.

In case (3), *RESTART* means a transfer in the restarting configuration, i.e., entering the initial state and placing the head on the first item of the list (containing ¢).

– *Cycles, tails:* Any computation of a restarting automaton M is composed from certain phases. A phase called a *cycle* starts in a restarting configuration, the head moves to the right along the input list until the restart operation is performed, and M is resumed in a new restarting configuration. M deletes at least once during a cycle – i.e., the new phase starts on a shortened word. A phase of a computation called the *tail* starts in a restarting configuration, and the head moves to the right along the input list until one of the halting states is reached.

This immediately implies that any computation of any restarting automaton is finite (finishing in a halting state).

In general, a restarting automaton is *nondeterministic*, i.e., there can be two or more instructions with the same left-hand side (q, au). If this is not the case, the automaton is *deterministic*.

– *Accepting, recognizing:* An input *word w is accepted by M* if there is a(n) (*initial*) computation that starts in the initial configuration with $w \in \Sigma^*$ (bounded by sentinels ¢,$) on the list, and finishes in an *accepting configuration*, where the control unit is in one of the accepting states. $L(M)$ denotes the language consisting of all words accepted by M; we say that M *recognizes the language* $L(M)$.

– *RRWW-automata:* $RRWW$-automaton is a restarting automaton that makes exactly one *REWRITE*-instruction in each cycle. M can also rewrite once during any tail.

By RWW-automata we mean $RRWW$-automata that restart immediately after any *REWRITE*-instruction.

By RW-automata we mean $RRWW$-automata with empty working alphabets. RW-automata use only the input symbols in the rewriting, i.e., in all instructions of the form (2) above, the string v contains symbols from the input alphabet Σ only. We can see that all restarting configurations of an RW-automaton are also its initial configurations.

By RR-automata we mean RW-automata which use deleting without rewriting (i.e., in all instructions of the form (2) above, the string v can always be obtained by deleting some symbols from au).

By *RW*-automata (*R*-automata) we mean *RRW*-automata (*RR*-automata) which restart immediately after any performance of a *REWRITE*-instruction.

– *Reductions:* The notation $u \Rightarrow_M v$ means that there exists a cycle of M starting in the restarting configuration with the word u and finishing in the restarting configuration with the word v; thc relation $\Rightarrow_M^*$ is the reflexive and transitive closure of $\Rightarrow_M$. If $u \Rightarrow_M v$ holds, we say that u is (can be) reduced to v by M. If $u \Rightarrow_M^* v$ holds, we say that u is stepwise reduced to v by M. If both u, v are from $L(M)$, we say that the *reduction* $u \Rightarrow_M v$ *is correct*.

We often (implicitly) use the next obvious claim:

Claim 13.1 (The error preserving property) *Let* $M = (Q, \Sigma, \Gamma, k, I, q_0, Q_A, Q_R)$ *be an arbitrary RRWW-automaton, u, v be arbitrary input words from* Σ^*. *If* $u \Rightarrow_M^* v$ *and* $u \notin L(M)$, *then* $v \notin L(M)$. *[In the same way, if* $u \Rightarrow_M^* v$ *and* $v \in L(M)$, *then* $u \in L(M)$.*]*

Corollary 13.1 *For every RW-automaton M, $u \Rightarrow_M^* v$ and $v \in L(M)$ always implies $u \in L(M)$, since all its restarting configurations are initial.*

– *Monotonicity properties*: Any cycle Cyc performed by an $RRWW$-automaton M has a significant configuration $c_w = (¢u, q, v\$)$ in which it performs the rewriting step of the cycle Cyc. We call the number $|v|$ ($|v|$ means the length of the string v) the distance from the place of the rewriting in Cyc to the right sentinel. We denote it $D_r(Cyc)$.

We say that a sequence of cycles $Cyc = Cyc_1, Cyc_2, \cdots Cyc_n$ of an $RRWW$-automaton M is *monotonic* if the sequence $D_r(Cyc_1), D_r(Cyc_2), \ldots, D_r(Cyc_n)$ is monotonic, i.e., not increasing. If Cyc is the (maximal) sequence of cycles of a computation C, and Cyc is monotonic, we say that C is a monotonic computation. Notice that the tails are not considered for monotonicity.

We distinguish between three types of monotonicity of $RRWW$-automata:

By a *monotonic $RRWW$-automaton* we mean an $RRWW$-automaton where all its computations are monotonic.

By an *accepting-monotonic (a-monotonic) $RRWW$-automaton* we mean an $RRWW$-automaton where all the accepting computations are monotonic.

By a *general-monotonic (g-monotonic) $RRWW$-automaton* we mean an $RRWW$-automaton where for an arbitrary accepting computation on some word w there is an monotonic accepting computation on w.

– *Weak cyclic form*: We say that an $RRWW$-automaton M is in the *weak cyclic form* if any word from $L(M)$ longer than k (the length of its lookahead) can be accepted by M only after performing one cycle (and a tail) at least.

We introduce some further special forms of $RRWW$-automata. These forms represent certain levels of determinism of $RRWW$-automata.

Three types of det-MVR-forms: We say that an $RRWW$-automaton M is in the *det-MVR$_1$*-form if it has the following two properties:

1) For any couple (q, v), where q is a state of M and v is a string of its symbols (of the length at most k), there is at most one instruction of the form $(q, v) \to (q', MVR)$.

2) If M rewrites in a tail, the tail is a rejecting one. Each accepting tail ends when the lookahead window moves completely past the right sentinel.

M is in the *det-MVR$_2$*-form if it is in the *det-MVR$_1$*-form, and for any pair of its instructions of the form $(q, v) \to (q', REWRITE(u))$, $(q, v) \to (q_1, MVR)$ (a pair of instructions with the same left-hand sides) it holds that the move-instruction $(q, v) \to (q_1, MVR)$ cannot be used in a cycle. In other words, the place of rewriting on a given word w is fixed for all possible cycles by an $RRWW$-automaton M in the *det-MVR$_2$*-form.

M is in the *det-MVR$_3$*-form if it is in the *det-MVR$_2$*-form, and for any of its instructions of the form $(q, v) \to (q', REWRITE(u))$ there is no MVR-instruction with the same left-hand side (i.e., (q, v)) in M. This means, in other words, that an *RRWW*-automaton M in the *det-MVR*$_3$-form can nondeterministically choose between several rewriting steps only.

Lemma 13.1 *For any $RRWW$-automaton M there is an $RRWW$-automaton M_1 in the det-MVR$_1$-form such that $L(M) = L(M_1)$, and $u \Rightarrow_{M_1} v$ exactly if $u \Rightarrow_M v$.*

Proof. The construction of M_1 is based on two ideas:

1. On the well known construction of an equivalent deterministic finite automaton for a given nondeterministic one. In this way from M we get an automaton M_0 with the property that for any couple (q, v), where q is a state of M and v is a string of its symbols, there is at most one instruction of the form $(q, v) \to (q', MVR)$.

2. On the idea that M_1 simulating a phase Ph of M_0 decides on the place of rewriting given by Ph (if there is one) whether M_1 further simulates the Ph exactly (performs the rewriting and so on), or it transforms the Ph into a tail which does not use any rewriting. Naturally, this transformation must preserve the language recognized by M_0.

   In the latter case, M_1 will behave in the following way: it will simulate Ph using the moves to the right only, without performing the rewrite operation. If Ph finishes with accepting, M_1 accepts. In any other case M_1 rejects.

   In the case that Ph does not contain any rewriting, M_1 simulates Ph exactly.

□

Using the fact that $u \Rightarrow_{M_1} v$ exactly if $u \Rightarrow_M v$, the following consequence of Lemma 13.1 can be shown.

Corollary 13.2 *Let $X \in \{Y, mon\text{-}Y, a\text{-}mon\text{-}Y, g\text{-}mon\text{-}Y\}$, where $Y \in \{R$, RW, RWW, RR, RW, $RRWW\}$. For any X-automaton M there is an X-automaton M_1 in the det-MVR$_1$-form such that $L(M) = L(M_1)$, and $u \Rightarrow_{M_1} v$ if and only if $u \Rightarrow_M v$.*

Remark 13.1 *A slightly more complex situation concerns the other det-MVR-forms. We will show some results concerning these forms in the next section.*

Notation: For brevity, prefix *wcf-* denotes the weak cyclic form versions of $RRWW$-automata, similarly *mon-* the monotonic, *a-mon-* the *a*-monotonic, and *g-mon-* the *g*-monotonic versions. Recall that *det-MVR*$_i$, for $i \in \{1, 2, 3\}$, denotes the three special forms introduced above. $\mathcal{L}(A)$, where A is some class of automata, denotes the class of languages recognizable by automata from A. E.g. the class of languages recognizable by monotonic RR-automata in the weak cyclic form is denoted by $\mathcal{L}(wcf\text{-}mon\text{-}RR)$.

3. RESULTS

In the first part of this section we show that for any *RR*-automaton (*RRW*-, *RRWW*-automaton resp.) M with any type of monotonicity there exists an equivalent automaton M' of the same type as M in the weak cyclic form.

Theorem 13.1 *Let* $X \in \{Y, mon\text{-}Y, a\text{-}mon\text{-}Y, g\text{-}mon\text{-}Y\}$, *where* $Y \in \{RR, RW, RRWW\}$. *For any* X*-automaton* M *there is an* X*-automaton* M_1 *in the weak cyclic form such that* $L(M) = L(M')$. *I.e.,* $\mathcal{L}(X) = \mathcal{L}(wcf\text{-}X)$.

Proof. We will outline a common transformation (construction) for all the above mentioned types X of *RRWW*-automata. Let M be an X-automaton. Because of Corollary 13.2, we can suppose M being in the *det-MVR*$_1$-form with the lookahead window of the size k, for some $k > 0$.

Let us also suppose that *Rgl* is the set of words which can be accepted by M in a tail (without restart). Let m be the number of *MVR*-instructions of M, and $p = m + k$. Let us consider the following fact.

Fact I. For any words v, w, y such that $vwy \in Rgl$, $|w| \geqslant p$ there is $u_1, u_1 \neq \lambda$ such that $w = xu_1z$, and $vxzy \in Rgl$ for some x, z.

In the case that the *Rgl* is finite, M can be simulated by an automaton M' which works in the same way as M and has the lookahead window of a size greater than the length of any word from *Rgl* (and at least k).

In the case that *Rgl* is an infinite language we will construct M' with the lookahead window of the size $p + k + 1$. M' simulates M using the first k symbols of the lookahead window. The suffix part of the length $p + 1$ serves for performing reductions based on Fact I (which does not change the (non)membership of the reduced word to the language *Rgl*). We need to use the suffix greater than p, because we need to reduce using Fact I also with the right sentinel in the lookahead window.

We construct an X-automaton M' simulating M in such a way that if $u \Rightarrow_{M'} v$ by a cycle C' then

(r1) $Dr(C') > k + p$, and $u \Rightarrow_M v$, or

(r2) $Dr(C) > k + p$, both u and v are from *Rgl*, and there is a cycle C of M starting on the word u such that $Dr(C) = Dr(C')$, or

(r3) $Dr(C') \leqslant k + p$, $|u| \geqslant k + p$, and either u and v are from *Rgl*, or $u \Rightarrow_M v$.

Reductions of types (r1), (r3) fully ensure the transformation into the weak cyclic form and at the same time the simulation of M by M'. Reductions of the type (r2) are used to ensure that M' rewrites in all places where rewrites M to preserve the possible monotonicities.

M' accepts or rejects (immediately) in a tail the words of the length at most $p+k$. Let us assume that $|w| > p+k$. M' will perform either a cycle or a rejecting tail on the string w. M' works in the following four modes: M'_0, M'_1, M'_2, M'_3.

M'_0: Any phase (i.e. cycle or tail) of M' starts in the mode M'_0. M being in the mode M'_0 moves its lookahead window from the left-hand side to the right-hand side, simulating the moving part of a (nondeterministically choosen) phase C of M. This mode ends by reaching the place p_1 in which C rewrites or the right sentinel appears in the lookahead of M'.

If $Dr(C) > k+p$, and if it cannot use any MVR-instruction on p_1 then M' switches into the mode M'_1.

If $Dr(C) > k+p$, M can use (also) a *MVR*-instruction at the place p_1; then M' switches nondeterministically into the mode M'_1, or into the mode M'_2.

If the right sentinel appears in the lookahead of M' then M'_0 switches into the mode M'_3.

M'_1: The task of M'_1 is to perform the reductions of the type (r1). M'_1 continues in a plain simulation of C (performing its rewriting). If C ends by the restart, M' restarts. If C ends by rejecting, M' rejects. According to the *det-MVR*$_1$-form of M another possibility (an accepting) cannot occur.

M'_2: The task of M'_2 is to perform the reductions of the type (r2). M'_2 performs two subtasks at the same time in a parallel (and nondeterministic) way.

The first subtask continues in the simulation of C (not performing its rewriting), in order to check whether C is a cycle of M or not.

The second subtask simulates another branch (continuation) of C, denoted as Br. The simulation of Br serves two purposes: first for checking whether Br is an accepting tail or not; secondly for a detection and deletion of a nonempty substring u_1 in the first step of M'_2 by Fact I (see above).

If the phase Br is an accepting tail, simultaneously M'_2 performs some deleting, and the phase C is a cycle, then M' restarts. Otherwise M' rejects.

M'_3: The task of M'_3 is to perform the reductions of the type (r3). M'_3 nondeterministically chooses (having the right sentinel in the lookahead) one of the following three actions.

If the simulated phase C is recognized as an accepting tail, then due to Fact I M'_3 deletes some nonempty substring u_1 from the last p symbols of the word, and performs the restart.

If the simulated phase C is recognized as a cycle then M'_3 simulates this cycle exactly (performing also its rewriting).

If the simulated phase C is recognized as a rejecting tail then M' rejects.

Obviously, the construction we present can be applied to any X-automaton M and yields an X-automaton M'. In particular, it is not hard to see that the monotonicity of M causes the monotonicity of M'. We can also see that M' is in the weak cyclic form, and that $L(M) = L(M')$. □

In what follows, we are looking for a transformation of automata of the above mentioned classes into the *det-MVR*$_3$-form, and simultaneously into the weak cyclic form. This is possible for *RR-*,*RRW-* and *RRWW*-automata in the *det-MVR*$_2$-form.

Theorem 13.2 *The following equality holds for* $X \in \{\lambda, W, WW\}$*:*
$\mathcal{L}(det\text{-}MVR_2\text{-}mon\text{-}RRX) = \mathcal{L}(wcf\text{-}det\text{-}MVR_3\text{-}mon\text{-}RRX)$.

Proof. Let M be a $mon\text{-}RRX$-automaton in the *det-MVR*$_2$-form with the lookahead window of the size k. For the main part of the proof we take the slightly modified construction from the proof of Theorem 13.1 applied to a slightly modified M. We can suppose w.l.o.g. that M always halts with the head on the right sentinel.

When the automaton M is a *det-MVR*$_2$-RRX-automaton, then the automaton M' constructed according to the (non-modified) construction from the proof of Theorem 13.1 is also a *det-MVR*$_2$-RRX-automaton: There is only one place p_1 of rewriting for all phases of M' on a word w – this is the place where the mode M'_0 ends. The added property of M ensures that M' always rewrites on p_1 in the modes M'_1 and M'_2. It is not hard to see (with this assumption) that M' can be modified in such a way that it rewrites (deletes) also by rejecting in the mode M'_3. In this way mode M'_0 ends at the only place where (the modified) M' always rewrites, thus M' is in the *det-MVR*$_3$-form. The weak cyclic form and the monotonicity of the constructed RRX-automaton is obvious. □

Let us consider the following language $L_{RR} = L^1_{RR} \cup L^2_{RR} \cup L^3_{RR}$, where $L^1_{RR} = \{ww^R \mid w \in (cd + dc)^*\}$, $L^2_{RR} = \{wddw^R \mid w \in (cd + dc)^*\}$ and $L^3_{RR} = (cd)^*(dc)^*a$.

Lemma 13.2 $L_{RR} \in \mathcal{L}(wcf\text{-}mon\text{-}RR)$.

Proof. A $wcf\text{-}mon\text{-}RR$-automaton M_{RR} recognizing L_{RR} can be sketched as follows.

M_{RR} contains two (sub)modules M_1, M_2.

M_{RR} at the beginning of any phase nondeterministically chooses one of the two modules M_1, M_2.

M_1 serves for checking of the membership to L^1_{RR} or to L^2_{RR}. M_1 looks for a subword $v_1 = cddc$, or for $v_2 = dccd$, or for z in the form $(cddddc + dcddcd)$ at odd positions from the left end of the input word w of the form $(cd + dc)^*(\lambda + dd)(cd + dc)^*$.

Finding z, M_1 rewrites z into dd (using only deleting). After this rewriting M_1 checks, moving to the right, the form $(cd + dc)^*$ of the remainder of the word. If the check is positive M_{RR} restarts, otherwise it rejects. Finding v_1, or v_2, M_1 nondeterministically decides between rewriting and moving further to the right in the searching mode of M_1, described above. M_1 rewrites (by deleting c's) the chosen string v_1 (resp. v_2) into dd. Further it checks, moving to the right, the remainder of w, whether it has the form $(cd+dc)^*$. If the check is positive M_{RR} restarts, otherwise it rejects. M_1 directly accepts the empty word and dd.

M_2 checks the membership to L^3_{RR} in the following way: On an input word of the form $w = (cd)^*(dc)^*a$ of a length of at least three it performs a cycle consisting of deleting the first pair of symbols from w. Beside this deleting, during the cycle M_2 checks the form $(cd)^*(dc)^*a$. M_2 accepts the string a.

M_{RR} does not accept any string not mentioned above.

We can see that M_{RR} is a *wcf-mon-RR*-automaton recognizing L_{RR}.

We can also see that the place of rewriting in a cycle of M_{RR} on a word w is not determined by the input word of the cycle. □

Lemma 13.3 $L_{RR} \in \mathcal{L}(wcf\text{-}mon\text{-}RR) - \mathcal{L}(wcf\text{-}det\text{-}MVR_3\text{-}mon\text{-}RRW)$.

Proof. The first part of this lemma is proved by Lemma 13.2. The second part remains to be proved.

Let us suppose that some $wcf\text{-}det\text{-}MVR_3\text{-}mon\text{-}RW$-automaton M with the size of its lookahead k recognizes L_{RR}. Because of the weak cyclic form for any word of the form $w = (cd)^n(dc)^n$, where $n \geqslant k$, there is a correct reduction. Any correct reduction of w by M must be of the following form: $w \Rightarrow_M w_1$, where $w_1 = (cd)^{n-i}v(dc)^{n-i}$, for some i, v, $2i \leqslant k, |v| < k$. Let us also consider the word $ww^R = (cd)^n(dc)^n(cd)^n(dc)^n$ in a similar way. For this word there also exists a correct reduction. Any correct reduction of ww^R by M must be of the following form: $ww^R \Rightarrow_M w_2$, where $w_2 = (cd)^n(dc)^{n-j}z(cd)^{n-j}(dc)^n$, for some j, z, $2i \leqslant k, |z| < k$. But that contradicts the *det-MVR*$_3$-form of M (otherwise M could not choose nondeterministically between *MVR*- and *REWRITE*- steps in the middle of w). □

The following inequalites are a direct consequence of the previous assertions, and the equality follows from (a construction showed in) [5] .

Theorem 13.3 *The following proper inequality holds for* $X \in \{\lambda, W\}$*:*
$\mathcal{L}(wcf\text{-}det\text{-}MVR_3\text{-}mon\text{-}RRX) \subset \mathcal{L}(wcf\text{-}mon\text{-}RRX)$.

Further the following equality holds:
$\mathcal{L}(wcf\text{-}det\text{-}MVR_3\text{-}mon\text{-}RRWW) = \mathcal{L}(wcf\text{-}mon\text{-}RRWW) = CFL.$

4. CONCLUDING REMARK

We have concentrated on the classification of languages recognized by restarting automata in the weak cyclic form. In the future, we will also be interested in some other types of special forms which promise to serve as a suitable paradigm for grammar checking.

References

[1] Jančar, P.; F. Mráz; M. Plátek & J. Vogel (1998), On Monotony for Restarting Automata, in M. Kudlek, ed., *Proceedings of the MFCS'98 Workshop on Mathematical Linguistics*: 71-84, Bericht 213, Universität Hamburg.

[2] Kuboň, V. & M. Plátek (1994), A Grammar Based Approach to a Grammar Checking of Free Word Order Languages, in *Proceedings of the 15th International Conference on Computational Linguistics COLING 94*, II: 906-910. Kyoto.

[3] Marcus, S. (1997), Contextual Grammars and Natural Languages, in G. Rozenberg & A. Salomaa, eds., *Handbook of Formal Languages*, II: 215-235, Springer, Berlin.

[4] Păun, Gh. (1997), *Marcus Contextual Grammars*, Kluwer, Dordrecht.

[5] Plátek, M. (1999), Weak Cyclic Forms of Restarting Automata, in W. Thomas, ed., *Preproceedings of DLT 99*: 63-72, Aachen, extended abstract.

[6] Plátek, M. (1999), Weak Cyclic Forms of Restarting Automata, Technical Report, No 99/4, KTI MFF, Charles University, Prague.

Chapter 14

THE TIME DIMENSION OF COMPUTATION MODELS

Sheng Yu
Department of Computer Science
University of Western Ontario
London, ON Canada N6A 5B7
syu@csd.uwo.ca

Abstract The time dimension (temporal dimension) of computation models is studied. We argue that there is no time dimension associated with the traditional computation models such as Church's *lambda*-calculus, Turing machines, and partial recursive functions. We define Iterative Turing Machines that have the time dimension. We also study the structures of time and their relation to computation models.

1. INTRODUCTION

> What, then, is time? If no one asks me, I know: If I wish to explain it to one that asketh, I know not.
> *St. Augustine [Confessiones XI, c. XIV, xvii./Gale pp. 40]*

Many philosophers and scientists have been fascinated with time. Quite a few theories on the time dimension have been developed. However, "Every concept of time arises in the context of some (no doubt useful) human purpose and bears, inevitably and essentially, the stamp of that human intent." [8]. Here, our consideration of time is in the context of computation models. In particular, we consider the relation of the time dimension and its structure to various computation models.

Recently, quite a few leading researchers have questioned the adequacy of the traditional computation models for the new paradigms of interactive computing.

Robin Milner wrote in his Turing lecture [9] in 1993: "For the much smaller world of sequential computation, a common semantic framework is founded on the central notion of *mathematical function* and is formally expressed in a functional calculus — of which Alonzo Church's λ-calculus is the famous

C. Martin-Vide and V. Mitrana (eds.), *Where Mathematics, Computer Science, Linguistics and Biology Meet*, 161-172.

prototype. Functions are an essential ingredient of the air we breathe, so to speak, when we discuss the semantics of sequential programming. But for concurrent programming and interactive systems in general, we have nothing comparable".

Peter Wegner et al. explicitly challenged the traditional Turing machine model and the Church-Turing thesis, claiming that "TMs are too weak to express interaction of object-oriented and distributed systems" ([15]) and that Church's Thesis "is valid in the narrow sense that TMs express the behavior of algorithms, the broader assertion that algorithms capture the intuitive notion of what computers compute is invalid" [13].

In formal language theory, many researchers also felt the need to develop new models that would be able to represent interactions. Notable examples of those models include *traces* [5], *cooperating distributed grammar systems* (CD) and *parallel communicating grammar systems* (PC) [4, 3, 10]. Gheorghe Păun has been actively involved in introducing and developing the latter two systems.

In this article, we join the discussions on computation models for interactive systems.

Let us consider a simple model of interaction:

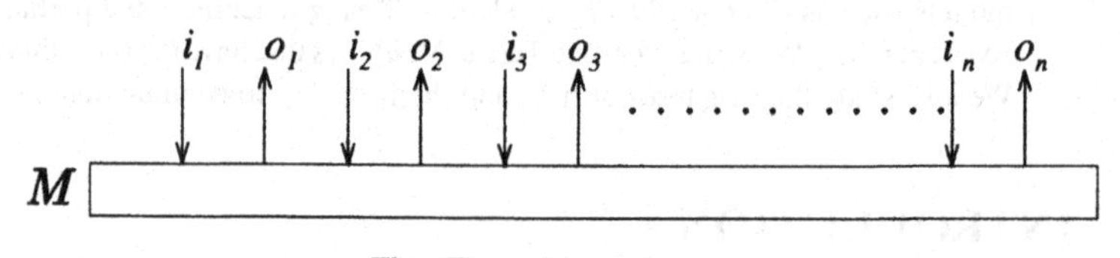

Figure 14.1 An interactive system.

Let M be an interactive machine, $i_1, \ldots, i_n$ be inputs to M and $o_1, \ldots, o_n$ outputs from M, where n is an arbitrary non-negative integer. The horizontal direction is the *time dimension*. We assume that the output o_i is a direct response by M for i_i and o_i may influence i_{i+1}. M is similar to the Sequential Interaction Machine (SIM) defined in [15] except that we DO NOT assume that the interaction sequences are infinite. We can consider M as an ATM banking machine.

Note that in the traditional functional model, a (partial) function f takes in an input and produces an output (or no output) in each computation, and that $f(x) = f(y)$ if $x = y$ and f is defined on x or y. Now, we consider whether M can be modelled with the traditional functional model. There are the following two possible ways that M might be modelled by a function f:

(1) *Multiple Calls*: A function f is called many times. At each call, f takes an input and (possibly) produces an output.

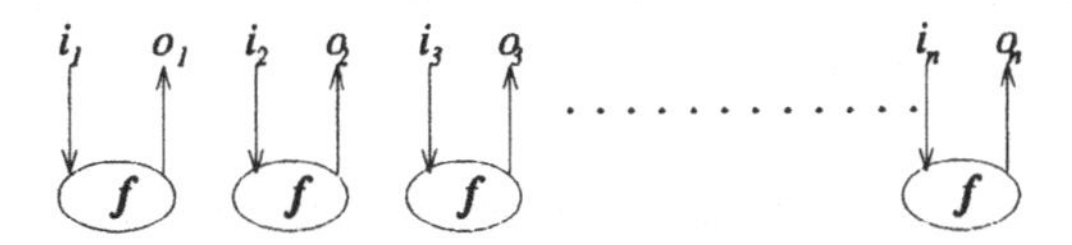

Figure 14.2 Modeling M by Multiple Call of f.

(2) *Multi-Value Input*: A function f takes an argument of the form $< i_1, o_1, i_2, o_2, \ldots, i_n, o_n >$ and (possibly) produces a Boolean value, i.e., $f(< i_1, o_1, i_2, o_2, \ldots, i_n, o_n >)$ is true if and only if the input is a valid sequence of interaction.

The first model "Multiple Calls" clearly cannot reply M simply because it is possible that $o_s \neq o_t$ even if $i_s = i_t$, but $f(i_s) = f(i_t)$ whenever $i_s = i_t$. M is history sensitive and time dependent, but f is not. Intuitively speaking, M ages and gets "wiser" in time, but f is always the same. Besides, f cannot represent the influence of o_i on i_{i+1}.

The second model "Multi-Value Input" can produce a correct answer for the question whether the sequence is a valid interaction sequence of M, but it does not model the computation of M. Imagine that if an ATM machine worked this way, then no customer would use it. The computation *requires* that o_1 is produced as a response to i_1; i_2 is entered after o_1 is produced and in the influence of o_1; o_2 is a direct response to i_2 but a result of all i_1, o_1, and i_2, and it influences i_3, etc. The main problem of the "Multi-Value Input" model is that it loses the relative timing of the input-output interaction. We say that it does not have the time dimension as M does.

In general, we say that there is no time dimension associated with the traditional computation models such as Church's λ-calculus, Turing machines, and partial recursive functions due to the following reasons:

(1) They are time independent. The output of a machine or a function is determined by its input without the influence of its history.

(2) Time, relative or absolute, is not part of the requirements of the computation carried out by those models.

The "time complexity" of a computation is only a measurement for the size or complexity of their internal or algorithmic structure rather than an external requirement of the computation.

The traditional models are adequate for representing the computations of mathematical functions and non-interactive business transactions, which were the main focus of early computer usage. However, computer programs have since been dealing with more and more complicated aspects of human life. The paradigms of computer programs are gradually approaching to paradigms of the real world, which are inherently concurrent, interactive, and "object-oriented".

To describe interactive or concurrent systems, the time dimension is not only needed but essential. In those systems, time is not only used to describe the complexity of a computation, it is (in the form of either relative or absolute time) an important part of the computation requirement. For example, in an interrogative system or a learning system, the relative timing of the inputs and outputs is essential. Such a system cannot be properly modelled by a function.

The traditional TM model is different from λ-calculus or partial recursive functions in its internal structure. TMs compute in the fashion of one transition at each time step. The time dimension is an innate characteristic of the TM model. So, the traditional TM model can be more naturally and easily extended to have the time dimension than λ-calculus and partial recursive functions can. The *Turing test* proposed by A. M. Turing himself [12] actually had extended his original model to include interaction. Recently, P. Wegner et al. proposed "interaction machines (IMs) as a stronger model that better captures computational behavior for finite interactive computing agents" [13, 14, 6]. IMs are also extensions of TMs.

In practice, computer programmers have long been programming for interactive software systems. Their natural instinct from their practice has told them that pure functional or logical languages are not adequate for interactive applications. The difference between the internal structures of TMs and other models in the time dimension gives a conceptual explanation to this natural instinct.

In the remaining part of this article, we give the definitions of Sequential Interactive Turing Machines and Multi-Stream Interactive Turing Machines in the next section. We study the structure of the time dimension and consider the point view of time versus the period view of time in Section 3. In Section 4, we discuss the relations between the structure of time and computation models. We conclude our discussion in Section 5.

2. INTERACTIVE TURING MACHINES

In this section, we define two types of interactive Turing machines, Sequential Interactive Turing Machines (SITMs) and Multi-Stream Interactive Turing Machines (MITMs). SITMs and MITMs are modifications of, respectively, SIMs and MIMs defined in [15]. However, in [15], the authors emphasize the *infinity* of interactions, and based on this they show that SIMs and MIMs are strictly more expressive than TMs. Instead, we emphasize the time dimension. We consider that the crucial difference between interactive machines and non-interactive machine or functions is their time dimension. We do not require SITMs or MITMs to have infinite interactions.

An SITM is similar to a multi-tape TM except that it has an interaction tape instead of an input tape (or an input/output tape). Physically, an SITM M, shown in the following figure, has a finite control, an interaction tape, n work tapes, $n \geqslant 1$, and $n+1$ heads, one on each tape. All tapes are one-way infinite tapes. The head on the interaction tape is called the interaction head, which can make only right moves or stationary moves. All other heads can move in either direction or stay put.

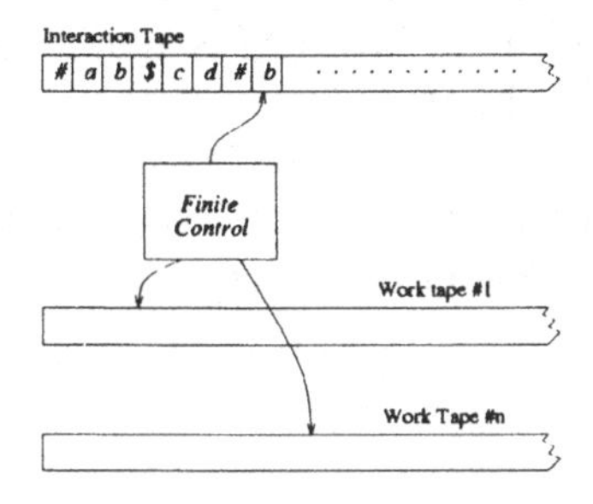

Figure 14.3 A sequential interactive Turing machine (SITM).

There are two special symbols that are used on the interaction tape to separate input and output strings: '#' marks the beginning of an input string and the end of an output string; and '$' marks the end of an input string and the beginning of an output string. The interaction head initially is placed on the first cell of the interaction tape. If initially '#' is on the first cell, M starts to read until the symbol '$'. Then it starts to write on the tape and ends the output with a '#'. Assume that an external source writes an input string following '#' and ends it with a '$'. M reads the input string and then writes an output string following the '$' symbol, etc. If initially '$' is on the first cell of the interaction tape, M starts to write on the tape and ends writing with a '#' symbol, and then starts to read, and repeat these reading and writing operations. M terminates normally if it enters a terminating state at the same time it finishes reading an input string or writing an output string.

Definition 14.1 *A sequential interactive Turing machine M is*

$$(I, O, Q, \Gamma, \gamma, \omega, s, \#, \$, B, T),$$

where I is the finite input alphabet; O is the finite output alphabet; Q is the non-empty finite set of states; $\#, \$, B$ are three special symbols, where $\#$ is *the delimiter preceding an input string or/and at the end of an output string, $ is the other delimiter that is ending an input string and preceding an output string, B is the blank symbol of the interaction tape; Γ is the finite work-tape alphabet; $\gamma : Q \times I \cup \{\#, \$\} \times \Gamma^k \rightarrow \{R, S\} \times (\Gamma \times \{L, R, S\})^k$ is the transition function of M in the reading phase; $\omega : Q \times \{B\} \times \Gamma^k \rightarrow (O \cup \{\#\} \times \{R, S\}) \times (\Gamma \times \{L, R, S\})^k$ is the transition function of M in the writing phase, where L (left), R (right), and S (stationary) are directions of the movement of a head; $s \in Q$ is the starting state of M; $T \subseteq Q$ is the set of terminating states.*

Note that M switches between the reading phase and the writing phase. It switches to the writing phase after reading a '$', and switches back to the reading phase after writing a '#'. When M is at the reading phase, it may wait at B for an input symbol to be ready.

An MITM is similarly defined as an SITM except that it may have more than one interaction tape. An MITM M, which has two interaction tapes, is shown in Figure 14.4, where the interior of M consists of the finite control and work tapes.

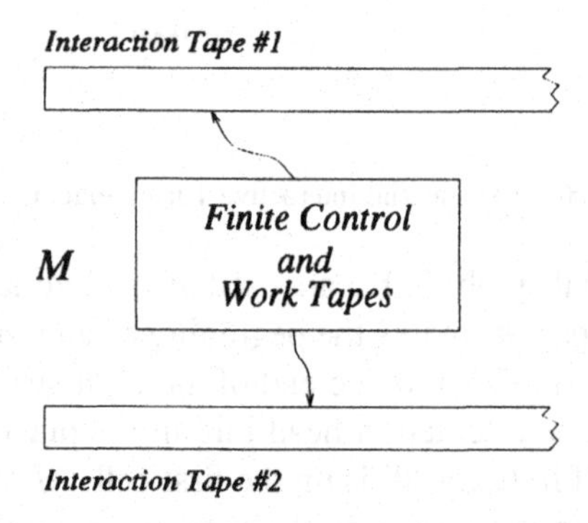

Figure 14.4 A multi-stream interactive Turing machine (MITM).

Note that there may be complicated synchronization constraints between the two interaction tapes of M.

Two MITMs M_1 and M_2 may share an interaction tape. See Figure 14.5 below. In this case, one of the MITMs needs to interchange the two special symbols '#' and '$' in its definition.

Based on these basic definitions, we can define the following relations between MITMs, informally. Let M_1 and M_2 be two MITMs.

- **Inheritance**
  We say that M_2 and M_1 have an inheritance relation if M_2 inherits all the properties of M_1, i.e., M_2 is exactly the same as M_1 except that it

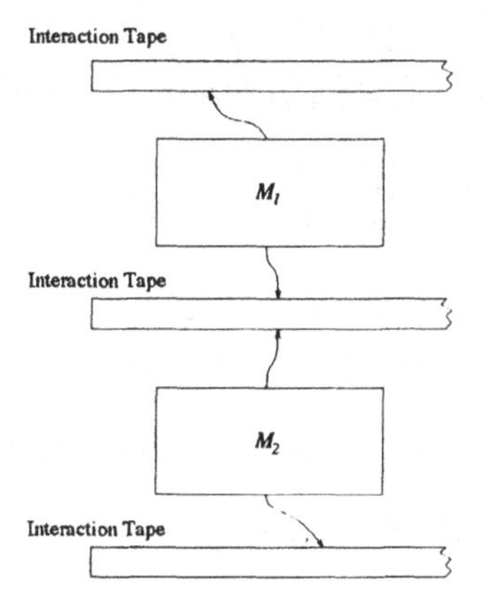

Figure 14.5 Two MITMs interact.

may have more interaction tapes. All the inherited interaction tapes of M_2 should work in exactly the same way as those of M_1.

- **Association**
  We say that M_1 and M_2 have an association relation if they share at least one interaction tape.

- **Aggregation**
  We say that M and M_1 have an aggregation relation if M_1 is a part of M. A MITM can be an aggregate of several MITMs. The aggregate MITM has all the interaction tapes of its component MITMs except those that are used for the interaction between two component MITMs, which become internal work tapes. This is illustrated in Figure 14.6, where M is an aggregate of M_1, and M_2, and M_3.

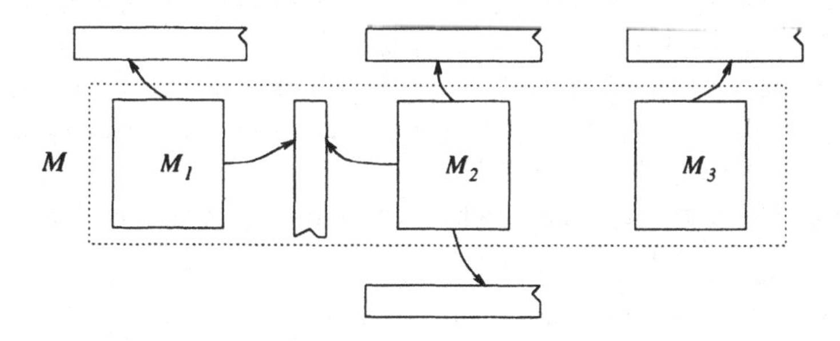

Figure 14.6 Aggregation of MITMs.

3. STRUCTURES OF TIME: POINT VIEW VERSUS PERIOD VIEW

We adopt the traditional temporal structure as the basic model of time, i.e., *time* consists of points ordered by a relation of precedence [2].

Definition 14.2 *A total point structure of time, $\mathcal{T}$, is a pair $(T, <)$, where T is a non-empty set and $<$ is a strict total order on T.*

By a strict total order, we mean a relation that satisfies the following properties:

(1) transitivity: $\forall t_a t_b t_c\ (t_a < t_b \wedge t_b < t_c \rightarrow t_a < t_c)$,

(2) irreflexivity: $\forall t_a\ (\neg(t_a < t_a))$,

(3) asymmetry: $\forall t_a t_b\ (t_a < t_b \rightarrow \neg t_b < t_a)$,

(4) linearity: $\forall t_a t_b\ (t_a = t_b \vee t_a < t_b \vee t_b < t_a)$.

Note that (3) is implied by (1) and (2) and, thus, can be omitted.

The total order in the above definition is often too rigid and strong a condition, which is unnecessary and superfluous for many applications. A partial order is often more satisfactory. In a partial ordering of time, two points with no *precedence* ($<$) relation are often interpreted as being concurrent.

Definition 14.3 *A partial point structure of time, $\mathcal{T}$, is a pair $(T, <)$, where T is a nonempty set and $<$ is a strict partial order on T.*

By a strict partial order, we mean a relation that satisfies transitivity and irreflexivity.

We can define a mapping that maps a partial point structure $\mathcal{T} = (T, <)$ into a set $\mathcal{S}$ of total point structures, $\mathcal{S} = \{\mathcal{T}_1, \ldots, \mathcal{T}_n\}$, such that for each $\mathcal{T}_i = (T, <_i)$, $1 \leqslant i \leqslant n$, if $t_a, t_b \in T$ and neither $t_a < t_b$ nor $t_b < t_a$ is defined in $\mathcal{T}$, then either $t_a <_i t_b$ or $t_b <_i t_a$ in $\mathcal{T}_i$.

For example, assuming that $T = \{t_a, t_b, t_c\}$, the partial point structure $(T, \{t_a < t_c, t_b < t_c\}$ can be mapped into a set of total point structures $\{(T, t_a < t_b < t_c), (T, t_b < t_a < t_c)\}$.

A special type of partial point structure, called multi-stream total point structure, is defined as follows:

Definition 14.4 *An n-stream total point structure $\mathcal{T}^n = (T_1, \ldots, T_n, <)$ is a partial point structure $(T_1 \cup \ldots \cup T_n, <)$ such that (1) $T_i \cap T_j = \emptyset$ for all $1 \leqslant i < j \leqslant n$, and (2) each $\mathcal{T}_i = (T_i, <_i)$ is a total point structure, where $<_i$ is $<$ restricted to T_i, $1 \leqslant i \leqslant n$.*

Intuitively, an n-stream total point structure can be considered as a time structure with a finite number of river-beds, not restricted to only one river-bed. The time structure in each river-bed is a total point structure.

Based on the above "point" views of time, we can define a basic "period" view of time. In this view, a period p is a pair of points $[p_s, p_t]$, where $p_s < p_t$, p_s is the starting point and p_t the terminating point of p. Two periods $p = [p_s, p_t]$ and $q = [q_s, q_t]$ may have one of the following 6 relations: (1) *precedes*: $p \lhd q$ if $p_t < q_s$, (2) *contains*: $p \subset q$ if $q_s < p_s$ and $p_t < q_t$, (3) *overlaps*: $p \bowtie q$ if $p_s < q_s < p_t < q_t$, and (1'), (2'), and (3') which are symmetric to (1), (2) and (3), respectively. Note that if we consider $p_s p_t$ and $q_s q_t$ are two words and $<$ is the catenation operator, then the 6 cases above are the results of the shuffle operation of $p_s p_t$ and $q_s q_t$.

In the following, we define a simple period view of time, in which the relation between two periods is either a 'precedes' relation or no relation defined. Two periods having no defined relation are usually interpreted as being concurrent, i.e., they can be in any of the above 6 cases.

Definition 14.5 *A simple period structure of time, $\mathcal{P}$, is a pair $(P, \lhd)$ where P is a nonempty set of periods and $\lhd$ is a partial order on P.*

More complicated period structures of time can be defined. For example, we can define $\mathcal{P} = (P, \lhd, \subset)$ or $\mathcal{P} = (P, \lhd, \subset, \bowtie)$. Period structures of time may also be defined based on the notion that a period is a set of points rather than two points. For example, the definition of periods in [2] is based on sets rather than pairs of points.

Many fine theories of concurrency have been introduced based on the point view of time. One outstanding example is the theory (theories) of traces [5].

In [16, 7, 11], synchronization expressions (SEs) have been introduced and studied, which are based on the simple period structure of time. An SE is an expression of processes that specify the synchronization constraints among those processes. The operators of SEs include: $\rightarrow$ (sequential), $\|$ (parallel), and $|$ (alternative). Processes are modeled as periods in a simple period structure of time. The $\rightarrow$ operator corresponds to the $\lhd$ relation, $\|$ corresponds to no relation defined, and $|$ corresponds to a substitution (one replacing the other). For example, SE $(a\|b) \rightarrow c$ corresponds to a simple period structure of time $(\{a, b, c\}, \{a \lhd c, b \lhd c\})$.

4. STRUCTURES OF TIME AND COMPUTATION MODELS

As we have discussed in the introduction, clearly, a function call in the traditional functional model can be considered as an instantaneous action or a point in the time dimension when viewed externally. In SITMs, the sequencing of the interactions is part of the (external) requirement of a computation. So, a computation of an SITM cannot be viewed as a point in time. It needs a linear dimension of time. However, a point structure of time is already adequate to model their interactions. For MITMs, we argue in the following that point structures of time is not adequate and a period structure is necessary.

Let us consider an MITM M which is an aggregate of M_1, M_2, and M_3, as shown in Figure 14.7. Note that for a read or write operation of an MITM on

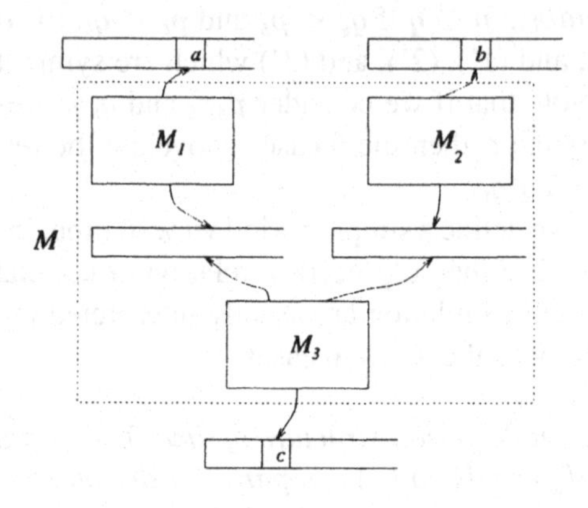

Figure 14.7 Synchronization constraints in MITM M.

an interaction tape, the machine may take k stationary moves before moving to the next cell. Then we say that the operation takes $k + 1$ steps. In general, we cannot assume that every read or write operation of an MITM takes only one step.

We assume that in M the operations of reading 'a', reading 'b', and writing 'c' have the following synchronization constraint: writing 'c' cannot proceed until both reading 'a' and reading 'b' finish. The latter two operations have no mutual time constraints. We also assume that the operations of reading 'a', reading 'b', and writing 'c' takes n_a, n_b, and n_c steps, respectively. In the following, we simply use a, b, and c to denote the three operations, respectively.

If we describe this situation with the point view of time, then we have either $a < b < c$ or $b < a < c$. In this view, we sequentialize the operations and cannot express that a and b can overlap in time. However, we can express this situation adequately with the period view of time.

We explain this more precisely in the following. We have two different situations with the period view of time (using the simple period structure of time we have defined):

(1) $\{a \lhd c,\ b \lhd c\}$

(2) $\{a \lhd b \lhd c\}$ or $\{b \lhd a \lhd c\}$.

The difference between (1) and (2) can be seen clearly if we map them into total sequential point structures of time below (we omit the operator $<$ for clarity):

(1) $a_s b_s a_t b_t c_s c_t$, $a_s b_s b_t a_t c_s c_t$, $a_s a_t b_s b_t c_s c_t$, $b_s a_s a_t b_t c_s c_t$, $b_s a_s b_t a_t c_s c_t$, or $b_s b_t a_s a_t c_s c_t$.

(2) $a_s a_t b_s b_t c_s c_t$, or $b_s b_t a_s a_t c_s c_t$.

However, if we use the point view of time, then we have

(1) $\{a < c,\ \ b < c\}$

(2) $\{a < b < c\}$ or $\{b < a < c\}$.

We would not be able to distinguish them. Therefore, the point view of time is not adequate for MITMs.

5. CONCLUSION

We have expressed our view on the time dimension in computation models. We argue that the traditional functional models do not have the time dimension associated with them. We define two types of interactive Turing machines, sequential interactive Turing machines and multi-stream interactive Turing machines. We consider two views of time, the point view and the period view, and two types of structures associated with them. We argue that the point view is appropriate for the SITM model but not adequate for the MITM model.

All the discussions in this paper are preliminary and mostly informal. It is clear that the main ideas behind the new computation models as well as the above discussions are interaction, concurrency, and object-orientation.

References

[1] Aalbersberg, I.J. & G. Rozenberg (1988), Theory of Traces, *Theoretical Computer Science*, 60: 1–82.

[2] Benthem, J. van (1991), *The Logic of Time*, Kluwer, Dordrecht.

[3] Csuhaj-Varjú, E.; J. Dassow; J. Kelemen & Gh. Păun (1994), *Grammar Systems. A Grammatical Approach to Distribution and Cooperation*, Gordon and Breach, London.

[4] Dassow, J.; Gh. Păun & G. Rozenberg (1997), Grammar Systems, in G. Rozenberg & A. Salomaa, eds., *Handbook of Formal Languages*, II: 155–213. Springer, Berlin.

[5] Diekert, V. & G. Rozenberg (1995), *The Book of Traces*, World Scientific, Singapore.

[6] Goldin, D. & P. Wegner (1999), Behavior and Expressiveness of Persistent Turing Machine, CS Technical Report, Brown University.

[7] Guo, L.; K. Salomaa & S. Yu (1994), Synchronization Expressions and Languages, in *Proceedings of the IEEE Symposium on Parallel and Distributed Processing*: 257–264.

[8] Lawrence, N. (1978), Levels of Language of Discourse about Time, in Fraser et al., eds., *The Study of Time*, III: 22–49. Springer, Berlin.

[9] Milner, R. (1993), Elements of Interaction — Turing Award Lecture, *Communications of the ACM*, 36.1: 78–89.

[10] Păun, Gh., ed., (1995), *Artificial Life. Grammatical Models*, The Black Sea University Press, Bucharest.

[11] Salomaa, K. & S. Yu (1999), Synchronization Expressions: Characterization Results and Implementation, in J. Karhumäki; H. Maurer; Gh. Păun & G. Rozenberg, eds., *Jewels are Forever*: 45–56. Springer, Berlin.

[12] Turing, A.M. (1950), Computing Machinery and Intelligence, *Mind*, 59.236: 433–460.

[13] Wegner, P. (1997), Why Interaction is More Powerful than Algorithms, *Communications of the ACM*, 40.5: 81–91.

[14] Wegner, P. (1998), Interactive foundations of Computing, *Theoretical Computer Science*, 192.2: 335–351.

[15] Wegner, P. & D. Goldin (1999), Interaction, Computability, and Church's Thesis, draft paper.

[16] Yu, S.; L. Guo; R. Govindarajan & P. Wang (1991), PARC Project: Practical Constructs for Parallel Programming Languages, in *Proceedings of the IEEE Fifteenth Annual International Computer Software and Applications Conference*: 183–189.

III

LANGUAGES AND COMBINATORICS

Chapter 15

AN INFINITE SEQUENCE OF FULL AFL-STRUCTURES, EACH OF WHICH POSSESSES AN INFINITE HIERARCHY

Peter R.J. Asveld
Department of Computer Science
Twente University of Technology
P.O. Box 217, 7500 AE Enschede, The Netherlands
infprja@cs.utwente.nl

Abstract We investigate different sets of operations on languages which result in corresponding algebraic structures, viz. in different types of full AFL's (full Abstract Family of Languages). By iterating control on ETOL-systems we show that there exists an infinite sequence $\mathcal{C}_m$ ($m \geqslant 1$) of classes of such algebraic structures (full AFL-structures): each class is a proper superset of the next class ($\mathcal{C}_m \supset \mathcal{C}_{m+1}$). In turn each class $\mathcal{C}_m$ contains a countably infinite hierarchy, i.e. a countably infinite chain of language families $K_{m,n}$ ($n \geqslant 1$) such that (i) each $K_{m,n}$ is closed under the operations that determine $\mathcal{C}_m$, and (ii) each $K_{m,n}$ is properly included in the next one: $K_{m,n} \subset K_{m,n+1}$.

1. INTRODUCTION

Usually, each newly introduced family of formal languages will be studied sooner or later with respect to its closure properties. In the early days of formal language theory, this meant that (non)closure under each known operation had to be established separately. Then one realised that some operations are more fundamental that others, and that some other operations can be expressed in these fundamental ones: they are "polynomials" over those fundamental operations. In short, a more algebraic view on this part of formal language theory emerged.

An important step in this algebraic approach to families of languages has been the introduction of the notion of full Abstract Family of Languages (full AFL),

C. Martin-Vide and V. Mitrana (eds.), Where Mathematics, Computer Science, Linguistics and Biology Meet, 175-186.

being a nontrivial family of languages closed under the operations: union, concatenation, Kleene $\star$, homomorphism, inverse homomorphism, and intersection with regular sets [9]. Similar as in ordinary algebra —where one went from groups to semigroups, rings, and fields— full AFL's gave rise to weaker structures (full trios, full semi-AFL's [9]) and more powerful ones: full substitution-closed AFL's [10], full super-AFL's [11] and full hyper-AFL's [1].

For each class $\mathcal{C}$ of these full AFL-like structures, it has been shown that $\mathcal{C}$ is not trivial in the sense that it does not solely consists of a few "isolated" language families but, to the contrary, that $\mathcal{C}$ is infinite. This latter fact is usually established by showing the existence of an infinite hierarchy, i.e. a countably infinite chain of language families K_n $(n \geqslant 1)$ such that (i) each K_n is closed under the operations that determine $\mathcal{C}$, and (ii) each K_n is properly included in the next one: $K_n \subset K_{n+1}$.

In this paper we show that by iterating control on ETOL-systems, as studied in [7, 8], we obtain an infinite sequence of full AFL-structures. Each class $\mathcal{C}_m$ $(m \geqslant 1)$ in this sequence is a proper superset of the next class: $\mathcal{C}_m \supset \mathcal{C}_{m+1}$. So the full AFL-structures in $\mathcal{C}_{m+1}$ are more powerful than those in $\mathcal{C}_m$. And each class $\mathcal{C}_m$ is nontrivial, since it contains an infinite hierarchy of language families $K_{m,n}$ $(n \geqslant 1)$, each of which is properly included in the next one: $K_{m,n} \subset K_{m,n+1}$. The proofs of these results heavily rely on the main results of [4] and [7]. Many properties of full substitution-closed AFL's, full super-AFL's and full hyper-AFL's (quoted in Section 5) have their counterparts for the classes $\mathcal{C}_m$; see Section 7.

The remaining part of this paper is organized as follows. Section 2 consists of Preliminaries. The definitions and properties of some generalized grammatical models (controlled K-iteration grammar, context-free K-grammar, regular K-grammar) are in Section 3 and Section 4, respectively. In Section 5 we recall the corresponding full AFL-structures. In Section 6 we quote two fundamental theorems which enable us to establish the main result in Section 7. Some concluding remarks are in Section 8.

2. PRELIMINARIES

We already mentioned the standard text [9] on full AFL's and related concepts. Some other books on formal language theory, like [12, 13, 15], also treat the relevant issues to the extent we use in this paper. For Lindenmayer or L systems we refer to [14].

Henceforth, Σ_ω denotes a countably infinite set of symbols. A *family of languages*, or a *family* for short, K is a set of languages L with $L \subseteq \Sigma_L^\star$ such that each Σ_L is a finite subset of Σ_ω. As usual, we assume that for each language L in the family K, the alphabet Σ_L is minimal, i.e. a symbol α belongs to Σ_L if and only if there exists a word w of L in which α occurs. A family K is called *nontrivial* if K contains a nonempty language L with $L \neq \{\lambda\}$, where λ denotes the empty word. We also assume that each family is closed under isomorphism.

We will use well-known families like FIN (family of finite languages), REG (regular languages), CF (context-free languages), as well as the family ONE of singleton languages: ONE $= \{\{w\} \mid w \in \Sigma_\omega^\star\}$, the family ALPHA of alphabets: ALPHA $= \{\Sigma \mid \Sigma \subset \Sigma_\omega, \Sigma \text{ is finite}\}$, and the family SYMBOL of singleton alphabets: SYMBOL $= \{\{\sigma\} \mid \sigma \in \Sigma_\omega\}$.

We often need the concept of substitution and a few of its generalizations (Definitions 15.1 and 15.9).

Definition 15.1 *Let K be a family and V an alphabet. A K-*substitution *is a mapping $\tau : V \to K$; it is extended to words over V by $\tau(\lambda) = \{\lambda\}$, and $\tau(\alpha_1 \ldots \alpha_n) = \tau(\alpha_1) \ldots \tau(\alpha_n)$ where $\alpha_i \in V$ $(1 \leqslant i \leqslant n)$, and to languages L over V by $\tau(L) = \bigcup\{\tau(w) \mid w \in L\}$. If K equals* FIN *or* REG, *τ is called a* finite *or a* regular substitution, *respectively.*

Given families K and K', let $\hat{\text{Sub}}(K, K')$ *be defined by* $\hat{\text{Sub}}(K, K') = \{\tau(L) \mid \tau \text{ is a } K'\text{-substitution}; L \in K\}$. *A family K is* closed *under K'-substitution if* $\hat{\text{Sub}}(K, K') \subseteq K$, *and K is* closed under substitution, *if K is closed under K-substitution.*

*$\tau : V \to K$ is a K-*substitution over *V if $\tau(\alpha) \subseteq V^\star$ for each $\alpha \in V$. A K-substitution τ over V is* nested, *if $\alpha \in \tau(\alpha)$ for each $\alpha \in V$.*

Definition 15.2 *A* prequasoid *K is a nontrivial family that is closed under finite substitution and under intersection with regular languages. For each family K, let $\Pi(K)$ denote the smallest prequasoid that includes K. A* quasoid *is a prequasoid that contains an infinite language.*

It is easy to see that each [pre]quasoid includes the smallest [pre]quasoid REG [FIN, respectively], whereas FIN is the only prequasoid that is not a quasoid; cf. [1, 2].

Definition 15.3 *A* full Abstract Family of Languages *or* full AFL *is a nontrivial family of languages closed under union, concatenation, Kleene $\star$, homomorphism, inverse homomorphism, and intersection with regular languages. A* full substitution-closed AFL *is a full AFL closed under substitution.*

Frequently, the following characterization of full AFL's is useful; cf. Theorem 15.3(1) below.

Proposition 15.1 [10, 9, 2] *A family K of languages is a full AFL if and only if K is a prequasoid closed under regular substitution (i.e.* $\hat{\text{Sub}}(K, \text{REG}) \subseteq K$*), and under substitution in the regular languages (i.e.* $\hat{\text{Sub}}(\text{REG}, K) \subseteq K$*).*

3. SOME GENERALIZED GRAMMARS

In this section we recall the definitions of some grammar types with a countably infinite number of rules rather than a finite number. These generalizations are based on the concepts of ETOL-system (Definition 15.4), controlled ETOL-system (15.5), context-free grammar (15.6), and non-self-embedding context-free grammar (15.7).

Definition 15.4 [1,2] *Let K be a family. A K-*iteration grammar *$G = (V, \Sigma, U, S)$ consists of an alphabet V, a terminal alphabet Σ ($\Sigma \subseteq V$), an initial symbol S ($S \in V$), and a finite set U of K-substitutions over V. The language $L(G)$ generated by G is defined by $L(G) = U^\star(S) \cap \Sigma^\star = \bigcup\{\tau_p(\ldots(\tau_1(S))\ldots) \mid p \geqslant 0;\ \tau_i \in U,\ 1 \leqslant i \leqslant p\} \cap \Sigma^\star$.*

The family of languages generated by K-iteration grammars is denoted by $H(K)$. For $m \geqslant 1$, $H_m(K)$ is the family generated by K-iteration grammars that contain at most m K-substitutions in U.

Definition 15.5 [1, 2] *Let Γ and K be a families of languages. A* Γ-controlled K-iteration grammar *or* (Γ, K)-iteration grammar *is a pair (G, M) that consists of a K-iteration grammar $G = (V, \Sigma, U, S)$ and a* control language *M, i.e. M is a language over U, and $M \in \Gamma$. The language $L(G, M)$ generated by (G, M) is defined by $L(G, M) = M(S) \cap \Sigma^\star = \bigcup\{\tau_p(\ldots(\tau_1(S))\ldots) \mid p \geqslant 0;\ \tau_i \in U,\ \tau_1 \ldots \tau_p \in M\} \cap \Sigma^\star$.*

The family generated by (Γ, K)-iteration grammars is denoted by $H(\Gamma, K)$. Similarly, $H_m(\Gamma, K)$ is the family generated by (Γ, K)-iteration grammars that contain at most m K-substitutions in U ($m \geqslant 1$).

Clearly, $H(K) = \bigcup_{m\geqslant 1} H_m(K)$ and $H(\Gamma, K) = \bigcup_{m\geqslant 1} H_m(\Gamma, K)$.

Example 15.1 *By taking concrete values for the parameter K we obtain some families of Lindenmayer languages; viz.* $H(\text{ONE}) = \text{EDTOL}$, $H_1(\text{ONE}) = \text{EDOL}$, $H(\text{FIN}) = \text{ETOL}$, *and* $H_1(\text{FIN}) = \text{EOL}$. *Readers unfamiliar with L systems are referred to [14]. Alternatively, they may view these equalities as definitions.*

Definition 15.6 [16, 2, 6] *Let K be a family. A* context-free K-grammar *G is a K-iteration grammar $G = (V, \Sigma, U, S)$ of which each substitution τ from U is a nested K-substitution over V; so $\alpha \in \tau(\alpha)$ for each $\alpha \in V$ and each $\tau \in U$.*

The family of languages generated by context-free K-grammars is denoted by $A(K)$. For $m \geqslant 1$, $A_m(K)$ is the family generated by context-free K-grammars that contain at most m K-substitutions in U.

Definition 15.7 [2, 5] *Let K be a family and let U be a finite set of nested K-substitutions over an alphabet V. Then U is called* not self-embedding *if* *for all* $u \in U^{\star}$ *and for all* α *in* V, *the implication* $w_1 \alpha w_2 \in u(\alpha) \Rightarrow (w_1 = \lambda$ *or* $w_2 = \lambda)$ *holds for all* $w_1, w_2 \in V^{\star}$.

A regular K-grammar $G = (V, \Sigma, U, S)$ *is a context-free* K*-grammar where* U *is a non-self-embedding set of nested* K*-substitutions over* V.

The family of languages generated by regular K*-grammars is denoted by* $R(K)$. *For each* $m \geqslant 1$, $R_m(K)$ *is the family generated by regular* K*-grammars that contain at most* m K*-substitutions in* U.

Example 15.2 *When we take* K *equal to* FIN, *we have* $A(\text{FIN}) = \text{CF}$ *and* $R(\text{FIN}) = \text{REG}$.

4. SOME PROPERTIES OF THESE GENERALIZED GRAMMARS

This section consists of some useful properties of the grammatical devices that we discussed in the previous section.

First, we remember that regular control does not extend the generating power of K-iteration grammars.

Theorem 15.1 [1, 2] *If* $K \supseteq \text{ONE}$, *then* $H(\text{REG}, K) = H(K)$.

The number of K-substitutions in a Γ-controlled K-iteration grammar can be reduced to two in case the parameters Γ and K satisfy some very simple conditions, since we have

Proposition 15.2 [1, 2] *Let* K *be a family with* $K \supseteq \text{SYMBOL}$.
(1) If Γ *is a family closed under* λ*-free homomorphism, then* $H_2(\Gamma, K) = H_m(\Gamma, K) = H(\Gamma, K)$ *for each* $m \geqslant 2$.
(2) For each $m \geqslant 2$, $H_2(K) = H_m(K) = H(K)$.

For [non-self-embedding] context-free K-grammars a reduction to a single, equivalent [non-self-embedding] K-substitution is possible.

Proposition 15.3 [2, 6, 5] *Let* K *be a family closed under union with languages from* SYMBOL. *If* $K \supseteq \text{SYMBOL}$, *then* $A_1(K) = A_m(K) = A(K)$ *and* $R_1(K) = R_m(K) = R(K)$ *for each* $m \geqslant 1$.

Comparing Propositions 15.2(2) and 15.3 reveals that providing regular or context-free K-grammars with control does not lead to interesting results.

We conclude this section with a few useful inclusion properties for which we need some additional terminology.

Definition 15.8 *A family* Γ *is closed under* left marking *[*right marking*] if for each language* L *in* Γ *with* $L \subseteq \Sigma^{\star}$ *for some* Σ, *and for each symbol* c *not in* Σ, *the language* $\{c\}L$ *[*$L\{c\}$*, respectively] belongs to* Γ*. And* Γ *is closed under* full marking *if* Γ *is closed under both left and right marking.*

Proposition 15.4 [1, 4] (1) *Let* Γ *be a family closed under right marking, and let* K *be a family with* $K \supseteq$ ONE*. Then* $\Gamma \subseteq H(\Gamma, K)$ *and* $K \subseteq H(\Gamma, K)$.
(2) *Let* Γ *be a family closed under (i) left or right marking, (ii) union or concatenation, and (iii) Kleene star. If* K *is a family with* $K \supseteq$ SYMBOL*, then* $H(K) \subseteq H(\Gamma, K)$.

Proposition 15.5 [2, 4, 5, 6] *Let* K *be a family closed under union with languages from* SYMBOL*. If* $K \supseteq$ SYMBOL*, then* $K \subseteq H(K)$*,* $K \subseteq A(K)$*, and* $K \subseteq R(K)$.

Proposition 15.6 [1, 2, 4, 5, 6] *Let* Γ *be a family closed under full marking. If the family* K *is a prequasoid, then so are the families* $R(K)$*,* $A(K)$*,* $H(K)$ *and* $H(\Gamma, K)$.

5. SOME FULL AFL-STRUCTURES

In Section 2 we already encountered full AFL's and full substitution-closed AFL's. For full AFL-like structures weaker than full AFL, we refer to [9]. The present section is devoted to structures stronger than full AFL, which are related to the generalized grammars of Section 3.

Definition 15.9 *A family* K *is closed under* iterated substitution *if for each language* L *from* K *with* $L \subseteq V^\star$ *for some alphabet* V*, and for each finite set* U *of* K*-substitutions over* V*, the language* $U^\star(L)$*, defined by*

$$U^\star(L) = \bigcup\{\tau_p \dots \tau_1(L) \mid p \geqslant 0,\ \tau_i \in U\ (1 \leqslant i \leqslant p)\},$$

belongs to K*. In case each substitution in* U *is nested, then* K *is called* closed under nested iterated substitution.

A full hyper-AFL *[1] is a full AFL closed under iterated substitution; a* full super-AFL *[11] is a full AFL closed under nested iterated substitution.*

Definition 15.10 *Let* K *be a family. By* $\hat{\mathcal{F}}(K)$ *[*$\hat{\mathcal{R}}(K)$*,* $\hat{\mathcal{A}}(K)$*, and* $\hat{\mathcal{H}}(K)$*] we denote the smallest full AFL [full substitution-closed AFL, full super-AFL, and full hyper AFL, respectively] that includes* K.

Theorem 15.2 [2, 4, 5, 6] *Let* K *be a family. Then* K *is a*
(1) *full substitution-closed AFL, if and only if* K *is a prequasoid and* $R(K) = K$.
(2) *full super-AFL, if and only if* K *is a prequasoid and* $A(K) = K$.
(3) *full hyper-AFL, if and only if* K *is a prequasoid and* $H(K) = K$.

Theorem 15.3 [2, 4, 5, 6] *Let K be a family. Then*
(1) $\text{Sûb}(\text{REG}, \text{Sûb}(\Pi(K), \text{REG})) = \text{Sûb}(\text{Sûb}(\text{REG}, \Pi(K)), \text{REG})$ *is a full AFL that includes K.*
(2) $R\Pi(K)$ *is a full substitution-closed AFL that includes K.*
(3) $A\Pi(K)$ *is a full super-AFL that includes K.*
(4) $H\Pi(K)$ *is a full hyper-AFL that includes K.*

Theorem 15.2(3) says that K is a full hyper-AFL if and only if $\Pi(K) = K$ and $H(K) = K$. Consequently, the smallest full hyper-AFL $\hat{\mathcal{H}}(K)$, that includes a family K, equals $\hat{\mathcal{H}}(K) = \bigcup\{w(K) \mid w \in \{\Pi, H\}^\star\}$ or, equivalently, $\hat{\mathcal{H}}(K) = \{\Pi, H\}^\star(K)$. According to Theorem 15.4 below, this infinite set of strings over $\{\Pi, H\}$ can be reduced to the single string $H\Pi$. Obviously, a similar remark applies to the other full AFL-structures.

Theorem 15.4 [2, 4, 5, 6] *Let K be a family of languages. Then* $\hat{\mathcal{F}}(K) = \text{Sûb}(\text{REG}, \text{Sûb}(\Pi(K), \text{REG})) = \text{Sûb}(\text{Sûb}(\text{REG}, \Pi(K)), \text{REG})$, $\hat{\mathcal{R}}(K) = R\Pi(K)$, $\hat{\mathcal{A}}(K) = A\Pi(K)$, *and* $\hat{\mathcal{H}}(K) = H\Pi(K)$.

Theorem 15.5 REG [REG, CF, ETOL, *respectively*] *is the smallest full AFL [full substitution-closed AFL, full super-AFL, full hyper-AFL].*

Each full hyper-AFL is a full super-AFL, and each full super-AFL is a full substitution-closed AFL. But none of the converse implications hold; cf. Theorem 15.5.

6. TWO FUNDAMENTAL RESULTS

This section contains two results (Theorems 15.6 and 15.7) that constitute the principal steps in obtaining the main result of this paper; cf. Section 7. The first one is a direct consequence of a more general statement from [4].

Theorem 15.6 [4] *Let Γ_1, Γ_2 and K be families of languages and let Γ_2 be closed under full marking, union or concatenation, and Kleene $\star$. If $K \supseteq$* ALPHA, *then* $H(\Gamma_1, H(\Gamma_2, K)) \subseteq H(\text{Sûb}(\Gamma_1, \Gamma_2), K)$.

Corollary 15.1 [4] (1) *Let Γ be a family of languages closed under full marking and under substitution that satisfies $\Gamma \supseteq$* REG. *If K is a family with $K \supseteq$* ALPHA $\cup$ ONE, *then* $H(\Gamma, H(\Gamma, K)) = H(\Gamma, K)$.
(2) *Let Γ be a family of languages that is closed under full marking, union, concatenation, and Kleene $\star$. If K is a family with $K \supseteq$* ALPHA $\cup$ ONE, *then* $H(H(\Gamma, K)) = H(\Gamma, K)$.

Corollary 15.1(2) has been used to show that certain families $H(\Gamma, K)$ are full hyper-AFL's. In particular, there exist infinite chains of full hyper-AFL's

[7, 8]; see also Theorem 15.7 below. In Section 7 we will use Corollary 15.1(1) to obtain related results.

Corollary 15.2 [4] *If* $K \supseteq$ ALPHA $\cup$ ONE, *then* $HH(K) = H(K)$.

In establishing results like Theorems 15.2(3), 15.3(4) and 15.4, Corollary 15.2 plays an important role.

The strictness of our infinite sequence of full AFL-structures, as well as the properness of our hierarchies rely on the following theorem that stems from a rich collection [7, 8] of similar hierarchies.

Theorem 15.7 [7, 8] *Let* $K_0 =$ REG *and* $K_{i+1} = H(K_i, \text{FIN})$ *for each* $i \geqslant 0$. *Then* $\{K_i\}_{i \geqslant 1}$ *is an infinite hierarchy of full hyper-AFL's, i.e.*
- *for each* $i \geqslant 1$, K_i *is a full hyper-AFL, and*
- *for each* $i \geqslant 1$, K_i *is properly included in* K_{i+1}: $K_i \subset K_{i+1}$.

7. AN INFINITE SEQUENCE OF FULL AFL-STRUCTURES

In this section we define a family of full AFL-structures (Definition 15.11). Then we show some properties of these full AFL-structures (Proposition 15.7, Theorems 15.8 and 15.9) and our main result (Theorem 15.10).

In this section we frequently write $H_\Gamma(K)$ instead of $H(\Gamma, K)$ in order to distinguish between the two arguments of $H(\Gamma, K)$: K will play the role of "ordinary" argument, whereas Γ is an additional parameter over which we proceed inductively (Theorems 15.7, 15.9 and 15.10).

In view of Theorem 15.2, the following definition is a natural extension of the notion of full hyper-AFL.

Definition 15.11 *Let* Γ *be a fixed family of languages. An arbitrary family* K *is a* full Γ-hyper-AFL *if* K *is a prequasoid with* $H(\Gamma, K) = K$, *or equivalently, with* $H_\Gamma(K) = K$. *For each family* K, *let* $\hat{\mathcal{H}}_\Gamma(K)$ *denote the smallest full* Γ*-hyper-AFL that includes* K.

Ordinary full hyper-AFL's are now obtained as a special instance of Definition 15.11, since Theorem 15.1 implies:

Proposition 15.7 *A family* K *of languages is a full hyper-AFL if and only if* K *is a full* REG*-hyper-AFL.*

Theorem 15.8 *Let* Γ *be a full substitution-closed AFL.*
(1) *Each full* Γ*-hyper-AFL is a full hyper-AFL.*
(2) *If* K *is a family, then* $H_\Gamma\Pi(K)$ *is a full* Γ*-hyper-AFL that includes* K.
(3) *For each family* K, $\hat{\mathcal{H}}_\Gamma(K) = H_\Gamma\Pi(K)$.

(4) $H_\Gamma(\mathrm{FIN})$ *is the smallest full* Γ*-hyper-AFL.*

Proof. (1) It suffices to show that $H(\Gamma, K) = K$ implies $H(K) = K$. Since $\Gamma \supseteq \mathrm{REG}$, we have by Propositions 15.5 and 15.4(2): $K \subseteq H(K) \subseteq H(\Gamma, K) = K$. Hence $H(K) = K$.

(2) The result follows from Propositions 15.6, 15.4(1), Corollary 15.1(1), and the fact that Γ is closed under substitution.

(3) By the inclusion $K \subseteq \hat{\mathcal{H}}_\Gamma(K)$ and the monotonicity of both H_Γ and Π, we have $H_\Gamma\Pi(K) \subseteq H_\Gamma\Pi\hat{\mathcal{H}}_\Gamma(K)$. According to Definition 15.11, this yields $H_\Gamma\Pi(K) \subseteq \hat{\mathcal{H}}_\Gamma(K)$. Now Theorem 15.8(2) and Proposition 15.4(1) imply that $H_\Gamma\Pi(K)$ is a full Γ-hyper-AFL that includes K. Hence we obtain that $\hat{\mathcal{H}}_\Gamma(K) = H_\Gamma\Pi(K)$.

(4) FIN is the smallest prequasoid, and Theorem 15.8(3). □

Compare Theorem 15.8(2), (3) and (4) with their "uncontrolled counterparts": Theorems 15.3(4), 15.4, and 15.5, respectively.

Theorem 15.9 *Let the family K be a prequasoid, let* $Q_0 = \mathrm{REG}$ *and* $Q_{i+1} = H(Q_i, K)$ *for each* $i \geqslant 0$*. Then for each* $i \geqslant 0$*,* Q_j *is a full* Q_i*-hyper-AFL provided that* $j > i$*.*

Proof. A simple proof by induction on i using Theorem 15.1, Propositions 15.4(1) and 15.6, and Corollary 15.1(2), yield the following facts:

(F1) Q_i is a full hyper-AFL for each $i \geqslant 1$, and

(F2) $Q_i \subseteq Q_j$ provided $j \geqslant i$.

Next we prove by induction on i ($i \geqslant 0$) that Q_j ($j > i$) is a full Q_i-hyper-AFL.

Basis ($i = 0$): We have to show that Q_j is a full Q_0-hyper-AFL for each $j \geqslant 1$. Since $Q_0 = \mathrm{REG}$ and each Q_j is a full REG-hyper-AFL if and only if Q_j is a full hyper-AFL (Proposition 15.7), the statement follows from (F1).

Induction step: Assume that for each $j > i$, Q_j is a full Q_i-hyper-AFL.

We have to show that each family Q_j with $j > i+1$ is a full Q_{i+1}-hyper-AFL.

Consider an arbitrary Q_j with $j > i + 1$; then $Q_j = H(Q_{j-1}, K)$. As $j - 1 > i$, the induction hypothesis implies that Q_{j-1} is a full Q_i-hyper-AFL. Now by Theorem 15.8(1) and Proposition 15.6, Q_j is a prequasoid.

So it remains to show that $H(Q_{i+1}, Q_j) \subseteq Q_j$, since the converse inclusion follows from Proposition 15.4(2) and (F1).

From the definition of Q_j and Theorem 15.6 respectively, we obtain

$$H(Q_{i+1}, Q_j) = H(Q_{i+1}, H(Q_{j-1}, K)) \subseteq H(\hat{\mathrm{Sub}}(Q_{i+1}, Q_{j-1}), K).$$

We already remarked that the induction hypothesis implies that Q_{j-1} is a full Q_i-hyper-AFL. By Theorem 15.8(1), Q_{j-1} is a full hyper-AFL and so Q_{j-1} is closed under substitution. As $j - 1 \geqslant i + 1$, we have $Q_{i+1} \subseteq Q_{j-1}$ by (F2), and consequently, $\hat{\text{Sub}}(Q_{i+1}, Q_{j-1}) \subseteq Q_{j-1}$. Hence we have $H(Q_{i+1}, Q_j) \subseteq H(\hat{\text{Sub}}(Q_{i+1}, Q_{j-1}), K) \subseteq H(Q_{j-1}, K) = Q_j$, which completes the induction. □

We are now ready for the main result.

Theorem 15.10 *Let* $K_0 = \text{REG}$ *and* $K_{m+1} = H(K_m, \text{FIN})$ *for* $m \geqslant 0$, *and let* C_m *be the class of all full* K_m*-hyper-AFL's. Then for each* $m \geqslant 1$,
(1) *the class* C_m *is a proper superset of* C_{m+1}: $C_m \supset C_{m+1}$,
(2) *the class* C_m *contains an infinite hierarchy of full* K_m*-AFL's, i.e. a countably infinite chain of language families* $K_{m,n}$ $(n \geqslant 1)$ *such that* (i) $K_{m,n}$ *is a full* K_m*-AFL, and* (ii) *for each* $n \geqslant 1$, $K_{m,n}$ *is properly included in the next one:* $K_{m,n} \subset K_{m,n+1}$.

Proof. (1) follows from Theorems 15.7, 15.9 (with $K = \text{FIN}$) and 15.8(4).
(2) For fixed m $(m \geqslant 1)$, we define $\{K_{m,n}\}_{n \geqslant 1}$ by $K_{m,n} = K_{m+n}$ for each $n \geqslant 1$. By Theorems 15.7 and 15.9 this is an infinite hierarchy of full K_m-hyper-AFL's. □

8. CONCLUDING REMARKS

We extended the finite sequence "full AFL, full substitution-closed AFL, full super-AFL, full hyper-AFL" to a countably infinite sequence of full AFL-structures, each of which possesses properties (Theorem 15.8) similar to those of the members of the initial, finite sequence (Theorems 15.3, 15.4 and 15.5). And each new class of full AFL-structures is nontrivial in the sense that it contains a countably infinite hierarchy (Theorem 15.10).

The concept of full AFL abstracts the regular languages in case they are characterized by nondeterministic finite automata and regular expressions. Full substitution-closed AFL's generalize the regular languages when they are viewed as the languages generated by non-self-embedding context-free grammars. Similarly, full super-AFL's, full hyper-AFL's and full Γ-hyper-AFL's correspond to context-free grammars, ETOL-systems and Γ-controlled ETOL-systems, respectively.

Actually, a full Γ-hyper-AFL is a full AFL closed under Γ-controlled iterated substitution. A family K is closed under *Γ-controlled iterated substitution*, if for each language L from K with $L \subseteq V^\star$ for some alphabet V, for each finite set U of K-substitutions over V, and for each language M over U from the family Γ, the language $M(L)$, defined by

$$M(L) = \bigcup\{\tau_p \dots \tau_1(L) \mid p \geqslant 0,\ \tau_i \in U\ (1 \leqslant i \leqslant p),\ \tau_1 \dots \tau_p \in M\},$$

belongs to K; cf. Definition 15.9.

We could take the obvious, next step: a family K is closed under *controlled iterated substitution* if K is closed under K-controlled iterated substitution. And a family K is a *full* $\top$*-hyper-AFL* if K is a prequasoid and $H(K, K) = K$.

Up to now there are only a few full $\top$-hyper-AFL's known. Of course there are the smallest full $\top$-hyper-AFL $K_\omega = \bigcup_{m \geqslant 1} K_m$ (cf. Theorem 15.7), and the largest effective one: the family RE of recursively enumerable languages (since we have $H(\text{RE}, \text{RE}) = \text{RE}$ by Church's thesis). A less trivial example is the family $\mathcal{L}_{\star\text{OI}}$ studied in [17]. In particular, it is an open question whether there exist infinitely many full $\top$-hyper-AFL's.

In this context we quote an interesting result (Proposition 15.8) for which we need some additional terminology. A family K is closed under *removal of right endmarker*, if for each language L ($L \subseteq \Sigma^\star$) and each symbol c with $c \notin \Sigma$, $L\{c\}$ in K implies L in K.

Proposition 15.8 [3] *Let K be a family closed under removal of right endmarker. If* $\text{DSPACE}(\log n) \subseteq K \subset \text{RE}$, *then K is not closed under controlled iterated (λ-free) substitution.*

(Originally, this statement has been formulated for λ-free substitutions only. Obviously, it applies to arbitrary substitutions as well.) Proposition 15.8 implies that $\text{DSPACE}(\log n)$ is neither a subfamily of K_ω nor of $\mathcal{L}_{\star\text{OI}}$.

On the other hand it is known that the hierarchy of Theorem 15.7, and consequently K_ω, is situated within the family of context-sensitive languages; see [3, 7, 8]. Therefore, the smallest elements in the classes $\mathcal{C}_m$ ($m \geqslant 1$), as well as the infinite hierarchies $\{K_{m,n}\}_{n \geqslant 1}$ ($m \geqslant 1$) (Theorem 15.10) are in the family of context-sensitive languages.

References

[1] Asveld, P.R.J. (1977), Controlled iteration grammars and full hyper-AFL's, *Information and Control*, 34: 248–269.

[2] Asveld, P.R.J. (1978), *Iterated Context-Independent Rewriting – An Algebraic Approach to Formal Languages*, Ph.D. Thesis, Department of Applied Mathematics, Twente University of Technology, Enschede.

[3] Asveld, P.R.J. (1980), Space-bounded complexity classes and iterated deterministic substitution, *Information and Control* 44: 282–299.

[4] Asveld, P.R.J. (1997), Controlled fuzzy parallel rewriting, in Gh. Păun & A. Salomaa, eds., *New Trends in Formal Languages — Control, Cooperation, and Combinatorics*: 49–70. Springer, Berlin.

[5] Asveld, P.R.J. (1999), The non-self-embedding property for generalized fuzzy context-free grammars, *Publicationes Mathematicae Debrecen*, 54: 553–573.

[6] Asveld, P.R.J. (2000), Fuzzy context-free languages — Part I: Generalized fuzzy context-free grammars, CTIT Technical Report No. 00-03, Enschede.

[7] Engelfriet, J. (1982), Three hierarchies of transducers, *Mathematical Systems Theory*, 15: 95-125.

[8] Engelfriet, J. (1985), Hierarchies of hyper-AFL's, *Journal of Computer and System Sciences*, 30: 86–115.

[9] Ginsburg, S. (1975), *Algebraic and Automata-Theoretic Properties of Formal Languages*, North-Holland, Amsterdam.

[10] Ginsburg, S. & E.H. Spanier (1970), Substitution in families of languages, *Information Sciences*, 2: 83–110.

[11] Greibach, S.A. (1970), Full AFL's and nested iterated substitution, *Information and Control*, 16: 7–35.

[12] Harrison, M.A. (1978), *Introduction to Formal Language Theory*, Addison-Wesley, Reading, Mass.

[13] Hopcroft, J.E. & J.D. Ullman (1979), *Introduction to Automata Theory, Languages, and Computation*, Addison-Wesley, Reading, Mass.

[14] Rozenberg, G. & A. Salomaa (1980), *The Mathematical Theory of L Systems*, Academic Press, New York.

[15] Salomaa, A. (1973), *Formal Languages*, Academic Press, New York.

[16] van Leeuwen, J. (1974), A generalization of Parikh's theorem in formal language theory, in J. Loeckx, ed., *Proceedings of the 2nd ICALP*: 17–26. Springer, Berlin.

[17] Vogler, H. (1988), The OI-hierarchy is closed under control, *Information and Computation*, 78: 187–204.

Chapter 16

TRELLIS LANGUAGES

Adrian Atanasiu
Department of Computer Science
Faculty of Mathematics
University of Bucharest
Str. Academiei 14, 70109 Bucharest, Romania
aadrian@pcnet.ro

Abstract We shall present here a new approach to languages representation, suggested by a technique widely used nowadays in the area of signal coding. Each word of the language is written in an array frame, where rows are words of regular languages and columns are equal length words of a finite language. Some simple properties of languages are presented, together with a "packing" method, which leads to an exprimation of words of a trellis language using a minimal number of symbols.

1. PRELIMINARY DEFINITIONS AND NOTATIONS

In recent years, the need to send messages by noisy channels with a high level of security (e.g. wireless phones) has led to using a technique which writes these messages in arrays; this frame allows one to control rows and columns of the message at the same time, and to correct a great number of errors succesfully ([3]). We use here the same idea in order to build a new class of languages – *trellis languages*; every word is arranged as an array frame, where rows are sequences of regular languages, and columns are equal length sequences of a finite language. This manner of representation allows, for example, a reduction (often significant) in the number of symbols in writing words from a trellis language.

The way in which the finite language (denoted by L_0) checks, by columns, the words of regular languages which forms sequences of a trellis language is quite similar to the original idea of grammar systems ([2]): a sequence is generated in sequence/parallel by several grammars, controlled by a coordinator/sinchronizing mechanism; here L_0 is a "coordinator" language, having the

C. Martin-Vide and V. Mitrana (eds.), Where Mathematics, Computer Science, Linguistics and Biology Meet, 187-198.

role of selecting the appropriate words in order to compose a sequence of the corresponding trellis language.

In this paper we suppose that the fundamental notions of formal languages are known. So, an *alphabet* V is a finite and non-empty set of symbols. V^* will be the set of all sequences (words) $\alpha = a_1 a_2 a_3 \ldots$ with symbols from V. The number of symbols from α is denoted by $|\alpha|$. The *empty word* $\epsilon \in V^*$ is a special sequence without symbols ($|\epsilon| = 0$). A *language* is a set $L \subseteq V^*$. Any language can be defined by a generative device (called *grammar*) or a recognition device (called *automaton*) (for other details, see [4]).

A hierarchy of languages can be established, depending on the complexity of these devices. So, L is a *context-free* language (*cfg* for short) only if it can be generated by a grammar $G = (V_N, V_T, S, P)$ where P contains only rules of type $A \longrightarrow \alpha$, $\alpha \in (V_N \cup V_T)^*$. By $\mathcal{L}_2$ we denote the class of context-free languages – in Chomsky classification.

A language L is *regular* if it can be generated by a grammar in which P contains only rules of type $A \longrightarrow aB|a$, $a \in V_T \cup \{\epsilon\}$. The class of all regular languages is denoted by $\mathcal{L}_3$. Finite automata $M = (Q, V, \delta, q_0, F)$ are usual recognition devices for these languages. Among all finite automata which accept a regular language L, the automaton with a minimal number of states is called the *minimal automaton.* We shall use here an assertion concerning regular languages, namely the "*Pumping Lemma*":

Let L be a regular language. Then, there exists an integer $n > 0$ with the property $\forall\, \alpha \in L$, $|\alpha| \geqslant n$; the decomposition $\alpha = uvw$ holds, where:

(*i*) $v \neq \epsilon$;
(*ii*) $|uv| \leqslant n$;
(*iii*) $uv^iw \in L \ \forall i \geqslant 0$.

2. CONSTRUCTION OF TRELLIS LANGUAGES

Definition 16.1 *Let $n \geqslant 1$ be an integer and $L_1, L_2, \ldots, L_n$ be regular languages. We also consider a finite language L_0 (called "coordinator language"), which contains only words of length n. One defines the trellis language $L_0(L_1, L_2, \ldots, L_n)$ as follows:*
if $w \in L_0(L_1, L_2, \ldots, L_n)$, $|w| = nk$, then

1. $w = \alpha_1 \alpha_2 \ldots \alpha_n$ *where*

   (a) $\alpha_i \in L_i$, $\quad 1 \leqslant i \leqslant n$;

   (b) $|\alpha_1| = |\alpha_2| = \ldots = |\alpha_n| = k$;

2. $\forall\, j = 1 \ldots, k$, $\quad \alpha_1(j)\alpha_2(j) \ldots \alpha_n(j) \in L_0$ *in which $\alpha_i(j)$ is the j-th symbol from the sequence α_i.*

The word w of the language $L_0(L_1, L_2, \ldots, L_n)$ can be arranged in an array $n \times k$ ($k \geqslant 0$)

$$w = \begin{array}{cccc} a_{11} & a_{21} & \ldots & a_{k1} \\ a_{12} & a_{22} & \ldots & a_{k2} \\ & & \vdots & \\ a_{1n} & a_{2n} & \ldots & a_{kn} \end{array}$$

where $\alpha_i = a_{1i}a_{2i}\ldots a_{ki} \in L_i$ ($1 \leqslant i \leqslant n$) and $a_{i1}a_{i2}\ldots a_{in} \in L_0$ ($1 \leqslant i \leqslant k$).

Every word w is obtained by concatenating the rows of this array; from this point of view, $L_0(L_1, L_2, \ldots, L_n) \subseteq L_1L_2\ldots L_n$. Knowing the total length of w (multiple of n), we can easily split this sequence into n subwords, the i-th subword being in L_i.

If the representation of w is obtained by concatenating the columns of this array, then a word from L_0^* results.

Example 16.1 *Let be* $L_1 = L_2 = \{a, b\}^*$.
If $L_0 = \{ab\}$, *then the language* $L_0(L_1, L_2) = \{a^n b^n | a, b \geqslant 0\}$ *is obtained;*
for $L_0 = \{aa, bb\}$ – $L_0(L_1, L_2) = \{\alpha\alpha | \alpha \in \{a, b\}^*\}$,
for $L_0 = \{ab, ba\}$ – $L_0(L_1, L_2) = \{\alpha\tilde{\alpha} | \alpha \in \{a, b\}^*\}$,
and – finally – for $L_0 = \{aa, ab, ba, bb\}$ – $L_0(L_1, L_2) = \{\alpha | \alpha \in \{a, b\}^*, |\alpha| = 2p\}$.

By $\mathcal{TL}_n$ we shall denote the class of all languages of type $L_0(L_1, L_2, \ldots, L_n)$ ($n \geqslant 1$). Some assertions obviously result from the Definition 16.1. For example:

1. $L_0(L_1, L_2, \ldots, L_n)L_0'(L_1', L_2', \ldots, L_n') = L_0L_0'(L_1, L_2, \ldots, L_n, L_1', L_2', \ldots, L_n')$
2. $L_0(L_1, L_2, \ldots, L_n) \cup L_0'(L_1, L_2, \ldots, L_n) \subseteq (L_0 \cup L_0')(L_1, L_2, \ldots, L_n)$.
3. $\bigcup_{\alpha \in L_0} \{\alpha\}(L_1, L_2, \ldots, L_n) \subseteq L_0(L_1, L_2, \ldots, L_n)$
4. $L_0(L_1 \cup L_1', L_2, \ldots L_n) = L_0(L_1, L_2, \ldots, L_n) \cup L_0(L_1', L_2, \ldots, L_n)$.
5. $L_0 \subseteq L_0' \implies L_0(L_1, L_2, \ldots, L_n) \subseteq L_0'(L_1, L_2, \ldots, L_n)$
6. $L_1 \subseteq L_1' \implies L_0(L_1, L_2, \ldots, L_n) \subseteq L_0(L_1', L_2, \ldots, L_n)$

As a remark, the reciprocal inclusion from (3) is not true; e.g., for $L_0 = \{ab\}$, $L_0' = \{ba\}$, $L_1 = L_2 = \{a, b\}^*$, we have $abba \in (L_0 \cup L_0')(L_1, L_2)$, but $abba \notin L_0(L_1, L_2) = \{a^n b^n | n \geqslant 1\}$, $abba \notin L_0'(L_1, L_2) = \{b^n a^n | n \geqslant 1\}$.

Proposition 16.1 $\mathcal{TL}_1 = \mathcal{L}_3$;

Proof. Let L be a regular language over an alphabet V. We shall construct the trellis language $L_0(L_1)$ as follows: $L_1 = L$, $L_0 = V$. The relation $L_0(L_1) = L$ is obvious.

The reciprocal implication is immediate ($w \in L_0(L_1) \implies w \in L$). □

Proposition 16.2 *$\mathcal{TL}_n$ $(n \geqslant 2)$ and $\mathcal{L}_2$ are incomparable.*

Proof. One can't find an integer n for which the language $\{\alpha \in \{a,b\}^* | N_a(\alpha) = N_b(\alpha)\}$ can be expressed as a trellis language.

Moreover, $\{\alpha\alpha | \alpha \in \{a,b\}^*\}$ is a trellis language (Example 16.1), but it is not context-free. □

Proposition 16.3 *Let be $L_0 \subset L_0'$. In this case,*
$L_0(L_1, L_2, \ldots, L_n) = L_0'(L_1, L_2, \ldots, L_n) \Longrightarrow (L_0' \setminus L_0)(L_1, L_2, \ldots, L_n) = \emptyset$.

Proof. Let us suppose there is $w \in (L_0' \setminus L_0)(L_1, \ldots, L_n)$; therefore $w = \alpha_1\alpha_2 \ldots \alpha_n$, $\alpha_i \in L_i$, and $\alpha_1(j)\alpha_2(j) \ldots \alpha_n(j) \in L_0' \setminus L_0$, $j = 1, 2 \ldots, |w|/n$. It results that $w \in L_0'(L_1, \ldots, L_n) \setminus L_0(L_1, \ldots, L_n)$. But, from $L_0(L_1, \ldots, L_n) = L_0'(L_1, \ldots, L_n)$ we have $L_0'(L_1, \ldots, L_n) \setminus L_0(L_1, \ldots, L_n) = \emptyset$, contradiction. □

Unfortunately, the reciprocal implication is not true. For example, if $L_0 = \{ab\}$, $L_0' = \{ab, bc\}$, $L_1 = ab^*$, $L_2 = bc^*$, then $(L_0' \setminus L_0)(L_1, L_2) = \{bc\}(L_1, L_2) = \emptyset$ but $L_0(L_1, L_2) = \{ab\}$, $L_0'(L_1, L_2) = \{ab^{n+1}c^n | n \geqslant 0\} \neq L_0(L_1, L_2)$.

3. GENERATIVE DEVICES FOR TRELLIS GRAMMARS

As a generative method for designing trellis languages, a regular grammar can be used (in order to generate the tabelar form for w by columns); it will be defined as follows:

For every i $(1 \leqslant i \leqslant n)$, L_i is generated by grammar $G_i = (V_N^i, V_T^i, S_i, P_i)$, where the elements of P_i are of type $A \longrightarrow aB|a$, $A, B \in V_N^i$, $a \in V_T^i$.

Let $V_N = V_N^1 \times V_N^2 \times \ldots \times V_N^n$, $V_T = \bigcup_{i=1}^{n} V_T^i$, $S = (S_1, S_2, \ldots, S_n)$; P contains only rules of type:

$$(A_1, A_2, \ldots, A_n) \longrightarrow \begin{bmatrix} a_1 \\ a_2 \\ \vdots \\ a_n \end{bmatrix} (B_1, B_2, \ldots, B_n) \Longleftrightarrow \begin{cases} A_i \longrightarrow a_i B_i \in P_i \\ i = 1, 2 \ldots, n \\ \\ a_1 a_2 \ldots a_n \in L_0 \end{cases}$$

$$(A_1, A_2, \ldots, A_n) \longrightarrow \begin{bmatrix} a_1 \\ a_2 \\ \vdots \\ a_n \end{bmatrix} \Longleftrightarrow \begin{cases} A_i \longrightarrow a_i \in P_i \\ i = 1, 2 \ldots, n \\ a_1 a_2 \ldots a_n \in L_0 \end{cases}$$

Proposition 16.4 *The grammar $G = (V_N, V_T, S, P)$ with all the components we defined above generates an array representation of the language $L_0(L_1, L_2, \ldots, L_n)$.*

The same language $L_0(L_1, \ldots, L_n)$ can be also generated in its sequential form using a matrix grammar, defined as follows:

Let $G_i = (V_N^i, V_T^i, S_i, P_i)$ be n regular grammars with $L(G_i) = L_i$ $(1 \leqslant i \leqslant n)$; we shall suppose (without loss of generality) that the nonterminal sets are disjoint pairwise, and each set P_i of rules verifies conditions:

- P_i contains only rules of type $A \longrightarrow aB|\epsilon$;
- There is only one nonterminal $X_i \in V_N^i$ cu $X_i \longrightarrow \epsilon$.

The grammar $G = (V_N, V_T, S, P)$ has $V_N = \{S\} \cup \bigcup_{i=1}^{n} V_N^i$, $V_T = \bigcup_{i=1}^{n} V_T^i$, and all productions are partitioned in

- $m_0 = [S \longrightarrow S_1 S_2 \ldots S_n]$,
- $m = [A_1 \longrightarrow a_1 B_1, A_2 \longrightarrow a_2 B_2, \ldots, A_n \longrightarrow a_n B_n]$

  for all words $a_1 a_2 \ldots a_n \in L_0$ with property $A_i \longrightarrow a_i B_i \in P_i$ $(1 \leqslant i \leqslant n)$; if there are several sets of such distinct rules, new matrices

  will be generated for all variants;
- $m_1 = [X_1 \longrightarrow \epsilon, X_2 \longrightarrow \epsilon, \ldots, X_n \longrightarrow \epsilon]$ pentru $X_i \longrightarrow \epsilon \in P_i$, $(1 \leqslant i \leqslant n)$.

We define the derivation for this grammar in the usual way:

1. First, the matrix of rules m_0 is applied. Also, the last matrix used in derivation is m_1;

2. Let $\beta_1 A_1 \beta_2 A_2 \ldots \beta_n A_n$ be a sentential intermediate form. We are looking for a matrix with rules of type $[A_1 \longrightarrow a_1 B_1, A_2 \longrightarrow a_2 B_2, \ldots A_n \longrightarrow a_n B_n]$; if such a matrix doesn't exist, the derivation has failed; otherwise, the direct derivation is built:

$$\beta_1 A_1 \beta_2 A_2 \ldots \beta_n A_n \Longrightarrow \beta_1 a_1 B_1 \beta_2 a_2 B_2 \ldots \beta_n a_n B_n.$$

Example 16.2 *Let us consider regular languages* $L_1 = \{a^n | n \geqslant 1\}$, $L_2 = \{b^{2n} | n \geqslant 1\}$, $L_3 = \{c^{3n} | n \geqslant 1\}$. *The three corresponding grammars will have sets of rules*

$P_1 = \{S_1 \longrightarrow aS_1 | a\};$
$P_2 = \{S_2 \longrightarrow bA, \quad A \longrightarrow bS_2 | b\};$
$P_3 = \{S_3 \longrightarrow cB, \quad B \longrightarrow cC, \quad C \longrightarrow cS_3 | c\}.$

Let us consider the coordinator language $L_0 = \{abc\}$.

The regular grammar we define for the array representation of $L_0(L_1, L_2, L_3) = \{a^{6n} b^{6n} c^{6n} | n \geqslant 1\}$ *has the rules*

$$(S_1, S_2, S_3) \longrightarrow \begin{bmatrix} a \\ b \\ c \end{bmatrix} (S_1, A, B), \quad (S_1, A, B) \longrightarrow \begin{bmatrix} a \\ b \\ c \end{bmatrix} (S_1, S_2, C)$$

$$(S_1, S_2, C) \longrightarrow \begin{bmatrix} a \\ b \\ c \end{bmatrix} (S_1, A, S_3), \quad (S_1, A, S_3) \longrightarrow \begin{bmatrix} a \\ b \\ c \end{bmatrix} (S_1, S_2, B)$$

$$(S_1, S_2, B) \longrightarrow \begin{bmatrix} a \\ b \\ c \end{bmatrix} (S_1, A, C),$$

$$(S_1, A, C) \longrightarrow \begin{bmatrix} a \\ b \\ c \end{bmatrix} (S_1, S_2, S_3) \Big| \begin{bmatrix} a \\ b \\ c \end{bmatrix}$$

The set of nonterminals only contains these 6 *elements used in the rules we wrote above;* (S_1, S_2, S_3) *is the start symbol and* $V_T = \{a, b, c\}$.

A derivation in this grammar (for instance, for $w = a^6 b^6 c^6$*) is*

$$(S_1, S_2, S_3) \Longrightarrow \begin{bmatrix} a \\ b \\ c \end{bmatrix} (S_1, A, B) \Longrightarrow \begin{bmatrix} aa \\ b \\ cc \end{bmatrix} (S_1, S_2, C) \Longrightarrow$$

$$\begin{bmatrix} aaa \\ bbb \\ ccc \end{bmatrix} (S_1, A, S_3) \Longrightarrow \begin{bmatrix} aaaa \\ bbbb \\ cccc \end{bmatrix} (S_1, S_2, B) \Longrightarrow \begin{bmatrix} aaaaa \\ bbbbb \\ ccccc \end{bmatrix} (S_1, A, C)$$

$$\Longrightarrow \begin{bmatrix} aaaaaa \\ bbbbbb \\ cccccc \end{bmatrix}.$$

The matrix grammar which generates this language has rules:

$m_0 = [S \longrightarrow S_1 S_2 S_3], \quad m_1 = [X_1 \longrightarrow \epsilon, \; X_2 \longrightarrow \epsilon, \; X_3 \longrightarrow \epsilon],$

$m_2 = [S_1 \longrightarrow aS_1,\ S_2 \longrightarrow bA,\ S_3 \longrightarrow cB]$, $m_3 = [S_1 \longrightarrow aS_1,\ A \longrightarrow bS_2,\ B \longrightarrow cC]$,
$m_4 = [S_1 \longrightarrow aS_1,\ S_2 \longrightarrow bA,\ C \longrightarrow cS_3]$, $m_5 = [S_1 \longrightarrow aS_1,\ A \longrightarrow bS_2,\ S_3 \longrightarrow cB]$,
$m_6 = [S_1 \longrightarrow aS_1,\ S_2 \longrightarrow bA,\ B \longrightarrow cC]$, $m_7 = [S_1 \longrightarrow aS_1,\ A \longrightarrow bS_2,\ C \longrightarrow cS_3]$,
$m_8 = [S_1 \longrightarrow aX_1,\ A \longrightarrow bX_2,\ C \longrightarrow cX_3]$

The derivation for the word $a^6b^6c^6$ will be in this case:

$S \Longrightarrow S_1S_2S_3 \Longrightarrow aS_1bAcB \Longrightarrow a^2S_1b^2S_2c^2C \Longrightarrow a^3S_1b^3Ac^3S_3 \Longrightarrow a^4S_1b^4S_2c^4B \Longrightarrow a^5S_1b^5Ac^5C \Longrightarrow a^6X_1b^6X_2c^6X_3 \Longrightarrow a^6b^6c^6.$

The coordinator language L_0 and the regular form of the generative grammar defined in Proposition 16.4 allows us to obtain the usual results of decidability concerning trellis languages.

Theorem 16.1 *The membership, emptiness and infinity problems are decidable for $\mathcal{TL}_n$.*

Proof. This results from "regular" representation of corresponding grammars; using similar algorithms we construct decision problems for $\mathcal{L}_3$. □

4. RESUMES

Two of the most important current applications of cryptography consist of message authentification and electronic signatures (for details, see [5]). Both of them are using the notion of the *resume* message.

A resume $Re(\alpha)$ of the message α is a representation of this message, but with a significantly shorter length. The basic condition which must be accomplished is injectivity: for two messages α, β of the same length,

$$Re(\alpha) = Re(\beta) \quad \Longrightarrow \quad \alpha = \beta.$$

The surjectivity condition: "α can be found by knowing $Re(\alpha)$ and $|\alpha|$" is advisable, but not necessary in *resume* constructions.

In this section we will present a method of *resume* construction for trellis languages which accomplishes both conditions.

Let V be an (nonempty) alphabet. For two words $\alpha, \beta \in V^*$, $\alpha \neq \beta$, $|\alpha| = |\beta|$, we define

$$p_{\alpha,\beta} = max\{k|\alpha = \gamma ax,\ \beta = \gamma by, |\gamma| = k-1,\ a, b \in V, a \neq b\}.$$

Remark: If $\alpha \in V^+$ is a fixed and nonempty sequence and $L_\alpha = \{\beta \in V^+|\beta \neq \alpha, |\beta| = |\alpha|\}$, then $p_{\alpha,\beta}$ can be considered a mapping from L_α onto $\{1, \ldots, n\}$.

For an arbitrary language $L \subseteq V^*$ and a positive integer s, one defines

$$p_L(s) = \begin{cases} 0 & \text{if } L \text{ contains only one word of length } s \\ \infty & \text{if } L \text{ has no words of length } s \\ max\{p_{\alpha,\beta} | \alpha, \beta \in L,\ \alpha \neq \beta,\ |\alpha| = |\beta| = s\} \end{cases}$$

Remark: $p_L(s)$ depends on the length s of the words from the language L. For example, for the regular language L, defined by finite automaton

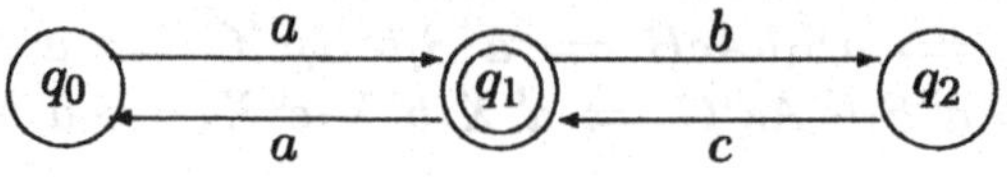

we have $p_L(2k+1) = 2k$ (hence $p_L(1) = 0$) and $p_L(2k) = \infty$.

Lemma 16.1 *If* $L \subseteq L'$ *then* $\forall\, s \geqslant 0, \quad p_L(s) < \infty \Longrightarrow p_L(s) \leqslant p_{L'}(s)$.

The proof is obvious.

Let $L_1, L_2, \ldots, L_n \in \mathcal{L}_3$ be languages generated by regular grammars $G_i = (V_N^i, V_T^i, S_i, P_i)$, and let us consider $L_0 \subset V^*$ to be a finite non-empty language, with all words of length n. Next, we shall suppose only the variant $L_0(L_1, L_2, \ldots, L_n) \neq \emptyset$.

For every $i = 1, 2, \ldots, n$ we will define the sets

$V_i = \{a \in V | \exists\, \alpha \in L_0,\ \alpha = \beta a \gamma,\ |\beta| = i-1\}$,

$R_i = \{A \longrightarrow aB \in P_i | a \in V_i\}$.

From this construction, some properties are easily proved:

- $\forall\, i\ (1 \leqslant i \leqslant n), \quad R_i \neq \emptyset$ (otherwise $L_0(L_1, L_2, \ldots, L_n) = \emptyset$).
- For every i, the regular grammar $G_{0i} = (V_N^i, V_i, S_i, R_i)$, has $L_{0i} = L(G_{0i}) \subseteq L_i$ and $L_0(L_1, L_2, \ldots, L_n) = L_0(L_{01}, L_{02}, \ldots, L_{0n})$.
- Languages L_{0i} are "commanded" by words from L_0; here we point out the coordinator role of L_0, which selects from each language L_i the useful sequences for the trellis language $L_0(L_1, L_2, \ldots, L_n)$.

Let $L_0(L_1, L_2, \ldots, L_n)$ be a trellis language over the alphabet V and L_{0i} regular languages defined in the previous section. Then:

1. $p_{L_0}(n)$ is a constant of language. Moreover, $s \neq n \Longrightarrow p_{L_0}(n) = \infty$.
2. $\forall\, s \geqslant 0,\ p_{L_{0i}}(s) \leqslant p_{L_i}(s)\ (1 \leqslant i \leqslant n)$ (Lemma 16.1).
3. $card(V_i) = 1 \Longrightarrow \forall\, s \geqslant 0,\ p_{L_{0i}}(s) \in \{0, \infty\}$.

In the following only languages L_{0i} will be used; thus we will denote them with L_i $(1 \leqslant i \leqslant n)$, without loss of generality.

Theorem 16.2 *For* $\alpha \in V^*$, $|\alpha| = ns$, *we define* $N(s) = \sum_{i=1}^{p_{L_0}(n)} p_{L_i}(s)$. *Then a resume* $Re(\alpha)$ *with* $N(s)$ *symbols, can be built for* α.

Proof. Because $p_{L_0}(n) \leqslant n$, the definition of $N(s)$ is correct.
Using the equality $|\alpha| = ns$, α can be arranged as an array with n rows and s columns

$$\alpha = \begin{matrix} a_{11} & a_{21} & \dots & a_{s1} \\ a_{12} & a_{22} & \dots & a_{s2} \\ & & \vdots & \\ a_{1n} & a_{2n} & \dots & a_{sn} \end{matrix}$$

Because $\forall i \ (1 \leqslant i \leqslant n) \quad a_{1i}a_{2i} \dots a_{si} \in L_i$, the first $p_{L_i}(s)$ symbols from row i are sufficiently in order to rebuild the whole word from L_i. Considering now the array columns, from characters of the first $p_{L_0}(s)$ rows, we can determine the words from $L_{p_n(L_0)+1} \dots, L_n$. We shall now construct the sequence

$$\beta = a_{11} \dots a_{p_s(L_1)1} a_{12} \dots a_{p_{L_2}(s)2} \dots a_{p_{L_{p_{L_0}(p_n)}}(s)) p_{L_0}(n)}.$$

It has $N(s)$ $(N(s) \leqslant ns)$ symbols; knowing β, s and $p_s(L_i)$ $(1 \leqslant i \leqslant p_n(L_0))$, it is easy to rebuild α. Therefore $Re(\alpha) = \beta$. □

Corollary 16.1 $\dfrac{N(s)}{ns} > 1 \iff L_0(L_1, L_2, \dots, L_n)$ *has no words of length* ns.

Example 16.3 *Let us consider the languages* $L_1 = a^*$, $L_2 = a^* + b^*$, $L_3 = a^*(b+c)$. *It is easy to see that* $p_{L_1}(s) = 0$, $p_{L_2}(s) = 1$, $p_{L_3}(s) = s$, $\forall\, s \geqslant 1$. *The language* $L_0 = \{aaa, aab, aba, abb, aac, abc\}$ *is the only one for which these languages are minimal. It has* $p_{L_0}(3) = 3$.

A word $\alpha = \alpha_1\alpha_2\alpha_3 \in L_0(L_1, L_2, L_3)$, $|\alpha| = 3s$ *has the resume* $Re(\alpha) = first_1(\alpha_2)\alpha_3$ *of length* $s+1$, *hence we can keep all the information by reducing three times its length.*

Using the sequence $\beta = Re(\alpha)Re(\alpha)Re(\alpha)$ *instead of* α, *some advantages occur. For example:*

1. *From* $|\beta| = 3s + 3$, *one obtains* $|\alpha| = 3s$. *First* $s+1$ *symbols of* β *forms* $Re(\alpha) = x\alpha_3$ $(x \in \{a, b\}, \ \alpha_3 \in L_3)$. *Therefore* $\alpha = \gamma_1\gamma_2\gamma_3$, *where* $\gamma_1 = a^s$, $\gamma_2 = x^s$, $\gamma_3 = \alpha_3$.

2. *By using error-correcting algorithms, characteristic of Coding theory ([1]),* $Re(\alpha)$ *can be obtained from* β, *even up to* 66% *of symbols from* β *are*

perturbated ("erased"). This is a remarkable "holographic" property of these resumes.

As we have seen, the limit $p_L(n)$ of a regular language L plays an important role in *resume* construction. If the number $p_L(n)$ of values for which $p_L(n) \neq \infty$ is finite, then the length of L's words resumes is significantly smaller than the lengths of words. A characterization criterion of this number of values $p_L(n)$ is given by the following theorem:

Theorem 16.3 *Let us consider the language $L \in \mathcal{L}_3$ and let be the set $\mathcal{N}_L = \{m \in \mathcal{N} | p_L(m) < \infty\}$. Then*

$$\lim_{n\to\infty,\, n\in\mathcal{N}_\mathcal{L}} \frac{p_L(n)}{n} = 0 \iff \exists\ M_L \text{ with } p_n(L) \leqslant M_L,\ \forall\ n \in \mathcal{N}_L.$$

Proof. The set of integers n with the property $p_L(n) = 0$ forms a substring with elements from $\mathcal{N}_L$; for these elements, L has only one word with length equal to n, and all assertions are obviously verified.

Let us consider now an integer n for which exists $\alpha, \beta \in L,\ \alpha \neq \beta,\ |\alpha| = |\beta| = n$ and $p_{\alpha,\beta} = p_L(n) = max\{p_{x,y} |\ x, y \in L,\ |x| = |y| = n\} = k$.

Therefore $\alpha = a_1 a_2 \dots a_n,\ \beta = b_1 b_2 \dots b_n$ with $a_i = b_i\ (i < n),\ a_k \neq b_k$.

" $\Longrightarrow$": Let $M = (Q, V, \delta, q_0, F)$ be a finite deterministic automaton which accepts L, hence there will be a (single) path $(q_0, q_1, \dots, q_{k-1})$ with $\delta(q_i, a_{i+1}) = q_{i+1}\ (0 \leqslant i \leqslant k-2),\ \delta(q_{i-1}, a_i) \neq \delta(q_{i-1}, b_i)$, denoted by common prefixes of length $k-1$ from α and β.

In this path, all automaton states are distinct. Otherwise, one can write $\alpha = uvwx,\ \beta = uvwy,\ v \neq \epsilon,\ |x| = |y|,\ u, v, w$ with properties from Pumping lemma; moreover, x, y differ in their first position. Then $\forall\ s \geqslant 0,\ \alpha_s = uv^s wx,\ \beta_s = uv^s wy \in L$. We shall have $n_s = n + (s-1)|v| \in \mathcal{N}_L,\ p_{n_s} = n + (s-1)|v| - |x|$ with $|v|, |x|$ constant. Applying the limit, one obtains

$$\lim_{s\to\infty} \frac{p_L(n_s)}{n_s} = \lim_{n\to\infty} \frac{n + (s-1)|v|}{n + (s-1)|v| - |x|} = 1 \neq 0,$$

in contradiction to the hypothesis.

Therefore, in the common path, the first k states are distinct; $k \leqslant card(Q)$ results. By considering now the minimal finite automaton M which accepts L, and writing $M_L = card(Q)$, the proof is complete.

" $\Longleftarrow$": Similarly. □

The Theorem 16.3 has several corolaries, such as:

Corollary 16.2 *The series $(a_n)_{n\in\mathcal{N}_\mathcal{L}},\ a_n = \dfrac{p_L(n)}{n}$ is convergent if and only if $p_n(L)$ verifies Theorem 16.3. In this case its limit is 0.*

Corollary 16.3 *The series $(p_L(n))_{n\in\mathcal{N}_\mathcal{L}}$ is a finite union of periodical sub-series.*

5. SOME PROBLEMS CONCERNING THE CONSTRUCTION OF TRELLIS LANGUAGES

In general, for trellis language construction, there are two important problems concerning $N(s)$, namely:

1. Let us consider the languages $L_0, L_1, \ldots, L_n$ with the required properties; how can a trellis arrangement of them be found, minimising $N(s)$ $(s \geqslant 0)$?
2. There are $L_1, L_2, \ldots, L_n \in \mathcal{L}_3$; how can a finite language L_0 which minimises $N(s)$ $(s \geqslant 0)$ be built?

In order to solve these problems, a rearrangement of languages $L_1, L_2, \ldots, L_n$ is used; the goal is to minimise the limit $p_{L_0}(n)$, through an eventual interchange of symbols from L_0. Thus, let

$L_0 = \{a_{11}a_{21}\ldots a_{n1}, a_{12}a_{22}\ldots a_{n2}, \ldots, a_{1p}a_{2p}\ldots, a_{np}\}$, $1 \leqslant p \leqslant [card(V)]^n$.

If $\pi \in S_n$ is an arbitrary interchange of n elements, then we can define

$\pi(L_0) = \{a_{\pi(1)1}a_{\pi(2)1}\cdots a_{\pi(n)1}, a_{\pi(1)2}a_{\pi(2)2}\cdots a_{\pi(n)2}, \ldots, a_{\pi(1)p}a_{\pi(2)p}\cdots, a_{\pi(n)p}\}$.

We shall choose $L_0' = \pi(L_0)$ the language with specific property

$$p_{L_0'}(n) = \min_{\sigma\in S_n} p_{\sigma(L_0)}(n)$$

Starting with this coordinator language and the permutation π, the regular languages $L_{01}, \ldots, L_{0n}$ can be defined, and then the trellis language given by Definition 16.1 is built.

Because this problem of $N(s)$ minimization also depends on the length ns of the words, a supplementary "weight" rearrangement of languages L_i on $p_{L_i}(s)$ $(1 \leqslant i \leqslant n,\ s \geqslant 1)$ will be necessary.

We shall present the results concerning these problems in a forthcoming paper.

References

[1] Adamek, J. (1991), *Foundation of Coding*, Wiley - Interscience, New York.

[2] Csuhaj-Varjú, E.; J. Dassow; J. Kelemen & Gh. Păun (1994), *Grammar Systems*, Gordon and Breach, London.

[3] Honary, B. & G. Markarian (1997), *Trellis Decoding of Block Codes*, Kluwer, Dordrecht.

[4] Hopcroft, J.E. & J.D. Ullman (1979), *Introduction to Automata Theory, Languages and Computation*, Addison-Wesley, Reading, Mass.

[5] Stinton, D. (1995), *Cryptographie, Théorie et pratique*, Intern. Thomson Publ., France.

Chapter 17

PICTURES, LAYERS, DOUBLE STRANDED MOLECULES: ON MULTI-DIMENSIONAL SENTENCES

Paolo Bottoni
Department of Computer Science
Pictorial Computing Laboratory
University of Rome "La Sapienza"
Via Salaria 113, 00198 Roma, Italy
bottoni@dsi.uniroma1.it

Abstract We discuss three examples of multi-dimensional sentences, to show problems deriving from a straightforward extension of one-dimensional notions, such as shuffle. A conjecture is proposed postulating the irreducibility of pictures to strings.

1. INTRODUCTION

In the last few years, the field of formal languages has been revitalised by the impact of: new communication media – especially exploiting pictures; new programming paradigms – especially distributed ones; and new computational paradigms – usually parallel ones. The formal language community has kept pace with these developments by providing formal models to study the ways in which sentences defining or resulting from computations in these new models can be generated, transformed, or recognised.

An interesting aspect of these studies is that in several cases they have to take into account the spatial arrangement of the symbols occurring in the sentences. Moreover, in these cases a special symbol, usually deemed *transparent* or *blank*, is required to act as a place-holder to be later filled by a symbol in a terminal alphabet, in order to maintain the spatial organisation of the sentence.

In this paper we explore the consequence of considering dimensionality and spatial arrangements in the formal definition of languages. Therefore we are interested in sentences which have a dimensional and spatial organisation, and in

C. Martin-Vide and V. Mitrana (eds.), Where Mathematics, Computer Science, Linguistics and Biology Meet, 199-209.

defining operations which preserve this organisation. This brings us to the observation that when we depart from the one-dimensionality of the usual strings, the single operation of concatenation is not sufficient to construct all possible sentences from a finite set of generators.

On the other hand, a multidimensional sentence can be expressed in terms of arrangements of constituting strings. This suggests a study of how usual operations defined on linear strings, such as concatenation, shuffle, insertion, etc. can be extended to the new structures. It appears that, at least in the cases considered here, we obtain different results according to whether we apply the operations to the constituting strings and then try to reconstruct a sentence according to the required organisation, or we try to directly define a multi-dimensional extension of the operation.

Paper outline. In Section 2, we first introduce three models of multidimensional sentence, the first two (pictures and layers) defined by the author in cooperation with Gheorghe Păun and other co-authors [1], [3], [2], the third being the well-known double stranded sequences at the basis of DNA computing [6]. In Section 3. we show the consequences of multidimensionality when trying to extend the *shuffle* operation to these multi-dimensional structures. In Section 4., we adapt the recently proposed notion of *shuffle with trajectory* [5] to these structures. Finally, we draw some conclusions and discuss future work and open problems.

2. MULTIDIMENSIONAL WORDS

Let $[\cdot] : \mathbf{N} \to \mathcal{P}(\mathbf{N})$ be the function defined by $[n] = \{1, \ldots, n\}$ if $n \geqslant 1$, $\{0\}$ otherwise. In the sequel, notation $[n]$ is a shorthand for $\{1, \ldots, n\}$. Given an alphabet of symbols V, we call V' the set $V \cup \{\tau\}$, where τ is a designated symbol, called the *transparent* symbol. For standard notions and notation on formal languages, we refer to [7].

Pictures. A *picture* π on V is a function $\pi : [\overline{r}] \times [\overline{c}] \to V'$ where $\overline{r}$ and $\overline{c}$ are two integers respectively called *height* and *width* of the picture. Given a picture π, the function $\pi_i^r : [\overline{c}] \to V'$ is defined by $\pi_i^r(j) = \pi(i,j)$, and π_i^r denotes the i-th *row* of π. Analogously $\pi_i^c : [\overline{r}] \to V'$ is defined by $\pi_i^c(j) = \pi(j,i)$ and π_i^c denotes the i-th *column* of π. An element $(r, c, \pi(r,c))$ is called a *pixel* of π. If $\pi(r,c) = \tau$ the pixel is said to be *transparent*, otherwise *visible*. A *pictorial language* PL on V is a subset of $\Pi(V)$, the set of all pictures on V (i.e. without transparent symbols).

Let $\pi : [\overline{r}] \times [\overline{c}] \longrightarrow V'$ be a picture, and r_1, r_2, c_1, c_2 four integers with $0 \leqslant r_1 \leqslant r_2 \leqslant \overline{r}$ and $0 \leqslant c_1 \leqslant c_2 \leqslant \overline{c}$. We say that the function $\pi_s : [\overline{r_2}] \times [\overline{c_2}]$ defined by $\pi_s(i,j) = \pi(i,j)$ if $r_1 \leqslant i \leqslant r_2$ and $c_1 \leqslant j \leqslant c_2$, τ otherwise, is a *subpicture* of π (in symbols $\pi_s \lhd \pi$). If $r_1 = c_1 = 1$, π_s is called a *prefix* of π.

The two operations $shift$ and sup allow the generation of all pictures in any PL by composing elementary images from a finite set, including the special images *row*:$[1] \times [0] \longrightarrow \{\tau\}$ and *col*:$[0] \times [1] \longrightarrow \{\tau\}$ [1].

Let π_1 and π_2 be two pictures $\pi_i : [\overline{r_i}] \times [\overline{c_i}] \longrightarrow V'$, for $i = 1, 2$. $shift(\pi_1, \pi_2) = \pi_3 : [\overline{r_1} + \overline{r_2}] \times [\, overline c_1 + \overline{c_2}] \rightarrow V'$ is defined as

$$\pi_3(r,c) = \begin{cases} \pi_2(r - r_1, c - c_1) & \text{if } \overline{r_1} < r \leqslant \overline{r_1} + \overline{r_2} \wedge \overline{c_1} < c \leqslant \overline{c_1} + \overline{c_2} \\ \tau & \text{otherwise} \end{cases}$$

$sup(\pi_1, \pi_2) = \pi_3 : [max(\overline{r_1}, \overline{r_2})] \times [max(\overline{c_1}, \overline{c_2})] \rightarrow V'$ is defined as

$$\pi_3(r,c) = \begin{cases} \pi_1(r,c) & \text{if } \pi_1(r,c) \in V \\ \pi_2(r,c) & \text{if } ((r,c) \notin [\overline{r_1}] \times [\overline{c_1}] \vee \pi_1(r,c) = \tau) \wedge \pi_2(r,c) \in V \\ \tau & \text{otherwise} \end{cases}$$

The picture $\Lambda : \{0\} \times \{0\} \rightarrow \{\tau\}$ (i.e. the picture with no pixel) is the neutral element for sup and the left neutral element for $shift$.

Layers. For two strings $x, y \in V'^*$ we denote by $[x, y]$ the two-level sequence obtained by placing x over y, justifying to left and completing the shortest string with occurrences of τ; $[x, y]$ is called a *layered string*. Given a layered string $[x, y]$, any symbol $x(i) \in V$ is said to be *observable*. A symbol $y(i) \in V \cup \{\tau\}$ is *observable* if and only if $x(i) = \tau$. If $y(i) \in V$, then $y(i)$ is also *visible*. Given a layered string, the operation $\diamond : V'^* \times V'^*$ is defined as $(x \diamond y)(i) = x(i)$ if $x(i) \in V$, $y(i)$ if $| x | < i \leqslant | y | \vee x(i) = \tau$, i.e. it is the one-dimensional version of sup. The unique string $(x \diamond y)$ is said to be *associated* with the layered string $[x, y]$. Properties of layered strings and of languages generated by grammars on them were studied in [2].

DNA sequences. Let $\rho \subseteq V \times V$ be a symmetrical relation, let (V, V) denote the set $\{(a, b) \mid a, b \in V, (a, b) \in \rho\}$. We use the notation (w_1, w_2) to indicate a pair of strings on V such that $| w_1 |=| w_2 |= k$ and $(w_1(i), w_2(i)) \in (V, V)$ for $i \in [k]$. We call $WK_\rho(V)$ the set of all such strings.

Besides these "complete" sentences, one also has the sets $L_\rho(V)$, $R_\rho(V)$ and $LR_\rho(V)$, of sentences obtained by concatenating sequences of pairs (x, τ) or (τ, x) to the left, to the right, and to both left and right of sentences in $WK_\rho(V)$, respectively. Then $W_\rho(V) = L_\rho(V) \cup R_\rho(V) \cup LR_\rho(V)$ [6].

A *DNA sequence* is a sentence in $WK_\rho(V)$. A sentence in $W_\rho(V)$ is a *sequence with sticky ends*. For a survey on languages on these sequences, see [6].

3. SHUFFLE ON MULTI-DIMENSIONAL SENTENCES

The *shuffle* operation maps pairs of strings into sets of strings $ш : V^* \times V^* \longrightarrow \mathcal{P}(V^*)$ and for $x, y \in V^*$ the *shuffle* of x, y is the set

$$x \sqcup\!\sqcup y = \{x_1 y_1 \ldots x_n y_n \mid x = x_1 \ldots x_n, y = y_1 \ldots y_n, x_i, y_i \in V^*\}$$

An equivalent recursive definition is: $(au \sqcup\!\sqcup bv) = a(u \sqcup\!\sqcup bv) \cup b(au \sqcup\!\sqcup v)$; $(u \sqcup\!\sqcup \lambda) = (\lambda \sqcup\!\sqcup u) = u$. For sets we have $L_1 \sqcup\!\sqcup L_2 = \bigcup_{u \in L_1, v \in L_2} u \sqcup\!\sqcup v$.

The set of strings $x \sqcup\!\sqcup y$ is characterised by the following properties: 1) each string in $x \sqcup\!\sqcup y$ uses all the symbols in both x and y; 2) the linear orders on occurrences of symbols in the two strings are maintained in all the strings in $x \sqcup\!\sqcup y$; 3) no other symbol is used. As a result of points 1) and 3) above, the length of all the strings w in $x \sqcup\!\sqcup y$ is the same and $\mid w \mid = \mid x \mid + \mid y \mid$.

In the next subsections we study under which conditions two-dimensional versions of shuffle maintain these properties for the three models of *pictures* (Section 3.1), *layers* (Section 3.2), and *DNA molecules* (Section 3.3). Conversely, we discuss how the relaxation of either of these properties produces new definitions.

3.1 PICTURES

If π_1 and π_2 are two pictures, the three properties for the linear shuffle operation can be maintained only if the following conditions hold:

1. $\overline{r_1} = \overline{r_2}$ or $\overline{c_1} = \overline{c_2}$. Let σ be the dimension in which the sizes are equal, ρ the other direction, i.e. $\{\sigma, \rho\} = \{r, c\}$.

2. for each picture π in $\pi_1 \sqcup\!\sqcup \pi_2$, for each string π_i^ρ in π, π_i^ρ is also a string π_{1j}^ρ or π_{2j}^ρ for some j.

3. For each picture π in $\pi_1 \sqcup\!\sqcup \pi_2$, for each string π_i^σ in π, $\pi_i^\sigma \in \pi_{1i}^\sigma \sqcup\!\sqcup \pi_{2i}^\sigma$.

The need to maintain the linear ordering of the occurrence of symbols also in direction ρ, as expressed in the condition 2 above, requires, however, that all the decompositions of the σ strings be consistent along the ρ direction, i.e. for each $j \in [\overline{\sigma}]$, $\pi_{kj}^\sigma = w_1^{k,j} \ldots w_s^{k,j}$, and $\mid w_i^{k,m} \mid = \mid w_i^{k,n} \mid$ for all $m, n \in [\overline{\sigma}]$, $i \in [s]$ and $k \in \{1, 2\}$. If $\overline{r_1} = \overline{r_2}$ and $\overline{c_1} = \overline{c_2}$, then the set $\pi_1 \sqcup\!\sqcup \pi_2$ is built by taking the union of the two sets constructed as described above for each direction.

Basically, this definition derives from considering a picture as the row (column) concatenation of row (column) strings, and the shuffle of two pictures results from the row (column) concatenation of shuffles of row (column) strings.

In this way, however, we define shuffle as a partial function, which can operate only on some pairs of pictures. The way to obtaining a total function again is to relax some of the properties of shuffle valid for the linear case. This can be achieved in two ways.

Not using all symbols. In this case one considers the maximal prefixes of the pictures having the same size in at least one direction. Formally, given the pictures π_1, π_2, we consider their respective restrictions ξ_1 and ξ_2 of size $\overline{r_{max}} \times \overline{c_{min}}$, and χ_1 and χ_2 of size $\overline{r_{min}} \times \overline{c_{max}}$, where $\overline{\sigma_{max}} = \max(\overline{\sigma_1}, \overline{\sigma_2})$ and $\overline{\sigma_{min}} = \min(\overline{\sigma_1}, \overline{\sigma_2})$ for $\sigma \in \{r, c\}$. To these two pairs of pictures one applies the shuffle operation as defined above and takes the union of the resulting sets.

Using new symbols. In this case one considers the pictures of minimal size sufficient to accommodate both π_1 and π_2 (i.e. one considers pictures of size $\max(\overline{r_1}, \overline{r_2}) \times \max(\overline{c_1}, \overline{c_2})$) and uses transparent symbols to fill the newly created positions in all possible ways. One then takes the union of all the shuffle of pairs in the resulting sets, using the definition of shuffle above for both directions.

In these cases we still regard a picture as the concatenation in one direction of strings oriented in the orthogonal direction. The final extension one could propose, to take into account the bidimensional nature of pictures, is therefore:
Not maintaining order in both directions. Any of the definitions above can be modified so as to eliminate the constraint $| w_i^{k,m} | = | w_i^{k,n} |$.

Such a situation, where the immediate extension of an operation from one to two dimensions allows the definition of only partial functions, is analogous to the case of concatenation. Indeed, the pair of operations proposed in the literature as horizontal and vertical versions of concatenation, while allowing the generation of all pictures in $\Pi(V)$, suffer from being formed by partial functions [8] (for a survey, see [4]). Our proposal of the pair *shift* and *sup* is motivated by having a pair of total functions able to generate all pictures in $\Pi(V)$. Moreover, in [3], it has been shown how this pair allows the expression of the pair formed by horizontal and vertical concatenation, realising a definite economy of notation.

In any case, the fact that the sufficiency of a pair of operations to generate $\Pi(V)$ has been proved for different pairs, while no single operation has been defined to this effect, suggests the following conjecture.

Conjecture 17.1 *There cannot exist a single binary operation on pictures (i.e. a function* $\gamma : \Pi(V) \times \Pi(V) \longrightarrow \Pi(V)$*) such that it has the following properties: 1)* γ *is total; 2)* γ *is associative; 3)* Λ *is the neutral element for* γ*; 4)* γ *generates* $\Pi(V)$ *starting from a finite set of generators.*

Stated otherwise, proving this conjecture would amount to establishing that $\Pi(V)$ is not a free monoid.

In the linear case, the shuffle operation is inherently connected to concatenation and the definitions above have somehow preserved such a connection. We now propose a notion of two-dimensional shuffle deriving from the operations of *shift* and *sup* which instead appear as more natural operations for pictures.

Let $S_{\pi_2}(\pi_1) = \{\pi \mid \exists r \in [\overline{r_2}], c \in [\overline{c_2}]$ s.t. $\pi = \mathit{shift}(z, \pi_1)$, with $z : [\overline{r}] \times [\overline{c}] \longrightarrow \{\tau\}\}$. $S_{\pi_2}(\pi_1)$ is therefore the set of all pictures obtained by shifting π_1 according to some subpicture of π_2.

Let $Q(\pi) = \{\pi_s \mid \pi_s \lhd \pi\}$, i.e. the set of all subpictures of π. A set $\Pi \subset Q(\pi)$ is said to be a *partition* of π (in symbols $\Pi \bowtie \pi$) if and only if $sup(\Pi) = \pi$ and $\sum_{\rho \in \Pi} \mid \rho \mid_V = \mid \pi \mid_V$. We can now define $\pi_1 ш_t \pi_2$. Let us first construct a set F of pictures of the form $\mathit{shift}(\pi_a, \pi_b)$ with $\pi_a \in Q(\pi_1)$ and $\pi_b \in \Pi_2$ for some $\Pi_2 \bowtie \pi_2$, taking each π_b only once. Analogously, we build a set S of pictures of the form $\mathit{shift}(\pi_c, \pi_d)$ with π_c a substructure of some $\pi_b \in \Pi_2$ and $\pi_d \in \Pi_1$ for some $\Pi_1 \bowtie \pi_1$, taking each π_b only once as a source for π_c. Finally, one takes the superposition of pictures of the form $sup(\pi_f, \pi_s)$ with $\pi_f \in F$ and $\pi_s \in S$ and such that if $\pi_f = \mathit{shift}(\pi_a, \pi_b)$, $\pi_s = \mathit{shift}(\pi_c, \pi_d)$ with $\pi_c \lhd \pi_b$.

In this case property 1) of one-dimensional shuffle is preserved. Property 2 is replaced by the weaker property that for each picture in $\pi_1 ш_t \pi_2$, if a pixel (i, j, v) appears in π_1 or π_2, the coordinates of the corresponding pixel (k, l, v) are such that $i \leqslant k$ and $j \leqslant l$. Finally, property 3) is replaced by the weaker property $\mid \pi_1 ш_t \pi_2 \mid_V = \mid \pi_1 \mid_V + \mid \pi_2 \mid_V$. Consequently, differently from the previous versions of shuffle introduced above, the size of the pictures in the shuffle set can vary. More precisely, the shuffle set contains pictures for each size from $(\overline{r_1} + \overline{r_2}, \max(\overline{c_1}, \overline{c_2}))$ to $(\max(\overline{r_1}, \overline{r_2}), \overline{c_1} + \overline{c_2},)$, through $(\overline{r_1} + \overline{r_2}, \overline{c_1} + \overline{c_2})$. It can be noted that these versions of properties 2) and 3) reduce to the original properties for the one-dimensional case.

Each version of shuffle acting on individual rows or columns considered above, such that it preserves the property of using all symbols from π_1 and π_2, produces a proper subset of the pictures produced by $ш_t$. However, $ш_t$ is forced to produce elements not in $\Pi(V)$, but in $\Pi(V')$ and is able to produce elements in $\Pi(V)$ only if π_1 and π_2 have the same size in at least one dimension. Even in this case, it can produce elements in $\Pi(V')$. Hence, $ш_t$ alone is not a possible generating function for $\Pi(V)$.

The negative result above holds also for the case when π_1 and π_2 are pictures with only one row. Even in this case, $ш_t$ can produce pictures with two rows. This brings to an important remark: contrary to the previous definitions of shuffle based on management of constituting strings, $ш_t$ does not reduce to linear shuffle when applied to one-row pictures. The same holds for the *shift* operation. Hence, even though there exists a one-to-one correspondence between one-row pictures (and one-column pictures as well) and strings, strings must be considered as defined only in one dimension, i.e. symbols are not considered to have a size in the vertical direction.

3.2 LAYERS

One can consider a layered string as the row concatenation of strings and obtain two different definitions of shuffle according to whether one considers the resulting vertical strings as unit elements or one treats the horizontal strings as independent objects.

In the first case, equality of positions in the two layers is seen as a form of synchronisation inherent in the layered string structure. Otherwise, the strings in each layer can be independently shuffled and then the layers reconstructed.

In the *synchronised* case, $[x_1, y_1] ш_s [x_2, y_2] = \{w \mid w = \alpha_1\beta_1\alpha_2\beta_2 \dots \alpha_n\beta_n, \alpha_1 \dots \alpha_n = [x_1, y_1], \beta_1 \dots \beta_n = [x_2, y_2]\}$. In the *independent* case, $[x_1, y_1] ш_i [x_2, y_2] = \{(w, z) \mid w \in x_1 ш x_2$ and $z \in y_1 ш y_2\}$.

One can then apply the operation $\diamond$ to the sets resulting from synchronised or independent shuffle. We indicate such a composition with $(\diamond, ш_\alpha)$ for $\alpha \in \{s, i\}$. Conversely, $(ш, \diamond)$ indicates that $\diamond$ is first separately applied to two layered strings and then $ш$ is applied to the resulting strings. We call *continuity* the property owned by a layered string $[x, y]$ such that $x \diamond y \in V^*$, (i.e. no pair (τ, τ) exists in $[x, y]$.

The relations between layered versions of shuffle and superposition are summarised by the following theorem.

Theorem 17.1 *The following properties hold:*

1. $(\diamond, ш_s)([x_1, y_1], [x_2, y_2]) \subset (\diamond, ш_i)([x_1, y_1], [x_2, y_2])$.

2. *if* $[x_1, y_1]$ *and* $[x_2, y_2]$ *are both continuous,* $[x_1, y_1] ш_s [x_2, y_2]$ *is continuous.*

3. $(\diamond, ш_i)([x_1, x_2], [y_1, y_2])$ *is continuous only if* $| x_1x_2 |_\tau = 0$ *or* $| y_1y_2 |_\tau = 0$.

4. $(ш, \diamond)(L_1, L_2) = (\diamond, ш_s)(L_1, L_2)$ *for each pair of languages of layered strings* L_1, L_2.

Proof. The proof of 1 is immediate from the definition. For 2, if a layered string is continuous, then if $x(i) = \tau$, it must be $y(i) \neq \tau$. Since $x(i)$ and $y(i)$ are moved together by $ш_s$, the property is preserved. For 3., if an occurrence of τ, exists in both x_j and y_k, for $j, k \in \{1, 2\}$ then a string with $[\tau, \tau]$ can be generated by $ш_i$, which is not continuous. Hence, at least one of x_1x_2 and y_1y_2 must not contain τ. For 4., let $[x_1, y_1] \in L_1$, $[x_2, y_2] \in L_2$ be two layered strings. Then all elements $[w', w'']$ in $[x_1, y_1] ш_s [x_2, y_2]$ are either in $[x_1, y_1]$ or in $[x_2, y_2]$. Hence, one of w' or w'' is in $x_1 \diamond y_1$ or in $x_2 \diamond y_2$, and hence in $(ш, \diamond)([x_1, x_2], [y_1, y_2])$. □

The theorem above suggests to consider $ш_s$ rather than $ш_i$ as the natural extension of $ш$ to layered strings. On the other hand, layered strings propose a situation in which using two dimensions is essential to understand the sentence structure, but composition of sentences can occur only in one direction. In this case, the correct definition of *shift* would be *shift*$([x_1, y_1], [x_2, y_2])(i- \mid x_1 \mid) = [x_2(i), y_2(i- \mid x_1 \mid)]$ if $\mid x_1 \mid < i \leqslant \mid x_1 \mid + \mid x_2 \mid$, τ otherwise, while *sup* would produce layers of length $\max(\mid x_1 \mid, \mid x_2 \mid)$. On the whole, layered strings seem to be a model of $1\frac{1}{2}$-dimensional sentences, where the second dimension is important to define the resulting string, but not to build new strings.

3.3 DNA

Again, one can consider a DNA sequence as the row concatenation of two strings, constrained by the complementarity relations. In this case the three properties of linear shuffle are preserved only if the shuffle for DNA sequences is defined so as to move only complementary pairs of symbols in the two strands. Formally, given a DNA sequence (w_1^1, w_2^1), we define its shuffle with the sequence (w_1^2, w_2^2), as the set of pairs (w_1, w_2) with $w_1 = \alpha_1^1\alpha_1^2 \ldots \alpha_n^1\alpha_n^2$, $w_2 = \beta_1^1\beta_1^2 \ldots \beta_n^1\beta_n^2$, for some decomposition $w_1^1 = \alpha_1^1 \ldots \alpha_n^1$, $w_2^1 = \beta_1^1 \ldots \beta_n^1$ and $w_1^2 = \alpha_1^2 \ldots \alpha_n^2$, $w_2^2 = \beta_1^2 \ldots \beta_n^2$, and $(\alpha_i^k, \beta_i^k) \in WK_\rho(V)$, for $k \in \{1, 2\}$.

On the other hand one could produce independent decompositions of the two strings in the DNA sequence (by using endonucleases followed by denaturation), and recombine the sequences via reannealing in such a way that possible resulting violations of complementarity are resolved by generating new complementary symbols (by using polymerases).

Formally, given (w_1^1, w_2^1) and (w_1^2, w_2^2) as above, one obtains the set of pairs $(w_1, w_2) \in WK_\rho(V)$ with $w_1 = \gamma_1\alpha_1^1\gamma_2\alpha_1^2 \cdots \gamma_{(2\times n)-1}\alpha_n^1\gamma_{2\times n}\alpha_n^2\gamma_{(2\times n)+1}$, $w_1 = \delta_1\beta_1^1\delta_2\beta_1^2 \ldots \delta_{(2\times n)-1}\beta_n^1\delta_{2\times n}\beta_n^2\delta_{(2\times n)+1}$, where α_i^k and β_i^k derive from decompositions of w_l^k for $k, l \in \{1, 2\}$, $\alpha_i^k\beta_i^k \neq \lambda$ for each i, and $\gamma_r \neq \lambda \Rightarrow \delta_r = \lambda$ and $\delta_r \neq \lambda \Rightarrow \gamma_r = \lambda$, for $r \in \{1, (2 \times n) + 1\}$ (they can all be λ, anyway).

One could develop a reasoning similar to the case of layered strings to conclude that DNA sequences constitute a model of $1\frac{1}{2}$-dimensional sentences which are somehow dual with respect to layers, where the second dimension is important to build new strings (through sticky ends), but not to define the resulting string (which can be reconstructed only by looking at either the upper or lower sequence). Also in this case the "synchronised" version of shuffle would seem a more natural extension.

4. SHUFFLE ON TRAJECTORIES AND MULTI-DIMENSIONAL SENTENCES

A recent proposal removes the non determinism inherent in the definition of the shuffle operation, by prescribing the use of a *trajectory* to specify the decompositions of x and y [5]. A trajectory is a word on the alphabet $V =$

$\{u, r\}$, where u and r are the *versors* for the *up* and *right* directions respectively. The shuffle of x and y on the trajectory $t \in V^*$ is a partial function $\amalg_t$: $V^* \times V^* \longrightarrow V^*$ where $x \amalg_t y$ is defined if and only if $| x |=| t |_r$ and $| y |=| t |_u$. In this case, $(x \amalg_t y)(i) = x(k_1 + 1)$ if $t_i = r$ and $y(k_2 + 1)$ if $t_i = u$, where $k_1 =| t_1 \dots t_{i-1} |_r$ and $k_2 =| t_1 \dots t_{i-1} |_u$. This corresponds to viewing x and y as axes of a Cartesian reference system and a trajectory as a walk in the resulting (finite) space, picking up a symbol from x each time we move to the right and a symbol from y each time we go up.

In [5], the definition is extended to sets of trajectories, by considering the union of all shuffles on all trajectories in the set, and to sets of languages in the natural way. In [5], it is also shown how all the usual operations on strings can be defined in terms of shuffles on a set T of trajectories. For example, usual shuffle is obtained by considering $T = \{r, u\}^*$, while concatenation is obtained by taking $T = r^*u^*$.

This introduction of a two-dimensional notion in the world of linear strings is particularly interesting for this paper. The reader can easily extend the notion of shuffle on trajectories to the case of layered strings and DNA sequences, bearing in mind the differences deriving by using trajectories for selection of paired symbols in a synchronised way or of symbols from the constituent strings independently. We study how such a notion can be extended to pictures.

We develop the definition for the case of pictures with equal number of rows; the reader can easily extend it for pictures with equal number of columns, or of the same size. Let $D = \{f, s\}$ an alphabet. Given two pictures π_1 : $[\overline{r}] \times [\overline{c_1}] \longrightarrow V$ and $\pi_2 : [\overline{r}] \times [\overline{c_2}] \longrightarrow V$, we call a *trajectory picture*, a picture $tp : [\overline{r}] \times [\overline{c_1} + \overline{c_2}] \longrightarrow D$ such that $| tp_i^r |_f = \overline{c_1}$, $| tp_i^r |_s = \overline{c_2}$, for $i \in [\overline{r}]\}$.

The shuffle of π_1 and π_2 *on the trajectory picture tp* is a picture $\pi_1 \amalg_{tp} \pi_2$ s.t. $(\pi_1 \amalg_{tp} \pi_2)_i^r = {\pi_1}_i^r \amalg_{t_i} {\pi_2}_i^r$ with $t_i = tp_i^r$ for $i \in [\overline{r}]\}$.

One can notice how a trajectory can be applied also to the case of the operation $\amalg_t$ introduced in section 3.1, by considering a trajectory $tp : [\overline{r}] \times [\overline{c}] \longrightarrow D$, with $r_1 \leqslant r \leqslant r_1 + r_2$, $c_1 \leqslant c \leqslant c_1 + c_2$ s.t. $| tp |_f = | \pi_1 |$ and $| tp |_s = | \pi_2 |$. For pairs of pictures with the suitable sizes, horizontal and vertical concatenation result by considering trajectories where all fs precede all ss in the corresponding direction. One can generalise this definition by considering trajectories where pictures of all fs or all ss have the same sizes as pictures π_1 and π_2 respectively and are suitably shifted. An analogous of anticatenation is obtained by making ss precede fs.

5. CONCLUSIONS

The attempt to introduce versions of the shuffle operation on languages with inherent multi-dimensionality has shown how sentences in these languages cannot be simply considered as spatial arrangements of strings. Indeed the preservation of the spatial arrangement of the sentence in some situation forces one to violate either the linear order or the continuity or the integrity of the constituent strings. This is particularly evident in the case of pictures, but provokes interesting phenomena also in the case of layered strings and of DNA sequences.

On the other hand, the notion of shuffle with trajectories highlights how even operations on linear strings actually require the use of two dimensions in order to make space for the modifications. In order to realise shuffle or shuffle on trajectories in two dimensions, however, there is no need of a third dimension. Space can be obtained by simply shifting subpictures and then superposing them on other subpictures. This is important in that it suggests that a greater degree of parallelism can be achieved for two dimensions than for one dimension. In fact, in one direction, the need to reconstruct the order of the string limits the possibility of full parallelism, for instance in the realisation of the shuffle operation. These issues deserve further investigation of the efficiency and the feasibility of operations on multidimensional words.

Based on the problems in defining two-dimensional versions of concatenation and shuffle, this paper has proposed a conjecture to the effect that the set of pictures is not a free monoid, but derives from a free construction using two operations, to be exact. Even though this is somehow implicit in most literature on images, to the best of our knowledge, this has never been stated explicitly. It appears that versions of this conjecture can be adapted for different operations defined on string languages, such as mirror image, anticatenation, different versions of shuffle. The algebraic nature of the set of pictures remains to be explored.

Acknowledgments

I am indebted to D. Vaida for drawing my attention to [5]. I have had many fruitful discussions with G. Mauri, P. Mussio, Gh. Păun, and D. Vaida on the algebraic structure of two-dimensional languages.

References

[1] Bottoni, P.; M.F. Costabile; S. Levialdi & P. Mussio (1997), Defining visual languages for interactive computing, *IEEE Transactions on Systems, Man, and Cybernetics – A*, 27.6: 773–783.

[2] Bottoni, P.; G. Mauri; P. Mussio & Gh. Păun (1998), Grammars Working on Layered Strings, *Acta Cybernetica*, 13.4: 339–358.

[3] Bottoni, P.; G. Mauri; P. Mussio & Gh. Păun, On the Power of Pictorial Languages, to appear in *International Journal of Pattern Recognition and Artificial Intelligence*.

[4] Giammarresi, D. & A. Restivo (1997), Two-dimensional Languages, in G. Rozenberg & A. Salomaa, eds., *Handbook of Formal Languages*, Springer, Berlin.

[5] Mateescu, A.; G. Rozenberg & A. Salomaa (1998), Shuffle on trajectories: Syntactic constraints, *Theoretical Computer Science*, 197.1–2: 1–56.

[6] Păun, Gh; G. Rozenberg & A. Salomaa (1998), *DNA Computing. New Computing Paradigms*, Springer, Berlin.

[7] Rozenberg, G. & A. Salomaa, eds., (1997), *Handbook of Formal Languages*, Springer, Berlin.

[8] Siromoney, G.; R. Siromoney & K. Krithivasan (1972), Abstract families of matrices and picture languages, *Computer Graphics and Image Processing*, 1.

Chapter 18

TRANSDUCTION IN POLYPODES

Symeon Bozapalidis
Department of Mathematics
Aristotle University of Thessaloniki
GR 54006 Thessaloniki, Greece
bozapalidis@ccf.auth.gr

Abstract We discuss transductions into the framework of polypodes. We focus on how locally recognizable transductions constitute an extension of rational transductions in monoids and describe most of the significant operations on trees words, etc. A Nivat-like representation theorem is given.

1. INTRODUCTION

Polypodes are sets endowed with operations of the form $M \times M^n \to M$ ($n \geqslant 1$) complying with some natural axioms. Such algebraic models unify the theories of words, trees, graphs, etc. They constitute a natural extension of monoids giving rise to new various recognizability and rationality notions.

Polypodes seem to provide the necessary structure for establishing Eilemberg's theorem on varieties in the framework of trees. In [1] the reader will find these ideas developed in detail.

Transductions between polypodes are discussed in this paper. Our goal is not to give an exhaustive presentation on the topic but rather to indicate how recognizability theory can be applied in this direction. Almost all interesting operations on trees, words, etc, can be described by locally recognizable polypodic transductions. The last type of transductions in the tree case are represented by the most general bimorphisms, namely, by compositions such as:

$$T_\Gamma(X_n) \xrightarrow{\phi^{-1}} T_\Delta(X_n) \xrightarrow{-nR} T_\Delta(X_n) \xrightarrow{\psi} T_\Sigma(X_n)$$

with ϕ, ψ general tree homomorphisms and R recognizable forest. A Nivat-like theorem is given.

C. Martin-Vide and V. Mitrana (eds.), Where Mathematics, Computer Science, Linguistics and Biology Meet, 211-218.

2. POLYPODES: BASIC FACTS

An $n-$*polypodic* operation ($n \geqslant 1$) on a set M is a function of the form:

$$M \times M^n \to M \qquad (m, m_1, ..., m_n) \longmapsto m[m_1, ..., m_n].$$

Such an operation is associative whenever for all $m, m_i, m_j' \in M$ it holds

$$m[m_1, ..., m_n][m_1', ..., m_n'] = m\,[m_1[m_1', ..., m_n'], ..., m_n[m_1', ..., m_n']].$$

The $n-$*tuple* $(e_1, ..., e_n) \in M^n$ is a unit if for all $m, m_i \in M$ it holds $m[e_1, ..., e_n] = m$, $e_i[m_1, ..., m_n] = m_i \; \forall i$. A unit whenever it exists is unique.

An $n-$*polypode* is a set endowed with an associative polypode operation admitting a unit. The algebra of polypodes is canonically defined. Just a word for the subpolypode $pol(L)$ generated by a subset L of an $n-$*polypode* M. It is given by the following inductive steps:

- $L \cup \{e_1, ..., e_n\} \subseteq pol(L)$, $(e_1, ..., e_n)$ the unit of M,
- if $a \in L$ and $m_1, ..., m_n \in pol(L)$ then $a[m_1, ..., m_n] \in pol(L)$.

The free $n-$*polypode* generated by the $n-$*ranked* alphabet Γ is $T_\Gamma(X_n)$ with $X_n = \{x_1, ..., x_n\}$ as a set of variables. Its polypodic operation is the usual tree substitution operation. Examples of polypodes (exept $T_\Gamma(X_n)$) are:

- the set of words $(\Sigma \cup X_n)^*$ with word substitution (Σ an ordinary alphabet).
- the set $P(T_\Gamma(X_n))$ of forests over $\Gamma \cup X_n$ with OI-forest substitution.
- the set $Graph(EX_n)$ of finite directed graphs whose edges are labelled over $E \cup X_n$ (E finite set with $E \cap X_n = \emptyset$) having a distinguished edge as a root. Edge substitution is a polypodic operation for this.
- the set of unordered trees $UnT_\Gamma(X_n)$.
- the set $UnLabT_\Gamma(Xn)$ of unlabelled trees.
- the set $[A^n, A]$ of functions $f : A^n \to A$ with function composition as a polypodic operation (called clones by P. Cohn)

Various types of recognizability can be defined in polypodes; we refer the most important ones, global and local. A subset L of an n-polypode M is said to be globally recognizable if there exist a finite n-polypode N and a polypode morphisme $g : M \to N$ so that $L = g^{-1}(R)$ for some $R \subseteq N$

L is locally recognizable whenever there exist a finite n-ranked alphabet Γ and a polypode morphism $h : T_\Gamma(X_n) \to M$ so that $L = h(R)$, for some recognizable forest $R \subseteq T_\Gamma(X_n)$.

In finitely generated polypodes we have the inclusion $GREC \subseteq LREC$ which is appropriate when dealing with polypodes of trees, words, etc. Locally recognizable subsets in polypodes play a role corresponding to rational subsets in monoids and equational subsets in algebras; hence our interest in them.

Closure properties: The class LREC(M) of locally recognizable subsets of M is closed under finite union, intersection with globally recognizable subsets, and direct image of polypode morphisms. Further, if $L, L_1, ..., L_n \in LREC(M)$. then $L[L_1, ..., L_n], pol(L) \in LREC(M)$.

The class of locally recognizable forests in the closure of the class of recognizable forests through arbitrary tree homomorphisms and the class of locally recognizable languages is the yield of the class of locally recognizable forests.

Now, we are going to apply recognizability in polypodes to Transduction Theory. A transduction from the polypode M to the polypode N is just a multivalued function $f : M \to P(N)$. Its graph is the relation: $\#f = \{(m, n)/m \in M, n \in f(m)\}$. Call f locally (globally) recognizable whenever $\#f$ is a locally (globally) recognizable subset of the product polypode $M \times N$. A bimorphism is a triple (ϕ, R, ϕ') where ϕ, ϕ' are polypode morphisms with common domain: $N \xrightarrow{\phi} M$, $N \xrightarrow{\phi^{-1}} M'$ and $R \subseteq N$. Its graph is the relation $\{(\phi(t), \phi'(t))/t \in R\}$.

The lemma below is useful for comparing globally and locally recognizable transductions.

Lemma 18.1 *In any finitely generated $n-polypode$ M $GREC(M) \subseteq LREC(M)$ holds.*

Proof. By hypothesis there exists a finite ranked alphabet Γ and an epimorphism of polypodes $h : T_\Gamma(X_n) \to M$.

Now if $L \in GREC(M)$, then so is $h^{-1}(L) \subseteq T_\Gamma(X_n)$, since $GREC\,(T_\Gamma\,(X_n)) \subseteq LREC(T_\Gamma(X_n))$ the subset $h(h^{-1}(L)) = L$ is locally recognizable. □

In the case where both M, N are finitely generated by the finite lists A, B respectively, then $M \times N$ is generated by the pairs (a, e_i'), (e_j, b), $a \in A, b \in B$ with $(e_1, ..., e_n)$ and $(e_1', ..., e_n')$ standing for the units of M and N respectively. By applying, therefore, the previous lemma we get

Theorem 18.1 *Assume both M, N are finitely generated. Then the class of globally recognizable relations from M to N is included into that of locally recognizable relations.*

The inclusion above is actually strict when dealing with trees, words, etc. The next result often facilitates the detection of locally recognizable transductions.

Proposition 18.1 *Given n – polypodes M, N it holds:*

i) Every finite subset R of $M \times N$ (including $\varnothing$) is an LR transduction

ii) If $R \subseteq M \times N$ (i=1,2) are LR transductions then so is their union $R_1 \cup R_2$.

iii) If $R \subseteq M \times N$ is LR transductions, then so is its inverse $R^{-1} \subseteq N \times M$ (in the sense of relations).

iv) If $R, R_1, ..., R_n \subseteq M \times N$ are LR transductions then so is $R[R_1, ..., R_n]$ and

v) If $R \subseteq M \times N$ is an LR transduction then so is $pol(R)$.

3. REMARKABLE LR TRANSDUCTIONS

We shall now present a series of interesting LR transductions; actually almost all significant operations have these characteristics.

1. The diagonal $\Delta = \{(t,t) / t \in T_\Gamma(X_n)\}$, Γ finite is an LR tree transduction since $\Delta = pol\{(\gamma(x_1, ..., x_n), \gamma(x_1, ..., x_n))/\gamma \in \Gamma\}$. The same holds for any finitely generated n – *polypode*.

2. The branch transduction $br : T_\Gamma(X_n) \rightarrow P_{fin}(T_{br(\Gamma)}(X_n))$ is locally recognizable because its graph equals $pol\{(\gamma(x_1, ..., x_n), \gamma_i(x_i))/\gamma \in \Gamma\}$, Γ finite.

3. Let M be a finitely generated polypode and $h : M \rightarrow N$ a polypode morphism. Its graph is: $\#h = pol\{(a, h(a))/a \in A\}$ where A is a finite set generating M. Hence h is an LR transduction and thus so is its inverse. We conclude that $yield : T_\Gamma(X_n) \rightarrow X_n^*$ and $er : T_\Gamma(X_n) \rightarrow (\Gamma \cup X_n)^*$, (ereses parentheses and commas) are LR transductions.

Although the relation $\{(w, \hat{w})/w \in X_n^*\} \subseteq X_n^* \times X_n^*$ ($\hat{w}$ the mirror image of w) is known to be nonrational, it is locally recognizable since the function $w \rightarrow \hat{w}$ is clearly a polypode morphism.

4. Let $s, t \in T_\Gamma(X_n)$; we say that t is an initial subtree of s if there exists a factorization $s = t[t_1, ..., t_n]$, $t_j \in T_\Gamma(X_n)$. More generally, let $m, m' \in M$, where M is an n – *polypode* ; m is the initial **subelement** of m' if $m' = m[m_1, ..., m_n]$, for some $m_j \in M$. For instance in $(\Sigma \cup X_n)^*$ the word $u = \sigma_1 x_1 \sigma_1 x_1 x_2 \sigma_2$ is an initial subword of $w = \sigma_1\sigma_1\sigma_2 x_1 x_2 \sigma_1\sigma_1\sigma_2 x_1 x_2 \sigma_2 \sigma_2$ since $w = u[\sigma_1\sigma_2 x_1 x_2/x_1, \sigma_2/x_2]$, etc. The relation $R = \{(m, m')/m$ initial subelement of $m'\}$ is a LR transduction provided M is finitely generated.

Indeed, we have $R = \Delta[pol\{(e_1, a)/a \in A\}, ..., pol\{(e_n, a) /a \in A\}]$ where A is a finite set generating M and $(e_1, ..., e_n)$ is the unit of M.

5. Assume that M and N are finitely generated n – *polypodes* and $K \in LREC(M)$, and $L \in LREC(N)$. Then the relation $K \times L \subseteq M \times N$ is locally recognizable.

Denote by A, B generating subsets and by $(e_1, ..., e_n)$, $(e_1', ..., e_n')$ the units of M and N respectively. Manifestly the subsets $K \times e_1' = \{(k, e_1')/k \in$

$K\}$, $e_1 \times L = \{(e_1, l)/l \in L\}$ are locally recognizable. Therefore so is $K \times L = (K \times e_1)[(e_1 \times L), (e_2, e_2), ..., (e_n, e_n)]$ As an application we see that the LR constants, that is the transductions $t \to K$, $K \in LREC(M)$ are LR because their graphs have the form $M \times K$. Choosing a forest $K \in LREC(T_\Gamma(X_n)) - GREC(T_\Gamma(X_n))$ the transduction $t \to K$ is LR but not GR. Also the transduction $t \to t \cup K$, $K \in LREC(M)$ is LR because its graph is $\Delta \cup (M \times K)$.

6. A second order substitution from $T_\Gamma(X_n)$ to $T_\Delta(X_n)$ is a function of the form $H : \Gamma \to P(T_\Delta(X_n))$. It is inductively extended into a function (still denoted H):

$H : T_\Gamma(X_n) \to P(T_\Delta(X_n))$

via the inductive formulas

$H(x_i) = x_i \,, 1 \leqslant i \leqslant n$

$H(\gamma(t_1, ..., t_n)) = H(\gamma)[H(t_1), ..., H(t_n)]$.

H is termed *locally recognizable* (LR) whenever all forests $H(\gamma)$, $\gamma \in \Gamma$ are locally recognizable.

Fact : Each LR second order substitution $H : T_\Gamma(X_n) \to P(T_\Delta(X_n))$ is a LR transduction.

Indeed we observe that the graph of H is

$$pol(\bigcup_{\gamma\in\Gamma} \gamma \times H(\gamma))$$

and each of the sets $\gamma \times H(\gamma)$ is an LR relation of $T_\Gamma(X_n) \times T_\Delta(X_n)$.

7. Transductions with more than two arguments can be easily defined: they are subsets of products of polypodes $M_1 \times M_2 \times ... \times M_k$, $(k \succ 2)$.

An important example is the polypodic operation on a polypode M :

$$\underbrace{M \times M \times ... \times M}_{n-times} \to M, (m, m_1, ..., m_n) \to m[m_1, ..., m_n].$$

It can be viewed as a transduction with $n + 2$ arguments.

Proposition 18.2 *In the case where M is a finitely generated n – polypode, its polypodic operation is a locally recognizable transduction.*

Proof. Assume that A is a generating finite set of M and let us put

$R = pol\{(a, e_1, ..., e_n, a)/a \in A\}$

$R_1 = pol\{(e_1, a, e_2, ..., e_n, a)/a \in A\}$

....

$R_n = pol\{(e_1, ..., e_n, a, a)/a \in A\}$

with $(e_1, ..., e_n)$ standing for the unit of M. Then

$\{(m, m_1, ..., m_n, m[m_1, ..., m_n])/m, m_i \in M\} = R[R_1, ..., R_n]$

which proves our assertion. □

Remark 18.1 *The above result in the case of free monoids collapses to the well-known fact that concatenation is actually a rational transduction.*

Corollary 18.1 *Tree substitution* $(t, t_1, ..., t_n) \to t[t_1, ..., t_n]$, $t, t_i \in T_\Gamma(X_n)$ *is a* LR *transduction*

Corollary 18.2 *The word substitution operation*
$(w, w_1, ..., w_n) \to w[w_1, ..., w_n]$, $w, w_i \in (\Sigma \cup X_n)^*$ $n \geqslant 2$,
is an LR transduction.

Corollary 18.3 *The substitution operation of unordered (resp.unlabeled) trees is an LR transduction.*

Corollary 18.4 *The composition of polynomial (resp. affine) functions in* $[N^n, N]$ $(f, g_1, ..., g_n) \to f \circ (g_1, ..., g_n)$ *is an LR transduction.*

4. REPRESENTATION THEOREM

Lemma 18.2 *Assume that* $h : M \to M'$ *is an epimorphism of* $n-polypodes$; *then each locally recognizable subset* $L' \subseteq M'$ *can be written as* $L' = h(L)$, *for some* $L \in LREC(M)$.

Proof. By definition there exists a polypode morphism $a : T_\Gamma(X_n) \to M'$ (Γ finite) so that $L' = a(R)$, for some $R \in REC(T_\Gamma(X_n))$.

The surjectivity of h guarantees the existence of a polypode morphism b : $T_\Gamma(X_n) \to M$ rendering the next triangle commutative:

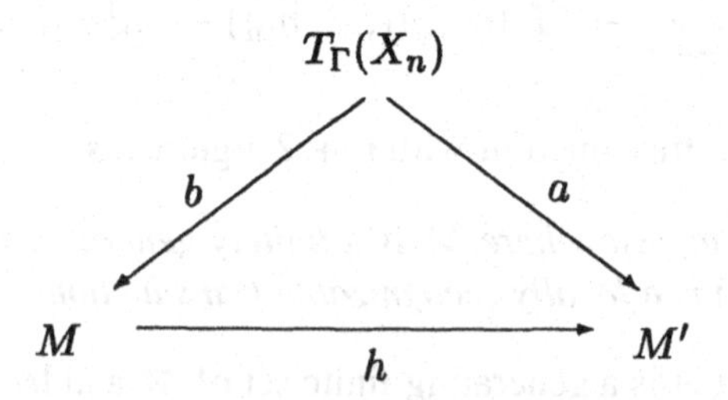

Now the subset $L = b(R)$ has the required properties. □

Next consider $n-ranked$ alphabets Γ, Δ and define the $n-ranked$ alphabet $\Gamma \vee \Delta = \{\langle\gamma, x_k\rangle / \gamma \in \Gamma, 1 \leqslant k \leqslant n\} \cup \{\langle x_k, \delta\rangle / \delta \in \Delta, 1 \leqslant k \leqslant n\}$

Two tree homomorphisms:

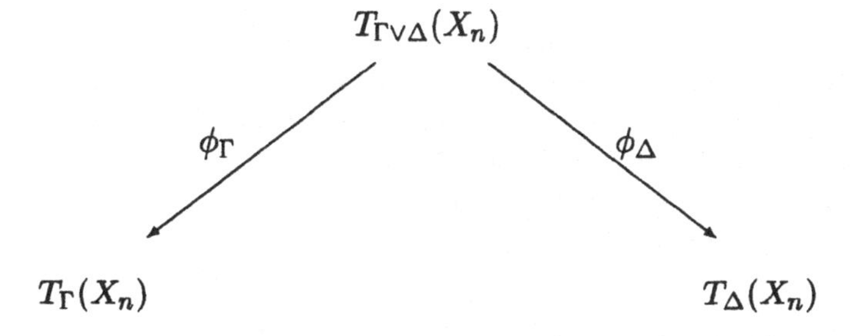

are obtained by setting

$\phi_\Gamma(\langle\gamma, x_k\rangle) = \gamma \quad , \quad \phi_\Delta(\langle\gamma, x_k\rangle) = x_k$

$\phi_\Gamma(\langle x_k, \delta\rangle) = x_k \quad , \quad \phi_\Delta(\langle x_k, \delta\rangle) = \delta$

They jointly derive a polypode morphism $\phi_{\Gamma,\Delta} : T_{\Gamma\vee\Delta}(X_n) \to T_\Gamma(X_n) \times T_\Delta(X_n)$ with $\phi_{\Gamma,\Delta}(t) = (\phi_\Gamma(t), \phi_\Delta(t))$, $t \in T_{\Gamma\vee\Delta}(X_n)$

Proposition 18.3 *$\phi_{\Gamma,\Delta}$ is surjective.*

A Nivat-like theorem follows:

Theorem 18.2 *Next conditions are mutually equivalent for a subset $R \subseteq T_\Gamma(X_n) \times T_\Delta(X_n)$:*

i) R is locally recognizable,

ii) There exists a locally recognizable forest $F \subseteq T_{\Gamma\vee\Delta}(X_n)$ such that $R = \{(\phi_\Gamma(t), \phi_\Delta(t))/t \in F\}$,

iii) There exists a finite n − ranked alphabet Σ with two alphabetic tree homomorphisms:

$\phi : T_\Sigma(X_n) \to T_\Gamma(X_n) \;\; , \;\; \psi : T_\Sigma(X_n) \to T_\Delta(X_n)$

and a locally recognizable forest $F \subseteq T_\Sigma(X_n)$ such that $R = \{(\phi(t), \psi(t))/t \in F\}$,

iv) There exists a finite n−ranked alphabet Σ with two tree homomorphisms $f : T_\Sigma(X_n) \to T_\Gamma(X_n)$, $g : T_\Sigma(X_n) \to T_\Delta(X_n)$ and a recognizable forest $F \subseteq T_\Sigma(X_n)$ such that $R = \{(f(t), g(t))/t \in F\}$.

Proof. $i) \Rightarrow ii)$ Apply the previous lemma to the epimorphism of polypodes $\phi_{\Gamma,\Delta} : T_{\Gamma\vee\Delta}(X_n) \to T_\Gamma(X_n) \times T_\Delta(X_n)$

$ii) \Rightarrow iii)$ Recall first that a tree homomorphism $h : T_\Sigma(X_n) \to T_\Gamma(X_n)$ is named *alphabetic* if for each symbol $\sigma \in \Sigma$ either $h(\sigma) \in \Gamma$ or $h(\sigma) = x_k \;\; (1 \leqslant k \leqslant n)$.

For instance both ϕ_Γ, ϕ_Δ are alphabetic. Now the proposed implication is immediate.

$iii) \Rightarrow iv)$ Since $F \subseteq T_\Sigma(X_n)$ is a locally recognizable forest, it is the image via a tree homomorphism $a : T_\Gamma(X_n) \to T_\Sigma(X_n)$ of a recognizable forest $L \subseteq T_\Sigma(X_n) : F = a(L)$. Then

$R = \{(\phi(t), \psi(t))/t \in F\} = \{(\phi(a(t)), \psi(a(t)))\ /t \in L\} =$

$= \{((\phi \circ a)(t), (\psi \circ a)(t)) \,/ t \in L\}.$

Thus we only have to take $f = \phi \circ a$ and $g = \psi \circ a$.

$iv \Rightarrow i)$ Apply the fact that each polypode morphism preserves locally recognizable subsets to the case of the morphism

$h : T_\Sigma(X_n) \to T_\Gamma(X_n) \times T_\Delta(X_n) \quad h(t) = (f(t), g(t))$

and to the recognizable forest L. □

Corollary 18.5 *In the transduction $t \to t \cap K$, K a locally recognizable forest is an LR transduction.*

Corollary 18.6 *The domain and codomain of an LR transduction are locally recognizable forests.*

References

[1] Theory of Polypodes I: Recognizability Seminar in Theor. Inf. TRI, Thessaloniki, 1999.

Chapter 19

SOME ALGEBRAIC PROPERTIES OF CONTEXTS AND THEIR APPLICATIONS TO CONTEXTUAL LANGUAGES

Rodica Ceterchi
Department of Computer Science
Faculty of Mathematics
University of Bucharest
Str. Academiei 14, R-70109 Bucureşti, Romania
rc@funinf.math.unibuc.ro

Abstract We present some elementary algebraic properties of contexts, and we investigate some relations between these properties and the generative power of the associated contextual grammars.

1. INTRODUCTION

The field of contextual languages has developed during the last three decades, becoming an important part of formal language theory (see [6]), which links this latter to the study of natural languages, as advocated in [2]. Some landmarks in its development are: the pioneering paper [1], which introduces the concept known today as *external context*, another pioneering paper [5], which introduces the concept of *internal context*, and the two monographs [3] and [4] published 15 years apart, which illustrate the dimension and the complexity of this field, as well as its extraordinary dynamics.

We believe that the richness of this field has not been exhausted, and that algebraic approaches to its study could be fruitful.

In the present paper we emphasize some elementary algebraic properties of contexts (in section 3), and begin in section 4 a study of the impact these properties have on the generative power of associated contextual grammars.

Neither the algebraic structures of section 3, nor their consequences presented in section 4, are exhaustive: on the contrary, they are only the tip of an iceberg, which, to the author's humble opinion, might deserve further investigation.

C. Martin-Vide and V. Mitrana (eds.), Where Mathematics, Computer Science, Linguistics and Biology Meet, 219-226.

2. PRELIMINARIES

Let V be an alphabet, V^* the free monoid generated by V, and let $\lambda \in V^*$ denote the empty word.

The following definitions come from [4]. A *contextual grammar without choice* is a triple $G = (V, A, C)$, where V is a finite alphabet, A is a finite language over V, called the set of *axioms*, and C is a finite subset of $V^* \times V^*$, called the set of *contexts*.

With respect to such a grammar, we can consider two derivation relations, the *external derivation relation* (introduced in [1]) defined by:

$$x \Longrightarrow_{ex} y \text{ iff } y = uxv, \text{ for a context } (u, v) \in C,$$

and the *internal derivation* relation (introduced in [5]) defined by:

$$x \Longrightarrow_{in} y \text{ iff } x = x_1x_2x_3, y = x_1ux_2vx_3 \text{ for any } x_1, x_2, x_3 \in V^*, (u, v) \in C$$

By $\overset{*}{\Longrightarrow}_\alpha$ we denote as usual the reflexive and transitive closure of $\Longrightarrow_\alpha$ for $\alpha \in \{ex, in\}$.

The *external contextual language* generated by G is:

$$L_{ex}(G) := \{y \in V^* \mid x \overset{*}{\Longrightarrow}_{ex} y, x \in A\}.$$

and in this case G is called *external contextual grammar* or **ecg** for short.

The *internal contextual language* generated by G is:

$$L_{in}(G) := \{y \in V^* \mid x \overset{*}{\Longrightarrow}_{in} y, x \in A\}.$$

and in this case G is called *internal contextual grammar* or **icg** for short.

A *contextual grammar with choice* is a quadruple $G = (V, A, C, \phi)$, where (V, A, C) is a contextual grammar without choice as before, to which a *choice* (or *selection*) *function* $\phi : V^* \to 2^C$ was added.

With respect to such a grammar the *external* and *internal derivation* relations on V^* are defined by:

$$x \Longrightarrow_{ex} y \text{ iff } y = uxv, \text{ for a context } (u, v) \in \phi(x),$$

$$x \Longrightarrow_{in} y \text{ iff } \quad x = x_1x_2x_3, \quad y = x_1ux_2vx_3 \text{ for any } x_1, x_2, x_3 \in V^* \text{ and } (u, v) \in \phi(x_2).$$

Their respective reflexive and transitive closures define the *external* and the *internal contextual language with choice* generated by G, by:

$$L_\alpha(G) := \{y \in V^* \mid x \overset{*}{\Longrightarrow}_\alpha y, x \in A\}, \text{ with } \alpha \in \{ex, in\}.$$

3. SOME ALGEBRAIC PROPERTIES OF CONTEXTS

In the following we will consider the maximal set of contexts, i.e. $V^* \times V^*$. Also, we will consider only the case of external contexts, although several considerations could be extended to inner contexts as well.

Let us note that in the case of external contexts, two succesive applications of the contexts $< u_1, v_1 >$ and $< u_2, v_2 >$ to a string x give us the succesive derivations

$$x \overset{<u_1,v_1>}{\Longrightarrow_{ex}} u_1 x v_1 \overset{<u_2,v_2>}{\Longrightarrow_{ex}} u_2 u_1 x v_1 v_2$$

which could be replaced by a one-step derivation using the context $< u_2 u_1, v_1 v_2 >$,

$$x \overset{<u_2u_1,v_1v_2>}{\Longrightarrow_{ex}} u_2 u_1 x v_1 v_2 \ .$$

This inspires us to define the following binary operation which we will call *composition of contexts*, by:

$$< u_1, v_1 > \cdot < u_2, v_2 > := < u_2 u_1, v_1 v_2 >$$

Note that $\cdot$ is catenation on the second component, and the mirror image of catenation (or anti-catenation) on the first component.

Proposition 19.1 *$(V^* \times V^*, \ \cdot, < \lambda, \lambda >)$ is a non-commutative monoid.*

We can also define two partial orders on $V^* \times V^*$:

$$< u, v > \leqslant_1 < u', v' > \quad \Longleftrightarrow \quad u \in Suf(u') \ \ and \ \ v \in Pref(v')$$
$$< u, v > \leqslant_2 < u', v' > \quad \Longleftrightarrow \quad u \in Pref(u') \ \ and \ \ v \in Suf(v')$$

The following result gives us alternative characterizations of the two orders in terms of composition of contexts.

Proposition 19.2 *The following are true:*

(1) $< u, v > \leqslant_1 < u', v' > \Longleftrightarrow$ *there exists a context* $< x, y >$ *such that* $< u, v > \cdot < x, y > = < u', v' >$

(2) $< u, v > \leqslant_2 < u', v' > \Longleftrightarrow$ *there exists a context* $< x, y >$ *such that* $< x, y > \cdot < u, v > = < u', v' >$

Proof. $u \in Suf(u')$ iff there exists an x such that $u' = xu$ (denote $x = u' \backslash u$), and $v \in Pref(v')$ iff there exists y such that $v' = vy$ (denote $y = v' \,/\, v$). For (1) take $< x, y > = < u' \setminus u, v' / v >$, and for (2) take $< x, y > = < u' / u, v' \setminus v >$. □

With the notations above, the following

$$\begin{aligned} <u,v> \rightarrow <u',v'> &:= <u' \setminus u, v' / v> \\ <u,v> \rightsquigarrow <u',v'> &:= <u' / u, v' \setminus v> . \end{aligned}$$

define $\rightarrow$ and $\rightsquigarrow$ as partial binary operations on $V^* \times V^*$. We have:

Proposition 19.3 *(1) For pairs of contexts* $<u,v>, <u',v'> \in V^* \times V^*$ *which satisfy* $<u,v> \leqslant_1 <u',v'>$ *we can define the binary operation* $<u,v> \rightarrow <u',v'> = <u' \setminus u, v' / v>$. *Its result satisfies the property that it is the biggest, w.r.t.* $\leqslant_1$, *element* $<a,b> \in V^* \times V^*$ *such that*

$$<u,v> \cdot <a,b> \leqslant_1 <u',v'> .$$

(2) For pairs of contexts $<u,v>, <u',v'> \in V^* \times V^*$ *which satisfy* $<u,v> \leqslant_2 <u',v'>$ *we can define the binary operation* $<u,v> \rightsquigarrow <u',v'> = <u' / u, v' \setminus v>$. *Its result satisfies the property that it is the biggest, w.r.t.* $\leqslant_2$, *element* $<a,b> \in V^* \times V^*$ *such that*

$$<a,b> \cdot <u,v> \leqslant_2 <u',v'> .$$

Another result which can be proved by straightforward calculations is the following:

Proposition 19.4 *Composition of contexts* $\cdot$ *is increasing to the right w.r.t.* $\leqslant_1$ *and increasing to the left w.r.t.* $\leqslant_2$.

These last two results tell us that the monoid $(V^* \times V^*, \ \cdot \ , <\lambda, \lambda>)$ is partially residuated to the right w.r.t. $\leqslant_1$ (the residual being the partial operation $\rightarrow$) and partially residuated to the left w.r.t. $\leqslant_2$ (the residual being the partial operation $\rightsquigarrow$).

4. BACK TO CONTEXTUAL LANGUAGES

In the case of **ecgs**, the set of contexts $C \subset V^* \times V^*$ is finite, but, due to the fact that we work with the symmetric and transitive closure of the one-step derivation, this is the same as working with the submonoid of $V^* \times V^*$ generated by C. Let us denote this submonoid by $\overline{C}$. It will contain $<\lambda, \lambda>$ and also all possible compositions of contexts in C.

Let $G = (V, A, C)$ be an **ecg**. For an axiom $a \in A$ we define

$$L_G(a) := \{x \in V^* | a \stackrel{*}{\Longrightarrow}_{ex} x\}$$

the set of words derived from a in G. The following is immediate.

Proposition 19.5 *For every* $a \in A$, $L_G(a)$ *is in bijective correspondence with* $\overline{C}$.

Proof. To a we associate $< \lambda, \lambda > \in \overline{C}$. To an $x \in L_G(a)$ of the form $x = u_n \dots u_1 a v_1 \dots v_n$, we associate $< u_1, v_1 > \cdot \dots \cdot < u_n, v_n > \in \overline{C}$. □

Since $L(G) = \bigcup_{a \in A} L_G(a)$, working with $L(G)$ implies, in a sense, working with $|A|$ copies of $\overline{C}$. We will make this sense more clear with the following definition and result.

Definition 19.1 *An axiom system is called* independent *iff for every pair of distinct axioms $a, b \in A$, $a \neq b$, we have $L_G(a) \cap L_G(b) = \emptyset$.*

Proposition 19.6 *If A is independent in $G = (V, A, C)$, then $L(G)$ is in bijective correspondence with the disjoint sum $\coprod_{i=1}^{|A|} \overline{C_i}$, with $C_i = C$ for every i.*

Proof. Since $\{L_G(a) | a \in A\}$ are pairwise disjoint, their union coincides with their disjoint sum, so we have $L(G) = \coprod_{a \in A} L_G(a)$. According to the previous result we have a bijective correspondence between $L_G(a)$ and $\overline{C}$ for every a, which will extend naturally to a bijective correspondence between $L(G)$ and $\coprod_{i=1}^{|A|} \overline{C_i}$, with $C_i = C$ for every i. □

In the following we will concentrate on **ecgs** with a single axiom, $A = \{a\}$, and the results can be extended to **ecgs** with independent axiom systems.

One advantage of identifying $L_G(a)$ with the submonoid $\overline{C}$ lies in the following fact: since the submonoid $\overline{C}$ does not have a unique set of generators, i.e. we can find (several) subsets $C' \subset V^* \times V^*$ such that $\overline{C'} = \overline{C}$, and the grammars attached to these subsets, $G' = (V, \{a\}, C')$, will have the same generative power as G, i.e. $L_{G'}(a) = L_G(a)$, we can address the issue of finding, for a given **ecg**, alternative sets of contexts, with the same generative power, but with certain desirable properties.

Among the desirable properties we can think about is the **minimality** (in the sense of cardinality) of the set of contexts. In other words, given an **ecg** $G = (V, \{a\}, C)$, can we find (and under what conditions) a minimal set C' of generators for $\overline{C}$, i.e. a minimal set of contexts C' with the same generative power as C?

In general, we will try to "construct" C' starting from C, and the question thus becomes: under what conditions can we eliminate elements from C, retaining the same generative power at the same time? An easy answer follows in the next result.

Proposition 19.7 *If C contains an element $< u, v >$ which is the composition of two other elements $< u_i, v_i > \in C$, $i = 1, 2$, not necessarilly distinct between them, but both distinct from $< u, v >$, then $C' = C \setminus \{< u, v >\}$ has the same generative power as C, i.e. $\overline{C'} = \overline{C}$.*

Proof. is trivial. □

It is not unexpected that we can effectively eliminate a context $< u, v >$ which is the composition of other contexts, and retain the same generative power, if we retain the elements which compose $< u, v >$. But what if not all the elements which compose $< u, v >$ belong to C? Can we find a way to compensate for the extraction of $< u, v >$ from C? The answer to this question involves an interplay between **ecgs without** and **with** choice, and is partially illustrated by the next results, which make use of the orders between contexts. In the following, we start with an *ecg* without choice $G = (V, \{a\}, C)$.

Proposition 19.8 *Let* $C = \{< u_i, v_i > | i = 1, 2\}$ *have two elements, such that* $< u_1, v_1 > \leqslant_1 < u_2, v_2 >$. *Denote* $< x, y > = < u_1, v_1 > \rightarrow < u_2, v_2 >$. *Suppose also that axiom a does not contain letters belonging to the components of contexts. Then the following are true:*

(1) The set of contexts $C' = C \setminus \{< u_2, v_2 >\} \bigcup \{< x, y >\}$ *has bigger generative power than C, i.e.* $\overline{C} \subset \overline{C'}$, *and the inclusion is strict if* $< x, y > \neq < u_1, v_1 >$.

(2) We can find a selection function ϕ *such that the selective ecg* $G' = (V, \{a\}, C', \phi)$ *has the same generative power as the initial* $G = (V, \{a\}, C)$.

Proof. (1) Recall that $< u_1, v_1 > \cdot < x, y > = < u_2, v_2 >$. Since $C' = \{< u_1, v_1 >, < x, y >\}$, it follows that $\overline{C} \subset \overline{C'}$. The fact that this inclusion is strict follows from the fact that $< x, y > \in C'$, but $< x, y > \notin C$, if $< x, y > \neq < u_1, v_1 >$.

(2) On $(V, \{a\}, C')$ add the selection function $\phi : V^* \rightarrow 2^{C'}$ given by the formulae:

$$\phi(a) = \{< u_1, v_1 >\}$$
$$\phi(u_1 w v_1) = C'$$
$$\phi(u_2 w v_2) = \{< u_1, v_1 >\}$$
$$\phi(w) = \emptyset, \quad for\ any\ other\ w \in V^*$$

The fact that $L_G(a) = L_{G'}(a)$ follows immediately, since the application of the context $< u_2, v_2 >$ in G can be replaced by two succesive applications of contexts in G', first $< u_1, v_1 >$, followed by the selected $< x, y >$. □

Remark 19.1 *We do* **not** *have a similar result for the order* $\leqslant_2$. *More precisely, take a set of contexts* $C = \{< u_i, v_i > | i = 1, 2\}$, *such that* $< u_1, v_1 > \leqslant_2 < u_2, v_2 >$, *denote* $< x, y > = < u_1, v_1 > \rightsquigarrow < u_2, v_2 >$, *and recall that* $< x, y > \cdot < u_1, v_1 > = < u_2, v_2 >$. *Take* $C' = \{< u_1, v_1 >, < x, y >\}$, *and it will be true that* $\overline{C} \subset \overline{C'}$, *the inclusion being strict if* $< x, y > \neq < u_1, v_1 >$. *In other words, we have a result similar to assertion (1) of the above proposition. But we can find no selection function* $\phi : V^* \rightarrow 2^{C'}$ *such that G and* $G' =$

$(V, \{a\}, C', \phi)$ have the same generative power, simply because words of the form xwy do not belong to $L_G(a)$.

For a fixed context $< u, v >$ let us denote

$$[< \lambda, \lambda >, < u, v >]_1 := \{< u', v' > \mid < u', v' > \leqslant_1 < u, v >\}$$

the closed interval w.r.t. $\leqslant_1$. (Such an interval will always be finite.)

Proposition 19.9 *Let $C = [< \lambda, \lambda >, < u, v >]_1$ be a closed interval w.r.t. $\leqslant_1$. Let $u, v \in V^*$ have the following letter-decompositions:*

$$\begin{aligned} u &= x_m \cdots x_2 x_1 \quad , \quad with\ x_i \in V\ \ for\ \ every\ \ i = \overline{1, m} \\ v &= y_1 y_2 \cdots y_n \quad , \quad with\ y_j \in V\ \ for\ \ every\ \ j = \overline{1, n}\ . \end{aligned}$$

Suppose also that the letters of a belong to $V \setminus \{x_i | i = \overline{1, m}\} \setminus \{y_j | j = \overline{1, n}\}$. Consider the set $C' = \{< x_i, \lambda > | i = \overline{1, m}\} \cup \{< \lambda, y_j > | j = \overline{1, n}\}$.
(1) C' has bigger generative power than C, i.e. $\overline{C} \subset \overline{C'}$.
(2) If we also consider the selection function $\phi : V^* \to 2^{C'}$ defined by:

$$\begin{aligned} \phi(a) &= \{< x_1, \lambda >, < \lambda, y_1 >\} \\ \phi(xwa) &= \{< \lambda, y_1 >\} \cup \{< x_{i+1}, \lambda > \mid x_i = x\} \\ \phi(awy) &= \{< x_1, \lambda >\} \cup \{< \lambda, y_{j+1} > \mid y_j = y\} \\ \phi(xwy) &= \{< x_{i+1}, \lambda > | x_i = x\} \cup \{< \lambda, y_{j+1} > \mid y_j = y\} \\ \phi(w) &= \emptyset\ ,\ for\ \ every\ \ other\ \ w \in V^*, \end{aligned}$$

then $G' = (V, \{a\}, C', \phi)$ has the same generative power as $G = (V, \{a\}, C)$.

Proof. The proof is by straightforward calculations, along the same lines as in the previous proposition. □

Remark 19.2 *Again, we cannot find a result similar to the assertion (2) above for the order $\leqslant_2$.*

We note, from these last two propositions, that certain modifications of the initial set of contexts, based on the order relation $\leqslant_1$, lead to the apparition of an interplay between **ecgs** without choice and **ecgs** with choice. Also, the last proposition shows "the power of one-sided contexts", which was also emphasized in [4].

As to the question "what is the advantage of working with C' instead of C?" (especially if we have to pay the price of using a selection function), the answer lies not with the minimization of the cardinality of the set of contexts, but with the minimization of the lengths of contexts in C' compared to those of the initial C.

References

[1] Marcus, S. (1969), Contextual grammars, *Revue Roumaine des Mathématiques Pures et Appliquées*, 14: 1525-1534.

[2] Marcus, S. (1997), Contextual grammars and natural languages, in G. Rozenberg & A. Salomaa, eds., *Handbook of Formal Languages*, II: 215-235, Springer, Berlin.

[3] Păun, Gh. (1984), *Contextual Grammars*, Publishing House of the Romanian Academy, Bucharest (in Romanian).

[4] Păun, Gh. (1997), *Marcus Contextual Grammars*, Kluwer, Dordrecht.

[5] Păun, Gh. & X.M. Nguyen (1980), On the inner contextual grammars, *Revue Roumaine des Mathématiques Pures et Appliquées*, 25: 641-651.

[6] Rozenberg, G. & A. Salomaa, eds., (1997), *Handbook of Formal Languages*, Springer, Berlin.

Chapter 20

ON FATOU PROPERTIES OF RATIONAL LANGUAGES

Christian Choffrut
L.I.A.F.A.
University of Paris 7
Tour 55-56, 1er étage, 2 Pl. Jussieu, 75251 Paris Cedex, France
cc@liafa.jussieu.fr

Juhani Karhumäki
Department of Mathematics and
Turku Centre for Computer Science
University of Turku
SF 20014 Turku, Finland
karhumak@cs.utu.fi

Abstract We consider so-called Fatou-type properties of rational languages and relations. For example, if the rational expression $E(X_1, \ldots, X_n)$ over arbitrary languages $X_1, \ldots, X_n$ represents a rational language, then each language X_i can be replaced by a rational language Y_i such that $E(X_1, \ldots, X_n) = E(Y_1, \ldots, Y_n)$ holds and moreover $X_i \subseteq Y_i$ for $i = 1, \ldots, n$. Some similar results for rational relations are established and an open problem on rational relations is stated.

1. INTRODUCTION

A classical result of Fatou, when translated in the terminology of formal series, states that if the coefficients of a $\mathbb{Q}$-rational series in one variable are in $\mathbb{Z}$ then the series is $\mathbb{Z}$-rational, [7]. That is to say, that $\mathbb{Z}$-rational series possess the *Fatou property* with respect to $\mathbb{Q}$. This has been subsequently extended to other semirings and for series in several variables, cf. e.g. [14]

*Supported by the Academy of Finland under the grant # 44087.

C. Martin-Vide and V. Mitrana (eds.), *Where Mathematics, Computer Science, Linguistics and Biology Meet*, 227-235.

or [2]. For example, it is well-known that $\mathbb{N}$-rational series possess the Fatou property with respect to $\mathbb{Q}_+$ but not with respect to $\mathbb{Z}$, see [6] or [14].

Though this paper can be viewed as a contribution to thinking on the Fatou property in a very strict sense, as explained in the conclusion, here we will understand "Fatou-type" properties in the more general way, loosely speaking, of those who claim that if something constructed from "complicated" parts is actually "simple" then it can be constructed from "simple" parts.

At this point and in order to explain more specifically what we are after, an illustration will suffice. It was observed in [5, Chap. VII] that whenever the product of two languages, say $X \cdot Y$, is rational, then there exist rational languages $\overline{X}$ and $\overline{Y}$ such that $\overline{X} \cdot \overline{Y} = X \cdot Y$. Moreover, $\overline{X}$ and $\overline{Y}$ are obtained in a monotone way in the sense that $X \subseteq \overline{X}$ and $Y \subseteq \overline{Y}$ hold. Later, this property was proved again several times, cf. for example [9] and [12].

The goal of this note is to consider the above property and its extensions in a systematic way. First we point out that the result holds in a much broader form; namely, if for some languages $X_1, \ldots, X_n$, the rational expression $E(X_1, \ldots, X_n)$ denotes a rational language, then the X_i's can be "saturated" in a unique way into subset-maximal rational languages $Y_1, \ldots, Y_n$ such that $E(Y_1, \ldots, Y_n)$ denotes the same language as $E(X_1, \ldots, X_n)$ and that $X_i \subseteq Y_i$ for all $i = 1, \ldots, n$. The use of the syntactic monoids makes the proof particularly simple.

Although in the above canonical construction from X_i to Y_i, Y_i is unique, this need not be true if the condition $X_i \subseteq Y_i$ is relaxed. Indeed, we give an example of a unary finite (and hence rational) language X having different maximal square roots, i.e., there exist two incomparable subset-maximal subsets Y and Z satisfying $Y^2 = Z^2 = X$.

Our main contribution is the generalization of this property to finite systems of rational expressions $E_j(X_1, \ldots, X_n)$ for $j = 1, \ldots, m$ over the common languages $X_1, \ldots, X_n$. This allows us to establish a Fatou property for finite transducers in the particular case where the transducer realizing a relation from a free monoid to another one has a deterministic underlying input automaton and arbitrary unary outputs for each transition. We call such transducers "input deterministic unary transducers", and we show for those that if the unary outputs associated with all inputs are rational then the outputs can be replaced, transition by transition, by rational languages. In proving the latter result we show that the equivalence problem for input deterministic unary transducers is decidable. This should be compared with two important undecidability results which state that if either one of the restrictions, i.e., input deterministic or unary, is relaxed, then the equivalence becomes undecidable, see [8,11] and [11]. In the conclusions we propose the Fatou problem for rational relations in its broadest sense.

2. PRELIMINARIES

In this section we recall necessary basics on automata theory, see [6], and we fix our notation.

2.1 RATIONAL SUBSETS

For a finite alphabet A we denote by A^* (resp. A^+) the free monoid (resp. free semigroup) generated by A. The identity of A^* referred to as the empty word, is denoted by 1. The *rational operations* on subsets of A^* (on languages) are the set union, the concatenation (product) and the star operation (also known as the Kleene star operation). The family of *rational* subsets of A^* is made up of those subsets of A^* which are obtained from singletons of A^* by a finite number of applications of the operations of union, concatenation and star, and is denoted by $\text{Rat}A^*$.

We recall that rational languages have several algebraic characterizations. A consequence of Kleene theorem establishing the equivalence between rational and *recognizable* (or *regular*) languages is that a language is rational if and only if there exist a finite monoid M, a morphism $\phi : A^* \to M$ and a finite subset $F \subseteq M$ such that $L = \phi^{-1}(F)$. The monoid M is said to *recognize* L. The *syntactic monoid* of L is the smallest monoid recognizing L. It is uniquely defined up to an isomorphism.

2.2 RATIONAL RELATIONS

We refer the interested reader to the monographs [1,6] for a more comprehensive exposition of the theory of rational relations. For our purpose the following will suffice.

A rational transducer is a tuple $\mathcal{T} = (Q, A, B, E, Q_-, Q_+)$ where Q is the finite set of *states*, $Q_- \subseteq Q$ is the set of *initial* states, Q_+ is the set of *final* states, A and B are the *input* and *output* alphabets and $E \subseteq Q \times A^* \times B^* \times Q$ is the finite set of *transitions*. A *path* is a sequence

$$p = (q_0, u_1, v_1, q_1), (q_1, u_2, v_2, q_2), \dots, (q_{n-1}, u_n, v_n, q_n) \qquad (20.1)$$

with $(q_{i-1}, u_i, v_i, q_i) \in E, i = 1, \dots, n$. Its *label* is the pair $(u_1 u_2 \dots u_n, v_1 v_2 \dots v_n)$ and $u_1 u_2 \dots u_n$ and $v_1 v_2 \dots v_n$ are respectively its *input* and *output* labels.

The path is *successful* if it leads from one initial state to some final state, i.e., $q_0 \in Q_-$ and $q_n \in Q_+$. The transducer $\mathcal{T}$ *defines* the relation $||\mathcal{T}|| \subseteq A^* \times B^*$ which is the set of labels of all successful paths. It is also said that $||\mathcal{T}||$ is the *behavior* of $\mathcal{T}$. The extension of the Kleene theorem to relations in $A^* \times B^*$ asserts that a relation is the behavior of a rational transducer if and only if it is *rational*, i.e. obtainable from singletons of $A^* \times B^*$ by a finite number of the operations of union, product and star.

The following notion is not standard but is quite natural. An *input deterministic transducer* is like a rational transducer except that Q_- is reduced to a unique state q_- and that the set of transitions E defines a partial function from $Q \times A$ into $\mathcal{P}(B^*) \times Q$ where $\mathcal{P}(B^*)$ is the power set of B^*. Observe that the behavior (defined in the obvious way) of an input deterministic transducer is not necessarily a rational relation.

A transducer, whether rational or not, is *output unary* whenever the output monoid has a single generator and can thus be identified to the additive monoid of the nonnegative integers.

2.3 RATIONAL EXPRESSIONS

A *rational expression* over a set $\{\xi_1, \ldots, \xi_n\}$ of variables is an element of free algebra with generators $\xi_1, \ldots, \xi_n$ and signature $\Sigma = \{+, \cdot, *\}$ where $+$ and $\cdot$ are two binary symbols and $*$ is a unary symbol. Since the symbols will be interpreted as the three rational operations, we adopt the infix instead of the prefix notation, e.g., $(\xi_1 + \xi_3^*) \cdot \xi_2$. The variables may be interpreted in any subalgebra of subsets of an arbitrary monoid closed under the rational operations but our preferred interpretations are $\mathcal{P}(A^*)$, $\mathcal{P}(A^* \times B^*)$, $\mathrm{Rat}(A^*)$ and $\mathrm{Rat}(A^* \times B^*)$.

3. A REDISCOVERED PROPERTY

In this section we consider the most basic Fatou property of rational languages stated in the next Theorem. It was first proved in [5] by means of quotients, later proved again among other things in [9] and finally explained in [12] through a direct construction based on automata.

Theorem 20.1 *Assume that L is rational and has the decomposition $L = X \cdot Y$ where $X, Y \subseteq A^*$. Then there exist two rational languages $\overline{X}$ and $\overline{Y}$ such that $L = \overline{X} \cdot \overline{Y}$, and moreover $X \subseteq \overline{X}$ and $Y \subseteq \overline{Y}$.*

We will prove a more general result. Intuitively, we will show that whenever some rational expression involving arbitrary languages represents a rational language, then each language occurring in the expression can be replaced by some rational language. The question is, how do we define these rational languages? If we deal with the star operation, it is well known that whenever X^* is rational then there exists a unique subset-minimal language Y such that $Y^* = X^*$ and that this language is rational. In other words, if we view the problem as trying to solve the equation $K = X^*$ for some fixed rational subset K, taking the subset-minimal solution seems to be a good candidate. However, this no longer holds if we consider the concatenation product. Indeed, we have $a^* = (a^n)^* \cdot (1 + a)^{n-1}$ and thus there are infinitely minimal (and rational) solutions below, say (a^*, a^*). But it is even worse with the third rational operation since the free monoid can be represented as the union of two nonrecursive sets which are the complements of one another. We are more lucky with maximal solutions as we will now prove.

We consider the set of n-tuples of subsets of a free monoid ordered componentwise, i.e., $(X_1, \ldots, X_n) \leqslant (Y_1, \ldots, Y_n)$ if and only if for each $i = 1, \ldots, n$, we have $X_i \subseteq Y_i$. Since the three rational operations are monotone in each variable, so is each rational expression $E(\xi_1, \ldots, \xi_n)$ interpreted as mapping from $(\mathcal{P}(A^*))^n$ to $\mathcal{P}(A^*)$.

Theorem 20.2 *[saturation] Let $E_j(\xi_1, \ldots, \xi_n)$ for $j = 1, \ldots, m$ be m rational expressions in the variables $\xi_1, \ldots, \xi_n$ and let $K_1, \ldots, K_m \subseteq A^*$ be rational subsets. If the system*

$$E_j(X_1, \ldots, X_n) = K_j, \; j = 1, \ldots, m \tag{20.2}$$

is satisfiable then it has maximal solutions. Furthermore, the components of all these maximal solutions are rational.

Observe that Theorem 20.1 corresponds to $E(\xi_1, \xi_2) = \xi_1\xi_2$.

Proof. For all $K_j = E_j(X_1, \ldots, X_n)$, $j = 1, \ldots, m$ let M_j be its syntactic monoid and $\phi_j : A^* \longrightarrow M_j$ its syntactic morphism. In particular, we have $K_j = \phi_j^{-1}(\phi_j(K_j)$. We let $\phi : A^* \longrightarrow M_1 \times \ldots \times M_m$ be the product morphism defined by $\phi(w) = (\phi_1(w), \ldots, \phi_m(w))$ and for all $i = 1, \ldots, n$ we set $F_i = \phi(X_i)$.

Denote by $\mathcal{M}$ the algebra whose elements are the subsets $\phi(Z)$ for some $Z \subseteq A^*$. Observe that $\mathcal{M}$ is closed under set union, concatenation and Kleene star and that we may interpret any rational expression in $\mathcal{M}$. We order the direct product $\mathcal{M}^n$ componentwise where on each component the ordering is the subset inclusion. Finally let $(G_1, \ldots, G_n) \in \mathcal{M}^n$ be a maximal element greater than or equal to $(F_1, \ldots, F_n)$ satisfying $E_j(G_1, \ldots, G_n) = E_j(F_1, \ldots, F_n)$ for all $j = 1, \ldots m$ where E_j is interpreted in the finite algebra $\mathcal{M}$.

Set $Y_i = \phi^{-1}(G_i)$ for all $1 \leqslant i \leqslant n$. Clearly, no n-tuple satisfying the m equalities of (20.2) is greater than $(Y_1, \ldots, Y_n)$. Indeed, if say $(Z_1, \ldots, Z_n)$ is greater than $(Y_1, \ldots, Y_n)$ then by the maximality of $(G_1, \ldots, G_n)$ and the monotonicity of the rational operations, we have $\phi(Z_i) = G_i$ for all $i = 1, \ldots, n$ and thus $Z_i \subseteq \phi^{-1}(G_i) = Y_i$.

At this point since all Y_i's are rational, we are left showing that indeed, the Y_i's satisfy (20.2). Since $X_i \subseteq Y_i$ holds and since the rational operations are monotone we have $K_j \subseteq E_j(Y_1, Y_2, \ldots, Y_n)$. Conversely, we have

$$\begin{aligned}\phi(E_j(Y_1, Y_2, \ldots, Y_n)) &= E_j(\phi(Y_1), \phi(Y_2), \ldots, \phi(Y_n)) \\ &= E_j(F_1, F_2, \ldots, F_n) = E_j(\phi(X_1), \phi(X_2), \ldots, \phi(X_n)) \\ &= \phi(E_j(X_1, X_2, \ldots, X_n)) = \phi(K_j)\end{aligned}$$

which proves the opposite inclusion and which completes the proof. □

The previous result shows that if the set of solutions of an equation of the form $K = E(X_1, \ldots, X_n)$, where K is a given rational subset, E a rational expression, and the X_i's are unknowns, is not empty, then it has maximal elements which are rational. Furthermore, there is a unique maximal solution above a given solution but the maximal solutions need not be unique. Indeed, set $X = \{0, 2, 3, 7, 10, 12, 14, 15\}$, $Y = \{0, 2, 3, 7, 12, 13, 14, 15\}$ and $Z = \{0 \leqslant i \leqslant 30 \mid i \neq 1, 8, 11, 23\}$. Then $X^2 = Y^2 = Z$ (the square is meant "additively") and X and Y are maximal with this property.

In [3] the expression $E(\xi) = \xi^2$ is studied in relation to free monoids and the extra condition assumed is that the product is unambiguous. It is shown that, given a rational subset K, the solution if it exists is unique i.e., there is at most one subset X such that $K = X^2$ holds. It is also proven that it is decidable whether or not the root is rational. Moreover, it is conjectured that this root is necessarily rational. Previous research in this direction was also done in [13].

4. AN APPLICATION TO INPUT DETERMINISTIC TRANSDUCERS

In this section we apply the above observations to the input deterministic transducers. We first start with a straightforward Lemma whose proof is left to the reader.

Lemma 20.1 *Let M be a commutative monoid. For all elements $x, \bar{x}, y, \bar{y}, u, \bar{u}, v, \bar{v}$ we have*

$$\left.\begin{array}{rcl} xy & = & \bar{x}\bar{y} \\ xuy & = & \bar{x}\bar{u}\bar{y} \\ xvy & = & \bar{x}\bar{v}\bar{y} \end{array}\right\} \Rightarrow xuvy = \bar{x}\bar{u}\bar{v}\bar{y}.$$

Given a transducer, whether rational or not, we define its *underlying automaton* as the finite automaton obtained by ignoring the output component of each transition. We may then speak of *useful* states which are visited in at least one succesful path. We say that a transducer is *trimmed* if all its states are useful. We have the following simple but interesting result:

Theorem 20.3 *[equivalence] Let $\mathcal{T}$ and $\mathcal{T}'$ be two input deterministic transducers with unary outputs and let M be the number of useful states (q, q') in the direct product of their underlying automata. If $||\mathcal{T}||(u) = ||\mathcal{T}'||(u)$ holds for all $u \in A^*$ of length less than $2M$ then $||\mathcal{T}|| = ||\mathcal{T}'||$.*

Proof. Assume by contradiction that $||\mathcal{T}||(u) \neq ||\mathcal{T}'||(u)$ for some shortest input $u \in A^*$. Clearly $|u| \geqslant 2M$. Consider the two unique successful paths in $\mathcal{T}$ and $\mathcal{T}'$ with input label u and assume they are traversed synchronously in $\mathcal{T}$ and $\mathcal{T}'$ when reading the different prefixes of u. The combined sequence thus obtained contains $|u|+1$ pairs of states of $Q \times Q'$ and thus some pair, say (q, q'), is visited at least three times. This means that the paths can be factored in $\mathcal{T}$ and in $\mathcal{T}'$ as

$$q_- \xrightarrow[\mathcal{T}]{u_1|X_1} q \xrightarrow[\mathcal{T}]{u_2|X_2} q \xrightarrow[\mathcal{T}]{u_3|X_3} q \xrightarrow[\mathcal{T}]{u_4|X_4} q_+$$

and

$$q'_- \xrightarrow[\mathcal{T}']{u_1|X'_1} q' \xrightarrow[\mathcal{T}']{u_2|X'_2} q' \xrightarrow[\mathcal{T}']{u_3|X'_3} q' \xrightarrow[\mathcal{T}']{u_4|X'_4} q'_+,$$

where u_2 and u_3 are nonempty and $q_+ \in Q_+$ and $q'_+ \in Q'_+$. By induction, we have the equalities $X_1X_4 = X'_1X'_4$, $X_1X_2X_4 = X'_1X'_2X'_4$ and $X_1X_3X_4 = X'_1X'_3X'_4$, which by the preceding lemma implies $X_1X_2X_3X_4 = X'_1X'_2X'_3X'_4$, a contradiction. □

Now, we are ready to consider a Fatou-type property of rational relations:

Theorem 20.4 *[Fatou] Let $\mathcal{T}$ be a trimmed input deterministic transducer with unary outputs such that $||\mathcal{T}||(u)$ is rational for each $u \in A^*$ of length less than $2N$, where N is the number of states of $\mathcal{T}$. Then $||\mathcal{T}||$ is rational.*

Proof. The proof follows the general "pumping property" pattern. Indeed, consider the outputs of the transitions as unknowns X_i for $i = 1, \ldots, m$, where m is the number of transitions, and for each input u of length $\ell < 2N$ consider the equality $||\mathcal{T}||(u) = X_{i_1} \ldots X_{i_\ell}$. We obtain thus a finite system of equalities indexed by all inputs of length less than ℓ, and we know, by Theorem 20.2, that it is possible to substitute a rational subset $X'_i \subseteq \mathbb{N}$ for each output label X_i in such a way that the finite system of equalities still holds. By the previous theorem, the rational transducer $\mathcal{T}'$ thus obtained defines the same relation as $\mathcal{T}$. This completes the proof. □

Despite of the simplicity of Theorem 20.3, it is theoretically interesting. Indeed, if either condition "input deterministic" or "unary outputs" is relaxed, the problem becomes undecidable as shown in [10,8] and [11], respectively.

As yet another consequence we obtain a decision result:

Corollary 20.1 *Let $\mathcal{F}$ be a family of subsets of $\mathbb{N}$ satisfying the following properties:*

i) *$\mathcal{F}$ is closed under set union and concatenation;*

ii) *There is an algorithm which, given an element $X \in \mathcal{F}$, determines whether or not X is rational.*

Then there exists an algorithm which, given an input deterministic transducer with outputs in $\mathcal{F}$, decides whether or not $||\mathcal{T}||$ is rational.

We conclude this section by noting that we can drop the assumptions on both the input determinism and the unarity in Theorem 20.4 if we assume that the domain of a transducer is finite, e.g. the transducer is *acyclic*:

Theorem 20.5 *Let $\mathcal{T}$ be an acyclic transducer such that for each $u \in dom(\mathcal{T})$ the output $||\mathcal{T}||(u)$ is rational. Then $||\mathcal{T}||$ is rational as well.*

Proof. Follows directly from Theorem 20.2. □

5. CONCLUSIONS

The original Fatou problem can be abstracted away by considering two semirings $K' \subseteq K$. Given a K-rational series whose coefficients are in K' is it true that it is K'-rational? The classical Fatou Theorem corresponds to the case when $K' = \mathbb{Z}$ and $K = \mathbb{Q}$.

We recall that a rational relation in $A^* \times B^*$ can be interpreted, in a natural way, as a rational series over A^* with coefficients in RatB^*, this is as an element of $\text{Rat}_{\text{Rat}B^*}(A^*)$ in terms of [6]. Consequently, the above problem can be formulated for rational transducers as a question of whether or not the Fatou property holds when K' is the semiring of all rational subsets of a free monoid B^* and K is the power set of all subsets of B^*. To our knowledge this important problem is still open.

In order to comment on this, we need to recall some background on rational series. It can be proven that given a semiring K, a K-series $s : A^* \to K$ is rational if and only if there exist an integer n and a triple (λ, μ, γ) where μ is matrix representation of A^* in the monoid $K^{n\times n}$ of square matrices of dimension n with entries in K, and $\lambda \in K^{n\times 1}$ and $\gamma \in K^{1\times n}$ are two matrices such that $s(u) = \lambda\mu(u)\gamma$ holds, see [2, p.8]. Now, consider the mapping σ of K onto $\{0, 1\}$ defined by $\sigma(k) = 0$ if and only if $k = 0$ and extend it to any $K^{r\times s}$ matrix where r and s are arbitrary integers. We say that two triples (λ, μ, γ) and $(\lambda', \mu', \gamma')$ have the same *support* if $\sigma(\lambda) = \sigma(\lambda'), \sigma(\gamma) = \sigma(\gamma')$ and for all $a \in A$, $\sigma(\mu(a)) = \sigma(\mu'(a))$ holds.

Now let K' be a subsemigroup of K. A second (and stronger) version of the Fatou question asks whether the condition that $s(u) = \lambda\mu(u)\gamma$ is in K' for all $u \in A^*$ implies that there is a triple $(\lambda', \mu', \gamma')$ with the same support as (λ, μ, γ) and with entries in K' such that $s(u) = \lambda'\mu'(u)\gamma'$ holds. A third (and even stronger) version assumes that there is a natural ordering on K and that the triple $(\lambda', \mu', \gamma')$ satisfies the further condition that for all positions $\lambda_i \leqslant \lambda'_i, \gamma_i \leqslant \gamma'_i$ and $\mu_{i,j}(a) \leqslant \mu'_{i,j}(a)$ hold. The reader can easily observe that Theorem 20.4 is a partial answer to the third version of the Fatou property with $K' = \text{Rat}B^*$ and $K = \mathcal{P}(B^*)$ and $\leqslant$ is the subset inclusion.

For more Fatou properties concerning some rational relations we refer the reader to [4].

References

[1] Berstel, J. (1979), *Transductions and context-free languages*, B. G. Teubner, Stuttgart.

[2] Berstel, J. & C. Reutenauer (1988), *Rational Series and Their Languages*, Springer, Berlin.

[3] Bertoni, A. & P. Massazza, On the square root of regular languages, submitted.

[4] Choffrut, C. (1992), Rational relations and rational series, *Theoretical Computer Science*, 98: 5–13.

[5] Conway, J.H. (1971), *Regular Algebra and Finite Machines*, Chapman and Hall, London.

[6] Eilenberg, S. (1974), *Automata, Languages and Machines*, Academic Press, New York.

[7] Fatou, P. (1904), Sur les séries entières à coefficients entiers, in *Comptes Rendues de l'Académie des Sciences de Paris*: 342–344.

[8] Ibarra, O.H. (1978), The solvability of the equivalence problem for ϵ-free ngsm's with unary input (output) alphabets and application, *SIAM Journal of Computing*, 7: 520–532.

[9] Kari, L. (1991), On insertions and deletions in formal languages, PhD thesis, Turku University.

[10] Lisovik, L.P. (1979), The identity problem of regular events over cartesian products of free and cyclic semigroups, *Doclady of Academy of Sciences of Ukraine*, 6: 480–497.

[11] Lisovik, L.P. (1997), The equivalence problem for finite substitutions on regular languages, *Doclady of Academy of Sciences of Russia*, 357: 293–301.

[12] Mateescu, A.; A. Salomaa & S. Yu (1998), On the composition of finite languages, Technical Report 222, Turku Centre for Computer Science.

[13] Restivo, A. (1978), Some decision results for recognizable sets of arbitrary monoids, in *Lecture Notes in Computer Science*, 62: 363–371, Springer, Berlin.

[14] Salomaa, A. & M. Soittola (1978), *Automata-Theoretic Aspects of Formal Power Series*, Springer, Berlin.

Chapter 21

MULTIPLE KEYWORD PATTERNS IN CONTEXT-FREE LANGUAGES

Pál Dömösi
Institute of Mathematics and Informatics
Lajos Kossuth University
Egyetem tér 1, H-4032 Debrecen, Hungary
domosi@math.klte.hu

Masami Ito
Faculty of Science
Kyoto Sangyo University
Kyoto 603, Japan
ito@ksuvx0.kyoto-su.ac.jp

Abstract A polynomial algorithm is shown to solve the following problem: Given a context-free language L and a finite list of nonempty words $w_1, \ldots, w_n$, let us decide whether or not there are words $z_0, \ldots, z_n$ having $z_0 w_1 z_1 \ldots w_n z_n \in L(G)$.

1. PRELIMINARIES

Being able to match various patterns of words is an essential part of computerized information processing activities such as text editing, data retrieval, bibliographic search, query processing, lexical analysis, and linguistic analysis. One of the simplest such questions is finding a (nonempty) keyword in a text string, a problem for which several algorithms have already been developed.

*This work was supported by the "Automata and Formal Languages" project of the Hungarian Academy of Sciences and Japanese Society for the Promotion of Science (No. 15), the Japanese Society for the Promotion of Science (No's. RC29818001, RC 29718009), Grant-in-Aid for Science Research 10044098 and 10440034, Ministry of Education, Science and Culture of Japan, and the Hungarian National Science Foundation (Grant No's. T019392 and T030140).

C. Martin-Vide and V. Mitrana (eds.), Where Mathematics, Computer Science, Linguistics and Biology Meet, 237-241.

A slightly more complicated problem is to find a consecutive appearance of a finite collection of nonempty words in a text string. But the problem becomes really complicated (or even undecidable) if we want to find an element of an infinite language with a consecutive appearance of a list of nonempty words as subwords.

A multiple keyword pattern is a finite list of nonempty words. The straightforward way to look for multiple keywords in a string is to apply a single-keyword pattern-matching algorithm consecutively once for each keyword component. This method can also be applied if we want to find an element of a finite set of strings to which we can match our multiple keyword. But we cannot use this idea when we want to decide whether or not there exists such a string in an infinite language. We will give a solution of this problem when the language is context-free.

In particular, we will consider the following problem. Given a finite ordered list of nonempty words $w_1, \ldots, w_n \in X^*$, and a context-free language $L \subseteq X^*$, let us decide whether or not there is a word $z \in L$ such that $z = z_0 w_1 z_1 \ldots w_n z_n$ for some words $z_0, \ldots, z_n \in X^*$. In other words, given a regular language of the form $R = X^* w_1 X^* w_2 \ldots X^* w_n X^*, \lambda \notin \{w_1, \ldots, w_n\}$ and a context-free language L, let us decide whether or not $R \cap L$ is empty or not.

It is well-known that the emptiness problem is decidable for context-free languages. (See, for example, [3].) On the other hand, it is clear that $R \cap L$ is a context-free language. Therefore, the discussed problem is decidable. We will show an algorithm to decide this problem. (We also note that R is defined as the shuffle ideal generated by the words $w_1, \ldots, w_n$ provided $w_1, \ldots, w_n \in X^*$.)

For all notions and notations not defined here, see [4, 7]. Consider an *alphabet* X and the *free monoid* X^* generated by X. λ denotes the *identity* of X^*, $X^+ = X^* \setminus \{\lambda\}$, and $| p |$ is the *length* of $p \in X^*$. The set of all *subwords* of any word p is denoted by $Sub(p)$. For any *language* $L \subseteq X^*$, we put $Sub(L) = \cup\{Sub(p) \mid p \in L\}$. In addition, we will denote by $|H|$ the cardinality of a given set H. Finally, we shall consider a grammar in the form $G = (V, X, S, P)$, where, in order, V and X are the sets of *variables* and *terminals*, respectively, S denotes the *start symbol*, and P is the set of *productions*. Moreover, $L(G)$ denotes the *language generated by* G.

The next statement can be derived directly from results in [2].

Theorem 21.1 *For any context-free grammar $G = (V, X, S, P)$ given in Chomsky normal form and $z \in L(G)$, if $|z| \geqslant |V|2^{|V|}\, e$ $(e > 0)$, and e positions of z are excluded, then z has the form $uvwxy$ where $|vx| > 0$, vx does not contain any excluded position, and $uv^i wx^i y$ is in $L(G)$ for all $i \geqslant 0$.*

2. MAIN RESULTS

We start with the following

Theorem 21.2 *Consider a context-free grammar $G = (V, X, S, P)$ and a word $z'_0 w_1 z'_1 \dots w_n z'_n \in L(G)$ with $\lambda \notin \{w_1, \dots, w_n\}$. There are words $z_1, \dots, z_n$ such that $z_0 w_1 z_1 \dots w_n z_n \in L(G)$ and $|z_0 w_1 z_1 \dots w_n z_n| < |V| 2^{|V|} |w_1 \dots w_n|$.*

Proof. Let $w_1, \dots, w_n$ be a fixed list of nonempty words and consider the shortest word p having $p = z'_0 w_1 z'_1 \dots w_n z'_n \in L(G)$ for some $z'_1, \dots, z'_n$. If $|p| < |V| 2^{|V|} |w_1 \dots w_n|$ then we are ready. Otherwise exclude all positions of $w_1, \dots, w_n$ in $z'_0 w_1 z'_1 \dots w_n z'_n$ and construct a decomposition $uvwxy$ of the word $z'_0 w_1 z'_1 \dots w_n z'_n$ such that $|vx| > 0$, and vx does not contain any excluded position. Then $uwy \in L(G)$; moreover, $uwy = z_0 w_1 z_1 \dots z_n w_n$ for some $z_1, \dots, z_n$ with $|uwy| < |uvwxy|$, a contradiction.

This ends the proof. □

Given a context-free grammar $G = (V, X, S, P)$, let us construct the grammar $G' = (V', X', S', P')$ such that $V' = V \cup \{\Omega\}$, $X' = X \cup \{\omega\}$, $P' = P \cup \{A \to A\Omega, A \to \Omega A : A \in V\} \cup \{\Omega \to \Omega\Omega, \Omega \to \omega\}$, where Ω and ω are new nonterminal and terminal symbols, respectively. The following statement is obvious.

Lemma 21.1 $L(G') = \{\omega^* x_1 \omega^* x_2 \dots \omega^* x_n \omega^* : x_1 \dots x_n \in L(G), x_1, \dots, x_n \in X\}$.

Next we prove

Theorem 21.3 *Given a context-free grammar $G = (V, X, S, P)$, let $w_1, \dots w_n$ be an arbitrary list of nonempty words. There are words $z_0, \dots, z_n$ with $z_0 w_1 z_1 \dots w_n z_n \in L(G)$ if and only if $z'_0 w_1 z'_1 \dots w_n z'_n \in L(G')$ holds for some words $z'_0, \dots, z'_n \in X'^* (= (X \cup \{\omega\})^*)$ with $|z'_i| = |V| 2^{|V|} |w_1 \dots w_n|$.*

Proof. First we suppose $z'_0 w_1 z'_1 \dots w_n z'_n \in L(G')$ for some words $z'_0, \dots, z'_n \in X'^* (= (X \cup \{\omega\})^*)$ with $|z'_i| = |V| 2^{|V|} |w_1 \dots w_n|$. Consider the words $z_i, i = 0, \dots, n$ such that for every $i = 0, \dots, n$, we omit all occurrences of the letter ω in z'_i. By Lemma 2.2, $z_0 w_1 z_1 \dots w_n z_n \in L(G)$.

Conversely, we now suppose $z_0 w_1 z_1 \dots w_n z_n \in L(G)$ for some $z_0, \dots, z_n$. By Theorem 21.2, we may assume $|z_0 w_1 z_1 \dots w_n z_n| < |V| 2^{|V|} |w_1 \dots w_n|$. Therefore, $|z_i| < |V| 2^{|V|} |w_1 \dots w_n|, i = 0, \dots, n$. Thus, let us define $z'_i = z_i \omega^{|V| 2^{|V|} (|w_1 \dots w_n| - |z_i|)}$. By Lemma 2.2, our conditions hold again.

The proof is complete. □

3. GENERALIZED CYK ALGORITHM

Now we are ready to describe our algorithm.

Using the well-known CYK-algorithm (Cocke-Younger-Kasami algorithm), on the basis of Theorem 21.3, in this part we give a cubic time algorithm to the

problem under discussion. For the description of our extended CYK algorithm, we will follow [3] and [4].
Let $G = (V, X, S, P)$ be a context-free grammar having Chomsky normal form and let $w_1, \ldots, w_n \in X^+$. Consider the grammar $G' = (V', X', S', P')$ such that $V' = V \cup \{\Omega\}$, $X' = X \cup \{\omega\}$, $P' = P \cup \{A \to A\Omega, A \to \Omega A : A \in V\} \cup \{\Omega \to \Omega\Omega, \Omega \to \omega\}$, where Ω and ω are new nonterminal and terminal symbols, respectively. Put

$$p = \omega^{|V|2^{|V|}|w_1 \ldots w_n|} w_1 \omega^{|V|2^{|V|}|w_1 \ldots w_n|} w_2 \ldots \omega^{|V|2^{|V|}|w_1 \ldots w_n|} w_n \omega^{|V|2^{|V|}|w_1 \ldots w_n|}.$$

For terminal words $z_0 w_1 z_1 \ldots w_n z_n \in L(G')$ having $|p| = |z_0 w_1 z_1 \ldots w_n z_n|$ all possible derivation trees can be arranged in a triangular matrix whose last diagonal column has $|p|$ cells and each other diagonal column has one less cell than the column just below it. This matrix is called a *recognition matrix.* The aim of the recognizing algorithm is to enter the appropriate variables of the grammar in the cells of the matrix. The matrix will be filled up along the diagonals column by column, from the bottom up where each column is filled up along from left to right. Each cell of the recognition matrix may contain zero or more variables. Given the above defined input word p, then the bottom column will be computed in such a way that $A \in V$ is entered in the cell $(i, 1)$ whenever $A \to a \in P$ and the ith letter of p is not ω. Moreover, all elements of V' $(= V \cup \{\Omega\})$ are entered in the cell $(i, 1)$ whenever the ith letter of p is ω. The next column will then be computed by entering $A \in V'$ in the cell $(i, 2)$ whenever $A \to BC \in P'$ for some $B, C \in V'$ and B occurs in the cell $(i, 1)$ while C occurs in the cell $(i + 1, 1)$. In general, the cell (i, j) for $i > j$ and $i + j = 2d + \mid p \mid +1$ will contain the variable A ,if and only if, $A \to BC$ is a rule of the grammar G' for some B and C such that B occurs in the cell (i, k) while C occurs in the cell $(i + k, j - k)$ for some k with $k = 1, \ldots, j - 1$. This computation makes certain that a variable A is entered in the cell (i, j) if and only if there exists a subtree representing the derivation of the portion of the input strings below the cell (i, j), that is, if and only if $P \overset{*}{\underset{G'}{\Rightarrow}} b_i \ldots b_j, b_i, \ldots, b_j \in X'(= X \cup \omega)$, where $b_i \ldots b_j$ is a subword of a word $p' \in L(G')$ of length $|p|$ having the form $z'_0 w_1 z'_1 \ldots w_n z'_n$ for some $z'_0, \ldots, z'_n \in X'^*$ $(= X \cup \{\omega\}$. Therefore, having finished the computation of the matrix, we have to check only whether S has been entered in the top part of the matrix having indices $(1, |p|)$. If so, then the language $L(G)$ has a word of the form $z_0 w_1 z_1 \ldots w_n z_n$; otherwise it does not, since all possible ways of deriving words of the form $z_0 w_1 z_1 \ldots w_n z_n, z_0 \ldots, z_n \in X^*$ have been taken into account during the computation of the matrix. (See Theorem 21.3.) Now we give a formal description of our algorithm.

begin

$p := \omega^{|V|2^{|V|}|w_1 \ldots w_n|} w_1 \omega^{|V|2^{|V|}|w_1 \ldots w_n|} w_2 \ldots \omega^{|V|2^{|V|}|w_1 \ldots w_n|} w_n$
$\omega^{|V|2^{|V|}|w_1 \ldots w_n|}$;

for i:=1 **to** $|p|$ **do**

if the ith symbol of p is ω then **do**

$V_{i,1} := V \cup \{\Omega\}$;

else do

$V_{i,1} := \{A \mid \exists a \in X$ suchthat $A \to a$ is a production in G and the ith symbolin p equals to a $\}$;

for $j := 2$ **to** $|p|$ **do**

for $i := 1$ to $|p| - j + 1$ **do**

begin

$V_{i,j} := \emptyset$;

for $k := 1$ **to** $j - 1$ **do**

$V_{i,j} := V_{i,j} \cup \{A \mid A \to BC$ is a production in G',B is in $V_{i,k}$ and C is in $V_{i+k,j-k}\}$;

end

$z_0 w_1 z_1 \ldots z_n w_n \in L(G)$ for some $z_0, \ldots, z_n \in X^*$iff $S \in V_{1,|p|}$.

end

References

[1] Dömösi, P. & M. Ito (1998), Characterization of languages by length of their subwords, in K.P. Shum; Y. Gao; M. Ito & Y. Fong, eds., *Proceedings of the International Conference on Semigroups and its Applications, Semigroups*: 117–129. Springer, Berlin.

[2] Dömösi, P.; M. Ito; M. Katsura & C. Nehaniv (1997), A new pumping property of context-free languages, in *Proceedings of DMTCS'96, Combinatorics, Complexity & Logic*: 187–193. Springer, Berlin.

[3] Hopcroft, J.E. & J.D. Ullman (1979), *Introduction to Automata Theory, Languages, and Computation*, Addison-Wesley, Reading, Mass.

[4] Révész, G. (1983), *Introduction to Formal Languages*, McGraw-Hill, New York.

[5] Salomaa, A. (1973), *Formal Languages*, Academic Press, New York.

Chapter 22

READING WORDS IN GRAPHS GENERATED BY HYPEREDGE REPLACEMENT

Frank Drewes
Department of Computer Science
University of Bremen
Postfach 330440, D-28334 Bremen, Germany
drewes@informatik.uni-bremen.de

Hans-Jörg Kreowski
University of Bremen
Postfach 330440, D-28334 Bremen, Germany
kreo@informatik.uni-bremen.de

Abstract The context-free syntax of programming languages like PASCAL and MODULA is often defined by means of syntax diagrams in a quite intuitive way. The syntactically correct programs are obtained by reading words along paths in syntax diagrams in a certain way. In this paper, we interpret hyperedge replacement graph languages as sets of syntax diagrams and investigate the string languages definable in this way.

1. INTRODUCTION

Walking along a path in, say, a directed and edge-labelled graph and picking up the labels of passed edges, one reads a word. This basic procedure is used in various contexts in the literature. First of all, if the considered graph is the state graph of a finite automaton then the words along paths from the initial state to some final state form the regular language accepted by the finite automaton. Secondly, syntax diagrams are often used to define the context-free syntax of programming languages like PASCAL and MODULA. The syntactically correct programs are obtained by reading words along paths in syntax diagrams in a certain way. Thirdly, a graph grammar may generate a language of string

C. Martín-Vide and V. Mitrana (eds.), Where Mathematics, Computer Science, Linguistics and Biology Meet, 243-252.

graphs, i.e., graphs consisting of single simple paths. By reading the words along these paths, a string language is obtained. In the case of hyperedge replacement grammars, a context-free type of graph grammar, it is known that the class of languages definable in this way subsumes the class of context-free languages, but contains non-context-free languages like $\{a^n b^n c^n \mid n \geqslant 1\}$ as well (see [5, 4, 1]).

In this paper, we employ and investigate the language generating mechanism of syntax diagrams in the more general setting of hyperedge replacement grammars. Such a grammar generates a set of (hyper-)graphs each of which has a sequence of distinguished, so-called external nodes. Reading the words along paths between two fixed external nodes, one gets a language of words. The class of languages specified in this way strictly contains the class of languages discussed in the third case above. The language $\{(a^n b^n)^m \mid m, n \geqslant 1\}$ is a separating example. Our main result is that the considered languages are characterized by a decomposition property which is based on the context freeness lemma for hyperedge replacement languages. As a consequence, it turns out that the membership problem of the new language class is in NP.

2. SYNTAX DIAGRAMS

Syntax diagrams are quite a popular concept. They are widely used as an intuitive means of defining the context-free syntax of programming languages. A syntactic feature represented by a nonterminal symbol A is associated with a syntax diagram $D(A)$ which may be seen as a directed graph the edges of which are labelled with terminal or nonterminal symbols. Moreover, every such diagram comes with two distinguished nodes *begin* and $\exists$. In the literature, one can encounter various ways of obtaining the language $L(D(A))$ of syntactically correct strings defined by $D(A)$. Two of these methods are discussed below.

First, let N and T be the sets of nonterminals and terminals, respectively. For every $A \in N$, $D(A)$ may be considered as the state graph of a finite automaton with *begin* as its initial and $\exists$ as its final state. Since the language $S(D(A))$ of sentential forms accepted by this automaton is regular, it can be specified by a regular expression $R(A)$. Now, the set of all pairs $(A, R(A))$ can be interpreted as a set P of context-free productions $A ::= R(A)$ in extended Backus-Naur form, yielding the syntax of a language in the traditional style. In this way, every nonterminal $A \in N$ determines a context-free grammar $G(A) = (N, T, P, A)$ whose generated language $L(G(A))$ consists of all syntactically correct strings with respect to the syntactic feature represented by A. Thus, one can define $L(D(A)) = L(G(A))$.

Second, one may use the pairs $A ::= D(A)$ for $A \in N$ as edge replacement rules. This allows one to derive syntax diagrams from syntax diagrams by replacing edges with nonterminal labels by their associated diagrams. In this way, one obtains a set $SD(A)$ of syntax diagrams for every $A \in N$, every derived diagram containing the two distinguished nodes *begin* and $\exists$. Now, $L(D(A))$ is obtained by reading the labels of the edges along paths from *begin* to $\exists$ in diagrams of $SD(A)$ if only paths with terminal labels are considered.

3. HYPEREDGE REPLACEMENT AND READABLE WORDS

In this section, we generalize the idea of reading words in syntax diagrams to sets of diagrams generated by hyperedge replacement grammars. For this, we first have to formalize some basic notations.

Given a natural number n, $[n]$ denotes the set $\{1, \ldots, n\}$. For a sequence $w = a_1 \cdots a_n \in S^*$ (where S is a set and $a_1, \ldots, a_n \in S$), $\mathbf{card}\, w$ denotes its length n, $[w]$ denotes the set $\{a_1, \ldots, a_n\}$, and $w(i) = a_i$ for $i \in [n]$. If $w = f(a)$, where $f: A \to S^*$ is a mapping and $a \in A$, then one may write $f(a, i)$ instead of $f(a)(i)$.

A *labelling alphabet* is a set C such that every $a \in C$, called a *label*, is given a type $type(a) \in \mathbb{N}$. A *hypergraph* H with labels in C is a quintuple $(V_H, E_H, att_H, lab_H, ext_H)$ such that

- V_H and E_H are finite sets of *nodes* and *hyperedges*, respectively,
- $att_H: E_H \to V_H^*$ is the *attachment function*, where $att_H(y)$ is a sequence without repetitions for every $y \in E_H$,
- $lab_H: E_H \to C$ is the *labelling function*, where $type(lab_H(y)) = \mathbf{card}\, att_H(y)$ for all $y \in E_H$, and
- $ext_H \in V_H^*$ is a sequence of nodes without repetitions, called the sequence of *external nodes*.

The *type* of a hypergraph H is given by $type(H) = \mathbf{card}\, ext_H$. Similarly, for every hyperedge $y \in E_H$ the *type* of y in H is $type_H(y) = \mathbf{card}\, att_H(y)$. Thus, $type_H(y) = type(lab_H(y))$ for all $y \in E_H$. A hypergraph is a *graph* if all its hyperedges are edges, i.e., hyperedges of type 2. The set of all hypergraphs with labels in C is denoted by $\mathcal{H}_C$. If C is irrelevant, the notation $\mathcal{H}$ may be used instead.

Hyperedge replacement is a context-free mechanism to hypergraphs from hypergraphs. Let $H \in \mathcal{H}$ and let $repl: E \to \mathcal{H}$ be a mapping, where $E \subseteq E_H$ and $type(repl(y)) = type_H(y)$ for all $y \in E$. Then $REPL(H, repl)$ denotes the hypergraph obtained from H by removing the hyperedges in E, adding disjointly each of the hypergraphs $repl(y)$, for $y \in E$ (without changing ext_H), and identifying $ext_{repl(y)}(i)$ with $att_H(y, i)$ for all $y \in E$ and $i \in [type_H(y)]$.

A *hyperedge replacement grammar* is a system $G = (N, T, P, Z)$ consisting of disjoint finite labelling alphabets N and T of *nonterminals* and *terminals*, a finite set P of *productions* $A ::= R$ such that $A \in N$, $R \in \mathcal{H}_{N \cup T}$, and

$type(A) = type(R)$, and an *axiom* $Z \in \mathcal{H}_{N \cup T}$. A hypergraph $H \in \mathcal{H}_{N \cup T}$ directly derives a hypergraph H', denoted by $H \Rightarrow_G H'$, if $H' = REPL(H, repl)$ for some $repl: E \to \mathcal{H}$, where $E \subseteq E_H$ and $lab_H(y) ::= repl(y) \in P$ for all $y \in E$. As usual, a derivation $H_0 \Rightarrow_G H_1 \Rightarrow_G \cdots \Rightarrow_G H_n$ of length $n \in \mathbb{N}$ (including the case $n = 0$) may be abbreviated by $H_0 \Rightarrow_G^n H_n$, or $H_0 \Rightarrow_G^* H_n$ if n does not matter. The *language generated by* G is the set

$$L(G) = \{H \in \mathcal{H}_T \mid Z \Rightarrow_G^* H\},$$

which is also called a *hyperedge replacement language*. Notice that all hypergraphs in a hyperedge replacement language have the same type, because they inherit their external nodes from the axiom.

It is often useful to assume that the axiom of a hyperedge replacement grammar is a single hyperedge. More precisely, for every label A of type n $A^\bullet$ denotes the hypergraph $(\{v_1, \ldots, v_n\}, \{y\}, att, lab, ext)$ with $att(y) = ext = v_1 \cdots v_n$ and $lab(y) = A$. Then, every hyperedge replacement grammar $G = (N, T, P, Z)$ can be transformed into a hyperedge replacement grammar G' generating the same language, such that the axiom of G' has the form $S^\bullet$ for some label S. Obviously, this can be done by adding a new nonterminal label S of type $type(Z)$ and the production $S ::= Z$.

Intuitively, hyperedge replacement is context-free because hyperedges are atomic items whose replacement does not affect the rest of the hypergraph. This is formalized by the so-called context-freeness lemma [4] which, once again, turns out to be very convenient.

Lemma 22.1 (context-freeness lemma) *Let* $G = (N, T, P, Z)$ *be a hyperedge replacement grammar,* $A \in N$, *and* $H \in \mathcal{H}$. *There is a derivation* $A^\bullet \Rightarrow_G^{n+1} H$ *if and only if* $H = REPL(R, repl)$ *for some production* $A ::= R$ *and a mapping* $repl: E_R \to \mathcal{H}$ *such that* $lab_R(y)^\bullet \Rightarrow_G^{n_y} repl(y)$ *for all* $y \in E_R$, *where* $n = \sum_{y \in E_R} n_y$.

As mentioned in the introduction, the purpose of this paper is to study the string languages obtained by walking along the paths in hypergraphs generated by a hyperedge replacement grammar, thereby reading the labels of the hyperedges being passed. The rest of this section is devoted to the formalization of these notions.

Let $H \in \mathcal{H}$ and $v, v' \in V_H$. A *path* from v to v' in H is a pair $p = (y_1 \cdots y_k, v_0 \cdots v_k)$ consisting of a sequence $y_1 \cdots y_k \in E_R^*$ and a sequence $v_0 \cdots v_k \in V_R^*$ for some $k \in \mathbb{N}$, subject to the following conditions:

- $v = v_0 \in [att_H(y_1)]$, $v' = v_k \in [att_H(y_k)]$, and
- for $i = 1, \ldots, k-1$, $v_i \in [att_H(y_i)] \cap [att_H(y_{i+1})]$.

The node v is called the *beginning* of p, denoted by $b(p)$, the node v' the *end* of p, denoted by $e(p)$, and the number k the *length* of p, denoted by $len(p)$.

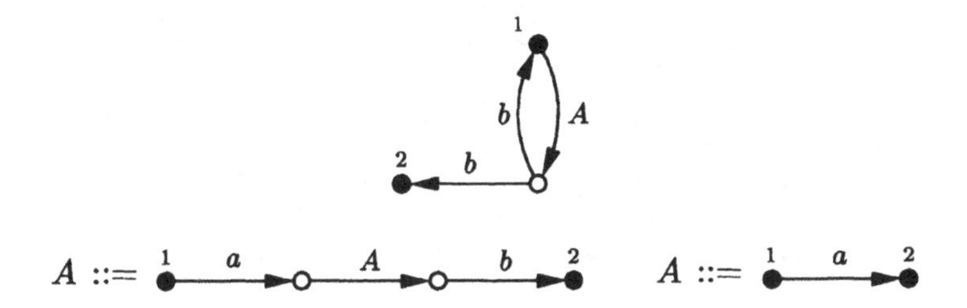

Figure 22.1 The axiom and the two productions considered in Example 22.1.

The path from v to v of length 0 is said to be *empty*. Given a second path $p' = (y'_1 \cdots y'_l, v'_0 \cdots v'_l)$ with $v_k = v'_0$, the *composition* of p and p' yields the path $pp' = (y_1 \cdots y_k y'_1 \cdots y'_l, v_0 \cdots v_k v'_1 \cdots v'_l)$.

By the definition of hypergraphs, the attachment of hyperedges is not allowed to contain repetitions. Therefore, for every path p as above and every $l \in [k]$ there are unique $i, j \in [type_H(y_l)]$ such that $v_{l-1} = att_H(y_l, i)$ and $v_l = att_H(y_l, j)$. Intuitively, the lth hyperedge on the path is entered through its ith attached node and left through the jth. In the following, the pair (i, j) is denoted by $num_p(l)$ (assuming that the hypergraph H referred to is clear from the context). The path p is *directed* if $num_p(l) = (i, j)$ implies $i < j$ for all $l \in [k]$.

Let H be a hypergraph and $i, j \in [type(H)]$. A path $p = (y_1 \cdots y_k, v_0 \cdots v_k)$ in H is an *(i, j)-passage* or just a *passage* through H if $v_0 = ext_H(i)$ and $v_k = ext_H(j)$. In this case, the word $lab_H(y_1) \cdots lab_H(y_k)$ is said to be *readable along p*. The set of all words which are readable along directed (i, j)-passages through H is denoted by $READ_{ij}(H)$. If $L \subseteq \mathcal{H}$ is a hyperedge replacement language consisting of hypergraphs of type 2, then the language $READ(L) = \bigcup_{H \in L} READ_{12}(H)$ is called an *hr readable language*.

Example 22.1 (hr readable language) *Let $G = (\{A\}, \{a, b\}, P, Z)$, where Z and P are as shown in Figure 22.1. (External nodes are depicted as filled dots, using numbers to indicate their order. Edges are drawn as arrows pointing from the first attached node to the second.) G generates all graphs of the form depicted in Figure 22.2 (where the cycle contains as many a-labelled edges as b-labelled edges). Thus, $READ(L(G)) = \{(a^n b^n)^m \mid n, m \geqslant 1\}$ is hr readable. Using nonterminals of a greater type one can easily modify G in such a way that, for some arbitrary but fixed $k \in \mathbb{N}$, the hr readable language $\{(a_1^n \cdots a_k^n)^m \mid m, n \geqslant 1\}$ is obtained.*

It should be mentioned that every hr readable language is of the form $READ(L)$ for some hyperedge replacement language L consisting of graphs. This is easy to see because, for a hypergraph H and $i, j \in [type(H)]$, $READ_{ij}(H)$ remains unchanged if every hyperedge $y \in E_H$ is replaced with

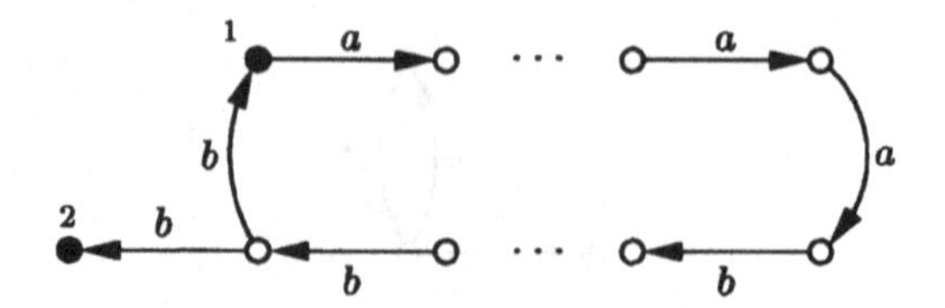

Figure 22.2 The sort of graphs generated in Example 22.1.

a $type_H(y)$-clique of edges labelled with $lab_H(y)$ (provided that the edges are directed in the appropriate way).

A graph of type 2 is called a *string graph* if it consists of a single $(1, 2)$-passage (and no further nodes and hyperedges). A string language is said to be an *hr string language* if it has the form $READ(L)$ for some hyperedge replacement language L consisting of string graphs. Hr string languages are studied in [5, 4] and in particular in [1], where they are characterized by means of tree-walking automata. Clearly, every context-free string language[1] is an hr string language. Moreover, every hr string language is an hr readable language by definition. As observed in [4], the language $\{a^n b^n c^n \mid n \geqslant 1\}$ is an hr string language, which means that the class of context-free languages is strictly contained in the class of hr string languages. Using the example above, one can show that, in turn, the hr readable languages form a proper superset of the hr string languages.

The argument makes use of the pumping lemma for hyperedge replacement languages (see [5, 4]). In order to keep things simple, consider a hyperedge replacement language L whose elements contain neither parallel hyperedges (i.e., hyperedges with identical attachment sequences) nor isolated nodes. For a hypergraph $G \in L$, let $size(G)$, the *size* of G, be the cardinality of E_G. Then the lemma says that there is a *pumping index* $p \in \mathbb{N}$ such that every hypergraph in L of size at least p can be written as $H_0[H[H_1]]$ (the definition of $[\ldots]$ is given below), where $1 < size(H) \leqslant p$, such that the hypergraph

$$H_0[\underbrace{H[\cdots[H}_{n \text{ times}}[H_1]]\cdots]]$$

is in L for each $n \in \mathbb{N}$. Here, H_0 and H are hypergraphs containing a distinguished hyperedge Y whose type equals the type of H and H_1. For all $G, G' \in \mathcal{H}$ with $Y \in E_G$ and $type_G(Y) = type(G')$, $G[G'] = REPL(G, repl)$ where $repl\colon \{Y\} \to \mathcal{H}$ maps Y to G'.

Assume that $\{(a^n b^n)^m \mid n, m \geqslant 1\}$ is an hr string language and let p be the pumping index of the corresponding string graph language. Consider the string graph representing $(a^n b^n)^m$. For large enough m, the hypergraph H in the

[1]Here, we restrict our attention to context-free languages L such that $\lambda \notin L$.

pumping lemma cannot contain edges from each of the $m-1$ components yielding substrings of the form ab^na. Thus, the string represented by $H_0[H_1]$ contains ab^na as a substring, which means that it equals $(a^nb^n)^k$ for some $k \in \mathbb{N}$. Since $size(H) > 1$ it follows that $k < m$. This shows that the difference in size between $H_0[H[H_1]]$ and $H_0[H_1]$ is at least $2n$, which, for large n, contradicts the assumption $size(H) \leqslant p$. Thus, we have obtained the following theorem.

Theorem 22.1 *The set of hr string languages is properly contained in the set of hr readable languages.*

4. CHARACTERIZATION

In this section, we exploit the context-freeness lemma in order to characterize the readable words of hyperedge replacement languages. Since readable words are closely related to passages, we begin by analyzing the composability of passages from passages.

Aiming at an application of the context-freeness lemma, one has to look at a hypergraph H of the form $H = REPL(R, repl)$ for some $R \in \mathcal{H}$ and $repl\colon E_R \to \mathcal{H}$. By the definition of hyperedge replacement, a passage through H can only contain edges of $repl(y)$ for some $y \in E_R$ if the passage enters and leaves $repl(y)$ through two of its external nodes. Consequently, every passage p can be decomposed into $p_1 \cdots p_n$ in such a way that there are hyperedges $y_1, \ldots, y_n \in E_R$ such that (a) $q = (y_1 \cdots y_n, b(p_1)b(p_2) \cdots b(p_n)e(p_n))$ is a passage in R and (b) for all $l \in [n]$, p_l is a $num_q(l)$-passage in $repl(y_l)$.

Obviously, this holds as well if $p_1, \ldots, p_n$ are required to be nonempty (i.e., $len(p_i) > 0$ for all $i \in [n]$), which implies $n \leqslant len(p)$. Conversely, it is clear that $p = p_1 \cdots p_n$ is a passage whenever q and $p_1, \ldots, p_n$ satisfy these conditions. This amounts to the following lemma.

Lemma 22.2 *Let $H = REPL(R, repl)$ for some hypergraph $R \in \mathcal{H}$ and a mapping $repl\colon E_R \to \mathcal{H}$, and let $i, j \in [type(H)]$. A path p is an (i,j)-passage in H if and only if there are*

- *an (i,j)-passage $q = (y_1 \cdots y_n, v_0 \cdots v_n)$ through R, where $n \leqslant len(p)$, and*
- *for every $l \in [n]$ a $num_q(l)$-passage p_l in $repl(y_l)$*

such that $p = p_1 \cdots p_n$.

Together with the context-freeness lemma, Lemma 22.2 yields a characterization of readable words. However, since two component passages p_i and p_j in the decomposition of a passage p may be different even if they stem from the same hyperedge, one cannot continue the decomposition recursively without speaking about sets of readable words. In fact, in order to avoid mixing up the words which stem from the same hyperedge, but connect different pairs of external nodes, one even has to deal with families of sets of words. This leads to the following definition.

Let $G = (N, T, P, Z)$ be a hyperedge replacement grammar, and consider some $A \in N \cup T$ with $type(A) = n$. A family $(W_{ij})_{i,j \in [n]}$ of sets of words is said to be *generated by* A if $A^\bullet \Rightarrow_G^* H$ for some $H \in \mathcal{H}_T$ such that $W_{ij} \subseteq READ_{ij}(H)$ for all $i, j \in [n]$. Notice that the case $A \in T$ is trivial: $(W_{ij})_{i,j \in [n]}$ is generated by $A \in T$ if and only if $W_{ij} \subseteq \{A\}$ for $i < j$ and $W_{ij} = \emptyset$ otherwise, because $A^\bullet$ is the only hypergraph which can be derived from $A^\bullet$. Another point which should be noticed is that a word w is in the hr readable language $READ(L(G))$ (assuming that $type(Z) = 2$) if and only if the family $(W_{ij})_{i,j \in [n]}$ with $W_{12} = \{w\}$ and $W_{ij} = \emptyset$ for $(i, j) \neq (1, 2)$ is generated by A. Thus, the word problem for hr readable languages turns out to be a special case of the question whether a family is generated by some $A \in N$ (since it can be assumed without loss of generality that $Z = S^\bullet$ for some $S \in N$).

Now, combining Lemma 22.2 with the context-freeness lemma and expressing everything in terms of generated families instead of passages, the main theorem of this paper is obtained.

Theorem 22.2 *Let $G = (N, T, P, Z)$ be a hyperedge replacement grammar and let $A \in N$ with $type(A) = n$. A family $(W_{ij})_{i,j \in [n]}$ is generated by A if and only if $W_{ij} = \emptyset$ for all $i, j \in [n]$ or there are*

(1) a rule $A ::= R$ in P,

(2) for every $y \in E_R$ a family $(W_{ij}^y)_{i,j \in [type_R(y)]}$ generated by $lab_R(y)$, and

(3) for every word $w \in W_{ij}$ ($i, j \in [n]$) an (i, j)-passage $q = (y_1 \cdots y_m, v_0 \cdots v_m)$ through R (for some $m \in \mathbb{N}$) such that $w = w_1 \cdots w_m$ for nonempty words $w_1, \ldots, w_m$ with $w_l \in W_{num_q(l)}^{y_l}$ for all $l \in [m]$.

For finite sets W_{ij} the theorem yields a nondeterministic decision procedure which determines whether $(W_{ij})_{i,j \in [n]}$ is generated by $A \in N \cup T$. As mentioned above, for $A \in T$ this can directly be verified. For $A \in N$, the input is accepted if $W_{ij} = \emptyset$ for all $i, j \in [n]$. Otherwise, the algorithm first "guesses" the production $A ::= R$ in (1) and, for all $i, j \in [n]$ and every $w \in W_{ij}$, the passage q and the decomposition into nonempty subwords $w_1, \ldots, w_m$ required in (3). Afterwards, the families $(W_{ij}^y)_{i,j \in [type_R(y)]}$ are collected: Every w_l is included in the set $W_{ij}^{y_l}$ for which $(i, j) = num_q(l)$ (all the guessed decompositions of all the words w contributing to the same families!). Finally, it is recursively verified that each family $(W_{ij}^y)_{i,j \in [type_R(y)]}$ obtained in this way is generated by $lab_R(y)$. Concerning termination, consider first the case where $m > 1$ for every decomposition in (3). In this case the maximum length of the considered words decreases in every step, so the algorithm terminates after a linear number of steps because no further decomposition is possible. Each step requires only linear time as the sum of the lengths of all words never increases. The case remains to be considered the case where the decomposition yields $w = w_1$, i.e., $m = 1$. Clearly, the algorithm can reject its input if this happens **card** N times in succession, because this indicates a loop. Thus, the running time remains quadratic as it is increased by at most a constant factor. Altogether, we obtain the following corollary.

Corollary 22.1 *The word problem for hr readable languages is in NP.*

5. CONCLUSION

In this paper, we have started to investigate hr readable languages—string languages which are obtained by interpreting context-free graph languages as sets of syntax diagrams (where context-freeness means that these languages are generated by hyperedge replacement grammars). The basic idea is to read words along directed paths in the generated graphs. Further studies, including the following aspects and questions, may help to estimate the significance of this approach.

(1) Theorem 22.1 shows that the class of hr string languages [1] is properly contained in the class of hr readable languages. For which classes of hyperedge replacement grammars do they coincide?

(2) Can the class of hr readable languages be characterized in terms of other language-generating or accepting devices?

(3) Under which conditions does Theorem 22.2 (in connection with the reasoning leading to Corollary 22.1) provide an efficient solution of the word problem? Is the general case NP-complete?

(4) Does the generating power increase if we use confluent node replacement (C-edNCE grammars, see [3]) instead of hyperedge replacement? In [2] these string languages are shown to be the output languages of nondeterministic tree-walking transducers.

References

[1] Engelfriet, J. & L. Heyker (1991), The string generating power of context-free hypergraph grammars, *Journal of Computer and System Sciences*, 43: 328–360.

[2] Engelfriet, J. & V. van Oostrom (1996), Regular description of context-free graph languages, *Journal of Computer and System Sciences*, 53: 556–574.

[3] Engelfriet, J. & G. Rozenberg (1997), Node replacement graph grammars, in G. Rozenberg, ed., *Handbook of Graph Grammars and Computing by Graph Transformation*, I: 1–94. World Scientific, London.

[4] Habel, A. (1992), *Hyperedge Replacement: Grammars and Languages*, Springer, Berlin.

[5] Habel, A. & H.J. Kreowski (1987), *Some structural aspects of hypergraph languages generated by hyperedge replacement:* 207–219. Springer, Berlin.

Chapter 23

REGULARLY CONTROLLED FORMAL POWER SERIES

Henning Fernau
Wilhelm-Schickard Institute of Informatics
University of Tübingen
Sand 13, D-72076 Tübingen, Germany
fernau@informatik.uni-tuebingen.de

Werner Kuich
Institute of Algebra and Computer Mathematics
Theoretical Informatics
Technical University of Wien
Wiedner Hauptstraße 8–10, A-1040 Wien, Austria
kuich@tuwien.ac.at

Abstract Regulated rewriting is one of the classical topics in formal language theory, see [3, 2]. This paper starts the research of regulated rewriting in the framework of formal power series, cf. [6, 7, 9]. More specifically, we model what is known as "free derivations" and "leftmost derivations of type 1" within context-free grammars controlled by regular sets in the language case. We show that the class which is the formal power series analogue of controlled free derivations forms a semiring containing the semiring of algebraic series, which in turn is characterized by the formal power series analogue of controlled leftmost derivations of type 1.

1. INTRODUCTION AND PRELIMINARIES

Our aim is to extend regulated rewriting, more specifically, context-free derivations controlled by regular control sets as introduced by Ginsburg and Spanier in [5], towards a formal power series framework. To this end, we as-

C. Martin-Vide and V. Mitrana (eds.), Where Mathematics, Computer Science, Linguistics and Biology Meet, 253-265.

sume the reader is familiar with both the basics of (algebraic) formal power series and of regulated rewriting, as seen in [3, 2, 6, 7].

We briefly recall the concept of an ω-continuous semiring. Consider a semiring $\langle A, +, \cdot, 0, 1\rangle$, where A is the underlying set of the semiring, $+$ denotes the addition with neutral element 0, and $\cdot$ the multiplication with neutral element 1. A semiring is called *naturally ordered* if the relation $\sqsubseteq$ defined by

$$a \sqsubseteq b \text{ iff there exists a } c \text{ such that } a + c = b$$

is a partial order. A semiring is called *complete* if arbitrary infinite sums of elements in A are well-defined. A complete naturally ordered semiring is called *ω-continuous* if, for all sequences $(a_i \mid i \in \mathbb{N}) \in A^{\mathbb{N}}$ and all $c \in A$, $\sum_{0 \leqslant i \leqslant n} a_i \sqsubseteq c$ implies $\sum_{i \in \mathbb{N}} a_i \sqsubseteq c$. Typical examples of ω-continuous semirings are: the *Boolean semiring* $\mathbb{B}$, $\mathbb{N}^{\infty} = \mathbb{N} \cup \{\infty\}$ with the usual addition and multiplication as well as with min as addition and $+$ as multiplication (the latter one is called the *tropical semiring*) and the formal languages $\langle \mathfrak{P}(\Sigma^*), \cup, \cdot, \emptyset, \{\varepsilon\}\rangle$. *In the rest of the paper, A denotes an ω-continuous commutative semiring.*

A *(formal) power series* is a mapping $s : \Sigma^* \to A$, where Σ is a finite alphabet. The image of a word $w \in \Sigma^*$ under s (called *coefficient*) is denoted by (s, w). Usually, formal power series are written as $s = \sum_{w \in \Sigma^*} (s, w) w$. Defining addition of two series s_1 and s_2 by $(s_1 + s_2, w) = (s_1, w) + (s_2, w)$ and multiplication by $(s_1 s_2, w) = \sum_{uv=w} (s_1, u)(s_2, v)$, the set of formal power series, denoted by $A\langle\langle \Sigma^* \rangle\rangle$, forms a semiring $\langle A\langle\langle \Sigma^* \rangle\rangle, +, \cdot, 0, \varepsilon\rangle$.

There are two notions which link formal languages and formal power series. The *support* of a series $s \in A\langle\langle \Sigma^* \rangle\rangle$ is $\operatorname{supp}(s) = \{w \in \Sigma^* \mid (s, w) \neq 0\}$. On the other hand, the *characteristic series* of a language $L \subseteq \Sigma^*$ is $\sum_{w \in L} w$.

The present work can only be seen as a first step in the development of a theory of regulated power series. Here, we basically provide the definitions of power series corresponding to the concept of regularly controlled grammars with free derivations and with leftmost derivations of type 1. Theorems 23.4 and 23.5 are shown by using constructions similar to the formal language case, but employing different proof methods.

2. REGULARLY CONTROLLED ALGEBRAIC SYSTEMS

Let Σ and Γ denote (finite) *alphabets* and let $Y = \{y_1, \dots, y_n\}$, $n \geqslant 1$ denote an *alphabet of variables*. We will work in the ω-continuous semiring

$$\langle (A\langle\langle \Gamma^* \rangle\rangle)\langle\langle \Sigma^* \rangle\rangle, +, \cdot, 0, \varepsilon\rangle$$

of (formal) power series of the form

$$r = \sum_{w \in \Sigma^*} (r, w)w, \text{ where } (r, w) \in A\langle\langle \Gamma^* \rangle\rangle .$$

Given a power series $r \in (A\langle\langle \Gamma^* \rangle\rangle)\langle\langle (\Sigma \cup Y)^* \rangle\rangle$, we define two mappings

$$\bar{r}, \hat{r} : ((A\langle\langle \Gamma^* \rangle\rangle)\langle\langle (\Sigma \cup Y)^* \rangle\rangle)^n \to (A\langle\langle \Gamma^* \rangle\rangle)\langle\langle (\Sigma \cup Y)^* \rangle\rangle$$

as follows. For $t = w_0 y_{i_1} w_1 \ldots w_{k-1} y_{i_k} w_k$, $w_j \in \Sigma^*$, $0 \leqslant j \leqslant k$, $k \geqslant 0$, and $r_1, \ldots, r_n \in (A\langle\langle \Gamma^* \rangle\rangle)\langle\langle (\Sigma \cup Y)^* \rangle\rangle$, we define

$$\bar{t}(r_1, \ldots, r_n) = \sum_{v_1, \ldots, v_k \in (\Sigma \cup Y)^*} ((r_{i_1}, v_1) ш \cdots ш (r_{i_k}, v_k)) w_0 v_1 w_1 \ldots w_{k-1} v_k w_k$$

and

$$\hat{t}(r_1, \ldots, r_n) = \sum_{v_1, \ldots, v_k \in (\Sigma \cup Y)^*} ((r_{i_1}, v_1) \ldots (r_{i_k}, v_k)) w_0 v_1 w_1 \ldots w_{k-1} v_k w_k .$$

Here ш denotes the shuffle product of power series in $A\langle\langle \Gamma^* \rangle\rangle$ (see [7, page 156]) and t is shorthand for the power series εt.

For $r, r_1, \ldots, r_n \in (A\langle\langle \Gamma^* \rangle\rangle)\langle\langle (\Sigma \cup Y)^* \rangle\rangle$ we define

$$\bar{r}(r_1, \ldots, r_n) = \sum_{t \in (\Sigma \cup Y)^*} (r, t)\bar{t}(r_1, \ldots, r_n)$$

and

$$\hat{r}(r_1, \ldots, r_n) = \sum_{t \in (\Sigma \cup Y)^*} (r, t)\hat{t}(r_1, \ldots, r_n) .$$

Observe that $\hat{r}(r_1, \ldots, r_n)$ is nothing more than the substitution of $r_1, \ldots, r_n$ into r which we also denote by $r[r_1|y_1, \ldots, r_n|y_n]$.

Theorem 23.1 *Let* $r \in (A\langle\langle \Gamma^* \rangle\rangle)\langle\langle (\Sigma \cup Y)^* \rangle\rangle$. *Then the functions* $\bar{r}$ *and* $\hat{r}$ *are* ω*-continuous.*

Proof. Due to the fact that the reasoning is completely analogous, we need only show $\bar{r}$ to be ω-continuous. Since the shuffle product is distributive over arbitrary sums, ш is an ω-continuous mapping and so is the k-fold shuffle product. Moreover, for $s_1 \in A\langle\langle \Gamma^* \rangle\rangle$ and $s_2 \in (A\langle\langle \Gamma^* \rangle\rangle)\langle\langle \Sigma^* \rangle\rangle$, the "scalar" product $s_1 \cdot s_2 \in (A\langle\langle \Gamma^* \rangle\rangle)\langle\langle \Sigma^* \rangle\rangle$ is ω-continuous. According to [8], Theorem 12 of Section 8.3, infinite sums of ω-continuous functions are again ω-continuous functions. One application of this theorem yields that $\bar{t}$ is an ω-continuous mapping. One more application of this theorem shows that $\bar{r}$, $r \in (A\langle\langle \Gamma^* \rangle\rangle)\langle\langle (\Sigma \cup Y)^* \rangle\rangle$, is an ω-continuous mapping. □

A *controlled system* (with variables $y_1, \ldots, y_n$) is a system of formal equations

$$y_i = \bar{r}_i, \quad 1 \leqslant i \leqslant n, \tag{23.1}$$

where each r_i is a formal power series in $(A\langle\langle\Gamma^*\rangle\rangle)\langle\langle(\Sigma \cup Y)^*\rangle\rangle$ with $Y = \{y_1, \ldots, y_n\}$. A *solution* to the controlled system (23.1) is given by $(\sigma_1, \ldots, \sigma_n) \in ((A\langle\langle\Gamma^*\rangle\rangle)\langle\langle\Sigma^*\rangle\rangle)^n$ such that

$$\sigma_i = \bar{r}_i(\sigma_1, \ldots, \sigma_n), \quad 1 \leqslant i \leqslant n.$$

A solution $(\sigma_1, \ldots, \sigma_n)$ is termed *least solution* iff $\sigma_i \sqsubseteq \tau_i$ for $1 \leqslant i \leqslant n$, where $(\tau_1, \ldots, \tau_n)$ is an arbitrary solution of (23.1).

It is often more convenient to write the controlled system (23.1) in matrix notation. Defining the two vectors

$$y = \begin{pmatrix} y_1 \\ \vdots \\ y_n \end{pmatrix} \quad \text{and} \quad r = \begin{pmatrix} r_1 \\ \vdots \\ r_n \end{pmatrix}$$

we can write our controlled system in the matrix notation $y = \bar{r}(y)$ or $y = \bar{r}$. A *solution* to $y = \bar{r}(y)$ is now given by $\sigma \in ((A\langle\langle\Gamma^*\rangle\rangle)\langle\langle\Sigma^*\rangle\rangle)^n$ such that $\sigma = \bar{r}(\sigma)$. A solution σ of $y = \bar{r}$ is termed *least solution* iff $\sigma \sqsubseteq \tau$ for any solution τ of $y = \bar{r}$. Here, the vector $r \in ((A\langle\langle\Gamma^*\rangle\rangle)\langle\langle(\Sigma \cup Y)^*\rangle\rangle)^n$ with ith component r_i, $1 \leqslant i \leqslant n$, induces a mapping

$$\bar{r} : ((A\langle\langle\Gamma^*\rangle\rangle)\langle\langle\Sigma^*\rangle\rangle)^n \to ((A\langle\langle\Gamma^*\rangle\rangle)\langle\langle\Sigma^*\rangle\rangle)^n$$

by $\bar{r}(s_1, \ldots, s_n)_i = \bar{r}_i(s_1, \ldots, s_n)$, where $1 \leqslant i \leqslant n$. The ith component of the value of $\bar{r}$ at $(s_1, \ldots, s_n) \in ((A\langle\langle\Gamma^*\rangle\rangle)\langle\langle\Sigma^*\rangle\rangle)^n$ is given by the value of the function $\bar{r}_i$ induced by the ith component of r at $(s_1, \ldots, s_n)$. Since all $\bar{r}_i$, $1 \leqslant i \leqslant n$, are ω-continuous, the mapping $\bar{r}$ is also ω-continuous.

Consider now a controlled system $y = \bar{r}$. Since the least fixed point of the mapping $\bar{r}$ can only be the least solution of $y = \bar{r}$, we achieve the following result when applying the fixed point theorem [10, Section 1.5]:

Theorem 23.2 *In $((A\langle\langle\Gamma^*\rangle\rangle)\langle\langle\Sigma^*\rangle\rangle)^n$, the least solution of a controlled system $y = \bar{r}$ exists and equals the least fixed point of $\bar{r}$, i.e.,*

$$\mathit{fix}(\bar{r}) = \sup(\bar{r}^i(0) \mid i \in \mathbb{N}).$$

Theorem 23.2 suggests how we can compute an approximation of the least solution of a controlled system $y = \bar{r}$. The approximation sequence $(\sigma^j \mid j \in \mathbb{N})$, where each $\sigma^j \in ((A\langle\langle\Gamma^*\rangle\rangle)\langle\langle\Sigma^*\rangle\rangle)^n$, associated to a controlled system $y = \bar{r}(y)$ is defined as follows:

$$\sigma^0 = 0, \quad \sigma^{j+1} = \bar{r}(\sigma^j), \quad j \in \mathbb{N}.$$

Clearly, $(\sigma^j \mid j \in \mathbb{N})$ is an ω-chain, and $\mathrm{fix}(\bar{r}) = \sup(\sigma^j \mid j \in \mathbb{N})$, i.e., we obtain the least solution of $y = \bar{r}$ by computing the least upper bound of its approximation sequence.

We have introduced controlled systems in order to model context-free grammars with control languages in the framework of formal power series. Let $G = (Y, \Sigma, P, y_1)$ be a context-free grammar, where the productions in P are labelled by symbols of Γ. Consider all productions with left-hand side y_i,

$$\gamma_1^i : y_i \to \alpha_1^i, \ldots, \gamma_{n_i}^i : y_i \to \alpha_{n_i}^i, \quad 1 \leqslant i \leqslant n,$$

where $\gamma_1^i, \ldots, \gamma_{n_i}^i \in \Gamma$ and $\alpha_1^i, \ldots, \alpha_{n_i}^i \in (\Sigma \cup Y)^*$. Write the rules in the form of a controlled system

$$y_i = \gamma_1^i \alpha_1^i + \cdots + \gamma_{n_i}^i \alpha_{n_i}^i \,.$$

Let $A = \mathbb{B}$ and consider the ith component σ_i, $1 \leqslant i \leqslant n$, of the least solution σ of this controlled system. Then, for $w \in \Sigma^*$, the power series $(\sigma_i, w) \in A\langle\langle\Gamma^*\rangle\rangle$ indicates all control words that govern a derivation from y_i to w, $1 \leqslant i \leqslant n$. If $((\sigma_i, w), \gamma) = 1$ and $\gamma \in \Gamma^*$, then w is derived from y_i by the control word γ. Due to space limitations, we defer the formal proof showing that the class of languages whose support is in $\langle \mathbb{B}^{\mathrm{rat,alg}}\langle\langle\Sigma^*\rangle\rangle, +, \cdot, 0, \varepsilon\rangle$ coincides with the class of languages generatable by regularly controlled context-free grammars with free derivations to a forthcoming paper. Similarly, left controlled algebraic systems correspond to regularly controlled context-free grammars with leftmost derivations of type 1 (see [2]).

Example 23.1 *(See [3, Example 2.2]) Consider*

$$G = (\{y_1, y_2, y_3\}, \{a, b, c\}, P, y_1),$$

where the rules in P are as follows: $\gamma_1 : y_1 \to y_2 y_3$, $\gamma_2 : y_2 \to a y_2 b$, $\gamma_3 : y_3 \to c y_3$, $\gamma_4 : y_2 \to ab$, *and* $\gamma_5 : y_3 \to c$. *The controlled system associated to this context-free grammar G is given by*

$$y_1 = \gamma_1 y_2 y_3, \quad y_2 = \gamma_2 a y_2 b + \gamma_4 ab, \quad \textit{and } y_3 = \gamma_3 c y_3 + \gamma_5 c.$$

Let $(\sigma^j \mid j \in \mathbb{N})$ be the approximation sequence of this controlled system, and let σ be its least upper bound. The second and third equation of this controlled system can be solved independently. Then

$$\sigma_2^j = \sum_{1 \leqslant i \leqslant j} (\gamma_2^{i-1} \cdot \gamma_4) a^i b^i, \quad \sigma_3^j = \sum_{1 \leqslant i \leqslant j} (\gamma_3^{i-1} \cdot \gamma_5) c^i,$$

and hence we have

$$\sigma_2 = \sum_{i \geqslant 1} (\gamma_2^{i-1} \cdot \gamma_4) a^i b^i, \quad \sigma_3 = \sum_{i \geqslant 1} (\gamma_3^{i-1} \cdot \gamma_5) c^i,$$

so that we derive

$$\sigma_1 = \sum_{i_2, i_3 \geqslant 1} (\gamma_1 \cdot (\gamma_2^{i_2-1} \cdot \gamma_4 \sqcup\!\sqcup\, \gamma_3^{i_3-1} \cdot \gamma_5)) a^{i_2} b^{i_2} c^{i_3} .$$

We define left controlled systems analogously. A *left controlled system* with variables $y_1, \ldots, y_n$ is a system of formal equations $y_i = \hat{r}_i$, $1 \leqslant i \leqslant n$, where each r_i is a power series in $(A\langle\langle\Gamma^*\rangle\rangle)\langle\langle(\Sigma \cup Y)^*\rangle\rangle$. A *solution* to such a system is given by $(\sigma_1, \ldots, \sigma_n) \in ((A\langle\langle\Gamma^*\rangle\rangle)\langle\langle\Sigma^*\rangle\rangle)^n$ such that

$$\sigma_i = \hat{r}_i(\sigma_1, \ldots, \sigma_n), \quad 1 \leqslant i \leqslant n .$$

Observe the only difference to controlled systems: $\bar{r}$ is replaced by $\hat{r}$. The remaining definitions (least solution, matrix notation, approximation sequence, etc.) are analogous to the respective definitions for controlled systems. We obtain accordingly:

Theorem 23.3 *The least solution of a left controlled system $y = \hat{r}$ exists in the semiring $((A\langle\langle\Gamma^*\rangle\rangle)\langle\langle\Sigma^*\rangle\rangle)^n$ and equals the least fixed point of $\hat{r}$, i.e.,*

$$\mathit{fix}(\hat{r}) = \sup(\hat{r}^i(0) \mid i \in \mathbb{N}) .$$

Left controlled systems are introduced in order to model context-free grammars with control languages for leftmost derivations in the framework of formal power series. Given a context-free grammar G whose rules are labelled by symbols of Γ, the ith component σ_i, $1 \leqslant i \leqslant n$, of the least solution σ of the associated left controlled system (with basic semiring $A = \mathbb{B}$) indicates the following: if, for $w \in \Sigma^*$, $\gamma \in \Gamma^*$, $1 \leqslant i \leqslant n$, $((\sigma_i, w), \gamma) = 1$, then w is derivable from y_i by a leftmost derivation with control word γ.

We continue to study Example 23.1, now under leftmost derivations. The second component σ_2 and the third component σ_3 of the least solution σ of the left controlled system which is associated to the context-free grammar G are the same as in the case of the associated controlled system. We only indicate the change to be in the first component σ_1, where we now have:

$$\sigma_1 = \sum_{i_2, i_3 \geqslant 1} \gamma_1 \gamma_2^{i_2-1} \gamma_4 \gamma_3^{i_3-1} \gamma_5 a^{i_2} b^{i_2} c^{i_3} .$$

We now restrict our systems in such a way that the right-hand sides of their equations are polynomials. A *controlled algebraic system* is a controlled system $y_i = \bar{r}_i$, $1 \leqslant i \leqslant n$, where each r_i is a polynomial in $(A\langle\Gamma^*\rangle)\langle(\Sigma \cup Y)^*\rangle$. A similar restriction on left controlled systems defines *left controlled algebraic systems.*

Let $y_i = \bar{r}_i$, $1 \leqslant i \leqslant n$, be a controlled (algebraic) system with least solution $(\sigma_1, \ldots, \sigma_n)$ and let s be a power series in $A\langle\langle\Gamma^*\rangle\rangle$. Then this controlled (algebraic) system, together with the *control power series s*, *generate* the following power series $\tau \in A\langle\langle\Sigma^*\rangle\rangle$:

$$\tau = \sum_{w \in \Sigma^*} h((\sigma_1, w) \odot s)w = \sum_{w \in \Sigma^*} (\sum_{v \in \Gamma^*} ((\sigma_1, w), v)(s, v))w,$$

where $h : A\langle\langle\Gamma^*\rangle\rangle \to A\langle\langle\Gamma^*\rangle\rangle$ is the morphism defined by $h(x) = \varepsilon$ for $x \in \Gamma$. Moreover, a power series in $(A\langle\varepsilon\rangle)\langle\langle\Sigma^*\rangle\rangle$ is identified with its isomorphic copy in $A\langle\langle\Sigma^*\rangle\rangle$. In an analogous way, a left controlled (algebraic) system and a *control power series* in $A\langle\langle\Gamma^*\rangle\rangle$ *generate* a power series in $A\langle\langle\Sigma^*\rangle\rangle$.

This generation of a power series models the generation of a language by a context-free grammar, where the allowed derivations (or leftmost derivations) must lie in a given control language (represented by the control power series s).

Proposition 23.1 *Assume that the controlled system $y_i = \bar{r}_i$, $1 \leqslant i \leqslant n$, with least solution σ and approximation sequence $(\sigma^j \mid j \in \mathbb{N})$, together with the control power series s, generate r. We define the sequence $(\eta^j \mid j \in \mathbb{N})$ by*

$$\eta^j = \sum_{w \in \Sigma^*} h((\sigma^j, w) \odot s)w\,.$$

Then, we can conclude that $r = \sup(\eta^j \mid j \in \mathbb{N})$.

A similar assertion holds in the case of left controlled systems.

Proof. Since both the Hadamard product (which is distributive over arbitrary sums) and morphisms are ω-continuous, we obtain

$$\sup(\eta^j \mid j \in \mathbb{N}) = \sum_{w \in \Sigma^*} h((\sup(\sigma^j, w)) \odot s)w = r\,.$$

□

Consider again our Example 23.1 from above, together with the control power series $s = \gamma_1(\gamma_2\gamma_3)^*\gamma_4\gamma_5$. Then, for $i_2, i_3 \geqslant 1$,

$$(\sigma_1, a^{i_2}b^{i_2}c^{i_3}) \odot s = \delta_{i_2,i_3}\gamma_1(\gamma_2\gamma_3)^{i_2}\gamma_4\gamma_5,$$

where δ is the Kronecker symbol. Hence, $\tau = \sum_{i \geqslant 1} a^i b^i c^i$. Observe that the power series $\gamma_1(\gamma_3\gamma_2)^*\gamma_4\gamma_5$ or $\sum_{i \geqslant 0} \gamma_1\gamma_2^i\gamma_3^i\gamma_4\gamma_5$ have the same function. Consider now the left controlled algebraic system given above using the control power series $s = \sum_{i \geqslant 0} \gamma_1\gamma_2^i\gamma_4\gamma_3^i\gamma_5$. Then, for $i_2, i_3 \geqslant 1$,

$$(\sigma_1, a^{i_2}b^{i_2}c^{i_3}) \odot s = \delta_{i_2,i_3}\gamma_1\gamma_2^{i_2}\gamma_4\gamma_3^{i_2}\gamma_5,$$

and hence $\tau = \sum_{i \geqslant 1} a^i b^i c^i$. As we will later see, it is no coincidence that we have not found a rational control power series in order to produce $\tau = \sum_{i \geqslant 1} a^i b^i c^i$ in a leftmost fashion, see Theorem 23.5.

A power series $r \in A\langle\langle \Sigma^* \rangle\rangle$ is called *regularly controlled algebraic series* iff there exists a controlled algebraic system and a rational power series in $A^{\mathrm{rat}}\langle\langle \Gamma^* \rangle\rangle$ that generate r. Similarly, *regularly left controlled algebraic series* are defined. The collection of all reguarly controlled (left controlled) algebraic series in $A\langle\langle \Sigma^* \rangle\rangle$ is denoted by $A^{\mathrm{rat,alg}}\langle\langle \Sigma^* \rangle\rangle$ ($A^{\mathrm{rat,lalg}}\langle\langle \Sigma^* \rangle\rangle$).

A power series $r \in \mathbb{N}^\infty\langle\langle \Sigma^* \rangle\rangle$ is called *unambiguous* iff $(r, w) \in \{0, 1\}$ for all $w \in \Sigma^*$.

In our running Example 23.1, the power series $\tau = \sum_{i \geqslant 1} a^i b^i c^i$ is a regularly controlled algebraic series. For $A = \mathbb{N}^\infty$, it is unambiguous. As we will see, τ is not a regularly left controlled algebraic series according to Theorem 23.5.

In fact, the control series can be used to disambiguate derivations. Consider the following example:

Example 23.2 *(see [6, Example 3.1]) The canonical $\mathbb{N}^\infty\langle\langle \{x\}^* \rangle\rangle$ left controlled algebraic system $y = \gamma_1 y^2 + \gamma_2 x$ pertaining to the context-free grammar*

$$(\{y\}, \{x\}, \{\gamma_1 = y \to y^2, \gamma_2 = y \to x\}, y)$$

has the solution (disregarding the control words)

$$\sum_{n \geqslant 0} C_n x^{n+1}, \quad \textit{where } C_n = \frac{(2n)!}{n!(n+1)!}, n \geqslant 0\,.$$

If we consider this to be a regularly left controlled algebraic series (controlled by $\gamma_1^* \gamma_2^*$), we get the solution

$$\sum_{n \geqslant 0} \gamma_1^n \gamma_2^{n+1} x^{n+1}\,.$$

3. CLOSURE PROPERTIES

A lot of closure properties are known for regularly controlled context-free languages (under free derivations), see [2, page 48] or [3, page 140]. In particular, standard constructions showing closure under union and concatenation (without formal proofs) are given in [2, Lemma 1.3.4] for the related formalism of so-called matrix grammars. The analogues in the formal power series case are proved in the following theorem.

Theorem 23.4 *$\langle A^{\mathrm{rat,alg}}\langle\langle \Sigma^* \rangle\rangle, +, \cdot, 0, \varepsilon \rangle$ is a semiring containing $A^{\mathrm{alg}}\langle\langle \Sigma^* \rangle\rangle$.*

Proof. We have to show that $\langle A^{\mathrm{rat,alg}}\langle\langle \Sigma^* \rangle\rangle, +, \cdot, 0, \varepsilon \rangle$ is a subsemiring, i.e., it is closed under $+$ and $\cdot$ and contains the corresponding neutral elements.

Let r and r' be in $A^{\text{rat,alg}}\langle\langle\Sigma^*\rangle\rangle$. Then there exist controlled algebraic systems $y_i = \bar{r}_i$, $1 \leqslant i \leqslant n$, and $y'_j = \bar{r}'_j$, $1 \leqslant j \leqslant m$ with least solutions σ and σ', respectively, and rational power series $s, s' \in A^{\text{rat}}\langle\langle\Gamma^*\rangle\rangle$ that generate r and r', respectively. We assume that both the sets of variables and the sets of labels are disjoint, that y_0 is a new variable and that γ, γ' are new labels in Γ.
(i) Closure under union: Consider the controlled algebraic system

$$\begin{aligned} y_0 &= \bar{r}_0(y_1, y'_1), \text{ where } r_0 = \gamma y_1 + \gamma' y'_1, \\ y_i &= \bar{r}_i, \quad 1 \leqslant i \leqslant n, \\ y'_j &= \bar{r}'_j, \quad 1 \leqslant j \leqslant m, \end{aligned}$$

and the rational power series $\gamma s + \gamma' s'$. We claim that $r + r'$ is generated by this controlled algebraic system and $\gamma s + \gamma' s'$.

Since the second and the third subsystem can be solved independently, it becomes clear that the y_0-component of the least solution of the controlled algebraic system equals

$$\bar{r}_0(\sigma_1, \sigma'_1) = \sum_{w \in \Sigma^*} (\gamma(\sigma_1, w) + \gamma'(\sigma'_1, w))w \,.$$

Hence, we obtain:

$$\begin{aligned} &\sum_{w \in \Sigma^*} h((\gamma(\sigma_1, w) + \gamma'(\sigma'_1, w)) \odot (\gamma s + \gamma' s'))w \\ &= \sum_{w \in \Sigma^*} h(\gamma((\sigma_1, w) \odot s))w + \sum_{w \in \Sigma^*} h(\gamma'((\sigma'_1, w) \odot s'))w \\ &= \sum_{w \in \Sigma^*} h((\sigma_1, w) \odot s)w + \sum_{w \in \Sigma^*} h((\sigma'_1, w) \odot s')w \\ &= r + r'. \end{aligned}$$

(ii) Closure under catenation: Consider the controlled algebraic system

$$\begin{aligned} y_0 &= \bar{r}_0(y_1, y'_1), \text{ where } r_0 = \gamma y_1 y'_1, \\ y_i &= \bar{r}_i, \quad 1 \leqslant i \leqslant n, \\ y'_j &= \bar{r}'_j, \quad 1 \leqslant j \leqslant m, \end{aligned}$$

and the rational power series $\gamma s s'$. We claim that rr' is generated by this controlled algebraic system and $\gamma s s'$. Since the second and the third subsystem can be solved independently, it is clear that the y_0-component of the least solution of the controlled algebraic system equals

$$\bar{r}_0(\sigma_1, \sigma'_1) = \sum_{w, w' \in \Sigma^*} \gamma((\sigma, w) \sqcup\!\sqcup (\sigma'_1, w'))ww' \,.$$

Hence, we obtain:

$$\begin{aligned}
&\sum_{w,w'\in\Sigma^*} h(\gamma((\sigma_1, w) \text{ш} (\sigma_1', w')) \odot \gamma s s')ww' \\
&= \sum_{w,w'\in\Sigma^*} h(((\sigma_1, w) \odot s)((\sigma_1', w') \odot s'))ww' \\
&= (\sum_{w\in\Sigma^*} h((\sigma_1, w) \odot s)w)(\sum_{w'\in\Sigma^*} h((\sigma_1', w') \odot s')w') \\
&= rr'.
\end{aligned}$$

(iii) Since $A^{\mathrm{alg}}\langle\langle\Sigma^*\rangle\rangle \subseteq A^{\mathrm{rat,alg}}\langle\langle\Sigma^*\rangle\rangle$, we have shown the claim to be true. □

Analogous results can be shown for leftmost derivations and with only small differences in the proofs. However, as shown in Theorem 23.5, the corresponding power series families are not "new" ones, therefore we refrain from giving details here.

We note that we conjecture that not all closure properties known for the language case (confer [2, page 48]) are transferable to the formal power series framework in full generality.

4. LEFTMOST DERIVATIONS

Theorem 23.5 $A^{\mathrm{rat,lalg}}\langle\langle\Sigma^*\rangle\rangle = A^{\mathrm{alg}}\langle\langle\Sigma^*\rangle\rangle$.

Proof. The construction in the proof of this theorem is analogous to the language case, see [4, Theorem 7].[1]

Consider a left controlled series $r \in A^{\mathrm{rat,lalg}}\langle\langle\Sigma^*\rangle\rangle$ generated by the left controlled algebraic system $y_i = \hat{r}_i$ with $1 \leqslant i \leqslant n$ together with a rational series $s \in A\langle\langle\Gamma^*\rangle\rangle$. Assume that s is given by a monoid morphism $\mu : \Gamma^* \to A^{m\times m}$, an initial state vector $I \in A^{1\times m}$ and a final state vector $F \in A^{m\times 1}$, i.e., $s = \sum_{v\in\Gamma^*}(I\mu(v)F)v$, see [1, Section I.5] or [9, Section II.2] for the definition of "automata" in this algebraic fashion. Extend μ in the usual manner to a semiring morphism $\mu : A\langle\langle\Gamma^*\rangle\rangle \to A^{m\times m}$.

Let $Z = \{y^i_{kj} \mid 1 \leqslant k, j \leqslant m, 1 \leqslant i \leqslant n\}$ be variables and define $m \times m$-matrices Y_i, $1 \leqslant i \leqslant n$, whose (k, j)th entry, $1 \leqslant k, j \leqslant m$, is the variable y^i_{kj}. Let

$$\rho : (\Sigma \cup Y)^* \to (\{\mathrm{diag}(w) \mid w \in \Sigma^*\} \cup \{Y_i \mid 1 \leqslant i \leqslant n\})$$

be the multiplicative monoid morphism defined by $\rho(x) = \mathrm{diag}(x)$, $x \in \Sigma$, and $\rho(y_i) = Y_i$, $1 \leqslant i \leqslant n$. Here, $\mathrm{diag}(w)$, $w \in \Sigma^*$, denotes the $m \times m$-diagonal matrix all of whose diagonal entries are equal to w.

[1] The proofs given in [2, Section 1.4] do not seem to be transferable as easily to the power series framework.

For $t \in (A\langle\langle\Gamma^*\rangle\rangle)\langle\langle(\Sigma \cup Y)^*\rangle\rangle$, define

$$\eta(t) = \sum_{\alpha \in (\Sigma \cup Y)^*} \mu((t, \alpha))\rho(\alpha) \in (A\langle\langle(\Sigma \cup Z)^*\rangle\rangle)^{m \times m}.$$

For later use, we compute, for $\sigma \in (A\langle\langle\Gamma^*\rangle\rangle)\langle\langle\Sigma^*\rangle\rangle$, the power series $I\eta(\sigma)F$:

$$\begin{aligned}
I\eta(\sigma)F &= I(\sum_{w \in \Sigma^*} \mu((\sigma, w))\mathrm{diag}(w))F \\
&= \sum_{w \in \Sigma^*} \sum_{v \in \Gamma^*} ((\sigma, w), v) I\mu(v)\mathrm{diag}(w)F \\
&= \sum_{w \in \Sigma^*} \sum_{v \in \Gamma^*} ((\sigma, w), v)(I\mu(v)F)w \\
&= \sum_{w \in \Sigma^*} \sum_{v \in \Gamma^*} ((\sigma, w), v)(s, v)w \\
&= \sum_{w \in \Sigma^*} h((\sigma, w) \odot s)w .
\end{aligned}$$

We now examine the effect of substituting power series and matrices of power series for variables and matrix variables, respectively. To this end, consider $\alpha = w_0 y_{i_1} w_1 \dots w_{k-1} y_{i_k} w_k$ with $w_j \in \Sigma^*$, $1 \leqslant j \leqslant k$, $k \geqslant 0$. We claim that, for $\sigma_i \in (A\langle\langle\Gamma^*\rangle\rangle)\langle\langle\Sigma^*\rangle\rangle$, $1 \leqslant i \leqslant n$,

$$\eta(\alpha[\sigma_i | y_i, 1 \leqslant i \leqslant n]) = \rho(\alpha)[\eta(\sigma_i) | Y_i, 1 \leqslant i \leqslant n]. \qquad (23.2)$$

We first compute the left-hand side of the equality (23.2):

$$\begin{aligned}
&\eta(\alpha[\sigma_i | y_i, 1 \leqslant i \leqslant n]) \\
&\quad = \eta(w_0 \sigma_{i_1} w_1 \dots w_{k-1} \sigma_{i_k} w_k) \\
&\quad = \sum_{v_1, \dots, v_k \in \Sigma^*} \mu((\sigma_{i_1}, v_1) \dots (\sigma_{i_k}, v_k))\rho(w_0 v_1 w_1 \dots w_{k-1} v_k w_k)
\end{aligned}$$

and now the right-hand side of the equality (23.2):

$$\begin{aligned}
&\rho(\alpha)[\eta(\sigma_i) | Y_i, 1 \leqslant i \leqslant n] \\
&\quad = \mathrm{diag}(w_0)\eta(\sigma_{i_1})\mathrm{diag}(w_1) \cdots \mathrm{diag}(w_{k-1})\eta(\sigma_{i_k})\mathrm{diag}(w_k) \\
&\quad = \sum_{v_1, \dots, v_k \in \Sigma^*} \mu((\sigma_{i_1}, v_1) \dots \mu((\sigma_{i_k}, v_k))\mathrm{diag}(w_0 v_1 w_1 \dots w_{k-1} v_k w_k) .
\end{aligned}$$

Since μ is a morphism, we have proved the equality (23.2).

Let $(\sigma^j \mid j \in \mathbb{N})$ be the approximation sequence of the left controlled system $y_i = \hat{r}_i$, $1 \leqslant i \leqslant n$. Let $(\tau^j \mid j \in \mathbb{N})$, $\tau^j \in (((A\langle\langle\Gamma^*\rangle\rangle)\langle\langle\Sigma^*\rangle\rangle)^{m \times m})^n$, be the approximation sequence of the algebraic system $Y_i = \eta(r_i)$, $1 \leqslant i \leqslant n$, in matrix notation. We claim that, for all $j \geqslant 0$, $\tau^j = \eta(\sigma^j)$ and prove it by induction on j. The case $j = 0$ is clear. Let now $j \geqslant 0$. Then we obtain

$$\begin{aligned}
\tau_i^{j+1} &= \sum_{\alpha \in (\Sigma \cup Y)^*} \mu((r_i, \alpha))\rho(\alpha)[\tau_k^j | Y_k, 1 \leqslant k \leqslant n] \\
&= \sum_{\alpha \in (\Sigma \cup Y)^*} \mu((r_i, \alpha))\rho(\alpha)[\eta(\sigma_k^j) | Y_k, 1 \leqslant k \leqslant n] \\
&= \sum_{\alpha \in (\Sigma \cup Y)^*} \mu((r_i, \alpha))\eta(\alpha[\sigma_k^j | y_k, 1 \leqslant k \leqslant n]) \\
&= \eta(\sum_{\alpha \in (\Sigma \cup Y)^*} (r_i, \alpha)\alpha[\sigma_k^j | y_k, 1 \leqslant k \leqslant n]) \\
&= \eta(\sigma_i^{j+1})
\end{aligned}$$

Denote $\tau_i = \sup(\tau_i^j \mid j \in \mathbb{N})$ and $\sigma_i = \sup(\sigma_i^j \mid j \in \mathbb{N})$, $1 \leqslant i \leqslant n$. Through Proposition 23.1 we infer that

$$\tau_i = \eta(\sigma_i), \quad 1 \leqslant i \leqslant n .$$

Consider now the algebraic system

$$y_0 = IY_1F, \quad Y_i = \eta(r_i), \quad 1 \leqslant i \leqslant n ,$$

where y_0 is a new variable. Then $I\tau_1 F$ is the y_0-component of the least solution. Hence,

$$I\tau_1 F = I\eta(\sigma_1)F = \sum_{w \in \Sigma^*} h((\sigma_1, w) \odot s)w = r$$

is in $A^{\mathrm{alg}}\langle\langle\Sigma^*\rangle\rangle$. □

Remark 23.1 *A similar argument shows that rational series left controlled by rational series characterize $A^{\mathrm{rat}}\langle\langle\Sigma^*\rangle\rangle$.*

References

[1] Berstel, J. & C. Reutenauer (1988), *Rational Series and Their Languages*, Springer, Berlin.

[2] Dassow, J. & Gh. Păun (1989), *Regulated Rewriting in Formal Language Theory*, Springer, Berlin.

[3] Dassow, J.; Gh. Păun & A. Salomaa (1997), Grammars with controlled derivations, in G. Rozenberg & A. Salomaa, eds., *Handbook of Formal Languages*, II: 101–154. Springer, Berlin.

[4] Fernau, H. (1998), Regulated grammars with leftmost derivation, in B. Rovan, ed., in *Proceedings of SOFSEM'98*, LNCS 1521: 322–331. Springer, Berlin. A revised and extended journal version will appear in *Grammars*, 3.1(2000).

[5] Ginsburg, S. & E.H. Spanier (1968), Control sets on grammars, *Mathematical Systems Theory*, 2: 159–177.

[6] Kuich, W. (1997), Semirings and formal power series, in G. Rozenberg & A. Salomaa, eds., *Handbook of Formal Languages*, I: 609–677. Springer, Berlin.

[7] Kuich, W. & A. Salomaa (1986), *Semirings, Automata, Languages*, Springer, Berlin.

[8] Manes, G.E. & M.A. Arbib (1986), *Algebraic Approaches to Program Semantics*, Springer, Berlin.

[9] Salomaa, A.K. & M. Soittola (1978), *Automata-Theoretic Aspects of Formal Power Series*, Springer, New York.

[10] Wechler, W. (1992), *Universal Algebra for Computer Scientists*, Springer, Berlin.

Chapter 24

FORBIDDEN SUBSEQUENCES AND PERMUTATIONS SORTABLE ON TWO PARALLEL STACKS

Tero Harju*
Mathematics Department
University of Turku
FIN-20014 Turku, Finland

Lucian Ilie†
Turku Centre for Computer Science (TUCS)
Lemminkäisenkatu 14 A, FIN-20520 Turku, Finland
luciliQutu.fi

Abstract We give a new combinatorial proof, based on stack graphs, for the characterization by forbidden subsequences of the permutations which are sortable on two parallel stacks. Some remarks about the permutations sortable on parallel queues are also made.

1. INTRODUCTION

Permutations that are sortable on two parallel stacks have been characterized by Even and Itai [5] using union graphs and circle graphs, see also Golumbic [7], and Rosenstiehl and Tarjan [10]. By [5], the circle graphs are very close to the permutations sortable on stacks. In Bouchet [3] the circle graphs are characterized by forbidden subgraphs with respect to local equivalence. For a

*This research was done while the authors visited Leiden University.
†Research supported by the Academy of Finland, Project 137358. On leave of absence from Faculty of Mathematics, University of Bucharest, Str. Academiei 14, R-70109, Bucharest, Romania.

C. Martin-Vide and V. Mitrana (eds.), Where Mathematics, Computer Science, Linguistics and Biology Meet, 267-275.

characterization using intersection graphs, see Fraysseix [6]. For permutations avoiding certain short sequences, we refer to Bóna [2] and Stankova [11].

Another characterization for these permutations, by forbidden subsequences, has been started by Tarjan [12], who gave some partial results on this topic, and completed by Pratt [9].

In this article we give a new proof for characterizing these permutations by forbidden subsequences. Our proof is totally different from the one by Pratt. It is based on a close investigation of stack graphs and we believe that it gives insight intohe problem from a different point of view.

We show that a permutation π is sortable on two parallel stacks if and only if π avoids (that is, does not contain a subsequence order isomorphic to) the permutations

$$\pi_k = 2(2k+1)41\ldots(2i)(2i-3)\ldots(2k+2)(2k-1)$$

and

$$\sigma_k = (2k+3)2(2k+5)41\ldots(2i)(2i-3)\ldots(2k+4)(2k+1)$$

for all odd integers $k \geqslant 1$. The permutation π_k is formed by shuffling the sequence $2, 4, \ldots, (2k+2)$ of even integers with the sequence $(2k+1), 1, 3, \ldots, (2k-1)$ of odd integers. The permutation σ_k is obtained similarly by shuffling the sequences $2, 4, \ldots, (2k+4)$ and $(2k+3), (2k+5), 1, 3, \ldots, (2k+1)$. For instance, $\pi_1 = 2341$, $\pi_3 = 27416385$, and $\sigma_1 = 5274163$, $\sigma_3 = 92(11)416385(10)7$.

2. DEFINITIONS

Let $\pi = (i_1, i_2, \ldots, i_n)$ be a permutation on $[1, n] = \{1, 2, \ldots, n\}$. It will be written more conveniently as $\pi = i_1 i_2 \ldots i_n$. For a $k \geqslant 1$, we say that π is *sortable on k parallel stacks*[1] if there is a way to obtain from π the identity $12\ldots n$ by the following procedure (see also Figure 24.1). We consider one by one the elements of the input π from left to right and two kinds of moves are allowed: (i) from the left of the input to the top of a stack and (ii) from the top of a stack to the right of the output.

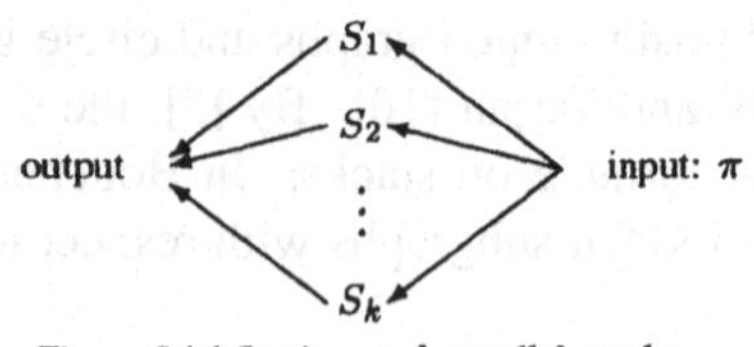

Figure 24.1 Sorting on k parallel stacks.

[1]This should not be confused with the two-stack sortable permutation of West [13] and Zeilberger [14]; the main difference is that here the stacks are used in parallel whereas in [13, 14] they are used sequentially.

As an example, we sort the permutation $\pi = 2431$ on two parallel stacks. We describe the sorting procedure by quadruples

$$(\text{output}, -\text{stack}_1, -\text{stack}_2, \text{input}),$$

where "–" marks the top of the respective stack. We have

$$(, -, -, 2431) \to (, -2, -, 431) \to (, -2, -4, 31) \to (, -2, -34, 1) \to$$
$$(, -12, -34,) \to (1, -2, -34,) \to (12, -, -34,) \to (123, -, -4,) \to$$
$$(1234, -, -,).$$

Thus, π can be sorted on two parallel stacks.

Let π be a permutation on $[1, n]$. We shall write $i \prec j$, if $\pi^{-1}(i) < \pi^{-1}(j)$, that is, if the element i comes before j in π. A sequence $j_1 j_2 \dots j_k$ is a *subsequence* of π if $j_1 \prec j_2 \prec \dots \prec j_k$ in π. A *factor* of π is a subsequence of consecutive elements, that is, a subsequence $i_k i_{k+1} \cdots i_{k+m}$ for some k and m. We write $i \in \sigma$, if the element i is in the subsequence σ.

A permutation σ on $[1, m]$ is *contained in* π, if π has a subsequence that is order isomorphic to σ. Also, π *avoids* a set of permutations Σ, if π does not contain any element of Σ. For instance, the permutation $\pi = 524163$ contains $\sigma = 3142$ in three ways; any of the subsequences 5263, 5163, and 4163 of π is order isomorphic to σ. It is easy to see that π avoids 2341.

It is known (see Knuth [8]) that a permutation π on $[1, n]$ avoids $\sigma = 231$ if and only if π is sortable on one stack.

For a permutation π on $[1, n]$, let $S[\pi]$ be the undirected graph on $[1, n]$ such that ij is an edge for $i < j$ if and only if there exists an $m < i$ with $i \prec j \prec m$. We call $S[\pi]$ the *stack graph* of π.

Let $\chi(G)$ be the chromatic number of an undirected graph G, that is, the smallest number of colours for the vertices of G such that no adjacent vertices get the same colour.

Denote by C_k a cycle of k vertices and edges without chords, that is, $C_k = (x_1, x_2, \dots, x_k)$, where the edges are the pairs x_1x_2, x_2x_3, ..., $x_{k-1}x_k$, x_kx_1. Now $\chi(C_k) = 2$, if k is even, and $\chi(C_k) = 3$, if $k \geqslant 3$ is odd. A cycle C_3 is called a *triangle*. A graph G is *triangle free*, if it has no triangles as subgraphs.

The following theorem is a "straightforward exercise" by Golumbic [7, p.236]; we agree with him.

Theorem 24.1 *The chromatic number $\chi(S[\pi])$ is the smallest nonnegative number k such that π can be sorted on k parallel stacks.*

3. PERMUTATIONS SORTABLE ON TWO PARALLEL STACKS

As a corollary to Theorem 24.1, we have

Theorem 24.2 *A permutation π is sortable on two parallel stacks if and only if its stack graph $S[\pi]$ is bipartite.*

In particular, for the permutation $\pi_1 = 2341$, the graph $S[\pi_1]$ consists of an isolated vertex 1 and of a triangle on the three vertices 2, 3, and 4. It is also clear that if a stack graph $S[\pi]$ contains a triangle, then π contains 2341, and therefore

Lemma 24.1 *A permutation π avoids* 2341 *if and only if $S[\pi]$ is triangle free. In particular, if π is sortable on two parallel stacks, then it avoids* 2341.

We notice that this lemma generalizes to: the maximal clique of $S[\pi]$ is of order at most k if and only if π avoids $23\ldots(k+2)1$.

We remark at this point that the latter claim in Lemma 24.1 is not true in converse: the permutation 5274163 is the shortest example that avoids 2341 but is not sortable on two parallel stacks. Indeed, it can be shown that every permutation π that avoids 2341 is sortable on five parallel stacks, but there exists such a π, which is not sortable on four parallel stacks. These assertions follow from results on circle graphs, see Ageev [1]. A graph G is a *circle graph* if it is (isomorphic to) the intersection graph of a set of chords of a (geometric) circle. By a theorem due to Even and Itai [5], a graph G is a circle graph if and only if it is isomorphic to a stack graph $S[\pi]$, when, in both, the isolated vertices are removed. It is worthwhile mentioning that the circle graphs, and thus the stack graphs, satisfy the perfect graph conjecture, see Buckingham and Golumbic [4]. According to this result, the chromatic number $\chi(G)$ equals $\omega(G)$, the maximum size of a clique in G, for all induced subgraphs G of $S[\pi]$ unless $S[\pi]$ contains a cycle C_k or its complement $\overline{C}_k$ for an odd $k \geqslant 5$. The (sub)graph $\overline{C}_k$ contains a triangle when $k \geqslant 7$, and $\overline{C}_5$ is isomorphic to C_5, and therefore many of our considerations reduce to the existence of odd cycles C_k in the stack graphs.

Let $\pi = i_1 i_2 \ldots i_n$ be a permutation on $[1, n]$. A permutation $\pi - i$ is obtained from π by *removing* an element $i \in [1, n]$, if i is removed, and each remaining element $j > i$ is replaced by $j - 1$. An element $i \in [1, n]$ is *removable*, if $\pi - i$ is not sortable on two parallel stacks. The permutation π is said to be *critical*, if it is not sortable on two parallel stacks, and it contains no removable elements.

Lemma 24.2 *For any critical permutation π, the stack graph $S[\pi]$ consists of a unique (induced) cycle $C = C_k$ for an odd $k \geqslant 3$, and of a set I of isolated vertices. Moreover, the isolated elements form an increasing subsequence $x_1 < x_2 < \ldots < x_t$ of π such that $x_{i+1} \geqslant x_i + 2$.*

Proof. By Theorem 24.2, $S[\pi]$ is not bipartite, and, as is well known, this is equivalent to saying that S contains a cycle C_k for some odd $k \geqslant 3$. Let

$C = C_k$ be such a cycle of $S[\pi]$ of the smallest length k. It is clear that C is an induced subgraph. Let E be the set of those elements m of $[1, n]$ such that there exist $i, j \in C$ with $i \prec j \prec m$ and $m < i < j$ (possibly $C \cap E \neq \emptyset$). Now $C \cup E = [1, n]$, since the subgraph induced by $C \cup E$ contains the odd cycle C as an induced subgraph, and π is critical by assumption (and elements not in $C \cup E$ are removable).

Let $i \notin C$, and assume on the contrary that there exists an edge ij in $S[\pi]$. Then necessarily there exists an $m < i$ such that either $i \prec j \prec m$ or $j \prec i \prec m$. Now, however, for any $x \prec y \prec i$ with $i < x < y$, also $x \prec y \prec m$ with $m < x < y$, and therefore the element i is removable, which contradicts the fact that π is critical. This shows that $[1, n] \setminus C$ consists of the isolated vertices in $S[\pi]$.

For two isolated elements i and j with $i \prec j$, if $i > j$, then, as in the above, i is removable; contradiction. Finally, if both i and $i + 1$ were isolated for some i, then, by the above, $i \prec i + 1$. Clearly, i would be then removable. □

Lemma 24.3 *If π is critical, then the isolated elements of π are exactly the elements i that satisfy the following* minimum condition: *for all j with $i \prec j$, $i < j$. Moreover, if $i \in I$, then $(i + 1) \prec i$.*

Proof. This first claim is clear from Lemma 24.2, since no element from C can satisfy the minimum condition. The second claim, $(i + 1) \prec i$, follows from this and Lemma 24.2. □

Lemma 24.4 *If a critical π avoids* 2341, *then it contains a factor* $2a41$ *for some* $a \geqslant 5$.

Proof. The element 1 is necessarily isolated, and therefore Lemma 24.3 implies that $2 \prec 1$ and $2 \in C$. Since the degree of 2 in $S[\pi]$ is two, there are exactly two elements a and b such that $2 \prec a \prec b \prec 1$ in π, and these are adjacent with 2, that is, $2a$ and $2b$ are edges of $S[\pi]$. By the assumption, $S[\pi]$ is triangle free, and hence ab is not an edge. Therefore $a > b$, and π has a factor $2ab1$.

We prove next that $4 \prec 1$. Assume the converse. By Lemma 24.2, either 3 or 4 is in C and we claim that 4 is in C whereas 3 is isolated. Indeed, if 3 is in C, then, necessarily, $3 \prec 1$. If $3 \prec 2$, then a cycle $(3, a, 2, b)$ of length four is created in $S[\pi]$, contradicting Lemma 24.2, and if $b = 3$, then 3 has degree one, which is a contradiction.

Therefore, we must have, using Lemma 24.2 again, $\pi = u2ab1v4cd3w$, where $d < c < a$ and $d < b$. Now, there must be a chain in $S[\pi]$ from d to a or b, say $(d, d_1, d_2, \ldots, d_r), r \geqslant 2$ such that $d_i \notin \{4, c\}$, for any $1 \leqslant i \leqslant r$ and $d_r \in \{a, b\}$. If $d_1 \prec d$, then $d_1 < d < c$ implies that d_1c is an edge and a cycle of length four, $(4, c, d_1, d)$, is obtained, contradicting Lemma 24.2. Thus necessarily $d_1 \in w$ and if $s \geqslant 2$ is minimal such that $d_s \notin w$, then $d_i < c$, for any $1 \leqslant i \leqslant s-1$, otherwise $d_i c$ is an edge. Therefore, since $d_s \neq c$, $d_s \in u2ab1v$ with $d_s < d_{s-1} < c$. A chain from d to c is obtained, thus contradicting Lemma 24.2. Consequently, $4 \prec 1$.

On the other hand, it is not the case that $4 \prec 2$, since then $4 < b < a$ ($b \neq 3$ as 3 cannot have degree one) and both $4a$ and $4b$ would be edges in $S[\pi]$, thus creating a cycle of length four in $S[\pi]$. Consequently, $4 \in \{a, b\}$ and, since $a > b$ and $b \neq 3$, we get $b = 4$. □

Theorem 24.3 *A permutation π on $[1, n]$ is sortable on two parallel stacks if and only if it avoids the permutations*

$$\pi_k = 2(2k+1)41 \dots (2i)(2i-3) \dots (2k+2)(2k-1) \quad (24.1)$$

$$\sigma_k = (2k+3)2(2k+5)41 \dots (2i)(2i-3) \dots (2k+4)(2k+1) \quad (24.2)$$

for all odd integers $k \geqslant 1$.

Proof. It is straightforward to verify that the stack graph $S[\pi_k]$ contains the cycle $C_{k+2} = (2, 4, \dots, 2(k+1), 2k+1)$ and $S[\sigma_k]$ contains the cycle $C_{k+4} = (2, 4, \dots, 2(k+2), 2k+3, 2k+5)$. Hence, if k is odd, then these cycles are of odd length, and accordingly $S[\pi_k]$ and $S[\sigma_k]$ are not bipartite. Therefore π_k and σ_k are not sortable on two stacks.

Suppose then that π is a critical permutation that avoids $\pi_1 = 2341$. By Lemma 24.4, π has a factor $2a41$, where $a \geqslant 5$, and thus 24 and $2a$ are edges of $S[\pi]$. Let the unique cycle C of $S[\pi]$ from Lemma 24.2 be of length k:

$$(b_1, b_2, \dots, b_{k-1}, a),$$

where $b_1 = 2$, $b_2 = 4$, and the last vertex a has an edge to b_1.

Let $b_1, b_2, \dots, b_m$ be the longest initial part such that $b_1 < b_2 < \dots < b_m$. Here $m \geqslant 2$. We have then

$$\pi = ub_1ab_2w_2b_3 \dots b_iw_ib_{i+1} \dots b_{m-1}w_{m-1}b_mw_m$$

for some factors u and w_i.

Let x_i be the minimum element in π after b_i, $1 \leqslant i \leqslant m$. Therefore $x_1 = x_2 = 1$, and $x_i < b_{i-1}$ for all $i \geqslant 2$ by the definition of stack graphs. Now for each $2 \leqslant i \leqslant m$,

$$x_i \in w_i. \quad (24.3)$$

Indeed, it is clear that $x_m \in w_m$. For $i < m$, if $b_{i+1} \prec x_i$, then $x_i < b_{i-1} < b_{i+1}$ would imply that $b_{i-1}b_{i+1}$ is an edge of $S[\pi]$, and thus there would be a triangle in $S[\pi]$ consisting of b_{i-1}, b_i and b_{i+1}.

Similarly, for $2 \leqslant i \leqslant m-1$,

$$y \in w_i \implies y < b_i, \tag{24.4}$$

since, otherwise, $b_i y$ is an edge of $S[\pi]$, and the degree of b_i would be at least three.

Claim. For any $2 \leqslant i \leqslant m-1$, the factor w_i contains no elements from C and, therefore, $w_i = x_i$.

Proof of Claim. Consider an i, $2 \leqslant i \leqslant m-1$, and assume that there exists an element of C in w_i. If b_{k-1} is in w_i, then $a < b_{k-1}$ and there exists a $z < a$ such that $b_{k-1} \prec z$. But then $z < a < b_{k-1} < b_i$ and $b_i \prec z$, which implies that ab_i is an edge, a contradiction since $i \leqslant k-2$. Therefore, there exists a $b_t \in w_i$ such that $b_{t+1} \notin w_i$. If $b_{t+1} \prec b_t$, then, $b_{t+1} < b_t$ and there exists an element $z < b_{t+1}$ such that $b_t \prec z$. By (24.4), $b_{t+1} < b_t < b_i$, and so the pair $b_{t+1}b_i$ is an edge of $S[\pi]$; a contradiction. Therefore $b_t \prec b_{i+1} \prec b_{t+1}$. Take an element z such that $b_{t+1} \prec z$ and $z < b_t$. Using (24.4), we obtain $b_t < b_i < b_{i+1}$, which implies that $b_t b_{i+1}$ is an edge, and thus $i = m-1$, $t = m+1$. Moreover, we have deduced that $b_{m+1} \in w_{m-1}$ and $b_{m+2} \in w_m$. Now $b_{m+2} < b_m$, since $b_m b_{m+2}$ is not an edge of $S[\pi]$ and $b_{m+1}b_{m+2}$ is. (In fact, we have even that $b_{m+2} < b_{m-1}$.)

Now, in order to complete the cycle C from b_{m+2} towards a, there necessarily exists an edge $b_s b_{s+1}$ connecting an element $b_s \in w_m$ with an element in the initial part $b_{s+1} \in u2a$. Indeed, if $b_s \in w_m$ and $b_{s+1} \in w_i$, for some $2 \leqslant i \leqslant m-1$, then $b_{s+1}b_m$ is an edge, a contradiction since $s+1 \geqslant m+2$. Therefore there is an x such that $b_s \prec x$ and $x < b_{s+1} < b_s$. Here $b_m < b_s$; for, otherwise, $x < b_{s+1} < b_m$ and thus $b_{s+1}b_m$ would be an edge but this is not possible, since $s+1 > m+1$.

Consequently, there are elements in the sequence $b_{m+3}b_{m+4}\dots b_{k-1}$ which are larger than b_m. Also, the first such element, say b_r, belongs to w_m and so do all the elements preceding it in the sequence, that is, $b_{m+3}, b_{m+4}, \dots, b_{r-1}$. Since $b_{r-1}b_r$ is an edge of $S[\pi]$, $z < b_{r-1} < b_m < b_r$ for some z with $b_r \prec z$. But this means that $b_m b_r$ is an edge of $S[\pi]$ which is a contradiction. Therefore $w_i = x_i$, for any $2 \leqslant i \leqslant m-1$. The Claim is proved.

We conclude from the Claim, that

$$\pi = u2a4x_2b_3x_3b_4x_4\dots b_{m-1}x_{m-1}b_m w_m.$$

In particular, $b_{m+1} \in u2a$.

If $b_{m+1} = a$, then we have completed the cycle C, and, in this case, $m = k-1$, u is empty, $w_m = x_m$, and $b_{k-2} < a < b_{k-1}$. Accordingly, $\pi = \pi_{k-2}$, since both sequences x_i and b_i are increasing.

Suppose then that $b_{m+1} \neq a$. In this case, $b_{m+1} \in u$ and $b_{m+1} < b_m$. If $a < b_m$, then, as ab_m is not an edge, we have also $a < x_m$ and there is $i, 1 \leqslant i \leqslant m-1$, such that $x_i < a < x_{i+1}$. As ab_i is not an edge, it follows that $a > b_i$ and from $b_i > x_{i+1}$ we get $a > x_{i+1}$, a contradiction. Therefore, $a > b_m$ and hence $b_{m+1}a$ is an edge and the cycle is thus completed: $m = k-2, u = b_{m+1}$, and again $w_m = x_m$. In this case, using again Lemma 24.3, $\pi = \sigma_{k-4}$. □

4. QUEUES AND STACKS

We consider in this section permutations sortable on parallel queues. Their definition is similar to the one for parallel stacks, just that the stacks are replaced by queues and elements are also allowed to be sent directly to the output (this is implicitly possible in the case of stacks), see also Figure 24.2.

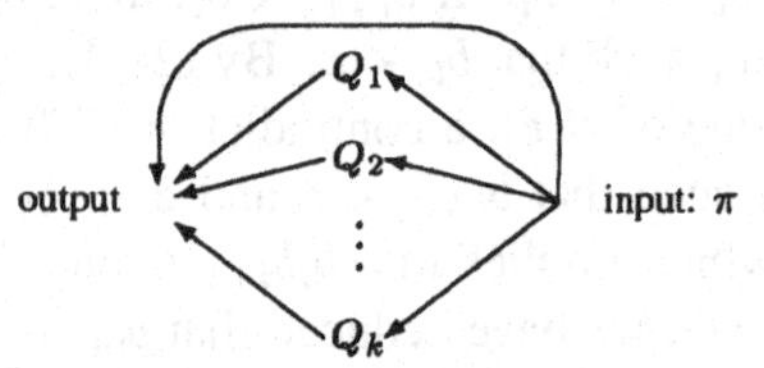

Figure 24.2 Sorting on k parallel queues.

A permutation π is sortable on k parallel queues if and only if the longest decreasing subsequence in π has the length $k+2$, see Golumbic [7, p. 166][2]. Therefore π is sortable on k parallel queues if and only if it avoids the permutation

$$\tau_k = (k+2)(k+1)k \ldots 1.$$

We note that in the permutations π_{i+2} and σ_i, for $i \geqslant 1$, the length of the longest decreasing subsequence is three (they do contain $\tau_1 = 321$), and therefore these permutations are sortable by two parallel queues, but not on a single queue. The remaining permutation $\pi_1 = 2341$ is sortable on one queue. We have thus

Theorem 24.4 *A permutation π is sortable on one queue and on two parallel stacks if and only if π avoids $\tau_1 = 321$ and $\pi_1 = 2341$.*

Also, it was shown by Stankova [11] that there are equally many permutations of any given length n that avoid 1234 and 4123, from which it follows that there are equally many permutations on $[1, n]$ that avoid 4321 and 2341. We therefore get:

[2]Notice that in [7] direct output of elements from the input is not allowed, in which case the characterization holds with the length of the longest decreasing subsequence $k+1$.

Theorem 24.5 *For any $n \geqslant 7$, the number of permutations of length n which are sortable on two parallel queues is strictly larger than the number of permutations of the same length which are sortable on two parallel stacks; that is, two queues are more powerful than two stacks.*

References

[1] Ageev, A. A. (1996), A triangle-free circle graph with chromatic number 5, *Discrete Mathematics*, 152: 295–298.

[2] Bóna, M. (1997), Permutations avoiding certain patterns, *Discrete Mathematics*, 175: 55–67.

[3] Bouchet, A. (1994), Circle graph obstructions, *Journal of Combinatorial Theory, Series B*, 60: 107–144.

[4] Buckingham, M.A. & M.C. Golumbic (1983), Partionable graphs, circle graphs, and the Berge strong perfect graph conjecture, *Discrete Mathematics*, 44: 45–54.

[5] Even, S. & A. Itai (1971), Queues, stacks, and graphs, in Z. Kohavi & A. Paz, eds., *Theory of Machines and Computation*: 71–76. Academic Press, New York.

[6] Fraysseix, H. de (1984), A characterization of circle graphs, *Eur. J. Combin.* 5: 223–238.

[7] Golumbic, M.C. (1980), *Algorithmic Graph Theory and Perfect Graphs*, Academic Press, New York.

[8] Knuth, D. (1973), *The Art of Computer Programming*, 1, Addison-Wesley, Reading, Mass.

[9] Pratt, V. (1973), Computing permutations with double-ended queues, parallel stacks and parallel queues. in *Proceedings of 5th STOC*: 268–277, Association for Computing Machinery.

[10] Rosenstiehl,P. & R.E. Tarjan (1984), Gauss codes, planar Hamiltonian graphs, and stack-sortable permutations, *J. Algorithms*, 5: 375–390.

[11] Stankova, Z. (1996), Classification of forbidden subsequences of lenght 4, *Eur. J. Combin.*, 17: 501–517.

[12] Tarjan, R. (1972), Sorting using networks of queues and stacks, *Journal of the Association for Computing Machinery*, 19.2: 341–346.

[13] West, J. (1993), Sorting twice through a stack, *Theoretical Computer Science*, 117: 303–313.

[14] Zeilberger, D. (1992), A proof of Julian West's conjecture that the number of two-stack-sortable permutations of length n is $2(3n)!/((n+1)!(2n+1)!)$, *Discrete Mathematics*, 102: 85–93.

Chapter 25

APPROXIMATE IDENTIFICATION AND FINITE ELASTICITY

Satoshi Kobayashi
Department of Information Sciences
Tokyo Denki University
satoshi@cs.uec.ac.jp

Yasubumi Sakakibara
Department of Information Sciences
Tokyo Denki University
yasu@cs.uec.ac.jp

Takashi Yokomori
Department of Mathematics
School of Education
Waseda University
1-6-1, Nishiwaseda, Shijuku-ku, Tokyo 169, Japan
yokomori@mn.waseda.ac.jp

Abstract In this paper, we investigate the upper-best approximate identifiability of a fixed target concept class using another fixed hypothesis space, and give a theorem which shows that this identification framework is closely related to an extended notion of finite elasticity. Further we implicitly give a method for enlarging the target concept class and for refining the structure of the hypothesis space.

1. INTRODUCTION

In the framework of identification in the limit introduced by Gold ([4]), the theory of approximately identifying a target concept outside the hypothesis space was first proposed by Mukouchi ([9]). In our previous work ([7]), we

C. Martin-Vide and V. Mitrana (eds.), Where Mathematics, Computer Science, Linguistics and Biology Meet, 277-286.

completely characterize the hypothesis spaces which can identify an approximate upper-best for *every* concept, and also show some interesting topological properties of such hypothetical spaces. The upper-best approximation of a given concept is defined as the minimum concept in the hypothesis space that contains the given concept, and the aim of the upper-best approximate identification is to identify the upper-best approximation of the given target concept at its limit.

In [7], we proved that the upper-best approximate identification of the class of all concepts is closely related to the finite elasticity of the hypothesis space. In this paper, we investigate the upper-best approximate identifiability of a fixed target concept class using another fixed hypothesis space, and prove that this identification framework is closely related to an extended notion of finite elasticity. Further we implicitly give a method for enlarging the target concept class and for refining the structure of the hypothesis space (Section 4.).

2. PRELIMINARIES

Let us consider a recursively enumerable set U, called a *universal set*. By $\subseteq$ ($\subset$) we denote the inclusion (proper inclusion) relation. A subset of U is called a *concept*, and a set of concepts is called a *concept class*. For any (possibly infinite) concept class $\mathcal{C}$, by $\cap\mathcal{C}$ and $\cup\mathcal{C}$, we denote the sets $\{x \mid \forall L \in \mathcal{C}\ x \in L\}$ and $\{x \mid \exists L \in \mathcal{C}\ x \in L\}$, respectively. For any concept class $\mathcal{C}$, the complementary class of $\mathcal{C}$ is defined to be $\{U - L \mid L \in \mathcal{C}\}$ and is denoted by $\mathcal{C}mp(\mathcal{C})$. By $\mathcal{F}in$ we denote the class of all finite concepts. By $\mathcal{C}o\mathcal{F}in$ we denote $\mathcal{C}mp(\mathcal{F}in)$. For any concept class $\mathcal{C}$, by $\mathcal{I}nt(\mathcal{C})$, we denote the smallest class containing $\mathcal{C}$ and closed under *finite* intersections. For any concept class $\mathcal{C}$, by $\mathcal{B}l(\mathcal{C})$, we denote the smallest class containing $\mathcal{C}$ and closed under the Boolean operations (union, intersection, and complement). In this paper, we assume that the Boolean closure of a given concept class contains both the intersection and the union of the empty concept class. Thus, for any concept class $\mathcal{C}$, $\mathcal{B}l(\mathcal{C})$ contains both $\emptyset$ ($= \cup\emptyset$) and U ($= \cap\emptyset$).

A class $\mathcal{C} = \{L_i\}_{i\in\mathbf{N}}$ of concepts is *an indexed family of recursive concepts* (or, *indexed family*, for short), iff there exists a recursive function $f : \mathbf{N} \times U \rightarrow \{0, 1\}$ such that

$$f(i, w) = \begin{cases} 1 & if\ w \in L_i, \\ 0 & otherwise. \end{cases}$$

where $\mathbf{N}$ denotes the set of positive integers.

3. APPROXIMATELY IDENTIFYING CONCEPTS

Let U be the universal set. Let $\mathcal{C}$ be a concept class and X be a concept (not always in $\mathcal{C}$). A concept $Y \in \mathcal{C}$ is called a *$\mathcal{C}$-upper-best approximation of a concept* X iff $X \subseteq Y$ and for any concept $C \in \mathcal{C}$ such that $X \subseteq C$,

$Y \subseteq C$ holds. A concept $Y \in \mathcal{C}$ is called a *$\mathcal{C}$-lower-best approximation of a concept X* iff $Y \subseteq X$ and for any concept $C \in \mathcal{C}$ such that $C \subseteq X$, $C \subseteq Y$ holds. By $\overline{\mathcal{C}}X$ ($\underline{\mathcal{C}}X$), we denote the $\mathcal{C}$-upper-best ($\mathcal{C}$-lower-best) approximation of a concept X. $\mathcal{C}_1$ has *upper-best approximation property (u.b.a.p)* relative to $\mathcal{C}_2$, written $\mathcal{C}_1 \overset{u.b.}{\to} \mathcal{C}_2$, iff for any concept X in $\mathcal{C}_2$, there exists a $\mathcal{C}_1$-upper-best approximation of X. Similarly, $\mathcal{C}_1$ has *lower-best approximation property (l.b.a.p.)* relative to $\mathcal{C}_2$, written $\mathcal{C}_1 \overset{l.b.}{\to} \mathcal{C}_2$, iff for any concept X in $\mathcal{C}_2$, there exists a $\mathcal{C}_1$-lower-best approximation of X. For any concept class $\mathcal{C}$ and a concept $X \subseteq U$, by $\mathcal{U}_{\mathcal{C}}(X)$ and $\mathcal{L}_{\mathcal{C}}(X)$, we denote the set $\{L \in \mathcal{C} \mid X \subseteq L\}$ and the set $\{L \in \mathcal{C} \mid L \subseteq X\}$, respectively.

We give below some propositions which will be used in the rest of the paper.

Proposition 25.1 *Let $\mathcal{C}_1, \mathcal{C}_2$ be concept classes.*
(1) $\mathcal{C}_1 \overset{u.b.}{\to} \mathcal{C}_2$ iff for any $L \in \mathcal{C}_2$, $\bigcap \mathcal{U}_{\mathcal{C}_1}(L) \in \mathcal{C}_1$. In this case, $\bigcap \mathcal{U}_{\mathcal{C}_1}(L) = \overline{\mathcal{C}}_1 L$ holds.
(2) $\mathcal{C}_1 \overset{l.b.}{\to} \mathcal{C}_2$ iff for any $L \in \mathcal{C}_2$, $\bigcup \mathcal{L}_{\mathcal{C}_1}(L) \in \mathcal{C}_1$. In this case, $\bigcup \mathcal{L}_{\mathcal{C}_1}(L) = \underline{\mathcal{C}}_1 L$ holds.

Proposition 25.2 *Let $\mathcal{C}$ be a concept class and A be any concept. Then, $\bigcup \mathcal{L}_{Cmp(\mathcal{C})}$*
$(U - A) = U - \bigcap \mathcal{U}_{\mathcal{C}}(A)$ holds.

Proposition 25.3 *Let $\mathcal{C}_1, \mathcal{C}_2, \mathcal{C}_3$ be concept classes.*
(1) $\mathcal{C}_1 \overset{u.b.}{\to} \mathcal{C}_2$ and $\mathcal{C}_1 \overset{u.b.}{\to} \mathcal{C}_3$ imply $\mathcal{C}_1 \overset{u.b.}{\to} \mathcal{C}_2 \cup \mathcal{C}_3$.
(2) $\mathcal{C}_1 \overset{u.b.}{\to} \mathcal{C}_3$ and $\mathcal{C}_2 \overset{u.b.}{\to} \mathcal{C}_3$ imply $\mathcal{I}nt(\mathcal{C}_1 \cup \mathcal{C}_2) \overset{u.b.}{\to} \mathcal{C}_3$.

Proof. (1) Immediate from the definition.
(2) Let L be any concept in $\mathcal{C}_3$, and define $L_1 = \overline{\mathcal{C}}_1 L$, $L_2 = \overline{\mathcal{C}}_2 L$, and $L_* = L_1 \cap L_2$. Note that $L_* \in \mathcal{I}nt(\mathcal{C}_1 \cup \mathcal{C}_2)$. We will show that L_* is $\mathcal{I}nt(\mathcal{C}_1 \cup \mathcal{C}_2)$-upper-best approximation of L. Note that $L \subseteq L_1$ and $L \subseteq L_2$ hold. Therefore, we have $L \subseteq L_*$.

Let us consider any concept $A \in \mathcal{I}nt(\mathcal{C}_1 \cup \mathcal{C}_2)$ such that $L \subseteq A$. Then, there are a finite sequence $L_{p_1}, \ldots, L_{p_k}$ of concepts in $\mathcal{C}_1$ and a finite sequence $L_{q_1}, \ldots, L_{q_l}$ of concepts in $\mathcal{C}_2$ such that $A = L_{p_1} \cap \cdots L_{p_k} \cap L_{q_1} \cap \cdots L_{q_l}$. We have that $L \subseteq L_{p_i}$ $(1 \leqslant i \leqslant k)$ and $L \subseteq L_{q_i}$ $(1 \leqslant i \leqslant l)$, which implies that $L_1 = \overline{\mathcal{C}}_1 L \subseteq L_{p_i}$ $(1 \leqslant i \leqslant k)$ and $L_2 = \overline{\mathcal{C}}_2 L \subseteq L_{q_i}$ $(1 \leqslant i \leqslant l)$. Therefore, we have $L_* = L_1 \cap L_2 \subseteq A$. This implies that $L_* \in \mathcal{I}nt(\mathcal{C}_1 \cup \mathcal{C}_2)$ is a $\mathcal{I}nt(\mathcal{C}_1 \cup \mathcal{C}_2)$-upper-best approximation of L. □

A positive presentation of a non-empty concept L is an infinite sequence $w_1, w_2, \ldots$ of the elements in L such that $\{w_1, w2, \ldots\} = L$. A positive presentation of an empty concept is an infinite sequence of the element $\# \notin U$. A negative presentation of a concept L is a positive presentation of the concept $U - L$. A learning algorithm is an algorithmic device which receives inputs from time to time and outputs positive integers from time to time. Let $\mathcal{C}$ be an indexed family. A learning algorithm M *$\mathcal{C}$-upper-best approximately identifies* a concept L in the limit from positive data iff for any positive presentation σ of L, M with σ outputs infinite sequence of integers, $g_1, g_2, ...$, such that $\exists n \geqslant 1 \, (\forall j \geqslant n \, g_j = g_n) \wedge (L_{g_n} = \overline{\mathcal{C}}L)$. A concept class $\mathcal{C}_1$ is *upper-best approximately identifiable in the limit* from positive data by an indexed family $\mathcal{C}_2$ iff there is a learning algorithm M which $\mathcal{C}_2$-upper-best approximately identifies every concept in $\mathcal{C}_1$ in the limit from positive data. The original definition of the identifiability from positive data ([1]) just corresponds to the case of $\mathcal{C}_1 = \mathcal{C}_2$. The lower-best approximate identifiability from negative data is defined in a similar manner.

4. APPROXIMATE IDENTIFICATION AND FINITE ELASTICITY

We introduce an extended notion of the finite elasticity ([13][8]). Let $\mathcal{C}$ be a concept class and L be any concept. $\mathcal{C}$ has *infinite elasticity* at L iff there is an infinite sequence $F_0, F_1, ...$ of finite subsets of L and an infinite sequence $L_1, L_2, ...$ of concepts in $\mathcal{C}$ such that (1) $\bigcup\{F_i \mid i \geqslant 1\} = L$ and (2) for any $i \geqslant 1$, $F_{i-1} \subset F_i$, $F_{i-1} \subseteq L_i$ and $F_i \not\subseteq L_i$ hold. A concept class $\mathcal{C}_1$ has *finite elasticity* relative to $\mathcal{C}_2$, written $\mathcal{C}_1 \stackrel{f.e.}{\to} \mathcal{C}_2$, iff for every concept $L \in \mathcal{C}_2$, $\mathcal{C}_1$ does not have infinite elasticity at L. In case of $\mathcal{C}_2 = 2^U$, this corresponds to the original definition of the finite elasticity, and we simply say that $\mathcal{C}_1$ has finite elasticity. It is easy to see that if $\mathcal{C}$ has finite elasticity, then $\mathcal{C}$ satisfies ACC.

Further, we must note that the above notion of the infinite elasticity of $\mathcal{C}$ at L was first equivalently introduced by [12] as the notion of the *infinite cross property* of L within $\mathcal{C}$, and this equivalence has been proved in [10].

We have the following interesting proposition.

Proposition 25.4 *Let $\mathcal{C}_1$, $\mathcal{C}_2$, and $\mathcal{C}_3$ be concept classes.*
(1) $\mathcal{C}_1 \stackrel{f.e.}{\to} \mathcal{C}_2$ and $\mathcal{C}_1 \stackrel{f.e.}{\to} \mathcal{C}_3$ imply $\mathcal{C}_1 \stackrel{f.e.}{\to} \mathcal{C}_2 \cup \mathcal{C}_3$.
(2) $\mathcal{C}_1 \stackrel{f.e.}{\to} \mathcal{C}_3$ and $\mathcal{C}_2 \stackrel{f.e.}{\to} \mathcal{C}_3$ imply $Int(\mathcal{C}_1 \cup \mathcal{C}_2) \stackrel{f.e.}{\to} \mathcal{C}_3$.

Proof. (1) Assume $\mathcal{C}_1 \stackrel{f.e.}{\not\to} \mathcal{C}_2 \cup \mathcal{C}_3$. Then, $\mathcal{C}_1$ has infinite elasticity at some concept $L \in \mathcal{C}_2 \cup \mathcal{C}_3$. Therefore, there is an infinite sequence $F_0, F_1, ...$ of finite subsets of L and an infinite sequence $L_1, L_2, ...$ of concepts in $\mathcal{C}_1$ satisfying the conditions of infinite elasticity at L. Since $L \in \mathcal{C}_2 \cup \mathcal{C}_3$, this implies that $\mathcal{C}_1$ does not have finite elasticity relative to either $\mathcal{C}_2$ or $\mathcal{C}_3$, which is a contradiction.
(2) Assume $Int(\mathcal{C}_1 \cup \mathcal{C}_2) \stackrel{f.e.}{\not\to} \mathcal{C}_3$. Then, $Int(\mathcal{C}_1 \cup \mathcal{C}_2)$ has infinite elasticity at some concept $L \in \mathcal{C}_3$. Therefore, there are an infinite sequence $F_0, F_1, ...$ of

finite subsets of L and an infinite sequence $L_1, L_2, \ldots$ of concepts in $Int(\mathcal{C}_1 \cup \mathcal{C}_2)$ satisfying the conditions of infinite elasticity at L. We can write $L_i = A_1^i \cap \cdots \cap A_{k(i)}^i$ for each $i \geqslant 1$, where $A_j^i \in \mathcal{C}_1 \cup \mathcal{C}_2$ $(1 \leqslant j \leqslant k(i))$. By $F_{i-1} \subseteq L_i$ $(i \geqslant 1)$, we have $F_{i-1} \subseteq A_j^i$ $(i \geqslant 1, 1 \leqslant j \leqslant k(i))$. By $F_i \not\subseteq L_i$ $(i \geqslant 1)$, we have for each $i \geqslant 1$, that some A_j^i does not contain F_i (for convenience, we say that A_j^i *breaks the elasticity*). Since $A_j^i \in \mathcal{C}_1 \cup \mathcal{C}_2$ $(i \geqslant 1, 1 \leqslant j \leqslant k(i))$, either $\mathcal{C}_1$ or $\mathcal{C}_2$ contains an infinite number of concepts A_j^i which breaks the elasticity. Without loss of generality, we can assume that $\mathcal{C}_1$ contains infinite number of such concepts, $A_{l_1}^{n_1}, A_{l_2}^{n_2}, \ldots$, where $n_i < n_{i+1}$ $(i \geqslant 1)$. Then, the sequences $F_0, F_{n_1}, F_{n_2}, \ldots$ and $A_{l_1}^{n_1}, A_{l_2}^{n_2}, \ldots$ lead to $\mathcal{C}_1 \overset{f.e.}{\not\to} \mathcal{C}_3$, which is a contradiction. □

Next, we introduce an extended notion of characteristic samples ([6]).

Let $\mathcal{C}$ be a concept class and consider a concept $L \in 2^U$. A finite subset F of L is called a *characteristic sample* of L relative to $\mathcal{C}$ iff for any $A \in \mathcal{C}$, $F \subseteq A$ implies $L \subseteq A$. Note that when we restrict L to a concept in $\mathcal{C}$, this definition coincides with the original notion of the characteristic sample ([1]).

Let $\mathcal{C}_1$ and $\mathcal{C}_2$ be concept classes. We say that $\mathcal{C}_1$ *has characteristic samples* relative to $\mathcal{C}_2$ iff for any $L \in \mathcal{C}_1$, there exists a characteristic sample of L relative to $\mathcal{C}_2$. The following lemma is useful.

Lemma 25.1 *$\mathcal{C}_1$ has characteristic samples relative to $\mathcal{C}_2$ iff $\mathcal{C}_2 \overset{f.e.}{\to} \mathcal{C}_1$.*

Proof. We first prove the only if direction of the claim. Suppose that there is a concept $L \in \mathcal{C}_1$ such that $\mathcal{C}_2$ has infinite elasticity at L. Let $F_0, F_1, \ldots$ be an infinite sequence of finite subsets of L and $L_1, L_2, \ldots$ be an infinite sequence of concepts in $\mathcal{C}_2$ satisfying the conditions of infinite elasticity at L. By the assumption, L has a characteristic sample $T \subseteq L$ relative to $\mathcal{C}_2$. Let n be the minimum integer j such that $T \subseteq F_j$. Then, we have for each $j > n$, $L \subseteq L_j$. This implies that for each $j > n$, $F_j \subseteq L \subseteq L_j$, which contradicts the condition of infinite elasticity at L.

To prove the converse direction, assume that there is a concept $L \in \mathcal{C}_1$ such that L does not have characteristic samples relative to $\mathcal{C}_2$. Let $w_0, w_1, \ldots$ be an enumeration of elements of L such that $w_i \neq w_j$ whenever $i \neq j$. Consider the following procedure:

stage 0 : $F_0 = \{w_0\}$;

stage i $(i \geqslant 1)$:

Find a concept $L_i \in \mathcal{C}_2$ such that $F_{i-1} \subseteq L_i$ and $L \not\subseteq L_i$;

Select an element e_i from $L - L_i$;

Set $F_i = F_{i-1} \cup \{w_i, e_i\}$;

Since L does not have characteristic samples, it holds that at the first step of each stage i ($i \geqslant 1$), we can find a concept L_i satisfying $F_{i-1} \subseteq L_i$ and $L \not\subseteq L_i$. (Otherwise, F_{i-1} should be a characteristic sample, a contradiction.) By $F_i - L_i \supseteq \{e_i\} \neq \emptyset$, we have $F_i \not\subseteq L_i$. Further, it is clear that $\bigcup\{F_i \mid i \geqslant 1\} = L$ holds. Thus, the sequences $F_0, F_1, ...$ and $L_1, L_2, ...$ satisfy the conditions of infinite elasticity at L. This is a contradiction. □

Let $e_1, e_2, ...$ be a fixed recursive enumeration of the universal set U. For any concept L, by $L^{(n)}$, we denote the set $L \cap \{e_1, ..., e_n\}$. It is clear that for any recursive concepts L_1 and L_2, and for any positive integer n, the inclusion $L_1^{(n)} \subseteq L_2^{(n)}$ is decidable.

Lemma 25.2 *Let $C_1 = \{L_i\}_{i \in \mathbf{N}}$ be an indexed family and C_2 be a concept class. C_2 is upper-best approximately identifiable in the limit from positive data by C_1 if $C_1 \overset{u.a.}{\rightarrow} C_2$ and $C_1 \overset{f.e.}{\rightarrow} C_2$.*

Proof. Let $w_1, w_2, ...$ be a positive presentation of a target concept $L_* \in C_2$ and consider the following learning algorithm:

stage 0 : $P_0 = \emptyset$;
stage n ($n \geqslant 1$):

$P_n = P_{n-1} \cup \{w_n\}$;

Select all concepts L_{p_i} in $\{L_1, ..., L_n\}$ such that $P_n \subseteq L_{p_i}$ and construct the set $Cns_n = \{L_{p_1}, ..., L_{p_{k_n}}\}$;

Select and output the minimum index g_n of concepts in Cns_n such that $L_{g_n}^{(n)}$ is a minimal concept in $\{L_{p_1}^{(n)}, ..., L_{p_{k_n}}^{(n)}\}$ with respect to $\subseteq$;

Claim A: There is a stage n_1 such that at any stage $n \geqslant n_1$, $L_* \subseteq L_{p_i}$ holds for any $L_{p_i} \in Cns_n$.

Proof of the Claim. By Lemma 25.1, L_* has a characteristic sample T relative to C_1. Let n_1 be the minimum integer n such that $T \subseteq P_n$. Then, the claim holds.

Let g_* be the minimum integer j of the concepts in C_1 such that $\overline{C_1} L_* = L_j$.
Claim B: There is a stage n_2 such that at any stage $n \geqslant n_2$, the learning algorithm always outputs g_*.

Proof of the Claim. Select all concepts L_{r_i} from $\{L_1, ..., L_{g_*-1}\}$ such that $L_* \subseteq L_{r_i}$ and construct the set $\{L_{r_1}, ..., L_{r_m}\}$. Then, for each L_{r_i} ($1 \leqslant i \leqslant m$), $L_{g_*} = \overline{C_1} L_* \subseteq L_{r_i}$ holds. By the minimality of g_*, for each L_{r_i} ($1 \leqslant i \leqslant m$), $L_{g_*} \subset L_{r_i}$ holds. Therefore, for each i ($1 \leqslant i \leqslant m$), we can choose the minimum integer s_i such that $L_{g_*}^{(s_i)} \subset L_{r_i}^{(s_i)}$. Let $n_2 = max\{s_1, ..., s_m, g_*, n_1\}$.

At any stage $n \geqslant n_2$, $L_{g_*} \in Cns_n$ holds since $n_2 \geqslant g_*$ and $L_* \subseteq L_{g_*}$. Further, at any stage $n \geqslant n_2$, $L_{g_*}^{(n)}$ is a minimal concept in $\{L_{p_1}^{(n)}, ..., L_{p_{k_n}}^{(n)}\}$, since for each L_{p_i} ($1 \leqslant i \leqslant k_n$), $L_* \subseteq L_{p_i}$ (by $n_2 \geqslant n_1$) and therefore $L_{g_*} = \overline{C_1} L_* \subseteq L_{p_i}$ hold.

It is only left to show that every $L_{p_i}^{(n)}$ with $p_i < g_*$ is not a minimal concept in $\{L_{p_1}^{(n)}, ..., L_{p_{k_n}}^{(n)}\}$. For any concept $L_{p_i}^{(n)}$ with $p_i < g_*$, by $n_2 \geqslant max\{s_1, ..., s_m\}$, $L_{g_*}^{(n)} \subset L_{p_i}^{(n)}$ holds. Therefore, every $L_{p_i}^{(n)}$ with $p_i < g_*$ is not a minimal concept.

This completes the proof of the lemma. □

Lemma 25.3 *Let C_1 be an indexed family and C_2 be a concept class such that $\mathcal{F}in \subseteq C_2$. C_2 is upper-best approximately identifiable in the limit from positive data by C_1 only if $C_1 \overset{u.a.}{\to} C_2$ and $C_1 \overset{f.e.}{\to} C_2$.*

Proof. By definition, it is clear that $C_1 \overset{u.a.}{\to} C_2$ is necessary for the upper-best approximate identification. Assume $C_1 \overset{f.e.}{\not\to} C_2$. Then, C_1 has infinite elasticity at some concept $L \in C_2$. Let $F_0, F_1, ...$ be an infinite sequence of finite subsets of L and $L_1, L_2, ...$ be an infinite sequence of concepts in C_1 satisfying the conditions of infinite elasticity at L. Let $w_0, w_1, ...$ be a recursive enumeration of elements of L. Let M be a learning algorithm which C_1-upper-best approximately identifies every concept in C_2 in the limit from positive data. For any finite sequence σ of elements of U, by $M(\sigma)$ we denote the last conjecture (integer) output by M when it is fed with σ. Consider the following procedure:

stage 0 : $\sigma_0 = w_0$;

stage i ($i \geqslant 1$):

Find a finite sequence σ_f of elements in L such that $M(\sigma_{i-1}) \neq M(\sigma_{i-1} \cdot \sigma_f)$;

$\sigma_i = \sigma_{i-1} \cdot \sigma_f \cdot w_i$;

At the first step of some stage i ($i \geqslant 1$), the procedure above cannot find any finite sequence σ_f satisfying the condition, and does not stop forever. Since otherwise, σ_∞ could be a positive presentation of L, and M with input σ_∞ does not converge to any integer, which is a contradiction.

Let n be the stage where the procedure cannot find any finite sequence σ_f satisfying the condition. Then, $L_{M(\sigma_n)}$ should be equivalent to $\overline{C_1} L$ since M C_1-upper-best approximately identifies L in the limit.

There is an F_m such that F_m contains all elements of σ_n, since σ_n is finite. Let δ be any positive presentation of the concept F_m. Then, M with input $\sigma_n \cdot \delta$ should converge to the index of the concept $\overline{C_1} F_m$ since $F_m \in C_2$ holds and $\sigma_n \cdot \delta$ is a positive presentation of F_m. Then, by definition of σ_n, $F_m \in C_2$ should be equivalent to $L_{M(\sigma_n)} = \overline{C_1} L$.

Note that, by the conditions of infinite elasticity, we have:

$$F_m \subseteq L_{m+1}, \quad (25.1)$$
$$F_{m+1} \not\subseteq L_{m+1}, \quad (25.2)$$
$$F_{m+1} \subseteq L. \quad (25.3)$$

By (25.1) and the definition of $\overline{\mathcal{C}}_1 F_m$, we have:

$$\overline{\mathcal{C}}_1 F_m \subseteq L_{m+1} \quad (25.4)$$

Therefore,

$$\begin{aligned}
\overline{\mathcal{C}}_1 L - \overline{\mathcal{C}}_1 F_m &\supseteq L - \overline{\mathcal{C}}_1 F_m && (\text{ by definition of } \overline{\mathcal{C}}_1 L\,) \\
&\supseteq F_{m+1} - \overline{\mathcal{C}}_1 F_m && (\text{ by (25.3) }) \\
&\supseteq F_{m+1} - L_{m+1} && (\text{ by (25.4) }) \\
&\neq \emptyset && (\text{ by (25.2) })
\end{aligned}$$

holds, which implies $\overline{\mathcal{C}}_1 L \neq \overline{\mathcal{C}}_1 F_m$, a contradiction. □

Theorem 25.1 *Let $\mathcal{C}_1$ be an indexed family and $\mathcal{C}_2$ be a concept class such that $\mathcal{F}in \subseteq \mathcal{C}_2$. Then, the following conditions are equivalent:*
(1) $\mathcal{C}_2$ is upper-best approximately identifiable in the limit from positive data by $\mathcal{C}_1$,
(2) $\mathcal{C}_1 \overset{u.a.}{\rightarrow} \mathcal{C}_2$ and $\mathcal{C}_1 \overset{f.e.}{\rightarrow} \mathcal{C}_2$.

Proof. By Lemma 25.2 and Lemma 25.3. □

By duality, we have the following.

Theorem 25.2 *Let $\mathcal{C}_1$ be an indexed family and $\mathcal{C}_2$ be a concept class such that $\mathcal{C}o\mathcal{F}in \subseteq \mathcal{C}_2$. Then, the following conditions are equivalent:*
(1) $\mathcal{C}_2$ is lower-best approximately identifiable in the limit from negative data by $\mathcal{C}_1$,
(2) $\mathcal{C}_1 \overset{l.a.}{\rightarrow} \mathcal{C}_2$ and $\mathcal{C}mp(\mathcal{C}_1) \overset{f.e.}{\rightarrow} \mathcal{C}mp(\mathcal{C}_2)$.

Theorem 25.3 *Let $\mathcal{C}_1$ and $\mathcal{C}_2$ be indexed families 1 and let $\mathcal{C}_3$ and $\mathcal{C}_4$ be concept classes such that $\mathcal{F}in \subseteq \mathcal{C}_3$ and $\mathcal{F}in \subseteq \mathcal{C}_4$. Consider any indexed family $\mathcal{C}_5$ for the concept class $\mathcal{I}nt(\mathcal{C}_1 \cup \mathcal{C}_2)$. Then, the following hold:*
(1) Assume that both $\mathcal{C}_3$ and $\mathcal{C}_4$ are upper-best approximately identifiable in the limit from positive data by $\mathcal{C}_1$. Then, $\mathcal{C}_3 \cup \mathcal{C}_4$ is upper-best approximately identifiable in the limit from positive data by $\mathcal{C}_1$.
(2) Assume that $\mathcal{C}_3$ is upper-best approximately identifiable in the limit from positive data by both $\mathcal{C}_1$ and $\mathcal{C}_2$. Then, $\mathcal{C}_3$ is upper-best approximately identifiable in the limit from positive data by $\mathcal{C}_5$.

Proof. By Proposition 25.3, Proposition 25.4 and Theorem 25.1 □

This theorem is important in the sense that it implicitly provides us with a method for enlarging the target concept class and for refining the mesh of the hypothesis space.

5. CONCLUSIONS

In this paper, we give a characterization of the upper-best approximate identifiability from positive data in case that the target concept class includes $\mathcal{F}in$. Further we implicitly give a method for enlarging the target concept class and for refining the structure of the hypothesis space. Characterize the upper-best approximate identifiability in a case where the target concept class does not include $\mathcal{F}in$ is still open. Further, also open to consideration is how to characterize the upper-best approximate identifiability from complete data.

References

[1] Angluin, D. (1980), Inductive inference of formal languages from positive data, *Information and Control*, 45: 117–135.

[2] Angluin, D. (1982), Inference of reversible languages, *Journal of the ACM*, 29: 741–765.

[3] Birkhoff, G. (1967), *Lattice Theory*, 3rd ed., American Mathematical Society.

[4] Gold, E. Mark (1967), Language identification in the limit, *Information and Control*, 10: 447–474.

[5] Kobayashi, S. & T. Yokomori (1994), Families of Noncounting Languages and Its Learnability from Positive Data, Technical Report CSIM 94-03, University of Electro-Communications, Tokyo.

[6] Kobayashi, S. & T. Yokomori (1995), Approximately Learning Regular Languages with respect to Reversible Languages: A Rough Set Based Analysis, in *Proceedings of the Second Annual Joint Conference on Information Sciences*: 91–94.

[7] Kobayashi, S. & T. Yokomori (1995), On Approximately Identifying Concept Classes in the Limit, in *Proceedings of the 6th International Workshop on Algorithmic Learning Theory*: 298–312. Springer, Berlin.

[8] Motoki, T.; T. Shinohara & K. Wright (1991), The correct definition of finite elasticity: corrigendum to identification of unions, in *Proceedings of the 4th Workshop on Computational Learning Theory*: 375–375.

[9] Mukouchi, Y. (1994), Inductive Inference of an Approximate Concept from Positive Data, in *Proceedings of the 5th International Workshop on Algorithmic Learning Theory*: 484-499. Springer, Berlin.

[10] Sato, M. (1995), Inductive Inference of Formal Languages, *Bulletin of Informatics and Cybernetics*, 27.1.

[11] Sato, M. & T. Moriyama (1994), Inductive Inference of Length Bounded EFS's from Positive Data, DMSIS-RR-94-2 Report, Department of Mathematical Sciences and Information Sciences, Univ. of Osaka Pref.

[12] Sato, M. & K. Umayahara (1992), Inductive inferability for formal languages from positive data, *IEICE Trans. Inf. & Syst.*, E75-D.4: 415–419.

[13] Wright, K. (1989), Identification of unions of languages drawn from an identifiable class, in *Proceedings of the 2nd Workshop on Computational Learning Theory*: 328–333.

Chapter 26

INSERTION OF LANGUAGES AND DIFFERENTIAL SEMIRINGS

Gabriel Thierrin
Department of Mathematics
University of Western Ontario
London, ON NGA 5B7, Canada
gab@csd.uwo.ca

Abstract The family of languages with addition or union and catenation as operations is a semiring with a partial order making it a semireticulated semigroup or gerbier. Furthermore, the operation of insertion of languages has properties similar to those related to the derivation of sum and product of functions. This suggests considering semirings and in particular gerbiers having some kind of abstract derivation. The investigation of these algebraic structures called differential semirings and gerbiers is the object of this paper.

1. INTRODUCTION

Let X^* be the free monoid generated by the alphabet X and let $\mathcal{F}(X^*)$ be the family of the languages over X. The standard operations on $\mathcal{F}(X^*)$ are the *union* $\cup$ (also denoted by $+$), the *intersection* $\cap$, the *catenation* . (also denoted by juxtaposition), the *catenation closure* or *iteration* $*$ and the *complementation*.

There are two other important operations on languages, *insertion* and *deletion*, that will be considered here.

Insertion. Let $L_1, L_2 \subseteq X^*$. The insertion of L_2 into L_1 is defined as:

$$L_1 \leftarrow L_2 = \{u \leftarrow v | u \in L_1, v \in L_2\}$$

where $u \leftarrow v = \{u_1 v u_2 | u = u_1 u_2\}$.

* Research partially supported by Grant R0504A01 of the Natural Sciences and Engineering Research Council of Canada.

C. Martin-Vide and V. Mitrana (eds.), Where Mathematics, Computer Science, Linguistics and Biology Meet, 287-296.

Deletion. The deletion of L_2 from L_1 is defined as:

$$L_1 \rightarrow L_2 = \{u \rightarrow v | u \in L_1, v \in L_2\}$$

where $u \rightarrow v = \{w \in X^* | u = w_1 v w_2, w = w_1 w_2, w_1, w_2 \in X^*\}$.

The definition of insertion and deletion given for a specific language can easily be extended to families of languages in the following way.

Let $\mathcal{T} = \{L_i | i \in I, L_i \subseteq X^*\}$ be a family of languages over X and let $L \subseteq X^*$. The insertion and deletion of L into $\mathcal{T}$ are defined respectively as:

$$\mathcal{T} \leftarrow L = \{L_i \leftarrow L | i \in I\}, \mathcal{T} \rightarrow L = \{L_i \rightarrow L | i \in I\}$$

These two operations of insertion and deletion have been studied extensively, in particular in [5]. They also play an important role in the field of computability, especially in relation with the topic of DNA computing. It can be shown [6] that a system using only these two operations can be constructed, a system that is equivalent to a Turing machine. Since these two operations can be realized biologically by DNA manipulations, this is an example showing the feasibility of universal DNA computers.

The operation of insertion in connection with union and catenation has the following interesting properties which are similar to some properties of derivation of functions related to addition and product of functions.

Let $L \subseteq X^*$ be a fixed language over the alphabet X and let $\rho(L_i) = L_i \leftarrow L$ for any language $L_i \subseteq X^*$. Then:

(i) $\rho_L(L_1 + L_2) = \rho_L(L_1) + \rho_L(L_2)$,

(ii) $\rho_L(L_1 L_2) = \rho_L(L_1) L_2 + L_1 \rho_L(L_2)$

Hence this operation ρ has properties similar that of derivation in calculus.

In abstract algebra, a *mapping* ρ, having the properties (i) and (ii) relatively to some operations $+$ an . of a given algebraic structure is generally called an *abstract derivation.* The corresponding algebraic structure is called a differential algebraic structure.

A well known example is the class of differential rings and in particular the ring of differentiable functions. Remark that many examples of differential rings exist, the corresponding abstract derivation having no relation at all with the usual derivation in calculus (See for example [3], [7]).

The family $\mathcal{F}(X^*, +, .)$ of languages over X with the operations of union $+$ and catenation . is not a ring because it fails to be a commutative group with respect to the addition $+$. However, by weakening the condition on $+$ by means of requiring that it is a commutative semigroup, $\mathcal{F}(X^*, +, .)$ is what is called a *semiring.* The mapping ρ, as defined above, makes $\mathcal{F}(X^*, +, .)$ a semiring with an abstract derivation and hence a differential semiring. This and many other examples motivate a general study of differential semirings in connection with the insertion operation in formal languages.

2. ABSTRACT DERIVATION AND DIFFERENTIAL SEMIRINGS

Let $\mathcal{F}(X^*)$ be the family of languages over X and let L be a fixed language. Define the following mappings ρ_L and τ_L of $\mathcal{F}(X^*)$ into $\mathcal{F}(X^*)$ as:

$$\rho_L(L_i) = L_i \leftarrow L, \quad \tau_L(L_i) = L_i \rightarrow L, \; L_i \subseteq X^*$$

Proposition 26.1 *The mapping* ρ_L defined on $\mathcal{F}(X^*)$ *by the insertion of a fixed language L has the following properties:*
(i) $\rho_L(L_1 + L_2) = \rho_L(L_1) + \rho_L(L_2)$
(ii) $\rho_L(L_1L_2) = \rho_L(L_1)L_2 + L_1\rho_L(L_2)$

Proof. (i) Immediate.
(ii) Clearly $\rho_L(L_1)L_2 + L_1\rho_L(L_2) \subseteq \rho_L(L_1L_2)$. Let $u \in \rho_L(L_1L_2)_L$. Then $u = wvw'$ with $ww' = u_1u_2 \in L_1L_2$, $v \in L$, $u_1 \in L_1$, $u_2 \in L_2$. If $|u_1| \leqslant |w|$, then the insertion of v is done through the word u_2 and $u = u_1u_2'vu_2''$ with $u_2'u_2' = u_2$. Hence $u_2'vu_2'' \in \rho_L(L_2)$ and $u \in L_1\rho_L(L_2)$. If $|u_1| > |w|$, then $|u_2| \leqslant w'$ and by a similar argument we get $u \in \rho(L_1)L_2$. This shows that $\rho_L(L_1L_2) \subseteq \rho_L(L_1)L_2 + L_1\rho_L(L_2)$. □

In general, we do not get similar results with τ_L. For example, let $X = \{a, b\}$, $L = \{a^2\}$, $L_1 = \{ba\}$ and $L_2 = \{ab\}$. Then:

$$\tau_L(L_1L_2) = \tau_L\{ba^2b\} = \{b^2\}, \tau_L(L_1) = \tau_L(L_2) = \emptyset$$

As will be seen later, many examples exist of families of languages that are preserved under insertions or mappings satisfying the above properties (i) and (ii). These properties are similar to the properties of the usual derivation in Calculus and in Differential Rings.

The previous results suggest considering a more general algebraic structure, including the preceding examples. This can be done by replacing rings by semirings. Also, taking into account the fact that addition of sets is an idempotent operation, we will also have to consider the case when the additive structure of the semiring is a semilattice.

A *semiring* $S(+, .)$ is a nonempty set with two binary operations $+$ and $.$ such that:

1. $S(+)$ is a commutative additive semigroup.
2. $S(.)$ is a multiplicative semigroup.
3. Distributive laws:

$$u.(v + w) = u.v + u.w, \quad (v + w).u = v.u + w.u$$

for every $u, v, w \in S$.

In general, juxtaposition is used instead of the symbol . .

If the operation . is commutative, then S is called a *commutative semiring*.

Remark. In general, *semirings* are defined with a *zero* element 0 and an *identity* element 1 (see for example [4]). However, since many interesting examples of semirings have no 0 or 1 , we consider semirings in the most general context here.

An important class of semirings is obtained by considering families of languages that are closed under union and catenation. In particular, the set $\mathcal{F}(X^*, +, .)$ of all the languages over the alphabet X is a semiring with the operations of union $+$ and catenation.

A *gerbier* $S(+,.)$ is a semiring with an idempotent addition $+$, i.e. $u+u=u$ for all $u \in S$, and hence $S(+)$ is a semilattice. This implies that the relation $\leqslant$ defined on S by

$$u \leqslant v \leftrightarrow u+v=v$$

is a partial order on S, compatible with the addition $+$ and the product . .

Remark. A *gerbier* is also called a *semireticulated semigroup*. However, we have adopted here the *French* terminology (See [1]) which is much more convenient.

Since union of languages is an idempotent operation, it follows then that semirings of languages with union as addition are also gerbiers.

A semiring $S(+,.)$ is called a *differential semiring* if there is a mapping $\delta : S \to S$ such that for all $u, v \in S$:

(i) $\delta(u+v) = \delta(u) + \delta(v)$;

(ii) $\delta(uv) = \delta(u)v + u\delta(v)$.

The mapping δ is called an *(abstract) derivation* of S.

Every differential ring is a differential semiring, but differential semirings can be quite different from the usual differential rings. This can be seen from the examples of differential semirings of languages given in the next section.

Let $S(X^*, +, ., \delta)$ be a differential semiring with derivation δ. If $0 \in S$, then $\delta(0) = 0$ and, if $1 \in S$, then $\delta(1) = 2\delta(1)$.

If S is a gerbier, then $u \leqslant v$ implies $\delta(u) \leqslant \delta(v)$ i.e. the *partial order* $\leqslant$ of S is *preserved* by the derivation δ. Indeed, we have $u \leqslant v$ if and only if $u + v = v$. Hence $\delta(u+v) = \delta(u) + \delta(v) = \delta(v)$ and hence $\delta(u) \leqslant \delta(v)$.

If S is a commutative gerbier and $n \geqslant 2$, then $\delta(u^n) = u^{n-1}\delta(u)$. This is true for $n = 2$ because $\delta(u^2) = u\delta(u) + \delta(u)u = 2u\delta(u) = u\delta(u)$.

Suppose it is true for $n-1$. Then

$$\delta(u^n) = \delta(u)u^{n-1} + u\delta(u^{n-1}) = \delta(u)u^{n-1} + uu^{n-2}\delta(u)$$

$$= 2u^{n-1}\delta(u) = u^{n-1}\delta(u)$$

3. DIFFERENTIAL SEMIRINGS OF LANGUAGES

Proposition 26.2 *Let* $\mathcal{F}(X^*, +, .)$ be the semiring of the languages over the alphabet X. Then, for every language $L \subseteq X^*$, $\mathcal{F}(X^*, +, ., \delta_L)$, where δ_L is the insertion relative to L, is a differential gerbier.

Proof. This follows immediately from Proposition 26.1 and the fact that $\mathcal{F}$ is a gerbier. □

We consider now the families of languages in the Chomsky hierarchy. These languages are generally defined in the context of a finite alphabet X and play a fundamental role in theoretical computer science. Let

$$\mathcal{C}(X^*) = \{F(X^*), R(X^*), CF(X^*), CS(X^*)\}$$

be, respectively, the set of the families of finite, regular, context-free and context-sensitive languages over a *finite* alphabet X.

Proposition 26.3 If X is a finite alphabet, then the family $\mathcal{T}(X^*, ., +, \delta_L)$ where $\mathcal{T}(X^*) \in \mathcal{C}(X^*)$ and δ_L is the insertion relative to a language $L \in \mathcal{T}(X^*)$ is a differential gerbier.

Proof. The family $\mathcal{T}(X^*)$ is closed under $+$ and . and the operation $+$ is commutative and idempotent. Hence $\mathcal{T}(X^*)$ is a gerbier. By a result of [LKT], $\mathcal{T}(X^*)$ is closed under insertion δ_L of a language $L \in \mathcal{T}(X^*)$. □

Proposition 26.4 Let $\mathcal{F}(X^*, +, ., \delta_a)$ be the family of languages over X, let $a \in X$ and $w \in X^*$. If δ_a is defined as:

$$\delta_a(T) = \{xwy | xay \in T\}, \;\; T \subseteq X^*$$

then $\mathcal{F}(X^*, +, ., \delta_a)$ is a differential gerbier.

Proof. It is immediate that $\delta(L_1 + L_2) = \delta(L_1) + \delta(L_2)$ and that $\delta(L_1)L_2 + L_1\delta(L_2) \subseteq \delta(L_1L_2)$. Let $u \in \delta(L_1L_2)$, i.e. $u = xwy$ with $xay \in L_1L_2$. This implies that $xay = x'x''ay$, $x' \in L_1$, $x''ay \in L_2$ or $xay = xay'y''$, $xay' \in L_1$, $y'' \in L_2$. In the first case, we have then $u = xwy = x'x''wy$ with $x' \in L_1$, $x''wy \in \delta_a(L_2)$, i.e. $u \in L_1\delta(L_2)$. Similarly for the second case, we get $u \in \delta(L_1)L_2$. Hence $\delta(L_1L_2) \subseteq \delta(L_1)L_2 + L_1\delta(L_2)$ and therefore $\delta(L_1L_2) = \delta(L_1)L_2 + L_1\delta(L_2)$. □

The previous proposition can easily be extended to the more general case where the derivation δ_a is replaced by the derivation δ_A defined by:

$$\delta_A(L) = \cup_{a \in A} \delta_a(L), \;\; A \subseteq X$$

where δ_a is replacing the letter $a \in A$ with a fixed word $w_a \in X^*$.

4. FIXED, ROBUST AND AMICABLE ELEMENTS

Let $G(+,.,\delta)$ be a differential gerbier. Relatively to the derivation δ, an element $u \in G$ is said to be:

(i) *fixed* if $\delta(u) = u$;

(ii) robust if $\delta(u) \leqslant u$;

(iii) *amicable if* $u \leqslant \delta(u)$.

Proposition 26.5 Let $G(+,.,\delta)$ be a differential gerbier and let $u, v \in G$. If u, v are fixed, robust or amicable, then their product uv is respectively fixed, robust or amicable.

Proof. If $\delta(u) = u$, $\delta(v) = v$, then:

$$\delta(uv) = \delta(u)v + u\delta(v) = uv + uv = uv$$

If $\delta(u) \leqslant u$, $\delta(v) \leqslant v$, then, from the compatibility of $\leqslant$ with the operations of G, we get:

$$\delta(u)v \leqslant uv, \quad u\delta(v) \leqslant uv$$

Hence:

$$\delta(uv) = \delta(u)v + u\delta(v) \leqslant uv + uv = uv$$

i.e. uv is robust.

If $u \leqslant \delta(u)$, $v \leqslant \delta(v)$, then, using a similar argument, we get $uv \leqslant \delta(uv)$, i.e. uv is amicable. □

Corollary 26.1 If not empty, the set of the fixed, robust or amicable elements of G *repectively* is a differential subgerbier of G.

Proof. The fact that these sets are subgerbiers of G follows immediately from the proposition.

To prove that they are diffential subgerbiers, we have to show that they are closed under the derivation δ. This is immediate for the fixed elements. If u is a robust element, then $\delta(u) \leqslant u$, $\delta(\delta(u)) = \delta^2(u) \leqslant \delta(u)$ and hence $\delta(u)$ is a robust element. A similar argument shows that the set of amicable elements is also closed under δ. □

The set of fixed, robust or amicable elements of a differential gerbier can be empty. For example, let X be a finite alphabet and let F be the family of the nonempty finite languages over X not containing the empty word 1. This family $F(+,.)$, closed under union $+$ and catenation ., is evidently a gerbier. Let $a \in X$ and let δ_a be the insertion relatively to the language $\{a\}$. Then $F(+,.,\delta_a)$ is a differential gerbier. However there is no fixed, no robust and no amicable element in this differential gerbier.

Consider now the differential gerbier $\mathcal{F}(X^*, +, ., \delta_L\}$. First let $L = \{a\}$ where $a \in X$. It is easy to see that the empty set $\emptyset$ is the only fixed element under the derivation δ_a. The language $A = a^+$ is a robust element under δ_a. Now let $L = \{1, a\}$. Then A becomes a fixed and hence a robust and amicable element under $\delta_{\{1,a\}}$.

The next proposition shows how to define a new derivation in a given differential gerbier in order to make every element an amicable element.

Proposition 26.6 Let $G(+, ., \delta)$ be a differential gerbier and let the mapping $\gamma : G \to G$ be defined by $\gamma(u) = u + \delta(u)$. Then:
(i) γ is a derivation of G;
(ii) for all $u \in G$, we have $u \leqslant \gamma(u)$.

Proof. (i) Clearly $\gamma(u + v) = \gamma(u) + \gamma(v)$. Furthermore:

$$\gamma(uv) = uv + \delta(uv) = uv + \delta(u)v + u\delta(v)$$

Since G is a gerbier, then the operation $+$ is commutative and idempotent. From $uv = uv + uv$, we get:

$$\gamma(uv) = uv + \delta(u)v + uv + u\delta(v) = (u + \delta(u))v + u(v + \delta(v))$$

$$\gamma(uv) = \gamma(u)v + u\gamma(v)$$

Hence γ is a derivation of $G(+, .)$.

(ii) We have $u + \gamma(u) = u + u + \delta(u) = u + \delta(u) = \gamma(u)$. Hence $u \leqslant \gamma(u)$.
□

A derivation δ such that $u \leqslant \delta(u)$ for every $u \in G$ is called a *stable derivation*. From the definition of $\leqslant$, it follows then that a derivation is stable if and only if $u + \delta(u) = \delta(u)$. A derivation δ is not necessarily stable. For example, this is the case for the differential gerbier $\mathcal{F}(X^*, +, ., \delta_a)$ of the languages over X where the derivation δ_a is the insertion of the language $L = \{a\}, a \in X$.

In the general case of $\mathcal{F}(X^*, +, ., \delta_L)$, the stable derivation γ_L as defined in the previous proposition is in fact the derivation δ_{L_1} where $L_1 = L \cup \{1\}$.

Proposition 26.7 Let $G(+, ., \delta)$ be a differential gerbier with a stable derivation δ. Then for every $u, v \in G$ and $n \geqslant 1$, we have:

$$\delta^n(uv) \leqslant \delta^n(u)\delta^n(v)$$

Proof. By induction on n. This is true for $n = 1$. We have $\delta(uv) = \delta(u)v + u\delta(v)$. Since δ is stable and $\leqslant$ is compatible with the product . in G, then $v \leqslant \delta(v)$ and hence $\delta(u)v \leqslant \delta(u)\delta(v)$. Similarly $u\delta(v) \leqslant \delta(u)\delta(v)$. Therefore:

$$\delta(uv) \leqslant \delta(u)\delta(v) + \delta(u)\delta(v) = \delta(u)\delta(v)$$

Suppose the proposition true for n. Then:

$$\delta^{n+1}(uv) = \delta(\delta^n(uv)) \leqslant \delta(\delta^n(u)\delta^n(v)) \leqslant \delta(\delta^n(u))\delta(\delta^n(v)) = \delta^{n+1}(u)\delta^{n+1}(v)$$

Hence the proposition is true for $n+1$. □

5. COMPLETE DIFFERENTIAL GERBIERS

A gerbier $G(+,.)$ is said to be *complete* if:

(i) every nonempty *subset* $A = \{u_i | i \in I\} \subseteq G$ has a *least upper bound* u noted $u = sup(A)$ or $u = \Sigma_{i \in I} u_i$;

(ii) $G(+,.)$ satisfies the *general distributivity laws:*

$$v.sup(A) = sup(v.A), \ \ sup(A).v = sup(A.v)$$

Clearly every finite gerbier is complete. Also the gerbier $\mathcal{F}(X^*,+,.)$ is complete.

A *complete differential gerbier* $G(+,.,\delta)$ is a differential gerbier such that:

(i) $G(+,.)$ is a complete gerbier;

(ii) the derivation δ commutes with *sup*, i.e. :

$$sup(\delta(A)) = \delta(sup(A)$$

for every subset $A \subseteq G$.

For example, $\mathcal{F}(X^*,+,.,\delta_L)$ is a complete differential gerbier.

Proposition 26.8 Let $G(+,.,\delta)$ be a complete differential gerbier with a stable derivation δ. For every element $u \in G$, there exists an element $\bar{u}$ such that:

(i) $u \leqslant \bar{u}$;

(ii) $\delta(\bar{u}) = \bar{u}$, i.e. $\bar{u}$ is a fixed element;

(iii) for every fixed element v with $u \leqslant v$, we have $\bar{u} \leqslant v$.

Proof. Let $\Delta = \{\delta^i(u) | i \geqslant 0\}$ with $\delta^0(u) = u$ and let $\bar{u} = sup(\Delta)$. Clearly $u \leqslant \bar{u}$. Then

$$\delta(\bar{u}) = \delta(sup(\Delta)) = sup(\delta(\Delta)) = sup(\delta^j(u) | j \geqslant 1)$$

Since $u + \delta(u) = \delta(u)$, then

$$sup(\delta^i(u) | i \geqslant 0) = sup(\delta^j(u)\} j \geqslant 1\}$$

Hence $\bar{u}$ $=\delta(\bar{u})$ and $\bar{u}$ is a fixed element.

From $u \leqslant v$, follows $\delta(u) \leqslant \delta(v) = v$ and, by induction, $\delta^i(u) \leqslant v$. Therefore $\bar{u} \leqslant v$. □

The gerbier $\mathcal{F}(X^*,+,.,\delta_L)$ is a complete differential gerbier. If $1 \in L$, the derivation δ_L is stable. For any language $T \subseteq X^*$, the language $\bar{T}$ is then the insertion closure of T (see [2]).

6. EXTENSIONS

If $S(+,.)$ is a semiring, then the derivation δ is a mapping of S into S satisfying the properties:

$$(i)\ \delta(u+v) = \delta(u) + \delta(v),\ \ (ii)\ \ \delta(uv) = \delta(u)v + u\delta(v)\ \ \forall u, v \in S$$

For semirings, particularly for semirings of languages, other specific properties of mappings can be considered, properties that are extensions of the properties of derivation. They are:

$$\rho(u+v) = \rho(u) + \rho(v),\ \ \rho(uv) = \rho(u)\rho(v) + \rho(u)v + u\rho(v) \qquad (26.1)$$

$$\rho(u+v) = \rho(u) + \rho(v),\ \rho(uv) = \rho(u)\rho(v) + \rho(u)v + u\rho(v) + uv \quad (26.2)$$

Remark that the second part of property (2) can be written as:

$$\rho(uv) = (\rho(u) + u)(\rho(v) + v)$$

We give now some examples, leaving a more detailed investigation to another paper.

Let $\mathcal{F}_+(X^*, +, ., \rho)$ be defined as:

(i) $\mathcal{F}_+(X^*, +, .)$ is the gerbier of the nonempty languages over X with $L \subseteq X^+$;

(ii) $\rho(L) = sub(L)$, where $sub(L)$ is the set of all nonempty subwords of words $w \in L$.

Then $\mathcal{F}_+(X^*, +, ., \rho)$ is a gerbier with the properties 26.1. Remark that $L \subseteq \rho(L)$.

Let $G(+, ., \tau)$ be a gerbier with a mapping τ satisfying the properties 26.1

$$\tau(RT) = \tau(R)\tau(T) + R\tau(T) + \tau(R)T$$

Suppose that for every $L \subseteq X^*$, we have $L \subseteq \tau(L)$. Then from $R \leqslant \tau(R)$, $T \leqslant \tau(T)$ follows $RT \leqslant \tau(R)\tau(T)$ and hence $RT + \tau(R)\tau(T) = \tau(R)\tau(T)$. Therefore:

$$\tau(RT) = \tau(R)\tau(T) + R\tau(T) + \tau(R)T + RT$$

and $G(+, ., \tau)$ is a gerbier with the properties 26.2.

References

[1] Dubreil-Jacotin, M.L; J. Lesieur & R. Croisot (1953), *Leçons sur la Théorie des Treillis, des Structures Algébriques Ordonnées et des Treillis Géométriques*, Cahiers Scientifiques XXI, Gauthier-Villars, Paris.

[2] Ito, M; L. Kari & G. Thierrin (1997), Insertion and deletion closure of languages, *Theoretical Computer Science*, 183: 3–19.

[3] Kaplansky, I. (1957), *An Introduction to Differential Algebra*, Actualités Scientifiques et Industrielles, 1251, Hermann, Paris.

[4] Kari, L. (1991), On insertion and deletion in formal languages, Ph.D. Thesis, University of Turku.

[5] Kari, L. & G. Thierrin (1996), Contextual insertions/deletions and computability, *Information and Computation*, 131: 47–61.

[6] Kuich, W. & A. Salomaa (1985), *Semirings, Automata, Languages*, Springer, Berlin.

[7] Ritt, J.F. (1950), *Differential Algebra,* American Mathematical Society, New York.

IV

MODELS OF MOLECULAR COMPUTING

Chapter 27

MOLECULAR STRUCTURES

Gabriel Ciobanu
Faculty of Computer Science
A.I. Cuza University
RO-6600 Iaşi, Romania
gabriel@info.uaic.ro

Abstract We introduce an abstract molecular structure, then we describe the DNA methylation by using a known calculus of the communicating concurrent systems, namely the π-calculus. Finally, we show that it is possible to have a well-defined interpretation of the π-terms by using our abstract molecular structures. In this way we introduce and study some abstract structures which are similar to the abstract machines in computer science (as Turing machines or automata), and suitable to express molecular interactions. We should remark that the considered interactions imply modifications too (e.g. DNA methylation); in order to express this aspect, we consider and study interactions with substitutions. Because the covalent bounds play a special role when we refer to DNA, our abstract molecular structures are systems with joined (shared) resources; they emphasize the use of some shared resources, as well as the resource transitions. Formally, we use some notions and results of concurrency theory, particularly π-calculus and multiset semantics.

1. INTRODUCTION

Any science is concerned with the real world. Usually a scientific theory consists of a mathematical formalism together with a way of relating such a formalism to the real world. To judge a theory, we should know its domain of applicability, and useful theories have limited domain of application. It is important to avoid confusion between a mathematical formalism and its corresponding physical reality. The mathematical formalism is simpler than physical reality; eliminating irrelevant details, science is the art of simplification ("A theory should be as simple as possible, but no simpler" - A. Einstein). Finally, the test of a scientific theory is how well it helps us to understand and manipulate the real world.

C. Martin-Vide and V. Mitrana (eds.), Where Mathematics, Computer Science, Linguistics and Biology Meet, 299-317.

We intend to present a "translation" of cell biology by using concepts, results, and terms from computer science. Taking into consideration the fact that biophysics is a translation of cell biology by using concepts, results and terms of physics, and that biochemistry is a translation of cell biology by using concepts, results and terms of chemistry, then an appropriate name for our field could be *biocomputing science*. Under such a name we can include the fundamental studies (structures, algorithms, results) related to DNA computing, molecular computing, and other biocomputing approaches.

In order to understand the cause-effect relations and mechanisms of DNA and molecular computing, we should identify, describe and study the most important operations over molecules. Looking at molecular reactions, interaction is a fundamental operation which appears in almost every process involving molecules, enzymes and proteins. As a further step, we should remark that some interactions imply modifications (e.g. methylation, acetilation, phosphorilation). In order to cover this aspect, we consider a more general case by studying interactions with substitutions. On the other hand, covalent bounds play a special role when we refer to DNA computing and to methylation. For all of these reasons, in this paper we present a formal description of molecular interaction using so-called *molecular structures*.

2. ABSTRACT MOLECULAR STRUCTURES

We introduce and study some systems (automata) with joined resources, called molecular structures. We want to use these structures in describing and understanding the molecular interactions, as well as in representing covalent structures for shared pair of electrons (common resources). This formalism emphasizes the use of some shared resources, as well as the resource transitions.

In order to define our structures, we start from two uncountable infinite mutually disjoint sets N and T of *nonterminals* and *terminals*. We shall use $x, y, z \ldots$ to range over N, and $a, b, c \ldots$ to range over T. The set of sources $R = N \cup T$ is their union, ranged over by $\alpha, \beta, \gamma \ldots$ We shall write $\tilde{x}$ for an enumeration of nonterminals $x_1, x_2, \ldots$, $\tilde{a}$ for an enumeration of terminals $a_1, a_2, \ldots$, and $\tilde{\alpha}$ for an enumeration of terminals or nonterminals $\alpha_1, \alpha_2, \ldots$ Given such an enumeration, its components are distinct, and the enumerated set is countable.

Definition 27.1 *Let I be a subset of* $\mathbf{N}$ *containing* 0. *A* **molecular structure** *is an automaton structure indexed by I, namely $MS = (S, r, \rightarrow, S_0)_I$, where*

- $S = (S_i)_{i \in I}$ *is a family of countable sets* $S_i \subset R$ *called* **states**;
- $r = (r_i)_{i \in I}$ *is a family of multisets* $r_i \in \mathcal{M}(S_i \times S_i)$ *representing the* **resources** *of the state* S_i *– (where* $\mathcal{M}(X)$ *denotes the set of all multisets over* X*);*

- $\rightarrow = (\rightarrow_{ij})_{i,j \in I}$ *is a family of* **transitions** $\rightarrow_{ij} \subseteq S_i \times S_j$ *over states;*
- S_0 *is the* **initial** *state.*

Molecular structures MS are in fact multiset transition systems, i.e. an extension of the notion of transition system [13] to multisets. We prefer to put in evidence the support set S_i for each multiset r_i.

The index set $I \setminus \{0\}$ is denoted by $I^{\star}$. We can associate a graphic representation to a molecular structure $MS = (S, r, \rightarrow, S_0)_I$. First we can associate a directed graph called state graph. The associated state graph of MS is $H_{MS} = (I, E)$, where $E = \{ (i,j) \in I \times I \mid \rightarrow_{ij} \neq \emptyset \}$. MS is called *quasi-finite* if every path in H_{MS} has a finite length, $|S_i|, |\rightarrow_{ij}|$ are finite, and $r_i(\alpha, \beta) < \omega$, $\forall \alpha, \beta \in S_i$, $\forall i, j \in I$. To simplify the notation, whenever F is a family of (multi)sets, we use the same F to denote the (multiset) union of the component (multi)sets.

The graphic representation of a molecular structure is given by a pseudo-graph. The nodes of this pseudo-graph are labeled injectively by elements of R. They are connected by directed edges and by hyperedges. Finally, the nodes connected by a hyperedge serve as the vertex set for some directed multigraph. More exactly, given a molecular structure $MS = (S, r, \rightarrow, S_0)_I$, the elements of the set S represent the labeled nodes, and the elements of the family S represent the hyperedges of the pseudo-graph; e.g. $S_0 = \{x, a, y\}$ is the hyperedge represented by an oval labeled with S_0 and connected through tentacles to the nodes labeled x, a, and y of the pseudo-graph, as in the following figure:

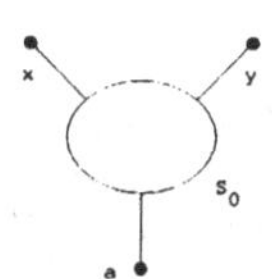

The elements of the family r are directed multigraphs. In particular, r_i is a directed multigraph having the vertex set S_i. This multigraph is represented inside the hyperedge labeled S_i; e.g. for the above S_0 and

$r_0(x, a) = r_0(a, a) = r_0(y, y) = 2$
$r_0(x, y) = r_0(y, x) = 1$
$r_0(\alpha, \beta) = 0$, otherwise

we have the following graphic representation:

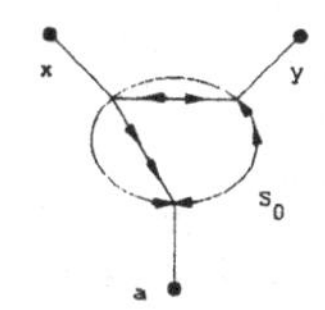

The elements of the set $\rightarrow$ are the directed edges of the pseudo-graph; e.g. $(\alpha, \beta) \in \rightarrow_{ij}$, denoted by $\alpha \rightarrow_{ij} \beta$, is the directed edge from the node labeled α to the node labeled β of the pseudo-graph, represented as in the figure:

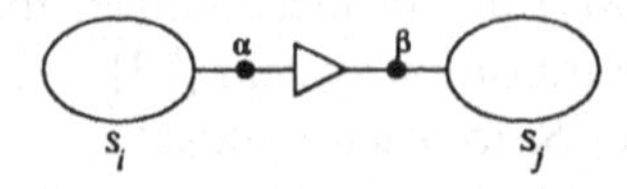

We don't want to distinguish between two molecular structures which differ from each other just by some irrelevant information. We shall identify them by a notion of isomorphism. We shall call *abstract molecular structure* such an isomorphism class.

Definition 27.2 *Let $MS = (S, r, \rightarrow, S_0)_I$ be a molecular structure. For every $i \in I$ we define a set* **interacting**(S_i) *of the potential interacting elements in S_i as the smallest set closed under the following rules*

- *if $(\alpha, \beta) \in_m r_i$, then $\alpha, \beta \in$* **interacting**(S_i)
- *if $\alpha \rightarrow_{ij} \beta$, then $\alpha \in$* **interacting**(S_i)
- *if $\alpha \rightarrow_{ki} \beta$, then $\beta \in$* **interacting**(S_i) .

Definition 27.3 *Let $MS = (S, r, \rightarrow, S_0)_I$ be an molecular structure. We define the molecular structure* **interacting**$(MS) = (S', r', \rightarrow', S'_0)_I$ *by*

- $S'_i =$ **interacting**(S_i), $i \in I$
- $r'_i = r_i|_{\textbf{interacting}(S_i) \times \textbf{interacting}(S_i)}$, $i \in I$
- $\rightarrow'_{ij} = \rightarrow_{ij}$, $i, j \in I$.

Definition 27.4 *Given A and B two molecular structures, and their corresponding molecular structures* **interacting**$(A) = (S^A, r^A, \rightarrow^A, S_0^A)_I$ *and* **interacting**$(B) = (S^B, r^B, \rightarrow^B, S_0^B)_J$, *we say that A and B are isomorphic if there exist two bijective mappings $\phi : S^A \rightarrow S^B$ satisfying $\phi(x) = x$ whenever $x \in N$, and $\sigma : I \rightarrow J$ satisfying $\sigma(0) = 0$ which fulfill the following conditions:*

- $\phi^\times(r_i^A) = r_{\sigma(i)}^B$, $i \in I$
- $\phi^\times(\rightarrow_{ij}^A) = \rightarrow_{\sigma(i)\sigma(j)}^B$, $i, j \in I$.

We use the notation $(\phi, \sigma) : A \cong B$, or simply $A \cong B$.

The conditions within the previous definition refer to resources and transitions. We can remark that these conditions are enough, and they ensure a suitable correspondence for states.

Proposition 27.1 *Let A and B be two molecular structures, and their corresponding molecular structures* **interacting**$(A) = (S^A, r^A, \rightarrow^A, S_0^A)_I$ *and*

interacting$(B) = (S^B, r^B, \rightarrow^B, S_0^B)_J$. *If* $(phi, \sigma) : A \cong B$*, then* $\phi(S_i^A) = S^B_{\sigma(i)}$*,* $\forall i \in I$*.*

The previously defined isomorphism is an equivalence.

Modulo this equivalence, we can rename the states of a molecular structure, remove elements which are not potential interacting elements, and rename the elements of T which appear within the molecular structure.

Abstract molecular structures have almost the same graphic representation as molecular structures. More exactly, the graphic form of an abstract molecular structure corresponding to a molecular structure $MS = (S, r, \rightarrow, S_0)_I$ is obtained starting from the graphic representation of MS. For every hyperedge S_i, a tentacle which leads to a vertex bearing a label that is not in **interacting**(S_i) is removed. The reached vertex is also removed, excepting the case when it is used by another hyperedge. All the labels from T together with the labels of hyperedges are removed.

When we describe the graphic representation of an abstract molecular structure, we did not explain how we identify the initial state after we have removed the labels of the hyperedges. In fact, we are working with a subset $\mathbf{MS}^\omega$ of the molecular structures where states form a directed rooted tree; the initial state is the root of this tree, and it is easy to identify it. In order to simplify our notation, from now on we use the notation $MS = (S, r, \rightarrow)_I$ instead of $MS = (S, r, \rightarrow, S_0)_I$, keeping in mind that S_0 is the initial state.

We define now the endomorphisms over molecular structures.

Definition 27.5 *Given a molecular structure* $A = (S, r, \rightarrow)_I$ *and a function* $f : R \rightarrow R$*, the image of* A *by* f *is a molecular structure* $fA = (S', r', \rightarrow')_I$ *defined by*

- $S'_i = f(S_i),\ i \in I$
- $r'_i = f^\times(r_i),\ i \in I$
- $\rightarrow'_{ij} = f^\times(\rightarrow_{ij}),\ i, j \in I$.

Substitutions are particular endomorphisms.

Definition 27.6 *Let* $\tilde{\alpha}$ *and* $\tilde{\beta}$ *be two enumerations from* R *such that* $|\tilde{\alpha}| = |\tilde{\beta}|$*. A substitution* $\{\tilde{\alpha}/\tilde{\beta}\} : R \rightarrow R$ *is given by* $\{\tilde{\alpha}/\tilde{\beta}\}(\gamma) =$ *if* $(\gamma = \beta_k)$*, then* α_k *else* γ*.*

We present some simple properties of substitution in the next lemma.

Lemma 27.1 *Let* $A = (S, r, \rightarrow)_I$ *be a molecular structure. Then*

1. $\{\tilde{\alpha}/\tilde{\beta}\}A = A$*, if* $\tilde{\beta} \cap S = \emptyset$
2. $\{\tilde{\alpha}/\tilde{\alpha}\}A = A$

3. $\{\tilde{\alpha}/\tilde{\beta}\}\{\tilde{\beta}/\tilde{\alpha'}\}A = \{\tilde{\alpha}/\tilde{\alpha'}\}A$, *if* $\tilde{\beta} \cap q = \emptyset$
4. $\{\tilde{\alpha}/\tilde{\beta}\}\{\tilde{\alpha'}/\tilde{\beta'}\}A = \{\tilde{\alpha'}/\tilde{\beta'}\}\{\tilde{\alpha}/\tilde{\beta}\}A$, *if* $\tilde{\beta} \cap \tilde{\beta'} = \tilde{\alpha} \cap \tilde{\beta'} = \tilde{\alpha'} \cap \tilde{\beta} = \emptyset$.

3. ABSTRACT MOLECULAR SYSTEMS

Definition 27.7 *The set* $\mathbf{MS}^\omega$ *of molecular systems is defined inductively by starting from two simple systems*

nil : $\mathbf{MS}^\omega$ *contains the molecular structure* $nil \stackrel{def}{=} (S, r, \rightarrow)_I$ *where* $I^\star = \emptyset$ *and*

- $S_0 = \emptyset$
- $r_0 = \emptyset$
- $\rightarrow_{00} = \emptyset$.

basic : *If* $x, y \in N$, *then* $\mathbf{MS}^\omega$ *contains the molecular structure* $1(x, y) \stackrel{def}{=} (S, r, \rightarrow)_I$ *where* $I^\star = \emptyset$ *and*

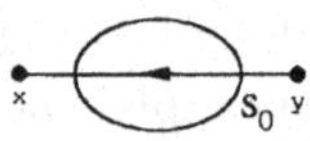

- $S_0 = \{x, y\}$
- $r_0 = \{\!\!\{ (y, x) \}\!\!\}$
- $\rightarrow_{00} = \emptyset$.

and using the following operations:

prefixing : *Let* $x, y \in N$, $A = (S, r, \rightarrow)_I \in \mathbf{MS}^\omega$, $a \in T \setminus q$, *and* $k \in \mathbf{N} \setminus I$. *If we note* $\{a/y\}A = (S', r', \rightarrow')_I$, *then* $\mathbf{MS}^\omega$ *contains the molecular structure* $(x, y)_{a,k}\, A \stackrel{def}{=} (S^1, r^1, \rightarrow^1)_K$ *where* $K = I \cup \{k\}$ *and*

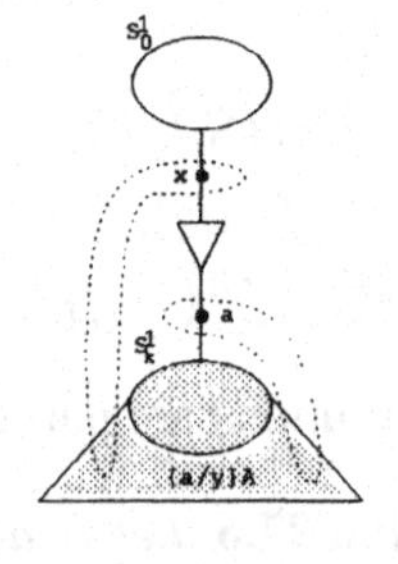

- $S^1_i = S'_i,\ i \in I^\star$
  $S^1_k = S'_0 \cup \{a\}$
  $S^1_0 = \{x\}$
- $r^1_i = r'_i,\ i \in I^\star$
  $r^1_k =_m r'_0$
  $r^1_0 = \emptyset$
- $\rightarrow^1_{ij} = \rightarrow'_{ij},\ i, j \in I^\star$
  $\rightarrow^1_{kj} = \rightarrow'_{0j},\ j \in I^\star$
  $\rightarrow^1_{0k} = \{(x, a)\}$
  $\rightarrow^1_{ij} = \emptyset$, *otherwise.*

restriction : *If* $A = (S, r, \rightarrow)_I \in \mathbf{MS}^\omega$, $\tilde{x} \subset N$, *and* $\tilde{a} \subset T \setminus q$ *such that* $|\tilde{x}| = |\tilde{a}|$, *then* $\mathbf{MS}^\omega$ *contains the molecular structure*

$(\tilde{x})_{\tilde{a}} A \stackrel{def}{=} \{\tilde{a}/\tilde{x}\}A$.

parallel composition : *Let* K *be a countable set. If* $A_k = (S^k, r^k, \rightarrow^k)_{I_k} \in \mathbf{MS}^\omega$, $k \in K$ *such that each of the families* $(T \cap S^k)_{k \in K}$ *and* $(I^\star_k)_{k \in K}$ *have* *mutually disjoint sets, then* $\mathbf{MS}^\omega$ *contains the molecular structure* $\otimes_{k\in K} A_k \stackrel{def}{=} (S, r, \rightarrow)_I$ *where* $I = \bigcup_{k\in K} I_k$ *and*

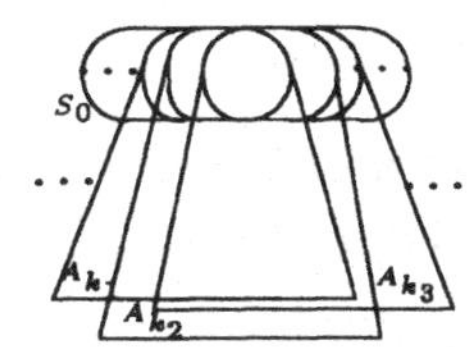

- $S_i = S_i^k,\ i \in I_k^\star,\ k \in K$
  $S_0 = \bigcup_{k\in K} S_0^k$
- $r_i = r_i^k,\ i \in I_k^\star,\ k \in K$
  $r_0 =_m \biguplus_{k\in K} r_0^k$
- $\rightarrow_{ij} = \rightarrow_{ij}^k,\ i, j \in I_k,\ k \in K$
  $\rightarrow_{ij} = \emptyset$, *otherwise.*

In the previous definition we have used (directly or indirectly) the following operations over multisets:

$\in_m$ If $r \in \mathcal{M}(X)$ and $a \in X$, then $a \in_m r$ if and only if $r(a) > 0$.

$=_m$ If $r_i \in \mathcal{M}(X_i)$, $i = 1, 2$, then $r_1 =_m r_2$ if and only if the following conditions are satisfied:

- $a \in_m r_1$ implies $a \in_m r_2$, and $r_1(a) = r_2(a)$, and
- $a \in_m r_2$ implies $a \in_m r_1$, and $r_1(a) = r_2(a)$.

$\uplus$ If $r_i \in \mathcal{M}(X_i)$, $i = 1, 2$, then $r_1 \uplus r_2 \in \mathcal{M}(X_1 \cup X_2)$ is defined by $(r_1 \uplus r_2)(a) = s_1(a) + s_2(a)$ for $a \in X_1 \cup X_2$, and $s_i \in \mathcal{M}(X_1 \cup X_2)$ such that $s_i =_m r_i$, $i = 1, 2$. Since the addition operation over $\mathbf{N}_\omega$ is associative, then the multiset union operation $\uplus$ is associative; it can be extended to countable sets of multisets.

In $\mathbf{MS}^\omega$ *we consider the subset* $\mathbf{MS}$ *of the quasi-finite molecular structures.*

Lemma 27.2 *Given a molecular structure A, if* $A \in \mathbf{MS}^\omega$, *then* $\mathbf{interacting}(A) = A$.

Lemma 27.3 *Let* $A, B \in \mathbf{MS}^\omega$ *such that* $A \cong B$. *If* $A \in \mathbf{MS}$, *then* $B \in \mathbf{MS}$.

Lemma 27.4 *If all constructions involved are valid, then*

1. $\{\tilde\alpha/\tilde\beta\}nil = nil$
2. $\{\tilde\alpha/\tilde\beta\}1(x, y) = 1(\{\tilde\alpha/\tilde\beta\}x, \{\tilde\alpha/\tilde\beta\}y)$
3. $\{\tilde\alpha/\tilde\beta\}(x, y)_{a,k}A = (\{\tilde\alpha/\tilde\beta\}x, y)_{a,k}\{\tilde\alpha/\tilde\beta\}A$, *if* $y, a \notin \tilde\alpha \cup \tilde\beta$
4. $\{\tilde\alpha/\tilde\beta\}(\tilde x)_{\tilde a}A = (\tilde x)_{\tilde a}\{\tilde\alpha/\tilde\beta\}A$, *if* $\tilde\alpha \cap \tilde x = \tilde\beta \cap \tilde x = \tilde\beta \cap \tilde a = \emptyset$
5. $\{\tilde\alpha/\tilde\beta\} \otimes_{k\in K} A_k = \otimes_{k\in K} \{\tilde\alpha/\tilde\beta\}A_k$.

Lemma 27.5 *Given a molecular structure* $A = (S, r, \rightarrow)$, *whenever the constructions involved are valid, we have*

1. $(x, y)_{a,k}A = (x, y')_{a,k}\{y'/y\}A$, *if* $y' \in N \setminus q$
2. $(\tilde x)_{\tilde a}A = (\tilde{x'})_{\tilde a}\{\tilde{x'}/\tilde x\}A$, *if* $\tilde{x'} \subset N \setminus (q \cup \tilde x)$.

Proposition 27.2 *If $A \in \mathbf{MS}^\omega$, then $\{\tilde{x}/\tilde{z}\}A \in \mathbf{MS}^\omega$ for any two enumerations $\tilde{x}, \tilde{z}$ from N such that $|\tilde{x}| = |\tilde{z}|$.*

The following result gives a characterization for **MS**.

Theorem 27.1 *Let $A = (S, r, \rightarrow, S_0)_I$ be a quasi-finite molecular structure. $A \in \mathbf{MS}^\omega$ if and only if for any $i, j \in I$, A satisfies the following conditions*

1. $|\rightarrow_{ij}| \leqslant 1$
2. *its state graph H_A is a directed rooted tree having 0 as root*
3. $\mathbf{interacting}(S_i) = S_i$
4. *if $\alpha \rightarrow_{ij} \beta$, then*
   (a) *$\beta \in T$ and*
   (b) *whenever $\beta \in S_k$ $(k \neq j)$, then there exists a jk-path in H_A.*

By $\mathbf{AMS}^\omega$ we denote the set of all abstract molecular systems, and by **AMS** the set of all abstract quasi-finite molecular systems. For every abstract molecular system $\mathcal{A}$, we associate the set $\lambda(\mathcal{A})$ of the nonterminals, and the number $\kappa(\mathcal{A})$ of the shared terminals of any molecular system $A \in \mathcal{A}$.

Definition 27.8 *Function $\lambda : \mathbf{AMS}^\omega \rightarrow 2^N$ is defined by $\lambda(\mathcal{A}) = S^A \cap N$, where $A \in \mathbf{MS}^\omega \cap \mathcal{A}$.*

Definition 27.9 *If $\mathcal{A} \in \mathbf{AMS}^\omega$ and $A \in \mathbf{MS}^\omega \cap \mathcal{A}$, then we define*

$$\kappa(\mathcal{A}) = |\bigcup_{\substack{i,j \in I_A \\ i \neq j}} S_i^A \cap S_j^A \cap T|$$

Definition 27.10 *We define the following operations upon $\mathbf{AMS}^\omega$:*

1. $(x, y)\mathcal{A} \stackrel{def}{=} [(x, y)_{a,k}A]_{\cong}$, $A \in \mathcal{A}$;
2. $(\tilde{x})\mathcal{A} \stackrel{def}{=} [(\tilde{x})_{\tilde{a}}A]_{\cong}$, $A \in \mathcal{A}$;
3. $\otimes_{k \in K} \mathcal{A}_k \stackrel{def}{=} [\otimes_{k \in K} A_k]_{\cong}$, $A_k \in \mathcal{A}_k$, $k \in K$.

Lemma 27.6

1. *If $A \cong B$, then $(x, y)_{a,k}A \cong (x, y)_{b,l}B$.*
2. *If $A \cong B$, then $(\tilde{x})_{\tilde{a}}A \cong (\tilde{x})_{\tilde{b}}B$.*
3. *If $A_k \cong B_k$ for every $k \in K$, then $\otimes_{k \in K} A_k \cong \otimes_{k \in K} B_k$.*

This lemma is not enough to prove that the previously introduced operations are well-defined. Next proposition comes to show the correctness of these definitions.

Proposition 27.3 *Let $x, y \in N$ and let $\tilde{x}$ be an enumeration in N. If $\mathcal{A}, \mathcal{A}_k \in$ **AMS**$^\omega$ where k belongs to the (countable) set K, then*

1. $(x, y)\mathcal{A} \in \mathbf{AMS}^\omega$;
2. $(\tilde{x})\mathcal{A} \in \mathbf{AMS}^\omega$;
3. $\otimes_{k \in K} \mathcal{A}_k \in \mathbf{AMS}^\omega$.

The following lemma is useful in the proof of the previous proposition. Moreover it also suggests that the requirement that some sets are uncountable is not essential.

Lemma 27.7 *Let $B \in \mathbf{MS}^\omega$ and let $S \subset T$, $L \subset \mathbf{N}^\star$ be two infinite countable sets. There exits $A \in \mathbf{MS}^\omega$ such that $S^A \cap T \subset S$, $I_A^\star \subset L$, and $A \cong B$.*

Definition 27.11 *If $x_1, x_2 \in N$, then the function $\{x_1/x_2\} : \mathbf{AMS}^\omega \to \mathbf{AMS}^\omega$ is defined by $\{x_1/x_2\}\mathcal{A} = [\{x_1/x_2\}A]_\cong$, where $A \in \mathbf{MS}^\omega \cap \mathcal{A}$.*

Lemma 27.8 *Let $A, B \in \mathbf{MS}^\omega$. If $A \cong B$, then $\{x_1/x_2\}A \cong \{x_1/x_2\}B$.*

Some results given for molecular systems are valid for abstract molecular systems too. We use the following notations: $\mathbf{nil} = [nil]_\cong$ and $\mathbf{1}(x, y) = [1(x, y)]_\cong$.

Lemma 27.9

1. $\{x_1/x_2\}\mathbf{nil} = \mathbf{nil}$;
2. $\{x_1/x_2\}\mathbf{1}(x, y) = \mathbf{1}(\{x_1/x_2\}x, \{x_1/x_2\}y)$;
3. $\{x_1/x_2\}(x, y)\mathcal{A} = (\{x_1/x_2\}x, y)\{x_1/x_2\}\mathcal{A}$, *if* $y \notin \{x_1, x_2\}$;
4. $\{x_1/x_2\}(x)\mathcal{A} = (x)\{x_1/x_2\}\mathcal{A}$, *if* $x \notin \{x_1, x_2\}$;
5. $\{x_1/x_2\} \otimes_{k \in K} \mathcal{A}_k = \otimes_{k \in K} \{x_1/x_2\}\mathcal{A}_k$.

Lemma 27.10

1. $(x, y)\mathcal{A} = (x, y')\{y'/y\}\mathcal{A}$, *if* $y' \notin \lambda(\mathcal{A})$;
2. $(x)\mathcal{A} = (x')\{x'/x\}\mathcal{A}$, *if* $x' \notin \lambda(\mathcal{A})$.

Lemma 27.11 *If $\mathcal{A} \in$ **AMS** and $A \in \mathbf{MS}^\omega \cap \mathcal{A}$, then $A \in$ **MS**.*

4. DYNAMICS OF THE MOLECULAR SYSTEMS

The static part of our structures uses concepts as those of states, transitions or resources derived from the formalisms that belong to the tradition of automata. The dynamics of our structures uses a *token-game* mechanism similar somehow to that of the Petri nets theory (which is a formalism for describing concurrent systems). However, our systems have a flexible structure, and in this way, a greater expressive power.

Definition 27.12 *Let* $A = (S^A, r^A, \rightarrow^A)_{I_A} \in \mathbf{MS}$. *If we have* $(\alpha_1, \gamma) \in_m r_0$ *and* $\gamma \rightarrow_{0k} \alpha_2$, *then* $A \mapsto [\{\alpha_1/\alpha_2\}B]_{\cong}$ *where* $B = (S^B, r^B, \rightarrow^B)_{I_B}$ *is given by* $I_B = I_A \setminus \{k\}$ *and*

- $S_j^B = S_j^A,\ j \in I_B^\star$,
  $S_0^B = S_0^A \cup S_k^A$;
- $r_j^B = r_j^A,\ j \in I_B^\star$,
  $r_0^B =_m (r_0^A - \{\!|(\alpha_1, \beta)|\!\})\ \uplus\ r_k^A$;
- $\rightarrow_{ij}^B = \rightarrow_{ij}^A,\ i, j \in I_B^\star$,
  $\rightarrow_{0j}^B = \rightarrow_{0j}^A \cup \rightarrow_{kj}^A,\ j \in I_B^\star$,
  $\rightarrow_{ij}^B = \emptyset$, *otherwise.*

The multiset equality $r_0^B =_m (r_0^A - \{\!|(\alpha_1, \beta)|\!\}) \uplus r_k^A$ is the essential part of the reduction mechanism. We can remark some similarity with the token-game of Petri nets. But what makes the difference is that in our structures the tokens are consumed. The resource $(\alpha_1, \beta) \in_m r_0$ – which corresponds to a token in Petri nets – is *consumed*, and not sent. Moreover, our systems have a flexible structure, and not a fixed one as that of Petri nets. After the resource of the initial state is consumed, this state, S_0, is *enriched* (by fusion) with the resources of the state S_k determined by the transition $\beta \rightarrow_{0k} \alpha_2$. $(\alpha_1, \beta) \in_m M_0$ represents an internal transition within the same state S_0, and $\beta \rightarrow_{0k} \alpha_2$ represents an external transition from S_0 to S_k; they correspond to the output process $\overline{x}z$, and to the input guard $x(y).P$ respectively.

Proposition 27.4 $\mapsto$ *is well-defined.*

We extend now $\mapsto$ to a relation $\Rightarrow$ over **AMS**.

Definition 27.13 *We define* $\mathcal{A} \Rightarrow \mathcal{B}$ *if* $A \mapsto \mathcal{B}$ *for some* $A \in \mathcal{A}$.

We present some results related to the new reduction relation $\Rightarrow$.

Lemma 27.12 *Let* $A, B \in \mathbf{MS}$. *If* $A \cong B$ *and* $A \mapsto \mathcal{A}$, *then* $B \mapsto \mathcal{A}$.

Lemma 27.13

1. *If* $\mathcal{A}_1 \Rightarrow \mathcal{A}_2$, *then* $(\tilde{x})\mathcal{A}_1 \Rightarrow (\tilde{x})\mathcal{A}_2$.
2. *If* $(\tilde{x})\mathcal{A}_1 \Rightarrow \mathcal{A}$, *then there exists* $\mathcal{A}_2$ *such that* $\mathcal{A}_1 \Rightarrow \mathcal{A}_2$ *and* $\mathcal{A} = (\tilde{x})\mathcal{A}_2$.

Lemma 27.14

1. $\mathbf{1}(x, z) \otimes (x, y)\mathcal{A} \Rightarrow \{z/y\}\mathcal{A}$.

2. *If* $\mathcal{A}_1 \Rightarrow \mathcal{A}_2$, *then* $\mathcal{A}_1 \otimes \mathcal{A} \Rightarrow \mathcal{A}_2 \otimes \mathcal{A}$ *and* $\mathcal{A} \otimes \mathcal{A}_1 \Rightarrow \mathcal{A} \otimes \mathcal{A}_2$.
3. *If* $\mathcal{A} \otimes \mathcal{B} \Rightarrow \mathcal{C}$, *then one of the following possibilities holds*

1a) $\mathcal{A} \Rightarrow \mathcal{A}'$
$\mathcal{C} = \mathcal{A}' \otimes \mathcal{B}$

1b) $\mathcal{B} \Rightarrow \mathcal{B}'$
$\mathcal{C} = \mathcal{A} \otimes \mathcal{B}'$

2a) $\mathcal{A} = \mathcal{A}_1 \otimes \mathbf{1}(x,z)$
$\mathcal{B} = (\tilde{u})((x,v)\mathcal{B}_1 \otimes \mathcal{B}_2)$
$\mathcal{C} = \mathcal{A}_1 \otimes (\tilde{u})(\{z/v\}\mathcal{B}_1 \otimes \mathcal{B}_2)$

2b) $\mathcal{A} = (\tilde{u})((x,v)\mathcal{A}_1 \otimes \mathcal{A}_2)$
$\mathcal{B} = \mathcal{B}_1 \otimes \mathbf{1}(x,z)$
$\mathcal{C} = \mathcal{B}_1 \otimes (\tilde{u})(\{z/v\}\mathcal{A}_1 \otimes \mathcal{A}_2)$

3a) $\mathcal{A} = (w)(\mathcal{A}_1 \otimes \mathbf{1}(x,w))$
$\mathcal{B} = (\tilde{u})((x,v)\mathcal{B}_1 \otimes \mathcal{B}_2)$
$\mathcal{C} = (w)(\mathcal{A}_1 \otimes (\tilde{u})(\{z/v\}\mathcal{B}_1 \otimes \mathcal{B}_2))$
where $\tilde{u}vw$ *is an sequence in* $N \setminus \lambda(\mathcal{A} \otimes \mathcal{B})$ *such that* $|\tilde{u}| \leqslant \kappa(\mathcal{B})$

3b) $\mathcal{A} = (\tilde{u})((x,v)\mathcal{A}_1 \otimes \mathcal{A}_2)$
$\mathcal{B} = (w)(\mathcal{B}_1 \otimes \mathbf{1}(x,w))$
$\mathcal{C} = (w)(\mathcal{B}_1 \otimes (\tilde{u})(\{z/v\}\mathcal{A}_1 \otimes \mathcal{A}_2))$
where $\tilde{u}vw$ *is a sequence in* $N \setminus \lambda(\mathcal{A} \otimes \mathcal{B})$ *such that* $|\tilde{u}| \leqslant \kappa(\mathcal{A})$

5. MOLECULAR INTERACTION IN CELL BIOLOGY

Molecules attract and repel each other. These attractions and repulsions arise from molecular interactions. A type of interaction occurs between molecules containing electronegative elements. For instance, a partially positive hydrogen atom of a molecule is attracted to the unshared pair of electrons of the electronegative atom of another molecule. This attraction is called a hydrogen bond. Hydrogen bonds are rather like glue between molecules. The carbon-halogen sigma bond is formed by the overlap of an orbital of the halogen atom and a hybrid orbital of the carbon atom. This is an example of shared resources which appear in molecular interactions.

J.D. Watson and F.H.C. Crick proposed a model for DNA which accounts for its behaviour. The Watson-Crick model proposed that the two polynucleotidic chains which run in opposite directions are not connected by covalent bonds, but associated by hydrogen bonding between the nitrogenous bases. The hydrogen bonds between DNA strands are not random, but they are specific between pairs of bases : guanine (G) is joined by three hydrogen bonds to cytosine (C), and adenine (A) is joined by two hydrogen bonds to thymine (T). These reactions are described as base pairing and the paired bases (G with C, A with T) are said to be complementary. In a perfect duplex of DNA, the strands are precisely complementary: only complementary sequences can form a duplex structure. Moreover, we consider particularly G-C pairs; these is because G-C base pairs have three hydrogen bonds, and therefore they are more stable then other base pairs which only have two hydrogen bonds. The stability of a double helix increase with the proportion of G-C base pairs. Therefore, a study of the Watson-Crick model leads to hydrogen bonds, i.e. to dipole-dipole interactions which occur between molecules containing a hydrogen atom bounded to nitrogen and/or oxygen. Hydrogen bonds are a glue between molecules and takes its name from the fact that the hydrogen atom is shared between the reacting groups. The formalism we introduce in this paper emphasizes the use of some shared resources, as well as dipole-dipole attraction (complementary communication) and resource transitions.

In order to have an example of how we intend to formalize DNA processes, we consider the DNA methylation. DNA methylation has been found in at least some members of almost every major biological species, from archaebacteria and viruses through to mammals and flowering plants. However *methylation is the only known covalent modification of DNA in eukaryotes*. Methylation shows how some of the bases incorporated into DNA can be chemically modified, and so provides a signal which marks a segment of DNA. Both prokaryotes and eukaryotes contain enzymes which methylate DNA. Methylation consists of substituting a hydrogen atom by a methyl group CH_3. Most of the methyl groups are found in CG doublets, and the majority of the CG sequences may be methylated. Usually the C residues on both strands of this short palindromic sequence [1] are methylated, giving the structure

$$\begin{matrix} 5' & {}^{m}CpG & 3' \\ 3' & GpC^{m} & 5' \end{matrix}$$

An important issue underlying the biology of DNA methylation is the catalytic mechanism of methyl transfer. The covalent addition of a methyl group to the C5 position of cytosine in the context of CpG dinucleotide is mediated by DNA methyltransferase enzyme. This mechanism is rather unusual: the target cytosine is extruded from the double helix into the active-site cleft in the enzyme. The distribution of methyl groups can be examined by taking advantage of restriction enzymes which cleave the sites containing the CG doublet. The methylation renders the target site resistant to restriction. This modification allows the bacterium to distinguish between its own DNA and any foreign DNA which lacks the characteristic modification pattern. The difference gives assistance to an invading foreign DNA in order to attack by restriction enzymes which recognize the absence of methyl groups at the appropriate sites. The isoschizomers are enzymes that cleaves the same target sequence in DNA, but have a different response to its state of methylation. The enzyme HpaII cleaves the sequence CCGG, but if the second C is methylated, the enzyme can no longer recognize the site. However, the enzyme MspI cleaves the same target site not taking account of the state of methylation of the second C. Thus MspI can be used to identify all the CCGG sequences, and HpaII can be used to deter-

[1] A palindrome is defined as a sequence of duplex DNA which is the same when either of its strands is read in a defined direction. For instance, the sequence given by the strands $5'GGTACC3'$ and $3'CCATGG5'$ are palindromic because when each strand is read in the direction $5'$ to $3'$, it generates the same sequence $GGTACC$.

mine whether they are methylated. In methylated DNA, the modified positions are not cleaved by HpaII.

6. A FORMAL DESCRIPTION OF DNA METHYLATION

We consider the full methylated DNA process, obtained by substitution of H by CH_3 on both strands of the structure

$$\begin{array}{ccc} 5' & CpG & 3' \\ 3' & GpC & 5' \end{array}$$

The language for the description of the methylation process is the π-calculus, a calculus introduced by Milner, Parrow, and Walker [11] as a formalism for describing mobile concurrent processes (see [9, 8]). In order to have a faithful representation of the methylatation process, we consider two types of "communication" channels: those corresponding to the hydrogen bonding between the complementary G-C pairs, and those corresponding to the interactions between the methyl group CH_3 and the corresponding cytosine. We represent these two types of channels by $||$, and $|$ respectively. Both these syntactic operations are forms of the unique interaction operator $|$ of π-calculus. We start with the two strands CpG and GpC which could be represented by $C(H).G \mid\mid G.C(H)$. In order to emphasize the presence of H which will be involved in the methylation process, we refine the expression of the two strands by $mt(H).\overline{hb}(H).C(H).G \mid\mid hb(H).G.C(H)$, where mt represents the methylation channel attached to Hydrogen, and hb represents the hydrogen bonding between the two DNA strands. Since G plays a passive role here, we could simplify this expression to

$$mt(H).\overline{hb}(H).C(H) \mid\mid hb(H).C(H).$$

By considering the interaction with the methyl group $\overline{mt}(CH_3)$, we can represent the process of methylation by the following expression:

$$\overline{mt}(CH_3) \mid mt(H).\overline{hb}(H).C(H) \mid\mid hb(H).C(H)$$

The most important reduction relation over processes of π-calculus is defined by the rule

$$(COM) \qquad \overline{x}(z).P \mid x(y).Q \to P \mid Q\{z/y\}$$

According to these reaction rules, we have the following representation of the full methylation process:

$$\overline{mt}(CH_3) \mid mt(H).\overline{hb}(H).C(H) \mid\mid hb(H).C(H) \to$$
$$\to \overline{hb}(CH_3).C(CH_3) \mid\mid hb(H).C(H) \to C(CH_3) \mid\mid C(CH_3).$$

This means that H was substituted by CH_3 on both strands of the starting structure, obtaining finally the structure

$$\begin{array}{ccc} 5' & {}^{m}CpG & 3' \\ 3' & GpC^{m} & 5' \end{array}$$

Therefore we provide an example how a version of π-calculus could be an appropriate specification (description) language for molecular processes. It would be useful to associate the previously introduced abstract molecular structures with the molecular process of methylation by giving a suitable interpretation of the specification language (π-calculus).

7. FROM π-CALCULUS TO ABSTRACT MOLECULAR SYSTEMS

We consider that interaction is established by a nondeterministic matching which dynamically binds "senders" to eligible "receivers". Even though there are many pairs which can satisfy the matching condition, only a single receiver gets the commitment of the sender. In process algebra, π-calculus models computing by reduction and interactive matching. From a semantic viewpoint, this kind of matching uses the "." prefix operator to symmetrically triggered input actions and complementary output actions. Both the sender and receiver offer their availability for communication, symmetrically narrowing their choice from a set of offered alternatives to a specific commitment. Similar mechanisms work in biology, chemistry, and computation. Thus we can model also the biological binding in DNA, where affinity between complementary pairs of nucleotides A-T and C-G determines the binding of DNA sequences, as well as chemical binding of positive and negative ions realized by availability to all ions of opposite polarity. Considering biological mechanisms in terms of their interaction patterns rather than state-transition rules can provide useful qualitative insights, and new forms of abstraction for computational models. The π-calculus uses names rather than values across send-receive channels, and a matching mechanism as the control structure for communication. Since variables may be channel names, computation can change the channel topology and process mobility is supported. Milner emphasized the importance of identifying the "elements of interaction", and his π-calculus extends the Church-Turing model to interaction by extending the algebraic elegance of lambda reduction rules to interaction between a sender and a receiver.

7.1 π-CALCULUS

The semantics of the π-calculus is usually given by a transition system of which the states are process terms. Concerning its expressiveness, it was shown that even a fragment of this calculus is enough to encode the λ-calculus.

An interaction may occur between two concurrent processes when one sends a name on a particular link, and the other is waiting for a name along the same link. An interaction is actually defined by a "sender" $\overline{x}z.P$ and a "receiver" $x(y).Q$, and it can be represented by the transition: $(\overline{x}z.P \mid x(y).Q) \rightarrow (P \mid Q[z/y])$. This is a synchronous interaction, where the send operation is blocking: an output guard cannot be passed without the simultaneous occurrence of an input action. In synchronous π-calculus, the interaction agents are just the (channel) names. We shall consider the names themselves as elementary agents, freely available for other processes waiting for them. This fits in very well with the idea of a molecular structure describing a process in π-calculus. In this model, the state of a system of concurrent processes is viewed as a multiset of processes in which some "molecules" can interact with each other according to some "reaction rules". In this setting, a particular molecule $\overline{x}z.P$ floating in the solution can interact with any molecule $x(y).Q$ floating in the same solution. The result of this interaction is a solution containing P and $Q[z/y]$.

We present the formal π-calculus framework by considering monadic π-calculus [9]. It is known that monadic π-calculus with only input guards has the full power of polyadic π-calculus, and that the output guarding can be defined in terms of input guarding [6, 1]. Therefore we don't use the output guards $\overline{x}z.P$, but only the process $\overline{x}z$ to denote the emission of a name z along a channel x ; in this way we use an asynchronous version of π-calculus. in fact, we use the asynchronous version of π-calculus [6], i.e. a fragment of the π-calculus with a particular form of replication, where there is no output prefixing and nondeterministic sum. Therefore we don't use the output guards $\overline{x}\langle z\rangle.P$, but only the output messages $\overline{x}\langle z\rangle$; an output message denotes the emission of a name z along a channel x. It is known that asynchronous π-calculus can simulate full π-calculus.

Let $\mathcal{X} \subset N$ be a infinite countable set of *names*. The elements of $\mathcal{X}$ are denoted by $x, y, z \ldots$ The terms of this formalism are called processes. The set of processes is denoted by $\mathcal{P}$, and processes are denoted by $P, Q, R \ldots$.

Definition 27.14 *The processes are defined over the set $\mathcal{X}$ of names by the following grammar*

$$P ::= 0 \mid \overline{x}\langle z\rangle \mid x(y).P \mid !x(y).P \mid \nu x P \mid P \mid Q$$

0 is the empty process. An input guard $x(y).P$ denotes the reception of an arbitrary name z along the channel x, behaving afterwards as $\{z/y\}P$. A replicated input guard $!x(y).P$ denotes a process that allows us to generate arbitrary instances of the form $\{z/y\}P$ in parallel, by repeatedly receiving names z along channel the x. The informal meaning of the restriction $\nu x P$ is that x is local in P. $P \mid Q$ represents the parallel composition of P and Q.

In $x(y).P$, the name y binds free occurrences of y in P, and in $\nu x P$, the name x binds free occurrences of x in P. By $fn(P)$ we denote the set of the names with free occurrences in P, and by $=_\alpha$ the standard α-conversion (like in λ-calculus).

Over the set of processes a structural congruence relation is defined; this relation provides a static semantics of some formal constructions.

Definition 27.15 *The relation* $\equiv \subset \mathcal{P} \times \mathcal{P}$ *is called structural congruence, and it is defined as the smallest congruence over processes which satisfies*

- $P \equiv Q$ *if* $P =_\alpha Q$
- $P \mid 0 \equiv P$, $P \mid Q \equiv Q \mid P$, $(P \mid Q) \mid R \equiv P \mid (Q \mid R)$, $!P \equiv P \mid !P$
- $\nu x 0 \equiv 0$, $\nu x \nu y P \equiv \nu y \nu x P$, $\nu x(P \mid Q) \equiv \nu x P \mid Q$ *if* $x \notin fn(Q)$.

Structural congruence deals with aspects related to the structure of processes. Dynamics is defined by a reduction relation.

Definition 27.16 *The reduction relation over processes is defined as the smallest relation* $\rightarrow \subset \mathcal{P} \times \mathcal{P}$ *satisfying the following rules*

(com)	$\overline{x}\langle z\rangle \mid x(y).P \rightarrow \{z/y\}P$
(par)	$P \rightarrow Q$ *implies* $P \mid R \rightarrow Q \mid R$
(res)	$P \rightarrow Q$ *implies* $(\nu x)P \rightarrow (\nu x)Q$
(str)	$P \equiv P'$, $P' \rightarrow Q'$, *and* $Q' \equiv Q$ *implies* $P \rightarrow Q$.

7.2 THE INTERPRETATION OF π-CALCULUS BY MOLECULAR STRUCTURES

We have used multisets and transition systems in order to provide a description of the molecular structures, and an operational semantics of the molecular interaction (systems). We intend to give now a "molecular" semantics for the asynchronous π-calculus, namely a new multiset semantics via the molecular systems. The reader could also find some results and comments concerning this interpretation in [3] and [5]. We define an interpretation from the asynchronous π-calculus to abstract molecular systems. The interpretation is defined on the structure of the π-terms by the following definition.

Definition 27.17 *The function of interpretation* $\mathcal{I} : \mathcal{P} \rightarrow$ **AMS** *is defined by*

- $\mathcal{I}(0) = \mathbf{nil}$
- $\mathcal{I}(\overline{x}\langle z\rangle) = \mathbf{1}(x, z)$
- $\mathcal{I}(x(y).P) = (x, y)\mathcal{I}(P)$
- $\mathcal{I}(!P) = \otimes_{n \geqslant 0} \mathcal{I}(P)$
- $\mathcal{I}(\nu x P) = (x)\, \mathcal{I}(P)$
- $\mathcal{I}(P_1 \mid P_2) = \mathcal{I}(P_1) \otimes \mathcal{I}(P_2)$.

We have the following results related to this interpretation.

Proposition 27.5 *The interpretation* $\mathcal{I} : \mathcal{P} \rightarrow$ **AMS** *is well-defined.*

Let $A = (S, r, \rightarrow)_I$ be a molecular structure of **MS**. We define the function *obs* from **AMS** to 2^N given by $obs([A]_{\simeq}) = \{\alpha \in N \mid (\beta, \alpha) \in_m r_0\}$, where

$A = (S, r, \rightarrow)_I \in$ **MS**. An observation predicate over **AMS** is defined by $\downarrow_x (\mathcal{A}) = true$ whenever $x \in obs(\mathcal{A})$, and $false$ otherwise. By $\sim_2$ we denote the strong barbed bisimulation over **AMS** defined in the same way as $\sim_1$, but using the observation predicate $\downarrow_x$, and the reduction relation $\Rightarrow$ over **AMS**.

Lemma 27.15 *Let P be a π-term, and $x \in N$. Then*

- $fn(P) = \lambda(\mathcal{I}(P))$,
- $|\lambda(\mathcal{I}(P))|$ *and* $\kappa(\mathcal{I}(P))$ *are finite,*
- $\downarrow_x (P) = true$ *if and only if* $\downarrow_x (\mathcal{I}(P)) = true$.

Proposition 27.6 *Let P be a π-term, and $x, y \in N$. Then $\mathcal{I}(\{x/y\}P) = \{x/y\}(\mathcal{I}(P))$.*

Proposition 27.7 *Let P_1, P_2 be two π-terms. If $P_1 \equiv P_2$, then $\mathcal{I}(P_1) = \mathcal{I}(P_2)$.*

Theorem 27.2 (operational correspondence) *Let P_1, P_2 be two π-terms, and $\mathcal{A} \in$* **AMS**.

1. *If $P_1 \rightarrow P_2$, then $\mathcal{I}(P_1) \Rightarrow \mathcal{I}(P_2)$.*
2. *If $\mathcal{I}(P_1) \Rightarrow \mathcal{A}$, then there exists P_2 such that $P_1 \rightarrow P_2$ and $\mathcal{A} = \mathcal{I}(P_2)$.*

Proof. Using lemmas 27.13, 27.14, and the previous propositions. The first part follows by induction on the derivation of $\rightarrow$. The second part is by induction on P_1. □

Theorem 27.3 (full abstraction) *Let P, Q be two π-terms.*

$$P \sim_1 Q \quad \textit{if and only if} \quad \mathcal{I}(P) \sim_2 \mathcal{I}(Q)$$

The last two results show that we have a good interpretation which does not lead to strange and undesirable situations. They offer the formal reason that our "molecular" semantics is interesting and correct. Therefore the compositional semantic mapping that associates a molecular structure to each process of the asynchronous π-calculus is well defined and makes sense.

8. CONCLUSION

Trying to understand the cause-effect relations and mechanisms of DNA and molecular computing, we should identify, describe and study the most important operations over molecules. Looking to the molecular reactions, interaction is a fundamental operation which appear in almost every process involving molecules, enzymes and proteins. This paper is devoted to molecular interaction.

First we introduce an abstract molecular structure, then we describe the DNA methylation by using a known calculus of the communicating concurrent systems, namely the π-calculus. Finally, we show that it is possible to have a well-defined interpretation of the π-terms by using abstract molecular structures. In this way we introduce and study some abstract structures which are similar to abstract machines in computer science (as Turing machines or automata), and also appropriate to express molecular interactions. We should remark that considered interactions imply also modifications (methylation); in order to express this aspect too, we consider and study interaction with substitution. Because covalent bounds play a special role when we refer to DNA, our abstract molecular structures are systems with joined (shared) resources; they emphasize the use of some shared resources, as well as the resource transitions.

Process algebras are considered as abstract specifications or concurrent programming languages, and they are based on compositionality. By our structures, we describe the details of the operational behaviour, but preserving also the compositionality and the expressive power of the π-calculus. We define an interpretation from asynchronous π-calculus to abstract molecular structures which is full abstract, i.e. two π-calculus terms are bisimilar if and only if their "molecular" interpretations are bisimilar. The equivalence used for this full-abstraction result is the barbed bisimulation. A connection between the reduction relation over π-terms and the reduction relation over their corresponding abstract molecular structures is given by an operational correspondence result.

Acknowledgments

I am grateful to Gheorghe Păun for his support and help, to Mihai Rotaru for our common work on π-calculus, and Bogdan Tănasă for discussions on DNA methylation. Thanks to Manfred Kudlek for reading the paper and detecting some spelling mistakes.

References

[1] Boudol, G. (1992), Asynchrony and the π-calculus, INRIA Research Report, 1702.

[2] Ciobanu, G. (1998), A molecular abstract machine, in Gh. Păun, ed., *Computing with Bio-Molecules. Theory and Experiments*: 61–79. Springer, Berlin.

[3] Ciobanu, G. (2000), Formal Description of the Molecular Processes, in Gh. Păun, ed., *Recent Topics in Mathematical and Computational Linguistics*, Publishing House of the Romanian Academy, Bucharest.

[4] Ciobanu, G. & M. Rotaru (1998), Faithful π-nets. A faithful graphic representation for π-calculus, *Electronic Notes in Theoretical Computer Science*

18.
http://www.elsevier.nl/locate/entcs/volume18.html.

[5] Ciobanu, G. & M. Rotaru (2000), A π-calculus machine, *Journal of Universal Computer Science*, 6.1, Springer, Berlin.
http://www.iicm.edu/jucs_6_1/a_pi_calculus_machine.

[6] Honda, K. & M. Tokoro (1991), An object calculus for asynchronous communication, in *Proceedings of the ECOOP'91*: 133–147. Springer, Berlin.

[7] Milner, R. (1989), *Communication and Concurrency*, Prentice-Hall, Englewood Cliffs, NJ.

[8] Milner, R. (1993), Elements of interaction, *Communications of the Association for Computing Machinery*, 36: 78–89.

[9] Milner, R. (1993), The polyadic π-calculus: a tutorial, in F.L. Bauer; W. Brauer & H. Schwichtenberg, eds., *Logic and Algebra of Specification*: 203–246. Springer, Berlin.

[10] Milner, R. (1996), Calculi for Interaction, *Acta Informatica*, 33.8: 707–737.

[11] Milner, R; J. Parrow & D. Walker (1992), A calculus of mobile processes, *Journal of Information and Computation*, 100: 1–77.

[12] Wegner, P. (1998) Interactive foundation of computing, *Theoretical Computer Science*.

[13] Winskel, G. & M. Nielsen (1995), Models for concurrency, in *Handbook of Logic in Computer Science*, IV. Oxford University Press, Oxford.

Chapter 28

A CHARACTERIZATION OF NON-ITERATED SPLICING WITH REGULAR RULES

Ray Dassen
Leiden Institute of Advanced Computer Science
University of Leiden
P.O. Box 9512, 2300 RA Leiden, The Netherlands

Hendrik Jan Hoogeboom
Leiden Institute of Advanced Computer Science
University of Leiden
P.O. Box 9512, 2300 RA Leiden, The Netherlands
hoogeboom@liacs.nl

Nikè van Vugt
Leiden Institute of Advanced Computer Science
University of Leiden
P.O. Box 9512, 2300 RA Leiden, The Netherlands
nvvugt@wi.leidenuniv.nl

Abstract The family $S(\mathsf{LIN}, \mathsf{REG})$ of languages obtained by (noniterated) splicing linear languages using regular rules does not coincide with one of the Chomsky families. We give a characterization of this family, and show that we can replace the regular rule set by a finite one.

1. INTRODUCTION

The molecular operation of cutting two DNA molecules with the help of restriction enzymes, and recombining (ligating) the halves into new molecules can be modelled within formal language theory as the *splicing* operation, as

C. Martin-Vide and V. Mitrana (eds.), Where Mathematics, Computer Science, Linguistics and Biology Meet, 319-327.

follows: two strings $x = x_1u_1v_1y_1$ and $y = x_2u_2v_2y_2$ can be spliced according to a splicing rule $r = (u_1, v_1, u_2, v_2)$ to produce another string $z = x_1u_1v_2y_2$:

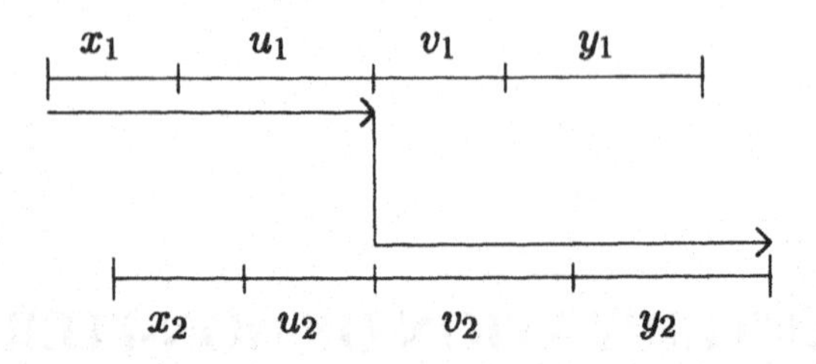

Here, the strings u_1v_1 and u_2v_2 represent the specific sites where restriction enzymes cut the DNA molecules. This operation was proposed by Head [3], long before the study of such operations became fashionable following Adleman's DNA implementation of an algorithm to solve the Hamiltonian Path Problem [1, 2].

The power of the splicing operation, with sets of rules classified within the Chomsky hierarchy, is investigated in [4, 8]. We reconsider the noniterated case of this operation, i.e., we consider the splicing operation as an operation on languages rather than as a language generating mechanism. In particular, we study the family of languages $S(\mathcal{F}, \mathsf{REG})$ that is obtained by splicing languages from a given family $\mathcal{F}$ using regular sets of rules. For families within the Chomsky hierarchy a precize characterization of $S(\mathcal{F}, \mathsf{REG})$ is known, except for the linear languages : $S(\mathsf{LIN}, \mathsf{REG})$ lies strictly in between linear and context-free languages.

Here we obtain an elementary characterization of $S(\mathcal{F}, \mathsf{REG})$ in terms of $\mathcal{F}$, and moreover, we show that the regular set of rules may be replaced by a finite set. For linear languages this then yields a new characterization of $S(\mathsf{LIN}, \mathsf{REG})$, Corollary 28.1.

In our final Section 5., we again try to reduce regular rule sets to finite ones, for a restricted form of splicing, where the rules may only be applied in a certain context [8, 6]. We do not fully succeed. For the case of 'increasing mode' which we consider, the reduction is implemented at the cost of losing words of length one from the generated language.

2. PRELIMINARIES

The empty word is denoted by λ.

A *generalized sequential machine* (gsm) is a finite state machine with additional output. It has a finite set of transitions of the form $(p, a, q, w) \in Q \times \Sigma \times Q \times \Delta^*$, where Q is the finite set of states, Σ and Δ are the input alphabet and output alphabet. Using such a transition, the machine may change from state p into state q, while reading the letter a on its input and writing the string w to its output. The gsm defines a relation in $\Sigma^* \times \Delta^*$, called a gsm mapping.

The *Chomsky hierarchy* is formed by the families FIN, REG, LIN, CF, CS, and RE, of finite, regular, linear, context-free, context-sensitive, and recursively enumerable languages. For a language family $\mathcal{F}$ we use $\mathcal{F} \oplus \mathcal{F}$ to denote finite unions of elements of $\mathcal{F}^2$, i.e., languages of the form $K_1 \cdot L_1 \cup \ldots \cup K_n \cdot L_n$, $n \geqslant 0$, with $K_i, L_i \in \mathcal{F}$. Assuming $\{\lambda\}$ and $\varnothing$ are elements of $\mathcal{F}$, then $\mathcal{F} \oplus \mathcal{F}$ equals $\mathcal{F}$ iff $\mathcal{F}$ is closed under union and concatenation. Hence $\mathcal{F} \oplus \mathcal{F} = \mathcal{F}$ for each Chomsky family, except LIN.

In the sequel, we need closure under gsm mappings and under union to be able to apply our constructions. Any family having these closure properties and containing all finite languages is called *friendly*. All Chomsky families, with the exception of CS, are friendly. We need a simple technicality.

Lemma 28.1 *A friendly family is closed under concatenation with symbols.*

Proof. Let K be an element of the friendly family $\mathcal{F}$. We show that $K\{a\} \in \mathcal{F}$. When $\lambda \notin K$, mapping K onto $K\{a\}$ can be performed by a gsm. If K contains λ, then observe that $K - \{\lambda\} \in \mathcal{F}$ as the intersection with a regular language can be computed by a gsm. Hence, $K\{a\} = (K - \{\lambda\})\{a\} \cup \{a\} \in \mathcal{F}$, by closure under union.

A symmetrical argumentation holds for $\{a\}K$. □

3. THE SPLICING OPERATION

We give basic notions and results concerning splicing and H systems, slightly adapted from [4].

Definition 28.1 *A* splicing rule *(over an alphabet V) is an element of* $(V^*)^4$. *For such a rule $r = (u_1, v_1, u_2, v_2)$ and strings $x, y, z \in V^*$ we write*

$$(x, y) \vdash_r z \quad \textit{iff} \quad x = x_1u_1v_1y_1,\ y = x_2u_2v_2y_2,\ \textit{and}$$
$$z = x_1u_1v_2y_2,\ \textit{for some}\ x_1, y_1, x_2, y_2 \in V^*.$$

We say that z is obtained by splicing *strings x and y* using *rule r.*

Definition 28.2 *An* H system *(or* splicing system*) is a triple $h = (V, L, R)$ where V is an alphabet, $L \subseteq V^*$ is the* initial language *and $R \subseteq (V^*)^4$ is a set of splicing rules, the* splicing relation. *The (*noniterated*)* splicing language *generated by h is defined as*

$$\sigma(h) = \{\, z \in V^* \mid (x, y) \vdash_r z \textit{ for some } x, y \in L \textit{ and } r \in R \,\}.$$

Usually a splicing rule $r = (u_1, v_1, u_2, v_2)$ is given as the string $\mathcal{Z}(r) = u_1\#v_1\$u_2\#v_2$ ($\#$ and $\$$ are special symbols not in V), i.e., $\mathcal{Z}$ is a mapping from $(V^*)^4$ to $V^*\#V^*\$V^*\#V^*$, that gives a *string representation* of each splicing rule. Now that the splicing relation R is represented by the language $\mathcal{Z}(R)$, we can consider the effect of splicing with rules from a certain family of languages : for instance, what is the result of splicing linear languages with finite sets of splicing rules?

Example 28.1 *Let* $L = \{a^n b^n \mid n \geqslant 1\} \cup \{c^n d^n \mid n \geqslant 1\}$, *thus* $L \in \mathsf{LIN}$. *Let* $h = (\{a, b, c, d\}, L, R)$ *be a splicing system with splicing relation* $R = \{\ (b, \lambda, \lambda, c)\ \}$ *consisting of a single rule. The language generated by h is* $\sigma(h) = \{\ a^{n_1} b^{m_1} c^{m_2} d^{n_2} \mid n_i \geqslant m_i \geqslant 1\ (i = 1, 2)\ \}$, *which is not in* LIN.

For any two families of languages $\mathcal{F}_1$ and $\mathcal{F}_2$, the family $S(\mathcal{F}_1, \mathcal{F}_2)$ of noniterated splicing languages (obtained by splicing $\mathcal{F}_1$ languages using $\mathcal{F}_2$ rules) is defined in the obvious way :

$$S(\mathcal{F}_1, \mathcal{F}_2) = \{\ \sigma(h) \mid h = (V, L, R) \text{ with } L \in \mathcal{F}_1 \text{ and } \mathcal{Z}(R) \in \mathcal{F}_2\ \}.$$

The families $S(\mathcal{F}_1, \mathcal{F}_2)$ are investigated in [7] and [8], for $\mathcal{F}_1$ and $\mathcal{F}_2$ in the Chomsky hierarchy. An overview of these results is presented in [4], from which we copy Table 28.1. As an example, the optimal classification within the Chomsky families of splicing LIN languages with FIN rules is $\mathsf{LIN} \subset S(\mathsf{LIN}, \mathsf{FIN}) \subset \mathsf{CF}$. It was shown in [5] that this table does not change when using the equally natural representation $u_1\#u_2\$v_1\#v_2$ instead of $u_1\#v_1\$u_2\#v_2$ for rule (u_1, v_1, u_2, v_2).

Additionally we will consider the family $S(\mathcal{F}, [1])$ of languages obtained by splicing $\mathcal{F}$ languages using rules of *radius* 1, i.e., for (u_1, u_2, u_3, u_4) we have $|u_i| \leqslant 1$ for $i = 1, 2, 3, 4$.

4. UNRESTRICTED SPLICING

We start by considering finite sets of rules.

Lemma 28.2 *Let* $\mathcal{F}$ *be a friendly family. Then* $S(\mathcal{F}, [1]) = S(\mathcal{F}, \mathsf{FIN}) = \mathcal{F} \oplus \mathcal{F}$.

Proof. The inclusion $S(\mathcal{F}, [1]) \subseteq S(\mathcal{F}, \mathsf{FIN})$ is immediate. We prove two other inclusions to obtain the result.

First we show $S(\mathcal{F}, \mathsf{FIN}) \subseteq \mathcal{F} \oplus \mathcal{F}$. Let $h = (V, L, R)$ be an H system with a finite number of rules, and with $L \in \mathcal{F}$.

Consider the rule $r = (u_1, v_1, u_2, v_2)$. When r is applied to strings $x_1 u_1 v_1 y_1$ and $x_2 u_2 v_2 y_2$, then only the substrings $x_1 u_1$ and $v_2 y_2$ are visible in the resulting

<table>
<tr><th>$\mathcal{F}_2$:</th><th>FIN</th><th>REG</th><th>LIN</th><th>CF</th><th>CS</th><th>RE</th></tr>
<tr><td>$\mathcal{F}_1$: FIN</td><td>FIN</td><td>FIN</td><td>FIN</td><td>FIN</td><td>FIN</td><td>FIN</td></tr>
<tr><td>REG</td><td>REG</td><td>REG</td><td>REG
LIN</td><td>REG
CF</td><td>REG
RE</td><td>REG
RE</td></tr>
<tr><td>LIN</td><td>LIN, CF</td><td>LIN, CF</td><td colspan="4" rowspan="4">RE</td></tr>
<tr><td>CF</td><td>CF</td><td>CF</td></tr>
<tr><td>CS</td><td colspan="2" rowspan="2"></td></tr>
<tr><td>RE</td></tr>
</table>

Table 28.1 The position of $S(\mathcal{F}_1, \mathcal{F}_2)$ in the Chomsky hierarchy.

string $x_1u_1v_2y_2$. We define two languages derived from the initial language following that observation: let $L_{\langle r} = \{xu_1 \mid xu_1v_1y \in L, \text{ for some } y \in V^*\}$, and let $L_{r\rangle} = \{v_2y \mid xu_2v_2y \in L, \text{ for some } x \in V^*\}$.

Observe that both $L_{\langle r}$ and $L_{r\rangle}$ can be obtained from L by a gsm mapping, and consequently these languages are in $\mathcal{F}$. Clearly, $\sigma(h) = \bigcup_{r \in R} L_{\langle r}L_{r\rangle}$, thus, $\sigma(h) \in \mathcal{F} \oplus \mathcal{F}$.

Second, we show $\mathcal{F} \oplus \mathcal{F} \subseteq S(\mathcal{F}, [1])$. Consider $K_1 \cdot L_1 \cup \ldots \cup K_n \cdot L_n$ with $K_i, L_i \subseteq V^*$ in $\mathcal{F}$, for some alphabet V. This union is obtained by splicing the initial language $\bigcup_{i=1}^n K_ic_i \cup \bigcup_{i=1}^n c_i'L_i$ with rules $(\lambda, c_i, c_i', \lambda)$, $i = 1, \ldots, n$, where the c_i, c_i' are new symbols. Note that the languages K_ic_i and $c_i'L_i$ belong to $\mathcal{F}$ by Lemma 28.1. □

The equality $S(\mathcal{F}, [1]) = S(\mathcal{F}, \mathsf{FIN})$ appears as [4, Lemma 3.10].

We can directly apply Lemma 28.2 to obtain the (known) characterizations of $S(\mathcal{F}, \mathsf{FIN})$ for the friendly families $\mathcal{F} =$ FIN, REG, CF and RE. The equality $S(\mathsf{LIN}, \mathsf{FIN}) = \mathsf{LIN} \oplus \mathsf{LIN}$ appears to be new, although the family $\mathsf{LIN} \oplus \mathsf{LIN}$ is hinted at in the proof of Theorem 3 of [7], where it is demonstrated that $S(\mathsf{LIN}, \mathsf{REG})$ is strictly included in CF.

Refining the above proof, we can extend it to regular rule sets.

Theorem 28.1 *Let $\mathcal{F}$ be a friendly family. Then $S(\mathcal{F}, \mathsf{FIN}) = S(\mathcal{F}, \mathsf{REG})$.*

Proof. By Lemma 28.2, it suffices to prove the inclusion $S(\mathcal{F}, \mathsf{REG}) \subseteq \mathcal{F} \oplus \mathcal{F}$.

Let $h = (V, L, R)$ be an H system with regular rule set, and initial language in $\mathcal{F}$. Assume $\angle(R) \subseteq V^*\#V^*\$V^*\#V^*$ is accepted by the finite state automaton $\mathcal{A} = (Q, \Sigma, \delta, q_{in}, F)$, with $\Sigma = V \cup \{\#, \$\}$.

Now, for $p \in Q$, let

$$\begin{aligned} L_{\langle p} = \{\, xu \mid & \; xuvy \in L, \text{ for some } x, u, v, y \in V^*, \\ & \text{such that } p \in \delta(q_{in}, u\#v) \,\}, \\ L_{p\rangle} = \{\, vy \mid & \; xuvy \in L, \text{ for some } x, u, v, y \in V^*, \\ & \text{such that } \delta(p, u\#v) \cap F \neq \varnothing \,\} \end{aligned}$$

Observe that both $L_{\langle p}$ and $L_{p\rangle}$ can be obtained from L by a gsm mapping. For example, the gsm computing $L_{\langle p}$ guesses the start of the segment u on its input, and simulates $\mathcal{A}$ on this segment (all the time copying its input to the output). At the end of u (nondeterministically guessed), it simulates the step of $\mathcal{A}$ on $\#$ and continues to simulate $\mathcal{A}$ on the input, without writing output, while checking whether state p is reached.

Some care has to be taken here. By definition, a gsm cannot use a λ-transition to simulate $\mathcal{A}$ on the additional symbol $\#$ that is not part of the input. As a solution, in its finite state the gsm may keep the values of both $\delta(q_{in}, u')$ and $\delta(q_{in}, u'\#)$ for the prefix u' of u that has been read.

Hence the languages $L_{\langle p}$ and $L_{p\rangle}$ are in $\mathcal{F}$.

We claim that $\sigma(h) = \bigcup_{(p,\$,q)\in\delta} L_{\langle p}L_{q\rangle}$, and consequently, $\sigma(h) \in \mathcal{F} \oplus \mathcal{F}$.

We prove the claim here in one direction : Assume $z \in L_{\langle p}L_{q\rangle}$ for some $(p, \$, q) \in \delta$. Then there exist $x_1, y_1, x_2, y_2, u_1, v_1, u_2, v_2 \in V^*$ such that $z = x_1u_1 \cdot v_2y_2$, $x_1u_1v_1y_1 \in L$, $p \in \delta(q_{in}, u_1\#v_1)$, $x_2u_2v_2y_2 \in L$, and $\delta(q, u_2\#v_2) \cap F \neq \varnothing$.

As $q \in \delta(p, \$)$, we conclude that $\delta(q_{in}, u_1\#v_1\$u_2\#v_2) \cap F \neq \varnothing$, and so $r = (u_1, v_1, u_2, v_2) \in R$. Hence, $z \in \sigma(h)$, as it is obtained by splicing $x = x_1u_1v_1y_1$ and $y = x_2u_2v_2y_2$ in L using r from R. □

Again, for most of the Chomsky families (including CS which is not friendly) the last result is implicit in Table 28.1. Here it is obtained through direct construction. We summarize the new results obtained for LIN.

Corollary 28.1 $S(\mathsf{LIN}, \mathsf{FIN}) = S(\mathsf{LIN}, \mathsf{REG}) = \mathsf{LIN} \oplus \mathsf{LIN}$.

5. RESTRICTED SPLICING

In this section we try to extend the result that a regular set of rules can be reduced to a finite set of rules (Theorem 28.1). We consider the setting where the general splicing operation $(x, y) \vdash_r z$ may only be applied in a certain context, as inspired by [8, 6].

We splice in *increasing mode* if the result z is as least as long as both inputs x and y. Formally,

$$(x, y) \vdash_r^{in} z \quad \text{iff} \quad (x, y) \vdash_r z \text{ and } |z| \geqslant |x|, |z| \geqslant |y|.$$

Example 28.2 *[6] Let $h = (\{a, b\}, L, R)$, where $L = ca^*b^*c \cup cb^*a^*c$ and $\mathcal{Z}(R) = \{ca^n\#b^nc\$c\#b^ma^mc \mid n, m \geqslant 1\} \in \mathsf{CF}$. Then $\sigma_{in}(h) = \{ca^nb^ma^mc \mid n, m \geqslant 1 \text{ and } n \leqslant 2m\}$, which is not context-free.*

With this restricted operation we define way the language $\sigma_{in}(h)$ for a splicing system h in the obvious. Thus, we consider the families $S_{in}(\mathcal{F}_1, \mathcal{F}_2)$, and we study the relation between $S_{in}(\mathcal{F}, \mathsf{FIN})$ and $S_{in}(\mathcal{F}, \mathsf{REG})$. We summarize the results concerning $S_{in}(\mathcal{F}_1, \mathcal{F}_2)$ obtained in [6].

Proposition 28.1 $S_{in}(\mathsf{REG}, \mathsf{CF}) - \mathsf{CF} \neq \varnothing$, $S_{in}(\mathsf{REG}, \mathsf{REG}) \subseteq \mathsf{REG}$, *and* $S_{in}(\mathsf{CS}, \mathsf{CF}) \subseteq \mathsf{CS}$.

Observe that $S_{in}(\mathcal{F}, \mathsf{FIN}) \subseteq S_{in}(\mathcal{F}, \mathsf{REG})$ by definition. We could not show the converse inclusion $S_{in}(\mathcal{F}, \mathsf{REG}) \subseteq S_{in}(\mathcal{F}, \mathsf{FIN})$. However, the families are *almost* equal, in the sense that for every language K_r in $S_{in}(\mathcal{F}, \mathsf{REG})$ there is a language K_f in $S_{in}(\mathcal{F}, \mathsf{FIN})$ such that K_r and K_f differ only by words of a maximum length of one.

Theorem 28.2 *Let $\mathcal{F}$ be a friendly family. Then $S_{in}(\mathcal{F}, \mathsf{FIN}) = S_{in}(\mathcal{F}, \mathsf{REG})$, almost (in the sense explained above).*

Proof. We develop some ideas from the proof of Theorem 28.1.

Let $h = (V, L, R)$ be an H system with regular rule set, and initial language in $\mathcal{F}$. We construct an H system with finite rule set that defines a language 'almost' equal to $\sigma(h)$. Assume $\mathcal{Z}(R) \subseteq V^*\#V^*\$V^*\#V^*$ is accepted by the finite state automaton $\mathcal{A} = (Q, \Sigma, \delta, q_{in}, F)$, with $\Sigma = V \cup \{\#, \$\}$, and $Q \cap \Sigma = \varnothing$. Assuming that the automaton is reduced (each state lies on a path from the initial to a final state) we can split the set of states into two disjoint subsets, $Q = Q_1 \cup Q_2$, such that Q_1 (Q_2) contains the states on a path before (after) the symbol \$ is read. Let ι be a new symbol.

First step. We construct a new initial language $L' \subseteq (V \cup Q \cup \{\iota\})^*$ from L as follows. For each $x, u, v, y \in V^*$, with $xuvy \in L$, and each $p \in Q_1$, $q \in Q_2$ we include in L' the following words, under the given constraints:

$xup\iota^k$ where	$p \in \delta(q_{in}, u\#v)$,	$\lvert vy \rvert \geqslant 1$, $k = \lvert vy \rvert - 1$.
xup	$p \in \delta(q_{in}, u\#v)$,	$\lvert vy \rvert = 0$.
$\iota^\ell qvy$	$\delta(q, u\#v) \cap F \neq \varnothing$,	$\lvert xu \rvert \geqslant 1$, $\ell = \lvert xu \rvert - 1$.
qvy	$\delta(q, u\#v) \cap F \neq \varnothing$,	$\lvert xu \rvert = 0$.

Observe that L' can be obtained from L by a gsm mapping, and consequently L' belongs to $\mathcal{F}$.

Let R' be the (finite) set of rules $\{ (\lambda, p, q, \lambda) \mid (p, \$, q) \in \delta \}$. Note that every rule (λ, p, q, λ) in R' corresponds to a (regular) set of rules $\{ (u_1, v_1, u_2, v_2) \mid p \in \delta(q_{in}, u_1 \# v_1), \delta(q, u_2 \# v_2) \cap F \neq \varnothing \}$ in R.

Let $h' = (V \cup Q \cup \{\iota\}, L', R')$.

It is easy to understand that $\sigma_{in}(h') \subseteq \sigma_{in}(h)$, following the construction of L' and R'. If $x' = x_1 u_1 p \iota^k$ and $y' = \iota^\ell q v_2 y_2$ in L', splice in the increasing mode to give $z = x_1 u_1 v_2 y_2$ using rule $r' = (\lambda, p, q, \lambda)$ in R', then there are strings $x = x_1 u_1 v_1 y_1$ and $y = x_2 u_2 v_2 y_2$ in L that splice to give z againusing rule $r = (u_1, v_1, u_2, v_2)$ in R. By construction, $|x| = |x'|$ (or $|x| = |x'| - 1$ when $|vy| = 0$) we know that $|x| \leqslant |x'|$, thus $|x'| \leqslant |z|$ implies $|x| \leqslant |z|$. Mutatis mutandis, this argument is also valid for y, y', so x and y splice in increasing mode, too.

In general, the reverse inclusion $\sigma_{in}(h) \subseteq \sigma_{in}(h')$ is not true. Assume, however, $z \in \sigma_{in}(h)$, obtained through $(x = x_1 u_1 v_1 y_1, y = x_2 u_2 v_2 y_2) \vdash_r^{in} z = x_1 u_1 v_2 y_2$, where $r = (u_1, v_1, u_2, v_2)$ in R.

By construction one finds $x' = x_1 u_1 p \iota^k$ and $y' = \iota^\ell q v_2 y_2$ in L' for suitable $k, \ell \in \mathbb{N}$, $p, q \in Q$. Consider $r' = (\lambda, p, q, \lambda)$ corresponding to r (as discussed above). Then $(x', y') \vdash_{r'} z$. This is increasing mode under the condition that

$$|v_1 y_1| + |v_2 y_2| \geqslant 1 \text{ and } |x_1 u_1| + |x_2 u_2| \geqslant 1.$$

This is seen as follows. If $|v_1 y_1| \geqslant 1$, then $|x'| = |x|$ and hence $|x'| \leqslant |z|$. Otherwise, if $|v_1 y_1| = 0$, then $x' = x_1 u_1 p$ is one symbol longer than x in the original splicing. However, $|x'| \leqslant |z|$ follows from the fact that $|v_2 y_2| \geqslant 1 = |p|$. An analogous argument holds for $|y'| \leqslant |z|$.

We conclude that $z \in \sigma_{in}(h')$, except when it can only be obtained using $|v_1 y_1| = |v_2 y_2| = 0$, i.e., $x = z = x_1 u_1$, $y = x_2 u_2$, and $r = (u_1, \lambda, u_2, \lambda)$, or $|x_1 u_1| = |x_2 u_2| = 0$, i.e., $x = v_1 y_1$, $y = z = v_2 y_2$, and $r = (\lambda, v_1, \lambda, v_2)$.

Second step. In order to accommodate almost all these cases we add additional strings to the initial language L'. Extend the alphabet by two copies of $V \times Q$, symbols which we will denote as $\langle a - p \rangle$, $\langle a + p \rangle$, $\langle q - a \rangle$, $\langle q + a \rangle$, where $a \in V, p \in Q_1$, and $q \in Q_2$.

For each $x, u, v, y \in V^*$, with $xu, vy \in L$, each $a \in V$, and each $p \in Q_1$, $q \in Q_2$ we add the following words to L', under the given constraints :

$w\langle a - p \rangle$	where	$p \in \delta(q_{in}, u\#)$,	$wa = xu$.
$\iota^k \langle q + a \rangle a$		$\delta(q, u\#) \cap F \neq \varnothing$,	$\lvert xu \rvert \geqslant 2,\ k = \lvert xu \rvert - 2$.
$a \langle a + p \rangle \iota^\ell$		$p \in \delta(q_{in}, \#v)$,	$\lvert vy \rvert \geqslant 2,\ \ell = \lvert vy \rvert - 2$.
$\langle q - a \rangle w$		$\delta(q, \#v) \cap F \neq \varnothing$,	$aw = vy$.

To R' add the set of rules $\{ (\lambda, \langle a-p\rangle, \langle q+a\rangle, \lambda), (\lambda, \langle a+p\rangle, \langle q-a\rangle, \lambda) \mid (p, \$, q) \in \delta \}$.

These new strings and new rules can only splice among themselves, and simulate most of the remaining splicings of the original system (with regular rule set). For instance, assuming $|x_1 u_1| \geqslant 2$, and writing $x_1 u_1 = wa$, the original system splices $(x_1 u_1, x_2 u_2) \vdash_r^{in} x_1 u_1 = z$, with $r = (u_1, \lambda, u_2, \lambda)$, iff the new system splices $(w\langle a{-}p\rangle, \iota^k \langle q{+}a\rangle a) \vdash_{r'}^{in} wa = z$ where $r' = (\lambda, \langle a{-}p\rangle, \langle q{+}a\rangle, \lambda)$ and $|\iota^k \langle q+a\rangle a| = |x_2 u_2|$ to ensure increasing mode.

Conclusion. Splicings we can *not* simulate have $v_1 y_1 = v_2 y_2 = \lambda$ to obtain $z = x_1 u_1$, or $x_1 u_1 = x_2 u_2 = \lambda$ to obtain $z = v_2 y_2$, in both cases with $|z| \leqslant 1$. The result is thus proved. □

The papers [6, 8] contain many other modes of restricted splicing. Many of these modes still lack a precise characterization. More specifically in connection with our investigations, it would be interesting to relate $S_\mu(\mathcal{F}, \mathsf{FIN})$ and $S_\mu(\mathcal{F}, \mathsf{REG})$ for each mode μ.

References

[1] Adleman, L.M. (1994), Molecular computation of solutions to combinatorial problems, *Science*, 226: 1021–1024.

[2] Dassen, J.H.M. & P. Frisco, eds., A bibliography of molecular computation and splicing systems. At url:
`http://www.liacs.nl/~pier/dna.html`.

[3] Head, T. (1987), Formal language theory and DNA: an analysis of the generative capacity of specific recombinant behaviours, *Bulletin of Mathematical Biology*, 49: 737–759.

[4] Head, T.; Gh. Păun & D. Pixton (1997), Language theory and molecular genetics: Generative mechanisms suggested by DNA recombination, in G. Rozenberg & A. Salomaa, eds., *Handbook of Formal Languages*, II, Springer, Berlin.

[5] Hoogeboom, H.J. & N. van Vugt (1998), The power of H systems: does representation matter?, in Gh. Păun, ed., *Computing with Bio-Molecules: Theory and Experiments*: 255–268. Springer, Singapore.

[6] Kari, L.; Gh. Păun & A. Salomaa (1996), The power of restricted splicing with rules from a regular language, *Journal of Universal Computer Science*, 2.4: 224-240.

[7] Păun, Gh. (1996), On the splicing operation, *Discrete Applied Mathematics*, 70: 57–79.

[8] Păun, Gh.; G. Rozenberg & A. Salomaa (1996), Restricted use of the splicing operation, *International Journal of Computer Mathematics*, 60: 17–32.

Chapter 29

UNIVERSAL AND SIMPLE OPERATIONS FOR GENE ASSEMBLY IN CILIATES

Andrzej Ehrenfeucht
Department of Computer Science
University of Colorado
Boulder, CO 80309-0347, USA
andrzej@cs.colorado.edu

Ion Petre
Turku Centre for Computer Science and
Department of Mathematics
University of Turku
FIN 20520 Turku, Finland
ipetre@cs.utu.fi

David M. Prescott
Department of Molecular, Cellular and Developmental Biology
University of Colorado
Boulder, CO 80309-0347, USA
prescotd@spot.colorado.edu

Grzegorz Rozenberg
Leiden Institute of Advanced Computer Science
Leiden University
Niels Bohrweg 1, 2333 CA Leiden, The Netherlands
rozenber@liacs.nl

C. Martin-Vide and V. Mitrana (eds.), Where Mathematics, Computer Science, Linguistics and Biology Meet, 329-342.

Abstract The way that ciliates transform genes from their micronuclear to their macronuclear form is very interesting (and unique), particularly from a computational point of view. In this paper, we describe the model of gene assembly in ciliates presented in [2], [4], and [3]. Moreover, we prove that the set of three operations underlying this model is universal, in the sense that it suffices for gene assembly from any micronuclear pattern. We also prove that the set of simple versions of these operations is not universal – this fact is interesting from the experimental point of view.

1. INTRODUCTION

DNA computing, initiated by L. Adleman in [1], is a fast growing interdisciplinary research area (see, e.g., [7]). The research in this area includes both DNA computing *in vitro* and DNA computing *in vivo*. Within the area of DNA computing *in vivo*, one investigates (the use of) living cells as computational agents. To this aim, one studies the computational aspects of DNA processing in living cells. A series of papers by L. Landweber and L. Kari (see, e.g., [5] and [6]) have brought to the attention of DNA computing community the computational power of DNA processing taking place in ciliates.

Ciliates are a very ancient group of organisms (their origin is estimated at about 2×10^9 years ago). They are a very extensive group of about 10000 genetically different organisms. A very interesting feature of ciliates is their nuclear dualism. A ciliate contains two kinds of nuclei: a germline nucleus (micronucleus) that is used in cell mating, and a somatic nucleus (macronucleus) that provides RNA transcripts to operate the cell. The DNA in the micronucleus is made of hundreds of kilobase pairs long with genes occurring individually or in groups dispersed along the DNA molecule, separated by long stretches of spacer DNA. The DNA molecules in the macronucleus are gene-size, on average about 2000 base-pairs long. As a matter of fact, a macronucleus results from a micronucleus, through an intricate transformation process. This process turns out to be fascinating (see, e.g, [8] and [9]), as it is from the computational point of view as well. In this paper, we present a model of this process, described in [2], [4], and [3]. We also prove that the set of three operations underlying this model is universal, in the sense that it suffices for gene assembly from any micronuclear pattern. Moreover, we prove that the set of simple versions of these operations is not universal. This is very intriguing, because the set of simple operations was used in [3] to provide possible assembly sequences for all experimentally determined micronuclear genes.

2. GENE ASSEMBLY IN CILIATES

The structure of a gene in a micronuclear chromosome consists of MDSs (macronuclear destined sequences) separated by IESs (internally eliminated

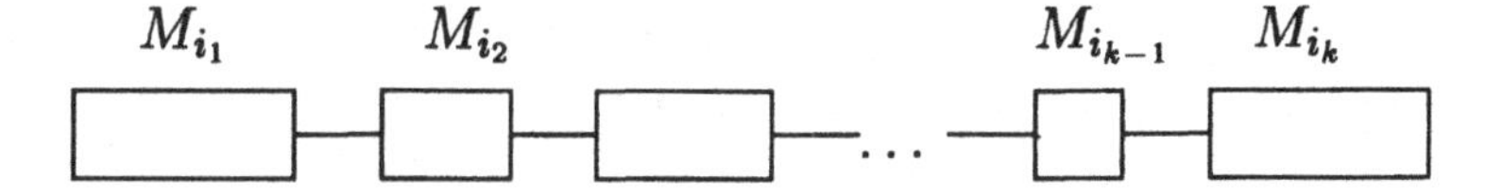

Figure 29.1 The structure of a gene in a micronuclear DNA.

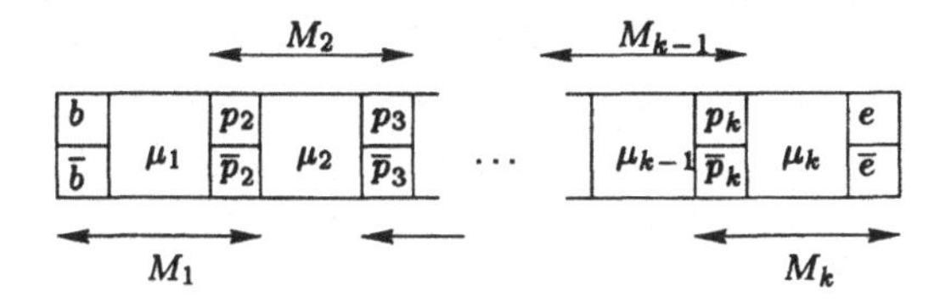

Figure 29.2 The structure of a macronuclear assembled gene.

sequences), as illustrated in Figure 29.1 – the rectangles represent MDSs and the connecting line segments between MDSs represent IESs. This structure is related to the structure of the macronuclear gene as follows. During the transformation process from micronucleus to macronucleus, the IESs become excised, and the MDSs are spliced in the orthodox order $M_1, M_2, \ldots, M_k$, which is suitable for transcription. Each MDS M_i has the structure

$$M_i = \left(\begin{matrix} p_i \\ \overline{p}_i \end{matrix}, \mu_i, \begin{matrix} p_{i+1} \\ \overline{p}_{i+1} \end{matrix} \right),$$

except for M_1 and M_k, which are of the form

$$M_1 = \left(\begin{matrix} b \\ \overline{b} \end{matrix}, \mu_1, \begin{matrix} p_2 \\ \overline{p}_2 \end{matrix} \right), \quad M_k = \left(\begin{matrix} p_k \\ \overline{p}_k \end{matrix}, \mu_k, \begin{matrix} e \\ \overline{e} \end{matrix} \right),$$

where b designates "beginning" and e designates "end". We refer to each $\begin{matrix} p_i \\ \overline{p}_i \end{matrix}$ as a *pointer*, and to μ_i as the *body* of M_i. Moreover, $\begin{matrix} p_i \\ \overline{p}_i \end{matrix}$ is called the *incoming pointer* of M_i, and $\begin{matrix} p_{i+1} \\ \overline{p}_{i+1} \end{matrix}$ is called the *outgoing pointer* of M_i. The pair $\left(\begin{matrix} p_i \\ \overline{p}_i \end{matrix}, \begin{matrix} p_i \\ \overline{p}_i \end{matrix} \right)$ is called *direct repeat pair* – these pairs help to guide the assembly of the macronuclear gene from its micronuclear form. As a matter of fact, each p_i is a sequence of nucleotides (and $\overline{p}_i$ is the inversion of p_i). On the other hand, b and e are not pointers – they simply designate the locations where an incipient macronuclear DNA molecule will be excised from macronuclear DNA.

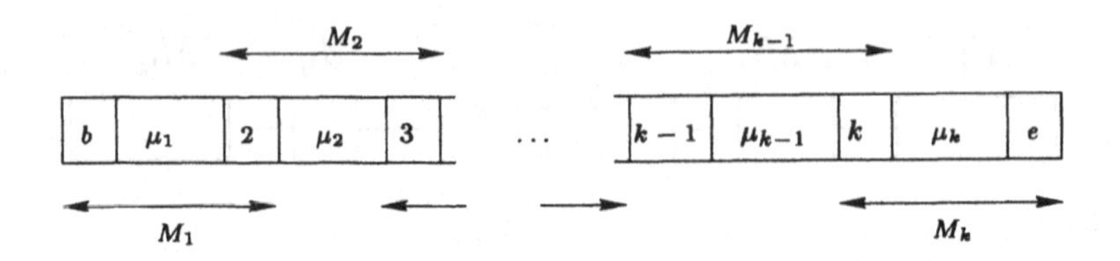

Figure 29.3 Simplified representation of an assembled gene.

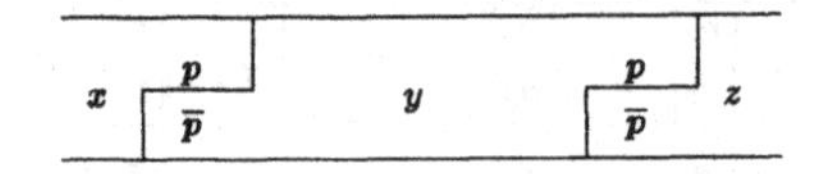

Figure 29.4 A direct repeat pattern (p, p).

In the macronucleus, the MDSs $M_1, M_2, \ldots, M_k$ are spliced in the way shown in Figure 29.2 – hence, M_j is "glued" with M_{j+1} on the pointer $\frac{p_{j+1}}{\overline{p}_{j+1}}$.

Since pointers are crucial for the considerations of this paper, we simplify our notations by using positive integers $2, 3, \ldots$ and $\overline{2}, \overline{3}, \ldots$ to denote pointers. Using the simplified notation, $M_i = (i, \mu_i, i+1)$, $\overline{M_i} = (\overline{i+1}, \overline{\mu}_i, \overline{i})$, and the structure of the assembled gene is represented as in Figure 29.3. Thus, we need the alphabets $\Delta = \{2, 3, \ldots\}$ and $\overline{\Delta} = \{\overline{2}, \overline{3}, \ldots\}$. We will use $\Pi = \Delta \cup \overline{\Delta}$ to denote the set of pointers. We will also use the extended sets $\Delta_{ex} = \Delta \cup \{b, e\}$, $\overline{\Delta}_{ex} = \overline{\Delta} \cup \{\overline{b}, \overline{e}\}$, and $\Pi_{ex} = \Pi \cup \{b, e, \overline{b}, \overline{e}\}$ to include the markers b and e and their inversions. We will use the "bar operator" to move from Δ_{ex} to $\overline{\Delta}_{ex}$ and the other way around. Hence, for $z \in \Pi_{ex}$, $\overline{z}$ is the *partner* of z from the "other half of Π_{ex}", where $\overline{\overline{z}} = z$. For any $z \in \Pi$, we call $\{z, \overline{z}\}$ *the pointer set* of z (and of $\overline{z}$).

It is argued in [4] and [3] that the gene assembly process in ciliates is accomplished through the use of the three following operations:

1. *(loop, direct repeat)-excision* (*ld-excision*, or just *ld*, for short),
2. *(hairpin, inverted repeat)-excision/reinsertion* (*hi-excision/reinsertion*, or just *hi*, for short), and
3. *(double loop, alternating direct repeat)-excision/reinsertion* (*dlad-excision/reinsertion*, or just *dlad*, for short).

We will now present each of these three operations in a uniform way by describing the domain of each of them, the kind of fold needed, and the "execution" of the operation.

1. The operation of *ld*-excision is applicable to molecules that have a direct repeat pattern (p, p) of a pointer p, i.e., it is of the form shown in Figure 29.4.

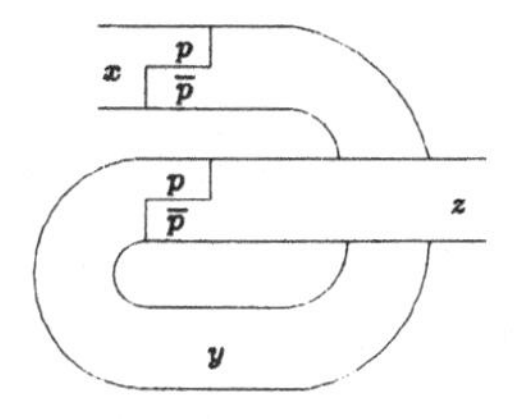

Figure 29.5 A loop aligned by the direct repeat pattern (p, p).

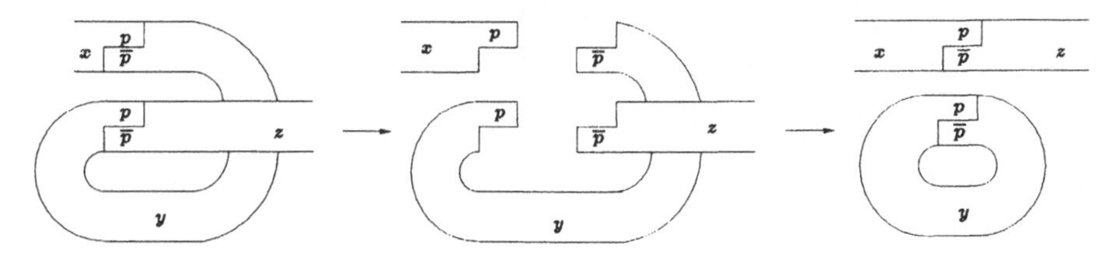

Figure 29.6 The *ld*-excision operation.

This molecule is folded into a loop aligned by the pair of direct repeats, as shown in Figure 29.5, and then, the operation proceeds as shown in Figure 29.6. The excision here involves staggered cuts and the operation yields two molecules, a linear one, and a circular one.

Moreover, if we require that the direct repeat pattern from Figure 29.5 is such that y consists of exactly one IES, then we call it a *simple direct repeat*, and we are dealing with *simple ld-excision*, or just *sld* for short.

It is important to note that if *ld*-excision is used in a process of assembly of a gene ν, then, with one exception, it must be simple, as otherwise an MDS would be excised, and ν could not be assembled, because the excised MDS (or MDSs) will be missing. The exceptional case is when the two occurrences of a pointer are on the two opposite ends of the molecule – we say then that we have a *boundary* application of the *ld*-excision. In this case, applying the *ld*-excision to p will lead to the assembly of the gene into a circular molecule (see [4]). As a matter of fact, if a boundary application is used in a successful assembly, then it can be used as the last operation.

2. The operation of *hi*-excision/reinsertion is applicable to molecules that have an inverted repeat pattern $(p, \overline{p})$ of a pointer p, i.e., it is of the form shown in Figure 29.7.

This molecule is folded into a hairpin aligned by the inverted repeat pair as shown in Figure 29.8, and then, the operation proceeds as shown in Figure 29.9. The excision here involves staggered cuts, but through reinsertion, the operation yields only one molecule.

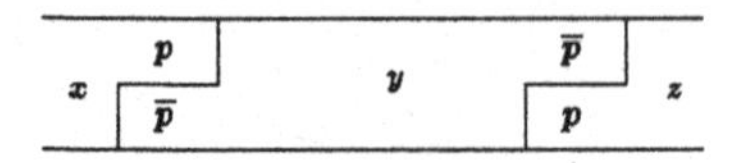

Figure 29.7 The inverted repeat pattern (p, p).

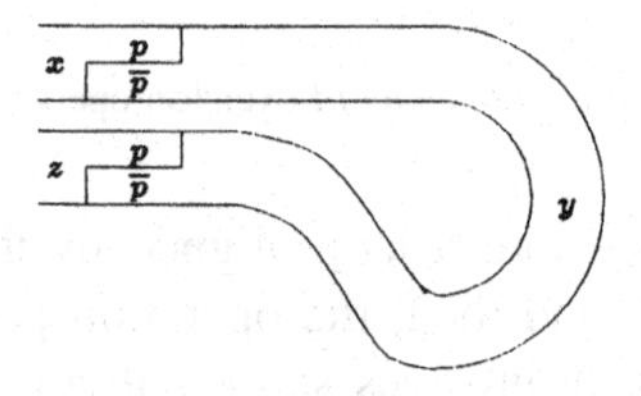

Figure 29.8 A loop aligned by the inverted repeat pattern $(p, \overline{p})$.

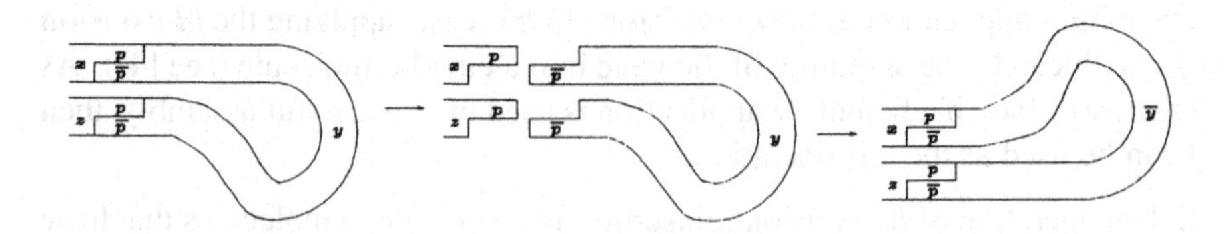

Figure 29.9 The *hi*-excision/reinsertion operation.

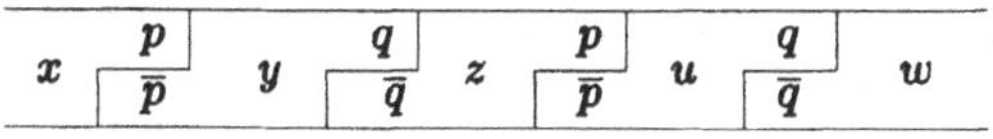

Figure 29.10 The alternating direct repeat pattern (p, q, p, q).

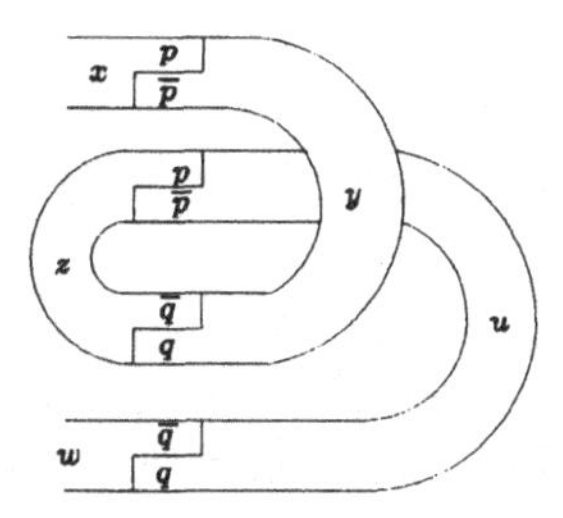

Figure 29.11 A double loop aligned by the alternating direct repeat pattern (p, q, p, q).

Moreover, if we require that the inverted repeat pattern from Figure 29.8 is such that y contains exactly one IES, then we are dealing with *simple hi-excision/reinsertion* operation, or just *shi*, for short.

3. The operation of *dlad*-excision/reinsertion is applicable to molecules that have an alternating direct repeat pattern (p, q, p, q), i.e., it is of the form shown in Figure 29.10. This molecule is folded into a double loop with one loop aligned by (p, p), and the other by (q, q), as shown in Figure 29.11. Then, the operation proceeds as shown in Figure 29.12. Again, the excision here involves staggered cuts, but because of the reinsertion, the operation yields one molecule only.

Moreover, if we require that the alternating direct repeat pattern from Figure 29.11 is such that one of y, u does not contain an IES, while the other one is an IES, then we are dealing with *simple dlad-excision/reinsertion* operation, or just *sdlad*, for short.

Note that a direct repeat pattern (p, p) has two occurrences from the pointer set $\{p, \overline{p}\}$: both of them are occurrences of p. An inverted repeat pattern $(p, \overline{p})$ has two occurrences from the pointer set $\{p, \overline{p}\}$: one of them is the occurrence of p, and the other one of $\overline{p}$. An alternating direct repeat pattern (p, q, p, q) has two occurrences from the pointer set $\{p, \overline{p}\}$, viz., two occurrences of p, and two occurrences from the pointer set $\{q, \overline{q}\}$, viz., two occurrences of q.

When a gene is assembled into its macronuclear form, it no longer has pointers anymore. On the other hand, its micronuclear precursor has pointers that flank IESs. During the process of assembly, as long as there are IESs, the assembly process is not yet completed. Thus, as long as an "intermediate molecule"

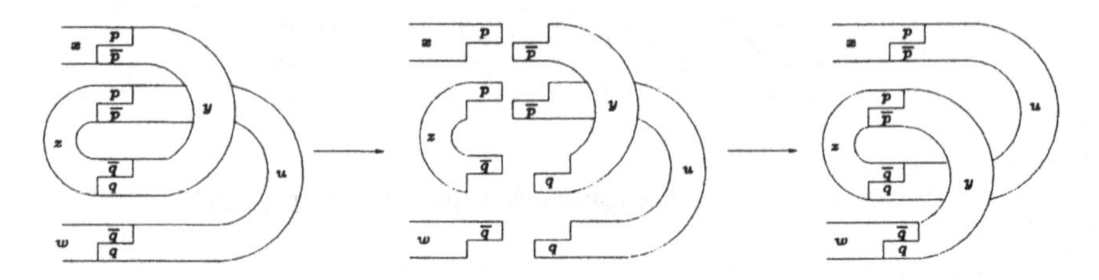

Figure 29.12 The *dlad*-excision/reinsertion operation.

contains pointers, the assembly process is not yet completed. This is the basic idea underlying the analysis of the assembly process by tracing of the pointers of the intermediate molecules. Here, our three operations, *ld*, *hi*, and *dlad*, are essential because each of them removes pointers. In particular:

1. The *ld* operation aligned on the direct repeat (p, p) removes (the occurrences from) the pointer set $\{p, \overline{p}\}$: one occurrence of p is now within the larger composite MDS, and one occurrence of p is within the excised loop together with an IES.
2. The *hi* operation aligned on the inverted repeat $(p, \overline{p})$ removes (the occurrences from) the pointer set $\{p, \overline{p}\}$: one occurrence of p is now a part of a larger composite MDS, and one occurrence of p is now within a larger composite IES.
3. The *dlad* operation aligned on the alternated direct repeat (p, q, p, q) removes (the occurrences from) the pointer sets $\{p, \overline{p}\}$ and $\{q, \overline{q}\}$: for both p and q, one occurrence is now within a larger composite MDS, and one occurrence is within a larger composite IES.

Thus, the process of gene assembly consists of a sequence of pointer set removals by our three operations.

The reader may have noticed that we do not distinguish carefully between a pointer p and its partner $\overline{p}$. From the computational point of view discussed in this paper, it really does not matter whether a given p is a pointer, and $\overline{p}$ its inversion, or whether $\overline{p}$ is a pointer and p its inversion. That is why all elements of Π are called pointers.

3. MDS STRUCTURES AND MDS DESCRIPTORS

From the point of view of the gene assembly process, the structural information about the micronuclear or an intermediate precursor τ of a macronuclear gene ν, can be given by the sequence of MDSs forming τ.

We will use the alphabet $\mathcal{M} = \{M_{i,j} \mid i, j \in \mathbb{N}^+ \text{ and } i \leqslant j\}$ to denote the MDSs of τ (we use $\mathbb{N}^+$ to denote the set of positive integers). The letters of the form $M_{i,i}$ may also be written as M_i, and they are called *elementary*, while the letters of the form $M_{i,j}$ with $i < j$, are called *composite*. Hence $\mathcal{M} = \mathcal{E} \cup \mathcal{C}$, where $\mathcal{E} = \{M_i \mid i \in \mathbb{N}^+\}$ and $\mathcal{C} = \{M_{i,j} \mid i,j \in \mathbb{N}^+ \text{ and } i < j\}$. The letters from $\mathcal{E}$ are used to denote the MDSs present in the micronuclear precursor of ν, referred to as *elementary* MDSs, while the letters from $\mathcal{C}$ are used to denote the *composite* MDSs formed in the process of gene assembly: $M_{i,j}$ denotes the composite MDS that results by splicing the elementary MDSs $M_i, M_{i+1}, \ldots, M_j$ by their pointers. To denote the inverted MDSs, we will use the alphabet $\overline{\mathcal{M}} = \{\overline{M}_{i,j} \mid M_{i,j} \in \mathcal{M}\}$; hence $\overline{\mathcal{E}} = \{\overline{M}_i \mid i \in \mathbb{N}^+\}$ and $\overline{\mathcal{C}} = \{\overline{M}_{i,j} \mid i,j \in \mathbb{N}^+ \text{ and } i < j\}$. Thus, the total alphabet we will use for denoting MDSs and their inversions is $\Theta = \mathcal{M} \cup \overline{\mathcal{M}}$. Clearly, Θ is infinite, but for each specific gene ν, the set of its elementary MDSs has only a finite number of elements, say k. They will be denoted by the letters from $\{M_1, \ldots, M_k\}$, where the orthodox order of the MDSs in the assembled macronuclear gene is $M_1, M_2, \ldots, M_k$ (we say then that ν is of MDS *size* k). Thus, we deal then with the finite alphabet $\Theta_k = \mathcal{M}_k \cup \overline{\mathcal{M}_k}$, where $\mathcal{M}_k = \{M_{i,j} \mid i,j \in \mathbb{N}^+ \text{ and } i \leqslant j \leqslant k\}$. To simplify the notation, we will assume that, unless clear otherwise, k is fixed for the sequel of this paper.

We generalize now the notion of the orthodox order of elements from $\mathcal{E}_k$ so that it includes also sequences containing elements of $\mathcal{C}_k$. Thus, a sequence $M_{i_1,j_1}, M_{i_2,j_2}, \ldots, M_{i_n,j_n}$, $n \leqslant k$, is *orthodox*, if $i_1 = 1$, $i_l = 1 + j_{l-1}$ for all $2 \leqslant l \leqslant n$ if $n \geqslant 2$, and $j_n = k$.

Let for $j \geqslant 1$, $\mathbf{com}_j$ be the operation defined on all finite sequences longer than j, such that the result of applying $\mathbf{com}_j$ to a sequence x is the commutation of the j-th and $(j+1)$-st element of x. We denote by COM the set of operations $\mathbf{com}_j$, for all $j \geqslant 1$.

Let for $j \geqslant 1$, $\mathbf{bar}_j$ be the operation defined on the set of all finite sequences over the alphabet Θ, longer than j, such that the result of applying $\mathbf{bar}_j$ to such a sequence x places the bar over the j-th element of x (here we assume that for each $X \in \Theta$, $\overline{\overline{X}} = X$). We denote by BAR the set of all operations $\mathbf{bar}_j$, for all $j \geqslant 1$.

We call then a sequence $X_1, \ldots, X_l$ over Θ_k a *real MDS structure*, if it is obtained from an orthodox sequence by a finite number of applications of operations fıom COM $\cup$ BAR.

As explained above, the tracing of pointers is crucial for the investigation of the assembly process. Here, the notion of a MDS descriptor turns out to be very handy. Intuitively speaking, given a sequence of MDSs corresponding to the micronuclear precursor of a gene or to an intermediate molecule, one denotes each MDS M by an ordered pair (p, q), and each $\overline{M}$ by an ordered pair $(\overline{q}, \overline{p})$, where (p, q) is the pair of pointers flanking M on its two ends; note that if an

MDS (elementary or composite) has the structure (p, μ, q), then its inversion has the structure $(\overline{q}, \overline{\mu}, \overline{p})$.

More formally, we define the mapping ψ on Θ_k by:

(i) $\psi(M_{1,k}) = (b, e)$, and $\psi(\overline{M}_{1,k}) = (\overline{e}, \overline{b})$,

(ii) $\psi(M_{1,i}) = (b, i+1)$, and $\psi(\overline{M}_{1,i}) = (\overline{i+1}, \overline{b})$, for all $1 \leqslant i < k$,

(iii) $\psi(M_{i,k}) = (i, e)$, and $\psi(\overline{M}_{i,k}) = (\overline{e}, \overline{i})$, for all $1 < i \leqslant k$,

(iv) $\psi(M_{i,j}) = (i, j+1)$, and $\psi(\overline{M}_{i,j}) = (\overline{j+1}, \overline{i})$, for all $1 < i \leqslant j < k$,

where b, e are reserved symbols (b stands for "beginning", and e for "end"). Then, for a sequence $X_1, \ldots, X_l$ over Θ_k,

$$\psi(X_1, \ldots, X_l) = \psi(X_1) \ldots \psi(X_l).$$

Thus, for example, for the sequence $x = M_{1,3}\overline{M}_{5,8}, M_9, M_{10,12}, \overline{M}_4$ and $k = 12$,

$$\psi(x) = (b, 4)(\overline{9}, \overline{5})(9, 10)(10, e), (\overline{5}, \overline{4}).$$

Let $\Pi_{ex,k} = \Delta_k \cup \overline{\Delta}_k \cup \{b, e, \overline{b}, \overline{e}\}$, where $\Delta_k = \{2, \ldots, k-1\}$ and $\overline{\Delta}_k = \{\overline{2}, \ldots, \overline{k-1}\}$, and let $\Pi^2_{ex,k}$ be the set of all ordered pairs over $\Pi_{ex,k}$. Then Γ_k is the subset of $\Pi^2_{ex,k}$ consisting of the following ordered pairs:

- (b, e), $(\overline{e}, \overline{b})$,
- (b, i), $(\overline{i}, \overline{b})$, for all $2 \leqslant i \leqslant k-1$,
- (i, e), $(\overline{e}, \overline{i})$, for all $2 \leqslant i \leqslant k-1$,
- (i, j), $(\overline{j}, \overline{i})$, for all $2 \leqslant i < j \leqslant k-1$.

A word over Γ_k is called an MDS *descriptor*.

Clearly, translating a sequence over Θ_k by ψ, we obtain a word over Γ_k. Then, a word δ over Γ_k is called a *realistic* MDS *descriptor* if $\delta = \psi(x)$, for some real MDS structure x (over Θ_k).

For two words u, v over an arbitrary alphabet Σ, we say that u is a *subword* of v if $v = v_1 u v_2$, for some words v_1, v_2 over Σ.

Recall that a sequence over Θ_k is a real MDS structure, if it is obtained from an orthodox sequence by a finite number of applications of operations from COM $\cup$ BAR. A word δ over Γ_k is *ordered* if $\delta = \psi(x)$, where x is an orthodox sequence over Θ_k - hence ordered words over Γ_k are translations by ψ of orthodox sequences over Θ_k. Thus, it is clear that a realistic MDS descriptor δ is ordered if and only if its first letter is of the form (b, i) for some i, its last letter is of the form (j, e) for some j, and for every subword $(x, y)(z, t)$ of δ, $y = z$.

Now, for each j, let $\mathbf{inv}_j$ be the operations defined on all words over Γ_k of length at least j, by:

$$\mathbf{inv}_j\left((x_1, y_1) \ldots (x_l, y_l)\right) \quad = \quad (x_1, y_1) \ldots (x_{j-1}, y_{j-1})(\overline{y}_j, \overline{x}_j)$$

$$(x_{j+1}, y_{j+1}) \ldots (x_l, y_l).$$

We denote by INV the set of all operations $\mathbf{inv}_j$, for $j \geqslant 1$. Clearly, the set of INV operations on words over Γ_k corresponds to the set of BAR operations on sequences over Θ_k. It is then straightforward to prove that an MDS descriptor is realistic if and only if it can be obtained from an ordered word over Γ_k by a finite number of applications of operations from COM $\cup$ INV.

Our basic molecule patterns: direct repeat, inverted repeat, and alternating direct repeat, are translated into the corresponding patterns of MDS descriptors in the obvious way (see [2]). Also, the *ld, hi,* and *dlad* operations on molecules (and their simple versions) are translated to $\vdash$, **hi**, and **dlad** operations on MDS descriptors (and their simple versions: **sld**, **shi**, and **sdlad**) in the obvious way. Thus, for example, for an MDS descriptor

$$\delta = (5,6)(\overline{4},\overline{3})(\overline{3},\overline{2})(6,7)(4,5)(b,2)(7,e),$$

we have:

$$\mathbf{sld}_{\overline{3}}(\delta) = (5,6)(\overline{4},\overline{2})(6,7)(4,5)(b,2)(7,e),$$

where $\overline{3}$ is the pointer involved in this application of **sld**,

$$\mathbf{hi}_2(\delta) = (5,6)(\overline{4},\overline{3})(\overline{3},\overline{b})(\overline{5},\overline{4})(\overline{7},\overline{6})(7,e),$$

where 2 is the pointer involved in this application of **hi**, and

$$\mathbf{dlad}_{5,7}(\delta) = (b,2)(4,6)(\overline{4},\overline{3})(\overline{3},\overline{2})(6,e),$$

where 5 and 7 are the pointers involved in this application of **dlad**.

In order to ensure that the use of realistic MDS descriptors as a framework for gene assembly is sound, one has to prove that the class of realistic MDS descriptors is closed with respect to the $\vdash$, **hi**, and **dlad** operations.

We will give the proof for the $\vdash$ and **hi** operations - the proof for **dlad** is completely analogous, but notationally more involved because one deals there with two pointer sets rather than one only. We will usc $\mathcal{F}$ to denote the class of all compositions of functions from COM $\cup$ INV.

Let δ_1 be a realistic MDS descriptor, and let **op** be an operation involving the pointer set $\{j, \overline{j}\}$, which is applicable to δ_1 - hence, $\mathbf{op}(\delta_1)$ is a word over Γ_k, with no occurrence from the pointer set $\{j, \overline{j}\}$.

Since δ_1 is a realistic MDS descriptor, there exists an ordered MDS descriptor ρ_1 and an $f \in \mathcal{F}$ such that $\delta_1 = f(\rho_1)$.

Since **op** involves the pointer set $\{j, \overline{j}\}$, $\rho_1 = \eta(i,j)(j,k)\zeta$ for some sequences η, ζ over Θ_k, and $\delta_1 = \alpha(x,y)\beta(u,v)\gamma$, for some $\alpha, \beta, \gamma \in \Gamma_k^*$ such that $f(\eta\zeta) = \alpha\beta\gamma$, and one of (x,y), (u,v) is in $\{(i,j),\ (\overline{j},\overline{i})\}$, while the other one is in $\{(j,k),\ (\overline{k},\overline{j})\}$.

We assume that the operation **op** satisfies the following condition W:

$\mathbf{op}(\delta_1) = \delta_2$, *where, for some* $f' \in \mathcal{F}$, δ_2 *results from* δ_1 *by inserting a pair* $(r,s) \in \{(i,k),\ (\bar{k},\bar{i})\}$ *somewhere in* $f'(\alpha\beta\gamma)$.

Let then $\rho_2 = \eta(i,k)\zeta$. Since ρ_1 is ordered, also ρ_2 is ordered. But $f'(\alpha\beta\gamma) = f'(f(\eta\zeta))$, where $f'f \in \mathcal{F}$, and so, because δ_2 results from δ_1 by inserting (r,s) somewhere in $f'f(\eta\zeta)$, $\delta_2 = g(\rho_2)$, for some $g \in \mathcal{F}$. Consequently, δ_2 is also realistic.

Since the operations $\vdash$ and **hi** satisfy the condition W given above, the class of realistic MDS descriptors is closed with respect to these operations.

4. THE ASSEMBLING POWER OF OPERATIONS

As illustrated in the last section, the application of the $\vdash$, **hi**, and **dlad** operations shortens the involved MDS descriptor. Hence, a successful assembly leads to an MDS descriptor without pointers. More formally, an MDS descriptor is *assembled* if it is of the form (b,e), or of the form $(\bar{e},\bar{b})$.

Let δ be a realistic MDS descriptor. For a subset S of $\{\vdash, \mathbf{hi}, \mathbf{dlad}\}$, and a sequence $\rho_1,\ldots,\rho_l$ of elements from S, we say that $\rho_1,\ldots,\rho_l$ is a *successful strategy* for δ, *based on* S, if $\rho_l \ldots \rho_1(\delta)$ is an assembled MDS descriptor. As a matter of fact, the corresponding assembled gene may end up on a circular or on a linear molecule – this is discussed in detail in [4].

We prove now that the set of our three operations on MDS descriptors, $\vdash$, **hi**, and **dlad**, is universal, i.e., any realistic MDS descriptor δ has a successful strategy based on this set of operations. Moreover, all operations are needed for the universality. First, however, we need two definitions.

For a realistic MDS descriptor $\delta = (x_1,x_2)\ldots(x_{l-1},x_l)$, and a pointer $p \in \Pi$, let x_i, x_j, with $1 \leqslant i < j \leqslant l$, be such that $\{x_i, x_j\} \subseteq \{p, \bar{p}\}$. Then, the *p-interval* of δ is the set of pointers $\{x_i, x_{i+1}, \ldots, x_j\}$. If the pointer p occurs twice in δ (i.e., if $x_i = x_j = p$), then we say that p is *negative*, and if both p and $\bar{p}$ occur in δ (i.e., if $\{x_i, x_j\} = \{p, \bar{p}\}$), then p is *positive* (in δ).

Theorem 29.1 *Any realistic MDS descriptor has a successful strategy based on the operations* $\{\vdash, \mathbf{hi}, \mathbf{dlad}\}$.

Proof. Recall that if the $\vdash$ operation is used within a successful strategy, then it must be simple or boundary.

We prove first that one of the three operations is always applicable to any realistic MDS descriptor. Let δ then be such a descriptor, and assume that neither $\vdash$, nor **hi** is applicable.

Since **hi** is not applicable, all pointers in δ are negative. Since $\vdash$ is not applicable, δ has no simple direct repeat pattern. Consequently, if a pointer

$p \in \Pi$ appears in δ, then the p-interval of δ must contain at least one other pointer.

Let p then be a pointer occurring in δ such that the number of pointers within the p-interval is minimal (no other pointer from δ has less pointers in its interval). Let q be a pointer that has an occurrence within the p-interval. Since all pointers in δ are negative, δ contains two occurrences of q. The other occurrence of q must be outside the p-interval, as otherwise the minimality of p is contradicted. Thus, **dlad** is applicable to δ either on the pattern (p, q, p, q), or on the pattern (q, p, q, p).

Since applying any of our three operations decreases the number of pointers, and applying any of them to a realistic MDS descriptor yields a realistic MDS descriptor again, the theorem holds. □

To ensure the existence of a successful strategy for all realistic MDS descriptors, one needs all three operations. In other words, for any proper subset S of $\{\vdash, \mathbf{hi}, \mathbf{dlad}\}$, there exist MDS descriptors which have no successful strategy based on S. Thus, for example, the MDS descriptor $(b, 2)(2, e)$ has no successful strategy if $\vdash$ is not in S, the MDS descriptor $(\overline{2}, \overline{b})(2, e)$ has no successful strategy if $\mathbf{hi} \notin S$, and $(b, 2)(3, e)(2, 3)$ has no successful strategy if $\mathbf{dlad} \notin S$.

When one of our three operations is applied to an MDS descriptor δ, the region of δ delimited by the pointers that are removed by the given application, may be of an arbitrary complexity, e.g., it can contain an arbitrary number of IESs. Simple operations were introduced in [3] in order to avoid such an arbitrary complexity. Thus, an application of either the **sld** or **shi** operations goes "across" one IES only, while an application of a **sdlad** exchanges (through recombination) one IES for one MDS. In this way, simple operations are easier to investigate. The price we pay is that the set of simple operations is no longer universal any more:

Theorem 29.2 *There exist realistic MDS descriptors that have no successful strategy based on the operations* $\{\mathbf{sld}, \mathbf{shi}, \mathbf{sdlad}\}$.

Proof. Consider the following realistic MDS descriptor

$$\delta = (b, 2)(4, 5)(7, 8)(2, 3)(5, 6)(8, 9)(3, 4)(6, 7)(9, e).$$

We prove that none of the three operations, **sld**, **shi**, **sdlad**, is applicable to δ.

Since δ does not contain a simple direct repeat pattern, **sld** is not applicable to δ.

Since for no pointer p from Π, δ contains an occurrence of $\overline{p}$, **shi** is not applicable to δ.

Since δ does not contain a simple alternating direct repeat pattern, **sdlad** is not applicable to δ. □

5. DISCUSSION

We have described in this paper the process of gene assembly in ciliates from the perspective of the model introduced in [4] and [3] (and elaborated in [2]). In particular, we have discussed the use of MDS descriptors for capturing the MDS structure of the micronuclear precursor and of any intermediate forms of the macromolecular gene. We proved that the set $\{\vdash, \mathbf{hi}, \mathbf{dlad}\}$ of operations underlying our model is universal, i.e., these operations can assemble any micronuclear MDS structure into a macronuclear gene. On the other hand, the set $\{\mathbf{sld}, \mathbf{shi}, \mathbf{sdlad}\}$ of the simple versions of these operations is not universal. This fact is especially interesting because it has been shown in [3] that all known (experimentally determined) MDS patterns of micronuclear genes can be assembled into their macronuclear form using *only* simple operations.

Acknowledgments

This work is supported by NIGMS Research Grant #GM 56161 to David M. Prescott. Ion Petre and Grzegorz Rozenberg gratefully acknowledge support by TMR Network GETGRATS.

References

[1] Adleman, L. (1994), Molecular computation of solutions to combinatorial problems, *Science*, 266: 1021–1024.

[2] Ehrenfeucht, A.; I. Petre, D.M. Prescott & G. Rozenberg (2000), String and graph reduction systems for gene assembly in ciliates, submitted.

[3] Ehrenfeucht, A.; D.M. Prescott & G. Rozenberg (1999), A formal model of DNA processing in Hypotrichous ciliates, submitted.

[4] Ehrenfeucht, A.; D.M. Prescott & G. Rozenberg (2000), Computational aspects of gene (un)scrambling in ciliates. In L. Landweber & E. Winfree, eds., *Evolution as Computation*: Springer, Berlin.

[5] Landweber, L.F. & L. Kari (1998), The evolution of cellular computing: nature's solution to a computational problem, in *Proceedings of the 4th DIMACS Meeting on DNA Based Computers*: 3–15, Philadelphia.

[6] Landweber, L.F. & L. Kari (2000), Universal molecular computation in ciliates, in L. Landweber & E. Winfree, eds., *Evolution as Computation*, Springer, Berlin.

[7] Păun, Gh.; G. Rozenberg & A. Salomaa (1998), *DNA Computing*, Springer, Berlin.

[8] Prescott, D.M. (1992), Cutting, Splicing, Reordering, and Elimination of DNA Sequences in Hypotrichous Ciliates, *BioEssays*, 14.5: 317–324.

[9] Prescott, D.M. (1992), The unusual organization and processing of genomic DNA in hypotrichous ciliates. *Trends in Genetics*, 8: 439–445.

Chapter 30

SEMI-SIMPLE SPLICING SYSTEMS

Elizabeth Goode
Department of Computer and Information Sciences
University of Delaware
Newark, DE 19706, USA
goode@mail.eecis.udel.edu

Dennis Pixton*
Department of Mathematics
Binghamton University
Binghamton, NY 13902-6000, USA
dennis@math.binghamton.edu

Abstract Generalizing a notion introduced by Păun and his coworkers, we introduce semi-simple splicing systems, in which all splicing rules have the form $(a, 1; b, 1)$ where a and b are single symbols. We find a simple graph representation of these systems, and from this representation we show that semi-simple splicing languages are reflexive splicing languages, that they contain constants, and that they are, in fact, strictly locally testable.

1. INTRODUCTION

In 1987 Tom Head [5] introduced the idea of *splicing* as a generative mechanism in formal language theory. This mechanism was popularized and extensively developed by Gheorghe Păun and his many coworkers, who recognized much of the subtlety and power of this mechanism and used it as a foundation for the rapidly developing study of DNA based computing. See, for example, [11] and [12]. (The most necessary definitions are reviewed briefly in section 2.)

*Research partially supported by DARPA/NSF CCR-9725021.

C. Martin-Vide and V. Mitrana (eds.), Where Mathematics, Computer Science, Linguistics and Biology Meet, 343-352.

However, one of Head's most basic questions is still unanswered. It was established early on that a finite splicing system would always produce a regular splicing language ([1], [13]), but not all regular languages can be generated in this way (a simple example is $(aa)^*$). Thus there remains the problem of precisely characterizing such splicing languages.

One of the first papers by Păun and his coworkers was [9], which introduced the notion of a *simple splicing system*. It is our goal here to give a simple description of the languages that result from a generalization of this notion, which we call a *semi-simple splicing system*. We are not interested in generalizing the results of [9], although our methods give simpler approaches even in the simple splicing case. Rather, we are interested in illustrating some approaches towards characterizing splicing languages in general.

Our main result is a characterization of semi-simple splicing languages in terms of certain directed graphs; this is carried out in section 3.. Using this, we verify a conjecture for all splicing languages in this special case: All semi-simple splicing languages must have a *constant* word. Constant words have been used by Head in recent constructions of splicing systems, and in fact our result shows that most factors of a semi-simple splicing language are constant, from which we derive two further conclusions using Head's results: Semi-simple splicing languages are *strictly locally testable*, and they can be generated by *reflexive* splicing systems.

It was only recently shown that there are splicing languages that are not reflexive. It is still unknown whether there are splicing languages that have no constants. Some of these results are in the first author's Ph.D. thesis [3]; related results will appear in [4].

2. REVIEW OF SPLICING THEORY

We work over a finite alphabet A. A *splicing rule over* A^* is a quadruple $r = (u, v; u', v')$ of strings in A^*. If x and y are strings in A^* then we say that the string z is a result of *splicing* x *and* y *using the rule* r if we can factor $x = x_1uvx_2$, $y = y_1u'v'y_2$ and $z = x_1uv'y_2$. If $I \subset A^*$ we write $r(I)$ for the set of all words z that can be obtained this way, with x and y in I.

A *splicing scheme* or *H scheme* is a pair $\sigma = (A, R)$ where A is the alphabet and R is a set of rules over A^*. We say σ is finite if R is a finite set.

Given a language I in A^* we define $\sigma(I)$ to be the union of all $r(I)$, where $r \in R$. We define *iterated splicing* as follows:

$$\sigma^0(I) = I, \quad \sigma^{k+1}(I) = \sigma(\sigma^k(I)) \cup \sigma^k(I), \quad \sigma^*(I) = \bigcup_{k=0}^{\infty} \sigma^k(I).$$

By a *splicing system* or *H system* we mean a pair (σ, I) where σ is a splicing scheme over A^* and $I \subset A^*$.

In this paper we shall require that both σ and I be **finite**. This is the original definition due to Head, but there have been many generalizations, which we shall not consider in this paper. We say L is a *splicing language* if $L = \sigma^*(I)$ for some finite splicing system (σ, I).

Following [9] we define a *simple* splicing scheme to be one in which all rules have the form $(a,\ 1\,;\ a,\ 1)$ for some $a \in A$ (we use 1 to refer to the empty string). We generalize this by defining a *semi-simple* splicing scheme to be one in which all rules have the form $(a,\ 1\,;\ b,\ 1)$ for $a,\ b \in A$.

We say a language L is a *simple splicing language* if $L = \sigma^*(I)$ for a finite simple splicing system (σ, I). Semi-simple and other varieties of splicing languages are defined similarly.

At the end of the paper we present an example of a semi-simple splicing language which is not a simple splicing language.

3. ARROW GRAPHS

We begin with a language $L \subset A^*$. We define $R(L)$ to be the set of all semi-simple rules $r = (a,\ 1\,;\ b,\ 1)$, for $a, b \in A$, such that $r(L) \subset L$. Let $\sigma = (A, R)$ be a splicing scheme for which $R \subset R(L)$. We shall consider the possibility that $L = \sigma^*(I)$ for some finite initial set I.

We augment A, L, σ with new symbols S and T not in A as follows: $\bar{A} = A \cup \{S, T\}$, $\bar{R} = R \cup \{(S,\ 1\,;\ S,\ 1),\ (T,\ 1\,;\ T,\ 1)\}$, $\bar{\sigma} = (\bar{A},\ \bar{R})$, and $\bar{L} = SLT$. This makes no significant difference:

Claim 30.1 *$\bar{\sigma}(\bar{L}) \subset \bar{L}$. Moreover, if $I \subset A^*$ then $\bar{\sigma}^*(SIT) = S\sigma^*(I)T$.*

Proof. This is immediate, since the extra rules involving S and T essentially do nothing: If x, y are in XA^*Y then splicing x, y using $(S,\ 1\,;\ S,\ 1)$ produces y, and splicing x, y using $(T,\ 1\,;\ T,\ 1)$ produces x. □

Now we develop a graphic representation for $\bar{L}$.

An *arrow* shall mean a triple $e = (a, w, a')$ in $\bar{A} \times A^* \times \bar{A}$. We normally write an arrow as $e = a \xrightarrow{w} a'$. Given such an arrow, we say a is the *initial vertex* of e and a' is the *terminal vertex* of e.

We define the *arrow graph* G of the pair $(\sigma,\ L)$ to be the following directed graph: The vertex set of G is $\bar{A}$, and the arrow $a \xrightarrow{w} a'$ is an *edge* of G from a to a' if there is a rule $(a,\ 1\,;\ b,\ 1)$ in $\bar{R}$ so that bwa' is a factor of $\bar{L}$. (A *factor* of a language M is a string y such that, for some strings x and z, $xyz \in M$.)

Note that G is an infinite graph, if L is infinite:

Claim 30.2 *$S \xrightarrow{w} T$ is an edge of G if and only if $SwT \in \bar{L}$.*

Proof. If $SwT \in \bar{L}$ then SwT is a factor of $\bar{L}$ and $(S,\ 1\,;\ S,\ 1)$ is in $\bar{R}$, so $S \xrightarrow{w} T$ is an edge. Conversely, if $S \xrightarrow{w} T$ is an edge then there is a rule

$(S, 1; b, 1)$ in $\bar{R}$ and bwT is a factor of $\bar{L}$. The only rule in $\bar{R}$ involving S is $(S, 1; S, 1)$, so $b = S$ and SwT is a factor of $\bar{L}$. Since $\bar{L} \subset SA^*T$ we conclude that $SwT \in \bar{L}$. □

We now introduce some algebraic structure on the arrow graph. If e_1 and e_2 are arrows then we say e_1 and e_2 are *adjacent* if the terminal vertex of e_1 is the initial vertex of e_2. If $e_1 = a_0 \xrightarrow{w_1} a_1$ and $e_2 = a_1 \xrightarrow{w_2} a_2$ are adjacent we define their *product*, written $e_1 \cdot e_2$, to be the arrow $a_0 \xrightarrow{w_1 a_1 w_2} a_2$.

Claim 30.3 (Closure) *If e_1, e_2 is an adjacent pair of edges of G, then $e_1 \cdot e_2$ is an edge of G.*

Proof. We write $e_1 = a_0 \xrightarrow{w_1} a_1$ and $e_2 = a_1 \xrightarrow{w_2} a_2$. Then there are rules $r_0 = (a_0, 1; b_0, 1)$ and $r_1 = (a_1, 1; b_1, 1)$ in $\bar{R}$ and words $x_0 b_0 w_1 a_1 y_1$ and $x_1 b_1 w_2 a_2 y_2$ in $\bar{L}$.

Using r_1 we can splice these words to produce $x_0 b_0 w_1 a_1 w_2 a_2 y_2$ in $\bar{L}$. Now $b_0 w_1 a_1 w_2 a_2$ is a factor of $\bar{L}$ and r_0 is a rule in $\bar{R}$, so $e_1 \cdot e_2 = a_0 \xrightarrow{w_1 a_1 w_2} a_2$ is an edge of G. □

As usual, a *path* in G is a sequence $\pi = \langle e_1, e_2, \ldots, e_n \rangle$ of edges so that each pair e_k, e_{k+1} is adjacent. In this case we write $e_k = a_{k-1} \xrightarrow{w_k} a_k$ and we say π is a path from a_0 to a_n. We use a nonstandard definition for the label of such a path:

$$\lambda(\pi) = a_0 w_1 a_1 w_2 \ldots a_{n-1} w_n a_n.$$

Of course we identify a single edge e with the path $\langle e \rangle$, so $\lambda(e)$ is also defined.

On the other hand, the multiplication on edges is obviously associative, so we can define unambiguously the product $e_1 \cdot e_2 \cdot \ldots \cdot e_n$ of the edges in a path π. The following is immediate from 30.3 and the definitions:

Claim 30.4 *If $\pi = \langle e_1, e_2, \ldots, e_n \rangle$ is a path in G then $e = e_1 \cdot e_2 \cdot \ldots \cdot e_n$ is an edge in G and $\lambda(e) = \lambda(\pi)$.*

Finally, we define the language of the graph G, written $L(G)$, to be the set of all labels of paths in G from S to T.

Claim 30.5 $L(G) = \bar{L}$.

Proof. According to 30.2, $\bar{L} \subset L(G)$. On the other hand, if π is a path in G from S to T then 30.4 produces an edge from S to T with the same label, so 30.2 also gives $L(G) \subset \bar{L}$. □

Now our goal is to replace G with a finite subgraph which still represents $\bar{L}$. For this we require a way to factor long edges. First we have a very simple partial factorization:

Lemma 30.1 (Prefix edge) *If $a \xrightarrow{ua'v} a''$ is an edge of G and $a' \in A$ then $a \xrightarrow{u} a'$ is an edge of G.*

Proof. This is trivial: Since $a \xrightarrow{ua'v} a''$ is an edge there are a rule $r = (a, 1; b, 1)$ in $\bar{R}$ and a word $z = xbua'va''y$ in $\bar{L}$. The rule r and the factor bua' of z establish that $a \xrightarrow{u} a'$ is an edge of G. □

We define an edge to be a *prime* edge if it cannot be written as the product of two edges. The following is proved by a standard induction on the length of the label of an edge:

Claim 30.6 *Every edge is the product of a sequence of prime edges.*

This factorization is unique. This is not necessary for our paper, so we omit the proof, but this uniqueness justifies our use of the term *prime edge* instead of *irreducible edge*.

We define the *prime arrow graph* G_0 to be the subgraph of G with the same vertex set but using only the prime edges of G. As an immediate consequence of 30.5 and 30.6 we have:

Claim 30.7 $L(G_0) = \bar{L}$.

The next result is the key observation about prime edges.

Claim 30.8 *Suppose $L = \sigma^*(I)$ for some subset $I \subset L$. If $a_0 \xrightarrow{w} a_1$ is a prime edge in G then w is a factor of I.*

Proof. For any edge $e = a_0 \xrightarrow{w} a_1$ there is a rule $(a_0, 1; b_0, 1)$ so that b_0wa_1 is a factor of $\bar{L} = \bar{\sigma}^*(\bar{I})$. Thus we can define an index $N(e)$ to be the minimal n such that, for some such rule, b_0wa_1 is a factor of $\bar{\sigma}^n(\bar{I})$. We shall prove the claim by induction on $N(e)$.

The claim is obviously true for all prime edges e with $N(e) = 0$. So let $e = a_0 \xrightarrow{w} a_1$ be a prime edge with $n = N(e) > 0$ so that the claim holds for all prime edges e' with $N(e') < n$. Then there is a rule $r_0 = (a_0, 1; b_0, 1)$ and a string z_0 in $\bar{\sigma}^n(\bar{I})$ which can be factored as $z_0 = x_0b_0wa_1y_1$. Since $n > 0$ there is a rule $r = (a, 1; b, 1)$ and strings $z = xay'$ and $z' = x'by$ in $\bar{\sigma}^{n-1}(\bar{I})$ so that z_0 is the result of splicing z and z' using r. That is,

$$xay = x_0b_0wa_1y_1.$$

There are four cases, depending on where the "a" appears in z_0:

Case 1: $|xa| \geqslant |x_0b_0wa_1|$: Then b_0wa_1 is a factor of xa, and hence of $z = xay'$, contradicting $N(e) = n$.

Case 2: $|x_0b_0| < |xa| < |x_0b_0wa_1|$: Then $w = uav$, so we have the prefix edge $e_1 = a_0 \xrightarrow{u} a$. Moreover, $y = va_1y_1$ so bva_1 is a factor of z'. Using this and the rule b we conclude that $e_2 = a \xrightarrow{v} a_1$ is an edge. Then the factorization $e = e_1 \cdot e_2$ contradicts primality.

Case 3: $|xa| = |x_0b_0|$: Then $y = wa_1y_1$. Thus $z' = x'by = x'bwa_1y_1$ so, using the rule r, we conclude that $e' = a \xrightarrow{w} a_1$ is an edge. Moreover, it is a prime edge. For, if not, then there are edges e_1 and e_2 with $e' = e_1 \cdot e_2$. We write $e_1 = a \xrightarrow{u} a'$ and $e_2 = a' \xrightarrow{v} a_1$, so $w = ua'v$. Then $e_1' = a_0 \xrightarrow{u} a'$ is a prefix edge of e, so $e = e_1' \cdot e_2$, contradicting primality of e.

Since $N(e') \leqslant n - 1$ the claim is true for e', and so w is a factor of I.

Case 4: $|xa| < |x_0b_0|$: Then b_0wa_1 is a factor of y, and hence of $z' = x'by$, contradicting $N(e) = n$. □

The following is our main result; it characterizes semi-simple splicing languages in terms of arrow graphs.

Theorem 30.1 *Suppose L is a language in A^*, R is a subset of $R(L)$, $\sigma = (A, R)$ and G_0 is the prime arrow graph for (σ, L). Then there is a finite $I \subset A^*$ such that $L = \sigma^*(I)$ if and only if G_0 is finite.*

Proof. It is immediate from 30.8 that G_0 is finite if I is finite.

Conversely, suppose that G_0 is finite. For each prime edge $e = a_0 \xrightarrow{w} a_1$ and each rule $(a_0, 1\,;\, b_0, 1)$ in $\bar{R}$ select, if possible, one word in $\bar{L}$ which has b_0wa_1 as a factor. Let $\bar{I} = SIT$ be the set of these strings. We claim that $\bar{L} = \bar{\sigma}^*(\bar{I})$, and hence $L = \sigma^*(I)$.

To see this, take any edge e in G from S to T, and let $e = e_1 \cdot e_2 \cdot \ldots \cdot e_n$ be its prime factorization, with $e_k = a_{k-1} \xrightarrow{w_k} a_k$. For $1 \leqslant k \leqslant n$ select a rule $r_k = (a_{k-1}, 1\,;\, b_{k-1}, 1)$ and a word z_k in $\bar{I}$ with $b_{k-1}w_ka_k$ as a factor. Note that $a_0 = b_0 = S$, so w_1 is a prefix of z_1. Similarly, w_n is a suffix of z_n. It is then clear that $z_1, z_2, \ldots, z_n$ can be spliced, using the rules $r_2, \ldots, r_n$ in this order, to produce $\lambda(e)$. □

Corollary 30.1 *A language $L \subset A^*$ is a semi-simple splicing language if and only if the prime arrow graph constructed from $(\hat{\sigma}, L)$, with $\hat{\sigma} = (A, R(L))$, is finite.*

Proof. Just notice that if $L = \sigma^*(I)$ for any $\sigma = (A, R)$ with $R \subset R(L)$ then $L = \hat{\sigma}^*(I)$. □

This corollary provides an algorithm for determining whether a regular language L is a semi-simple splicing language. First, to check whether $r = (a, 1\,;\, b, 1)$ is in $R(L)$ we have to decide whether $r(L) \subset L$, and this is algorithmically feasible since $r(L)$ is a regular language if L is regular; specifically, in terms of the quotient operation, $r(L) = (L(aA^*)^{-1})a((A^*b)^{-1}L)$. Further, if a, a' are in $\bar{A}$ we define $E(a, a')$ to be the set of w such that $a \xrightarrow{w} a'$ is an edge of G. This is a regular language, since it can be written as the union, over b for which $(a, 1\,;\, b, 1) \in R(L)$, of $(A^*b)^{-1}L(aA^*)^{-1}$. Moreover, the set $E_0(a, a')$, defined as above but for edges of G_0, is also regular, since it may be written as $E(a, a')$ minus the union of the products $E(a, a'')a''E(a'', a')$, and so it is algorithmically decidable whether $E_0(a, a')$ is finite.

For the applications in the next section we require the following:

Lemma 30.2 (Approximate factorization) *If G_0 is finite then there is an integer K with the following property: Suppose $e = a_0 \xrightarrow{xyz} a_1$ is an edge of G and $|y| \geqslant K$. Then there is a factorization $e = e' \cdot e''$ so that $e' = a_0 \xrightarrow{xu} a$ and $e'' = a \xrightarrow{vz} a_1$ (and hence $y = uav$).*

Proof. Let K_0 be the maximum length of the label of an edge of G_0, and set $K = K_0 - 1$.

Take e as in the statement of the lemma and let $e_1 \cdot e_2 \cdot \ldots \cdot e_n$ be its prime factorization. Let k be the last index for which $|\lambda(e_1 \cdot e_2 \cdot \ldots \cdot e_k)| \leqslant |a_0x|$. By the definition of K, $|\lambda(e_1 \cdot e_2 \cdot \ldots \cdot e_{k+1})| \leqslant |a_0xy|$. Then $e' = e_1 \cdot e_2 \cdot \ldots \cdot e_{k+1}$ and $e'' = e_{k+2} \cdot \ldots \cdot e_n$ have the desired properties. □

4. CONSTANTS AND REFLEXIVITY

Schützenberger [14] defined a *constant* of a language L to be a string $c \in A^*$ so that, for all strings x, y, x', y', we have xcy' in L if xcy and $x'cy'$ are in L. This has an obvious relation to splicing: c is a constant of L if and only if $r(L) \subset L$, where $r = (c, 1\,;\, c, 1)$. For simple splicing systems we can interpret the graph characterization of section 3. in terms of constants:

Theorem 30.2 *A language L in A^* is a simple splicing language if and only if there is K so that every factor of L of length at least K contains a symbol which is a constant of L.*

Proof. Let M be the set of all symbols in A which are constants of L, let $\tilde{R}$ be the set of all rules $(a, 1\,;\, a, 1)$ with $a \in M$, and let $\tilde{\sigma} = (A, \tilde{R})$. Then clearly $L = \sigma^*(I)$ for some simple splicing scheme σ if and only if $L = \tilde{\sigma}^*(I)$. Hence, by Theorem 30.1, L is a simple splicing language if and only if the prime arrow graph G_0 constructed from $(L, \tilde{\sigma})$ is finite. However, an arrow $a \xrightarrow{w} a'$ is an edge if and only if awa' is a factor of $\bar{L}$ and $a = S$ or $a \in M$. Hence an edge $a \xrightarrow{w} a'$ is a prime edge if and only if w does not contain any elements of M. Thus G_0 is finite if and only if there is a bound, K_1, on the lengths of factors of L that do not contain symbols in M, and the result follows with $K = K_1 + 1$. □

It is unknown whether every splicing language must have a constant; there are regular languages which have no constants. In the semi-simple case there are many constants:

Theorem 30.3 *If L is a semi-simple splicing language then there is a constant K so that every string in A^* of length at least K is a constant of L.*

Proof. Let K be the number given by approximate factorization (30.2). Suppose that $c \in A^*$ has length at least K and suppose xcy and $x'cy'$ are in L. Then approximate factorization applied to $e = S \xrightarrow{x'cy'} T$ gives $e = e' \cdot e''$, where $e'' = a \xrightarrow{d'y'} T$ and $c = dad'$. Now $S \xrightarrow{xcy} T$ is an edge and $xcy = xdad'y$ so we have the the prefix edge $\bar{e} = S \xrightarrow{xd} a$ (from 30.1). Hence $\bar{e} \cdot e'' = S \xrightarrow{xdad'y'} T$ is an edge, so $xdad'y' = xcy'$ is in L, as desired. □

McNaughton and Papert [10] defined the notion of *strict local testability*, and de Luca and Restivo [2] translated this definition to the following: A language L is strictly locally testable if and only if there is K so that every string in A^* of length at least K is a constant. Thus, as in [8], we have:

Corollary 30.2 *If L is a semi-simple splicing language then L is strictly locally testable.*

A splicing scheme $\sigma = (A, R)$ is called *reflexive* if whenever a rule $(u,\ v\,;\ u',\ v')$ is in R then $(u,\ v\,;\ u,\ v)$ and $(u',\ v'\,;\ u',\ v')$ are also in R. This condition was introduced in [6] as a necessary condition for a splicing scheme to actually represent DNA recombination. It was recently discovered that there are splicing languages that cannot be generated by reflexive splicing systems. An example is the set of all words on $\{a, b\}^*$ that have at most two b's. See [3] and [4].

Head has shown ([7, 8]) how the presence of sufficiently many constants in a language leads to a splicing structure for the language. Precisely, he proves that if L is a regular language and there is a finite set F of constants of L so that $L \setminus A^*FA^*$ is finite then L is a reflexive splicing language. Thus:

Corollary 30.3 *If (σ, I) is a semi-simple splicing system then there is a reflexive splicing system $(\hat{\sigma}, \hat{I})$ such that $\sigma^*(I) = \hat{\sigma}^*(\hat{I})$.*

We conclude with a very simple example of a semi-simple splicing language, illustrating our main results.

Let $A = \{a, b, c\}$, let R contain the single rule $(a,\ 1\,;\ b,\ 1)$, and let $I = \{abccab\}$. Following the arguments of 30.8 we can readily determine the prime arrow graph G_0; we find the following set of prime edges: $S \xrightarrow{abccab} T$, $S \xrightarrow{1} a$, $S \xrightarrow{abcc} a$, $a \xrightarrow{cc} a$, $a \xrightarrow{1} T$, $a \xrightarrow{ccab} T$. (There are other prime edges, but we shall ignore them, since they do not appear on any paths from S to T.) From this determination of G_0 it is easy to see that the splicing language is $L = \sigma^*(I) = \{a\} \cup \{a, ab\}(cca)^+\{1, b\}$. The reader may check that any rule $r = (x, 1; x, 1)$, with $x \in A$, may be applied to two copies of *abccab* to produce a string which is not in L; hence L is not a simple splicing language.

From the proof of Theorem 30.3 we see that every factor of L of length at least 8 is a constant of L, and we can follow the arguments of [7] to produce a reflexive splicing system whose language is L. In fact, our example is simpler: The shortest constant of L is *cc*, and this constant appears in all elements of L except the string a. Moreover, L is generated by a reflexive splicing scheme with only one rule, $(cc, 1; cc, 1)$.

References

[1] Culik II, K. & T. Harju (1991), Splicing semigroups of dominoes and DNAk, *Discrete Applied Mathematics*, 31: 261–277.

[2] DeLuca, A. & A. Restivo (1980), A characterization of strictly locally testable languages and its application to subsemigroups of a free semigroup, *Information and Control*, 44: 300–319.

[3] Goode, E. (1999), *Constants and Splicing Systems*, PhD thesis, Binghamton University.

[4] Goode, E. & Pixton, D., Syntactic monoids, simultaneous pumping, and H systems, in preparation.

[5] Head, T. (1987), Formal language theory and DNA: an analysis of the generative capacity of specific recombinant behaviors, *Bulletin of Mathematical Biology*, 49.6: 737–759.

[6] Head, T. (1992), Splicing systems and DNA, in G. Rozenberg & A. Salomaa, eds., *Lindenmayer Systems: Impacts on Theoretical Computer Science, Computer Graphics and Developmental Biology*: 371–383. Springer, Berlin.

[7] Head, T. (1998), Splicing languages generated with one sided context, in Gh. Păun, ed., *Computing with Bio-molecules–Theory and Experiments*: 269–282. Springer, Singapore.

[8] Head, T. (1998), Splicing representations of strictly locally testable languages, *Discrete Applied Mathematics*, 87: 139–147.

[9] Mateescu, A.; Gh. Păun; G. Rozenberg & A. Salomaa (1996), Simple splicing systems, *Discrete Applied Mathematics*.

[10] McNaughton, R. & S. Papert (1971), *Counter-Free Automata*, MIT Press, Cambridge, Mass.

[11] Păun, Gh. (1995), On the power of the splicing operation, *International Journal of Computer Mathematics*, 59: 27–35.

[12] Păun, Gh.; G. Rozenberg & A. Salomaa (1996), Computing by splicing, *Theoretical Computer Science*, 168.2: 321–336.

[13] Pixton, D. (1996), Regularity of splicing languages, *Discrete Applied Mathematics*, 69.1–2: 99–122.

[14] Schützenberger, M.P. (1975), Sur certaines operations de fermeture dans les langages rationnels, *Symposia Math.*, 15: 245–253.

Chapter 31

WRITING BY METHYLATION PROPOSED FOR AQUEOUS COMPUTING

Tom Head
Mathematical Sciences
Binghamton University
Binghamton, NY 13902-6000, USA
tom@math.binghamton.edu

Abstract The general concept of aqueous computing calls for molecules to be used as tablets on which bits can be written by physical or chemical procedures. Recently DNA plasmids have been written on successfully by using restriction enzymes and a ligase. Reading has been done from a single gel separation. Here we suggest writing on DNA molecules using methylating enzymes. A provisional reading procedure is suggested that uses interleaved sequences of cuts and gel separations.

1. INTRODUCTION

The present work has developed in the context of a sequence of articles that includes [8] [4] [6] [11] and [5]. For a full understanding of the present short article it may be necessary to consult one of the last four articles in this list. The volume by Gh. Păun, G. Rozenberg and A. Salomaa [10] provides a broad theoretical context for molecular computing and Păun's book [9] collects many interesting related articles.

The use of DNA plasmids in aqueous computing has recently been illustrated with successful wet lab computations [7] [6] [11]. The fundamental concept of this approach is the use ofmolecules in a manner that is analogous to the use of memory registers in a conventional computer. At a predetermined set of locations on the molecule bits are 'written' and later 'read' using biochemical processes. Here we continue to suggest the use of double stranded DNA molecules as the memory registers. But now we suggest that writing be done with methylase enzymes and that reading be done with restriction enzymes. *A*

C. Martin-Vide and V. Mitrana (eds.), Where Mathematics, Computer Science, Linguistics and Biology Meet, 353-360.

remarkable feature of this proposal is that no ligation is used at all. In [6] and [11] the cut and paste operations used required that circular DNA molecules (plasmids) be used as the memory registers. The present suggestions allow the choice of either linear or circular DNA molecules. Linear molecules have been chosen for this exposition.

2. THE CONCEPTUAL FRAMEWORK OF AQUEOUS COMPUTING: EXPRESSED FOR DNA

A double stranded DNA molecule D is chosen that has a base pair sequence adequate to allow the choice of an appropriate pattern of nonoverlapping subsegments of D for which the sequence of base pairs occurring in each of these segments occurs nowhere else in the molecule. We call these subsegments the *stations* of the molecule. The stations are chosen so that at each station a specific modification can be made at that station that does not alter D at any other location. Each station, in its original state, is interpreted as a representation of either the bit 0 or the bit 1, as desired. Once a station has been modified it is regarded as representing the opposite bit. Several NP-complete algorithmic problems admit solutions by an essentially uniform procedure - even when the station modification method used alters bits irreversibly [4] [6] [11]. All computations begin with an aqueous solution containing a vast number of the original molecules. (By recalling Avogadro's number, one may note that a nanomole, i.e. a billionth of a gram molecular weight, of a compound provides approximately 602 trillion molecules.)

Aqueous computations fall naturally into two phases. In the first (writing) phase the modifications are carried out at the stations as appropriate to the problem instance being solved. Once the first phase is complete the solution of the problem is represented in the molecular content of the remaining aqueous solution. The second (reading) phase consists of recovering the problem solution from the aqueous solution.

3. NEW SUGGESTIONS FOR WRITING AND READING

Assume now that, as stations of D, a sequence of enzyme sites has been chosen at which D may be cut. Recall that, for each restriction enzyme, there is a companion enzyme, a methylase, that acts at the same site as the restriction enzyme and covalently joins a methyl group to one of the bases in the site, usually a cytosine. The methyl group is adjoined in such a way that the corresponding restriction enzyme will no longer cut at this site. Here we accept methylation as an *irreversible* process. If the method of computation suggested here proves to merit continued investigation, then other methods of blocking enzyme sites may be developed, perhaps using proteins that adhere *reversibly* at the desired sites. Methylation is the method suggested here for 'writing' on molecules.

Once the writing by methylation is complete, 'reading' is required to recover the problem solution. The reading process suggested here involves a sequence of cutting operations using the restriction enzymes appropriate to each station. A sequence of gel separations is interleaved with the cuts. It is hoped that the gel separations can be replaced soon by a more rapid and efficient technology. These suggested forms of writing and reading are concisely communicated in Sections 4 and 5 by treating in detail a specific instance of the classical Boolean satisfiability (SAT) problem [2].

4. DECIDING SATISFIABILITY: WRITING BY METHYLATION

Let p, q, r, s be propositional variables and let p', q', r', s', respectively, be their negations. Thus, for example, setting p =TRUE, q =FALSE, r =TRUE, s =FALSE requires p' =FALSE, q' =TRUE, r' =FALSE, s' =TRUE. We outline a solution for the following:

Problem. Does there exist an assignment of truth values (TRUE or FALSE) to the propositional variables p, q, r, s for which each of the following four disjunctive clauses evaluates to TRUE: $p+q', q+r'+s', p'+r+s', p'+q+r+s$?

Starting at one end of the molecule D, associate the eight stations with the eight literals in the following order: $p, p', q, q', r, r', s, s'$. The crucial point here is that the station associated with each propositional variable must be adjacent to the station associated with its negation. (For convenience in the reading phase, we also assume that the numbers of base pairs between the tip of D and the station p, and between successive stations of D, are small compared to the number of base pairs between the station s' and the end of D.) We use the eight literals as the names of the eight stations with which they are associated. The writing phase of a biochemical procedure for answering the stated instance of the satisfiability problem follows:

Begin the computation with a test tube T_0 containing an aqueous solution that contains a vast number of the original molecules D. Regard the initial, necessarily nonmethylated, condition of each of the eight stations as representing the bit 0. *Interpret each occurrence of the bit* 0 *to mean that a truth assignment has not been made for the associated literal. Do not interpret* 0 *to mean FALSE.* Divide the contents of T_0 equally between test tubes T_1 and T_2. To T_1 add the methylase appropriate to methylate the station p. To T_2 add the methylase appropriate to methylate the station q'. Return the DNA in T_1 and T_2 to a new test tube T_0 to form an aqueous solution containing these partially methylated

DNA molecules. Each DNA molecule in T_0 now has either station p or station q' methylated. A literal is regarded as having been assigned the truth value TRUE if and only if its associated station has been methylated, i.e. set to 1. Thus each DNA molecule in T_0 now satisfies the first of the four clauses, $p+q'$. This completes the first of four writing (methylation) steps and it is typical of each writing step. One more such step is given in similar detail:

Divide the contents of T_0 equally among tubes T_1, T_2, and T_3. To T_1, T_2 and T_3 add the methylases appropriate to methylate the stations q, r' and s', respectively. Return the DNA molecules in T_1, T_2, and T_3 to a new tube T_0. Each DNA molecule in T_0 now has: (1) at least one of the two stations p and q' methylated, and (2) at least one of the three stations q, r' and s' methylated. Thus each DNA molecule in T_0 now satisfies the first two of the four clauses, namely $p + q'$ and $q + r' + s'$. Notice, however, that some DNA molecules in T_0 now have both the station q and the station q' methylated! Such molecules express the logical contradiction that both q and q' are TRUE. The occurrence of such undesired molecules is typical in the computational architecture suggested here. The writing of contradictions is not resisted. Instead, the contradictions are eliminated in the reading process.

Each of the two remaining clauses requires an additional writing step. In the first of these, T_0 is divided among three tubes T_1, T_2, T_3 in which p', r, and s', respectively, are set to TRUE with their DNA returned to T_0. In the last writing step T_0 is divided among four tubes T_1, T_2, T_3, T_4 in which p', q, r, and s, respectively, are set to TRUE with their DNA returned to T_0. These four steps, one for each clause, complete the writing phase of the computation. Each DNA molecule in T_0 now satisfies each of the four clauses - but among these molecules are those that express contradictions. Notice that any molecule in T_0 that has at least one literal in each of the four sets $\{p, p'\}$, $\{q, q'\}$, $\{r, r'\}$, and $\{s, s'\}$ that has NOT been set to TRUE (i.e., has not been methylated) provides at least one valid truth assignment for the variables $\{p, q, r, s\}$ that satisfies the four clauses. (When q' has been set to TRUE then q must be set to FALSE. However, if neither q nor q' has been set to TRUE, then q may be set to either TRUE or FALSE.) Consequently, a set of clauses is satisfiable if and only if after the writing phase has been completed - and all molecules expressing contradictions have been removed - residues of D still remain. Molecules that express contradictions are removed in the reading process.

5. DECIDING SATISFIABILITY: READING BY RESTRICTION AND SEPARATION

Recall that the writing phase terminated with the test tube T_0 containing molecules that satisfy all the clauses - although some certainly represent contradictions. To T_0 add the restriction enzyme that cuts at nonmethylated stations

p. Segments of distinct lengths may arise in T_0. Short segments, all of the same length, will arise from the tips cut from the nonmethylated stations p. Such short tip segments will be ignored completely from this point on. (In gel separations they move far away from the segments in which we are interested.) We are concerned with only two classes of DNA molecules: (1) molecules that were methylated at station p and therefore were not cut, and (2) residues of molecules that were not methylated at station p and therefore were cut by the enzyme. Molecules of type (1) will be referred to as *uncut* and those of type (2) as *residues*. Since the uncut are longer, as measured in base pairs (bps), than the residues, the two classes can be separated on a gel.
Motivating Remark: Some of the uncut molecules may also be methylated at p'; others not. Those that are also methylated at p' represent contradictions and must be eliminated. Those that are not methylated at p' do not represent contradictions and must be retained. The residue molecules are the residues of molecules that were not methylated at p. Therefore they arose from molecules that did not represent contradictions and they must be retained.

Make a gel separation of the DNA molecules in T_0. Transfer the DNA molecules in the band consisting of the uncut molecules to a new tube T_1. Transfer the DNA molecules in the band consisting of the residue molecules directly to a new tube T_0. To T_1 add the restriction enzyme that cuts at non-methylated stations p'. Make a gel separation of the DNA molecules in T_1. Discard the DNA molecules in the band consisting of the still uncut molecules (because they have arisen from DNA molecules that represented contradictions). Add to tube T_0 the DNA molecules from the band consisting of the residues of molecules produced by cutting at the station p'. (We ignore the short tips cut off at p'.) At this stage, T_0 contains molecules of at most two lengths: the residues from cuts at p and the residues from cuts at p'. In T_0 there is no molecule that is methylated at both p and p'. This completes the first step in a loop that consists of four virtually identical steps. The pair $\{p, p'\}$ has been treated in this first step. In the remaining three steps the pairs $\{q, q'\}$, $\{r, r'\}$, and $\{s, s'\}$ receive similar treatment which must be done in this specified order. Each of the remaining three steps is carried out by returning to the second paragraph of this Section, replacing $\{p, p'\}$ first by $\{q, q'\}$, then by $\{r, r'\}$, and finally by $\{s, s'\}$.

At the termination of the four step loop each DNA molecule appearing in test tube T_0 has arisen from one of the DNA molecules which has been successfully cut in at least one member of each pair of stations $\{p, p'\}$, $\{q, q'\}$, $\{r, r'\}$, and $\{s, s'\}$. Thus any remaining DNA molecule is the ultimate residue of a DNA molecule that represented a logically noncontradictory truth setting that satisfies all of the specified clauses. Consequently the final step of this computation is the determination of whether or not residues of D are now present in tube T_0. There is a setting of the truth values for $\{p, q, r, s\}$ for which the four clauses:

$p+q', q+r'+s', p'+r+s', p'+q+r+s$ evaluate to TRUE, if and only if residues of D remain in T_0. (Note: a DNA of appropriate length, which would not be cut at all, would be added early to measure any loss of DNA. Thus the final decision would be made by a gel separation that would test for the presence of a DNA band other than the band which must be displayed by the specially introduced test DNA.)

6. ADVANTAGES AND DISADVANTAGES OF THESE WRITING AND READING PROCEDURES

Among the advantages of these procedures are: (0) the positive features shared by all methods of aqueous computing: all computations begin with the same single molecular variety D and this D is grown in bacteria making purchase of DNA unnecessary; (1) the elimination of ligation from all aspects of the computation; and (2) the elimination of the use of restriction enzymes from the writing phase.

Among the disadvantages as compared to other forms of aqueous computing are: (1) the selectively methylated DNA molecules cannot be amplified while preserving their methylation patterns; and (2) the reading procedure answers the YES/NO question of satisfiability, but in the process it destroys the molecules that would allow one to specify specific truth assignments for which the clauses evaluate to TRUE.

Even in the face of disadvantage (1), we remain interested in exploring the blockage of restriction enzyme sites, by methylation or otherwise, as a writing tool. The reading procedure described here is the only one we have to offer at this time to accompany writing by methylation. We would like to replace this reading procedure by a procedure in which molecules having sufficiently closely neighboring methylated sites could be pulled from a solution without disturbing the remaining molecules. Such a procedure would not have disadvantage (2).

7. THE FUTURE

We would like to implement the computational scheme suggested here in a wet lab. We do not suggest that this scheme will scale up to provide efficient solutions to practical problems. We suggest that the aqueous pattern of computation continues to be worth exploration. We are interested in experimenting with its realizations in various media and with various writing and reading technologies. We find it plausible that the algorithmic procedures described in greatest generality in [5] and confirmed by wet lab implementations in [6] and [11] may be implementable in the coming decades using molecules or molecular scale devices dissolved or suspended in a fluid, probably water. Almost surely, any practical implementation would be realized in a capillary flow system, rather than in test tubes. Perhaps a technology that provides a variety of stations, each sensitive to electromagnetic radiation of specific frequencies, can be developed that will provide rapid writing and reading.

Acknowledgments

(1) The efforts that I have devoted to relating the computational and molecular sciences would have ended a decade ago if it were not for the encouragement provided by Gheorghe Păun's contagious enthusiasm for splicing theory which brought my earlier work [3] before a wide audience. The move from pure theory to the wet lab was, of course, stimulated by the exciting appearance of Leonard Adleman's paper [1]. When Takashi Yokomori directed me to [1], I recall emailing Gheorghe this message: "Since you liked splicing systems, you're going to LOVE this: DNA computing [1]!".

(2) Partial support for this research through NSF CCR-9509831 and DARPA/NSF CCR-9725021 is gratefully acknowledged.

References

[1] Adleman, L. (1994), Molecular computation of solutions of combinatorial problems, *Science*, 266: 1021– 1024.

[2] Garey, M.R. & D.S. Johnson (1979), *Computers and Intractability - A Guide to the Theory of NP-Completeness*, W.H. Freeman, San Francisco, Ca.

[3] Head, T. (1987), Formal language theory and DNA: an analysis of the generative capacity of specific recombinant behaviors, *Bulletin of Mathematical Biology*, 49: 737–759.

[4] Head, T. (1999), Circular suggestions for DNA computing, in A. Carbone; M. Gromov & P. Pruzinkiewicz, eds., *Pattern Formation in Biology, Vision and Dynamics*, World Scientific, London, to appear.

[5] Head, T., Biomolecular realizations of a well grounded parallel architecture, submitted.

[6] Head, T.; G. Rozenberg; R. Bladergroen; K. Breek; P.M.H. Lommerese & H. Spaink, The plasmid alternative for biomolecular computing, ms.

[7] Head, T.; M. Yamamura & S. Gal (1999), Aqueous computing: writing on molecules, in *Proceedings of the Congress on Evolutionary Computation 1999*: 1006–1010. IEEE Service Center, Piscataway, NJ.

[8] Ouyang, Q.; P.D. Kaplan; S. Liu & A. Libchaber (1997), DNA solution of the maximal clique problem, *Science*, 278: 446–449.

[9] Păun, Gh., ed. (1998), *Computing with Bio-molecules - Theory and Experiments*, Springer, Singapore.

[10] Păun, Gh.; G. Rozenberg & A. Salomaa (1998). *DNA Computing: New Computing Paradigms*, Springer, Berlin.

[11] Yamamura, M.; T. Head & S. Gal, Boolean satisfiability decided by DNA plasmids, ms. in progress.

Chapter 32

CONTEXT-FREE RECOMBINATIONS

Jarkko Kari
Department of Computer Science
15 MLH
University of Iowa
Iowa City, IA 52242, USA
jjkari@cs.uiowa.edu

Lila Kari
Department of Computer Science
University of Western Ontario
London, ON N6A 5B7, Canada
lila@csd.uwo.ca

Even if the rope breaks nine times, we must splice it back together a tenth time.
— Tibetan proverb

Abstract We address the issue of the computational power of a formal model ([3], [6]) for the guided homologous recombinations that take place during gene rearrangement in ciliates. Results in [7], [5] have shown that a generalization of this model that assumes context-controlled recombinations has universal computational power. Complementing results in [4], [11], [12], [15], we study properties of context-

*Research partially supported by Grant R2824A01 of the Natural Sciences and Engineering Research Council of Canada to L.K. and NSF Grant CCR 97-33101 to J.K.

C. Martin-Vide and V. Mitrana (eds.), Where Mathematics, Computer Science, Linguistics and Biology Meet, 361-375.

free recombinations and characterize the languages generated by context-free recombination systems. As a corollary, we obtain context-free recombinations which are computationally weak, being able to generate only regular languages. This is one more indicator that, most probably, the presence of direct repeats does not provide all the information needed for accurate splicing during gene rearrangement.

1. INTRODUCTION AND NOTATION

Ciliates are a diverse group of a few thousand types of unicellular eukaryotes (nucleated cells) that emerged more that 10^9 years ago [13]. Despite their diversity, ciliates still share two common features: the possession of a hair-like cover of cilia used for moving and food capture, and the presence of two nuclei [13]. The *micronucleus* is functionally inert and becomes active only during sexual exchange of DNA, while the active *macronucleus* contains the genes needed for the development of the ciliate. When two cells mate, they exchange micronuclear information and afterwards develop new macronuclei from their respective micronuclei.

In some of the few ciliates studied, the protein-coding segments of the genes (or *MDSs* for macronuclear destined sequences) are present also in the micronucleus interspersed with large segments of non-coding sequences (*IESs* for internally excised sequences). Moreover, these segments are present in a permuted order in the micronucleus. The function of the various eliminated sequences is unknown and moreover they represent a large portion of the micronuclear sequences: the *Oxytricha* macronucleus (average length of 2,200 basepairs per molecule) has $\sim$ 4 % of the DNA sequences present in the micronucleus whereas the *Stylonychia Lemnae* has $\sim$ 2 % ([13]).

As an example, the micronuclear actin I gene in *Oxytricha Nova* is composed of 9 MDSs separated by 8 IESs. The 9 MDSs are present in the permuted order 3-4-6-5-7-9-2-1-8, the proper order being defined by the 1 through 9 arrangement in the functional macronuclear gene [13]. Instructions for unscrambling the micronuclear actin I gene are apparently carried in the gene itself [13]. At the end of each ith MDS ($1 \leqslant i \leqslant 8$) is a sequence of 9 to 13 bp that is identical to a sequence preceding the $(i+1)$th MDS (which occurs somewhere else in the gene). In the model proposed in [13] the homologous recombination between repeats joins the MDSs in the correct order.

In the following we describe a formal system intended to model the guided homologous recombinations that take place during gene rearrangement. Before introducing the formal model, we summarize our notation. An alphabet Σ is a finite, nonempty set. A sequence of letters from Σ is called a string (word) over Σ and in our interpretation corresponds to a linear strand. The length of a word w is denoted by $|w|$ and represents the total number of occurrences of letters in the word. A word with 0 letters in it is called an empty word and is denoted by λ. The set of all possible words consisting of letters from Σ is denoted by Σ^*, and the set of all nonempty words by Σ^+. We also define circular words over Σ by declaring two words to be equivalent if and only if (iff) one is a cyclic permutation of the other. In other words, w is equivalent to w' iff they can be decomposed as $w = uv$ and $w' = vu$, respectively. Such a circular word $\bullet w$ refers to any of the circular permutations of the letters in w. Denote by $\Sigma^\bullet$ the set of all circular words over Σ.

For a linear word $w \in \Sigma^*$, $\text{Pref}(w) = \{x \in \Sigma^* | \; w = xv\}$, $\text{Suff}(w) = \{y \in \Sigma^* | \; w = uy\}$ and $\text{Sub}(w) = \{z \in \Sigma^* | \; w = uzv\}$.

For a circular word $\bullet w \in \Sigma^\bullet$, we define $\text{Pref}(\bullet w) = \text{Suff}(\bullet w) = \emptyset$ and

$$\text{Sub}(\bullet w) = \{x \in \Sigma^* | \; \bullet w = \bullet uxv, u, v \in \Sigma^*\}$$

as the set of prefixes, suffixes, respectively subwords of $\bullet w$.

For more notions of formal language theory the reader is referred to [14] With this notation we introduce several operations studied in [6], [7] in the context of gene unscrambling in ciliates.

Definition 32.1 *If $x \in \Sigma^+$ is a junction sequence then the recombinations guided by x are defined as follows:*

(i) $uxv + u'xv' \Rightarrow uxv' + u'xv$ (linear/linear), Figure 32.1,

(ii) $uxvxw \Rightarrow uxw + \bullet vx$ (circular/linear), Figure 32.2,

(iii) $\bullet uxv + \bullet u'xv' \Rightarrow \bullet\; uxv'u'xv$ (circular/circular), Figure 32.3.

Note that all recombinations in Definition 32.1 are reversible, i.e., the operations can also be performed in the opposite directions.

For example, operation (ii) models the process of intramolecular recombination. After x finds its second occurrence in $uxvxw$, the molecule undergoes a strand exchange in x that leads to the formation of two new molecules: uxw and a circular DNA molecule $\bullet vx$. Intramolecular recombination accomplishes the deletion of either sequence vx or xv from the original molecule $uxwxv$ and the positioning of w immediately next to ux. This implies that (ii) can be used to rearrange sequences in a DNA molecule thus accomplishing gene unscrambling.

The above operations are similar to the "splicing operation" introduced by Head in [2] and "circular splicing" and "mixed splicing" ([3], [15], [10], [11], [12]). [9], [1] and subsequently [16] showed that some of these models have the computational power of a universal Turing machine. (See [4] for a review.)

In [7] the above strand operations were generalized by assuming that homologous recombination is influenced by the presence of certain contexts, i.e., either the presence of an IES or an MDS flanking a junction sequence. The observed dependence on the old macronuclear sequence for correct IES removal

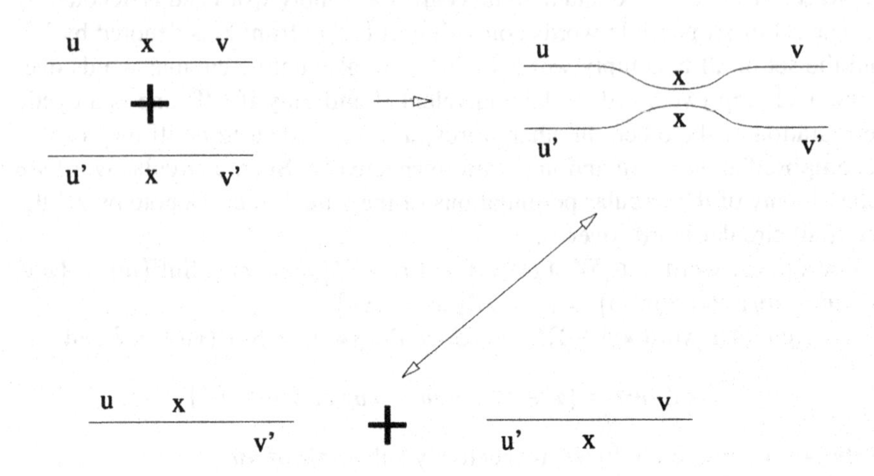

Figure 32.1 Linear/linear recombination.

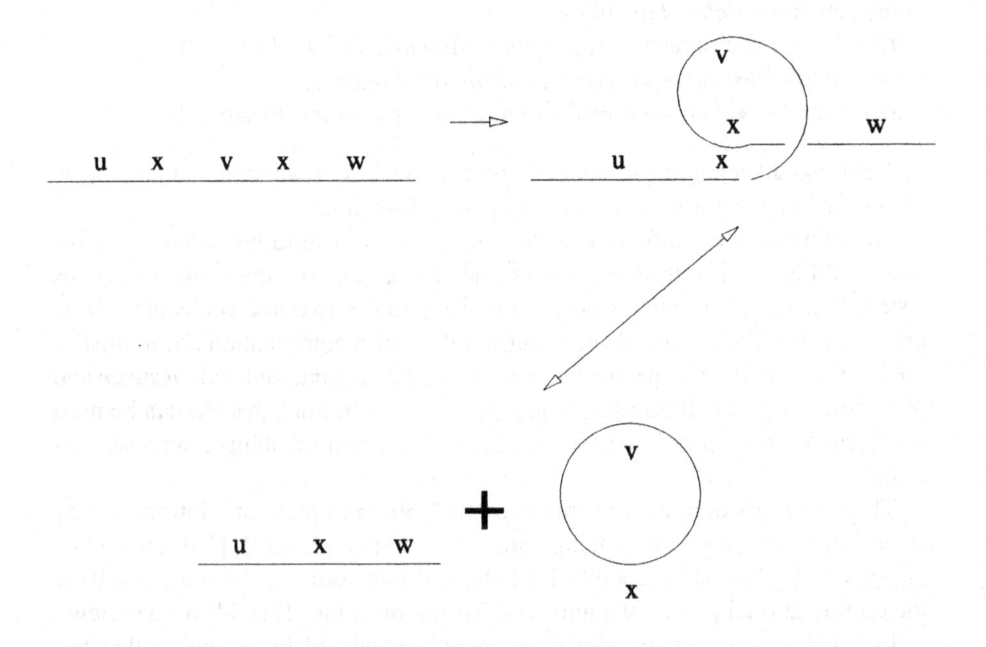

Figure 32.2 Linear/circular recombination.

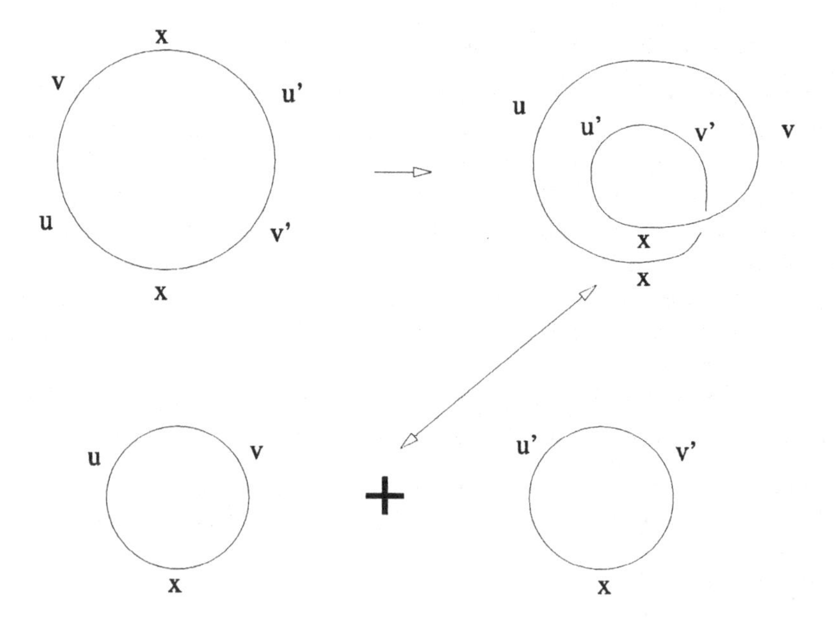

Figure 32.3 Circular/circular recombination.

in *Paramecium* suggests that this is the case ([8]). This restriction captures the fact that the guide sequences do not contain all the information for accurate splicing during gene unscrambling. In particular, in [7] we defined the notion of a *guided recombination system* based on operation (ii) and proved that such systems have the computational power of a Turing machine, the most widely used theoretical model of electronic computers.

We now consider the case where all types of recombinations (linear/linear, linear/circular, circular/circular) are allowed and moreover no context restrictions apply. We study properties of such recombinations. This study complements results obtained in [10], [11], [12], [15], [4] on linear splicing, circular splicing, self-splicing and mixed splicing. However, while Theorem 32.2 may follow from a result in [4] on the closure of an AFL under all splicings, Theorem 32.1 characterizes the language $L(R)$ of an arbitrary context-free recombination system with a possibily infinite set of junction sequences and arbitrary axiom sets.

Theorem 32.2 shows that the resulting rewriting systems are computationally weak having only the power to generate regular languages. This is one more indicator that, most probably, the presence of direct repeats does not provide all the information needed for accurate splicing during gene rearrangement.

2. CONTEXT-FREE RECOMBINATIONS

In our intuitive image of context-free recombinations we can view strings as cables or "extension cords" with different types of "plugs". Given a set of junction sequences J, each $x \in J$ defines one type of "plug". Strings, both linear and circular can then be viewed as consisting of "elementary" cables that only have plugs at their extremities. (A circular strand consists of elementary cables connected to form a loop.) A recombination step amounts to the following operations: take two connections using identical plugs (the connections can be in two different cables or in the same cable); unplug them; cross-plug to form new cables.

In view of Lemma 32.4 to be proved later, we will assume, without loss of generality, that all sets of plugs J are subword-free.

Definition 32.2 *Let $J \subseteq \Sigma^+$ be a set of plugs. We define the set of elementary cables (respectively left elementary cables and right elementary cables) with plugs in J as*

$E_J = (J\Sigma^+ \cap \Sigma^+ J) \setminus \Sigma^+ J \Sigma^+,$
$L_J = \Sigma^* J \setminus \Sigma^* J \Sigma^+,$
$R_J = J\Sigma^* \setminus \Sigma^+ J \Sigma^*.$

Note that an elementary cable in E_J is of the form $z_1 u = v z_2$ where $z_1, z_2 \in J$ are plugs. In other words, an elementary cable starts with a plug, ends with a plug, and contains no other plugs as subwords. The start and end plugs can overlap.

A left elementary cable is of the form wz, where $z \in J$ is a plug and wz does not contain any other plug as a subword. In other words, if we scan wz from left to right, z is the first plug we encounter.

Analogously, a right elementary cable is of the form zw where $z \in J$ is a plug and wz does not contain any other plug as a subword.

Definition 32.3 *For a set of plugs $J \subseteq \Sigma^+$ and a linear word $w \in \Sigma^+$, the set of elementary cables with plugs in J occurring in w is defined as*

$$E_J(w) = E_J \cap \mathit{Sub}(w),$$

while the set of left, respectively right, elementary cables occurring in w is

$$L_J(w) = L_J \cap \mathit{Pref}(w)$$
$$R_J(w) = R_J \cap \mathit{Suff}(w)$$

Note that $L_J(w)$ and $R_J(w)$ are the empty set or singleton sets.

Examples: If $\Sigma = \{a, b\}$ and $J = \{b\}$ then $L_J(aba) = ab$, $R_J(aba) = ba$, $E_J(aba) = \emptyset$. Also $L_J(ab) = ab$, $R_J(ab) = b$, $E_J(ab) = \emptyset$ and $L_J(ba) = b$, $R_J(ba) = ba$, $E_J(ba) = \emptyset$.

Definition 32.4 *For a set of plugs $J \subseteq \Sigma^+$ and a circular word $\bullet w \in \Sigma^\bullet$ we define the elementary cables occurring in $\bullet w$ as follows:*

(i) If $\exists x \in J \cap Sub(\bullet w)$ the elementary cables with plugs in J occurring in $\bullet w$ are defined as $E_J(\bullet w) = E_J(www)$, $L_J(\bullet w) = R_J(\bullet w) = \emptyset$,

(ii) If $J \cap Sub(\bullet w) = \emptyset$ then $E_J(\bullet w) = L_J(\bullet w) = R_J(\bullet w) = \emptyset$.

Examples. If $\Sigma = \{a, b\}$ and $J = \{aba, baa\}$ then $E_J(\bullet aba) = \{abaa, baaba\}$. If $\Sigma = \{a, b, c, d\}$ and $J = \{abc, bcdab\}$ then $E_J(\bullet abcd) = \{abcdab, bcdabc\}$.

From the above examples we see that in circular words start and end plugs are allowed to overlap.

In the definitions for elementary cables, left and right elementary cables can be easily generalized to languages. For a language $L \subseteq \Sigma^* \cup \Sigma^\bullet$,

$$E_J(L) = \bigcup_{w \in L} E_J(w), L_J(L) = \bigcup_{w \in L} L_J(w), R_J(L) = \bigcup_{w \in L} R_J(w)$$

The following two lemmas introduce some properties of elementary cables.

Lemma 32.1 *Given a set of plugs $J \subseteq \Sigma^+$,*

(i) If $u, v \in \Sigma^*$ and $u \in Sub(v)$ then $E_J(u) \subseteq E_J(v)$.

(ii) If $u \in Pref(v)$ then $L_J(u) \subseteq L_J(v)$.

(iii) If $u \in Suff(v)$ then $R_J(u) \subseteq R_J(v)$.

Proof. (i) $E_J(u) = E_J \cap \text{Sub}(u) \subseteq E_J \cap \text{Sub}(v) = E_J(v)$ as $\text{Sub}(u) \subseteq \text{Sub}(v)$.

(ii) and (iii) are proved analogously. □

Lemma 32.2 *If $x \in J$ is a plug then*

(i) $E_J(uxv) = E_J(\{ux, xv\})$,

(ii) $E_J(\bullet ux) = E_J(xux)$,

(iii) $L_J(uxv) \subseteq L_J(ux)$,

(iv) $R_J(uxv) \subseteq R_J(xv)$.

Proof. (i) Let $e = z_1u_1 = u_2z_2$ be an elementary cable in $E_J(uxv)$, $z_1, z_2 \in J$. If e is a subword of ux or of xv then $e \in E_J(ux)$ or $E_J(xv)$, respectively.

If e is not a subword of either ux or xv but e is a subword of uxv then necessarily x is a subword of e. Because x is a plug and e is an elementary cable, e must be a suffix of ux or a prefix of xv. In either case, e is a subword of ux or xv.

(ii) By definition, $E_J(\bullet ux) = E_J(uxuxux)$. By (i), $E_J(uxuxux) = E_J(\{ux, xuxux\}) = E_J(\{ux, xux, xux\})$, which by Lemma 32.1 equals $E_J(xux)$.

(iii) ux is a prefix of uxv therefore $L_J(ux) \subseteq L_J(uxv)$ by Lemma 32.1. Because x is a plug, both $L_J(ux)$ and $L_J(uxv)$ are nonempty and therefore they are equal singleton sets.

(iv) xv is a suffix of uxv therefore $R_J(xv) \subseteq R_J(uxv)$ by Lemma 32.1. Because x is a plug, both $R_J(xv)$ and $R_J(uxv)$ are nonempty and therefore they are equal singleton sets. □

The proposition below shows that recombination of cables does not produce additional elementary cables, i.e. the set of the elementary cables of the result strings equals the set of elementary cables of the strings entering recombination.

Proposition 32.1 *If $J \subseteq \Sigma^+$ is a set of plugs and $x \in J$ then*
(i) $E_J(uxvxw) = E_J(uxw) \cup E_J(\bullet vx)$,
(ii) $E_J(\{uxv, u'xv'\}) = E_J(\{uxv', u'xv\})$,
(iii) $E_J(\{\bullet uxv, \bullet u'xv'\}) = E_J(\bullet uxv'u'xv)$,
(iv) $L_J(uxvxw) = L_J(\{uxw, \bullet vx\})$,
(v) $L_J(\{uxv, u'xv'\}) = L_J(\{uxv', u'xv\})$,
(vi) $R_J(uxvxw) = R_J(\{uxw, \bullet vx\})$,
(vii) $R_J(\{uxv, u'xv'\}) = R_J(\{uxv', u'xv\})$.

Proof. (i) By Lemma 32.2, $E_J(uxvxw) = E_J(\{ux, xvxw\}) = E_J(\{ux, xvx, xw\}) = E_J(\{uxw, xvx\}) = E_J(\{uxw, \bullet vx\})$.

(ii) Similarly, $E_J(\{uxv, u'xv'\}) = E_J(\{ux, xv, u'x, xv'\}) = E_J(\{uxv', u'xv\})$.

(iii) $E_J(\bullet uxv, \bullet u'xv'\}) = E_J(\{\bullet xvu, \bullet xv'u'\}) = E_J(\{xvux, xv'u'x\})$ while

$E_J(\bullet uxv'u'xv) = E_J(\bullet xv'u'xvu) = E_J(xv'u'xvux) = E_J(\{xv'u'x, xvux\})$.

(iv) By Lemma 32.2, $L_J(uxvxw) = L_J(ux)$ while $L_J(\{uxw, \bullet vx\}) = L_J(uxw) = L_J(ux)$.

(v) By Lemma 32.2, $L_J(\{uxv, u'xv'\}) = L_J(\{ux, u'x\})$ while $L_J(\{uxv', u'xv\}) = L_J(\{ux, u'x\})$.

(vi) By Lemma 32.2, $R_J(uxvxw) = R_J(xw)$ while $R_J(\{uxw, \bullet vx\}) = R_J(xw)$.

(vii) By Lemma 32.2, $R_J(\{uxv, u'xv'\}) = R_J(\{xv, xv'\})$ while $R_J(\{uxv', u'xv\}) = R_J(\{xv', xv\})$. □

We are now ready to define the notion of a context-free recombination system. This is a construct whereby we are given a starting set of sequences and a list of junction sequences (plugs). New strings may be formed by recombinations among the existing strands: if one of the given junctions sequences is present, recombinations are performed as defined in Section 1. Recombinations are context-free, i.e., they are not dependent on the context in which the junctions sequences appear. The language of the system is defined as the set of all strands that can thus be obtained by repeated recombinations starting from the initial set.

Definition 32.5 *A context-free recombination system is a triple*

$$R = (\Sigma, J, A)$$

where Σ is an alphabet and $J \subseteq \Sigma^+$ is a set of plugs, while $A \subseteq \Sigma^+ \cup \Sigma^\bullet$ is the set of axioms of the system.

Given a recombination system R, for sets $S, S' \subseteq \Sigma^+ \cup \Sigma^\bullet$ we say that S derives S' and we write $S \Rightarrow_R S'$ iff there exists $x \in J$ such that one of the following situations holds:

(i) $\exists uxv, u'xv' \in S$ such that $uxv + u'xv' \Rightarrow uxv' + u'xv$ and $S' = S \cup \{uxv', u'xv\}$,

(ii) $\exists uxvxw \in S$ such that $uxvxw \Rightarrow uxw + \bullet vx$ and $S' = S \cup \{uxw, \bullet vx\}$,

(iii) $\exists uxw, \bullet vx \in S$ such that $uxw + \bullet vx \Rightarrow uxvxw$ and $S' = S \cup \{uxvxw\}$,

(iv) $\exists \bullet uxv, \bullet u'xv' \in S$ such that $\bullet uxv + \bullet u'xv' \Rightarrow \bullet uxv'u'xv$ and $S' = S \cup \{\bullet uxv'u'xv\}$,

(v) $\exists \bullet uxv'u'xv \in S$ such that $\bullet uxv'u'xv \Rightarrow \bullet uxv + \bullet u'xv'$ and $S' = S \cup \{\bullet uxv, \bullet u'xv'\}$.

Definition 32.6 *The language generated by a context-free recombination system R is defined as*

$$L(R) = \{w \in \Sigma^* \cup \Sigma^\bullet |\ A \Rightarrow_R^* S, w \in S\}$$

Lemma 32.3 *For any context-free recombination systems $R = (\Sigma, J, A)$ there exists a context-free recombination system $R' = (\Sigma, J', A)$ such that J' is subword-free and $L(R) = L(R')$.*

Proof. Let $J' = \{w \in J |\ \nexists u \in J, u \neq w$ such that $w = x'ux'', x', x'' \in \Sigma^*\}$. As $J' \subseteq J$, obviously $L(R') \subseteq L(R)$ as any recombination sequence using a plug in J' is a recombination according to R as well.

Conversely, let $x \in J \setminus J'$. There exist $y \in J', x', x'' \in \Sigma^*, y \neq x$ such that $x = x'yx''$.

Then,

(i) $uxv + u'xv' = ux'yx''v + u'x'yx''v' \Rightarrow_{R'} ux'yx''v' + u'x'yx''v = uxv' + u'xv$,

(ii) $uxvxw = ux'yx''vx'yx''w \Rightarrow_{R'} ux'yx''w + \bullet x''vx'y = uxw + \bullet vx'yx'' = uxw + \bullet vx$,

(iii) $uxw + \bullet vx = ux'yx''w + \bullet vx'yx'' \Rightarrow_{R'} ux'yx''vx'yx''w = uxvxw$,

(iv) $\bullet uxv + \bullet u'xv' = \bullet ux'yx''v + \bullet u'x'yx''v' \Rightarrow_{R'} \bullet ux'yx''v'u'x'yx''v = uxv'u'xv$,

(v) $\bullet uxv'u'xv = \bullet ux'yx''v'u'x'yx''v \Rightarrow_{R'} \bullet ux'yx''v + \bullet yx''v'u'x' =$
$\bullet uxv + \bullet u'x'yx''v' = \bullet uxv + \bullet u'xv'$.

Consequently, any derivation step in R can be simulated by a derivation step in R', i.e., $L(R) \subseteq L(R')$. □

As a consequence of the preceding lemma we may assume, without loss of generality, that a context-free recombination system has a subword-free set J of plugs. The following Lemma will aid in the proof of our main result.

Lemma 32.4 *Let $R = (\Sigma, J, A)$ be a context-free recombination system. Let $ux = x'u'$ start and end with plugs $x', x \in J$ where $u, u' \neq \lambda$. If ux satisfies $E_J(ux) \subseteq E_J(A)$, then there exist $\alpha, \beta \in \Sigma^*$ such that $\alpha ux\beta$ or $\bullet\alpha ux\beta$ is in $L(R)$.*

Proof. Induction on k, the number of occurrences of plugs in $ux = x'u'$.

Base case: $k = 2$. Then ux is elementary, $ux \in E_J(ux)$, therefore there exists an axiom $a \in A$ such that $ux \in E_J(a)$. If $a \in \Sigma^*$ is a linear word, then $a = \alpha ux\beta \in L(R)$. If $\bullet a \in \Sigma^\bullet$ is circular, as $\bullet a$ contains at least one plug, $\bullet a + \bullet a + \bullet a = \bullet aaa \in L(R)$. By definition $ux \in E_J(aaa)$ which implies $aaa = \alpha ux\beta$ and $\bullet aaa = \bullet\alpha ux\beta \in L(R)$.

Inductive step: Let $ux = x'u' = vyz$ where y is any plug in the middle, e.g. the second last plug.

Words vy and yz satisfy the conditions of the claim as they contain fewer than k plugs, so we may apply the inductive hypothesis to both of them. Consequently, $\alpha vy\beta$ or $\bullet\alpha vy\beta$ is in $L(R)$ and $\gamma yz\delta$ or $\bullet\gamma yz\delta$ is in $L(R)$.

One more recombination yields:
$\alpha vy\beta + \gamma yz\delta \Rightarrow \alpha vyz\delta + \gamma y\beta$ or
$\alpha vy\beta + \bullet\gamma yz\delta \Rightarrow \alpha vyz\delta\gamma y\beta$ or
$\bullet\alpha vy\beta + \gamma yz\delta \Rightarrow \gamma y\beta\alpha vyz\delta$ or
$\bullet\alpha vy\beta + \bullet\gamma yz\delta \Rightarrow \bullet\ \alpha vyz\delta\gamma y\beta$, respectively.

Note that in each of the four possible cases the result contains a linear or circular word in $L(R)$ that has $vyz = ux = x'u'$ as a subword, as required. □

The theorem below shows that a context-free recombination system characterized by a set of plugs J and a set of axioms A has the following property. Any cable that consists of elementary cables plugged together one after the other and that is either linear or circular can be obtained from the axioms using cross-plugging. Conversely, no other types of cables can be obtained from the axioms.

Theorem 32.1 *Let $R = (\Sigma, J, A)$ be a context-free recombination system. Then $L(R) = X$ where*

$X = \{w \in \Sigma^* \cup \Sigma^\bullet |$ either $E_J(w) = L_J(w) = R_J(w) = \emptyset$ and $w \in A$, or $E_J(w), L_J(w), R_J(w)$ are not all empty and $E_J(w) \subseteq E_J(A)$, $L_J(w) \subseteq L_J(A)$, $L_J(w) \subseteq L_J(A)\}$.

Proof. "$X \subseteq L(R)$"

Let $w \in X$. If $E_J(w) = L_J(w) = R_J(w) = \emptyset$ and $w \in A$, then $w \in L(R)$.

Assume now that $w \in \Sigma^*$ is a linear word such that $E_J(w), L_J(w), R_J(w)$ are not all empty and $E_J(w) \subseteq E_J(A)$, $L_J(w) \subseteq L_J(A)$, $R_J(w) \subseteq R_J(A)$.

If w contains only one plug $x \in J$ then $w = uxv$ and $L_J(w) = ux$, $R_J(w) = xv$. As $L_J(w) \subseteq L_J(A)$, there exists an axiom $a_1 \in A \cap \Sigma^*$ such that $a_1 = uxt$. As $R_J(w) \subseteq R_J(A)$, there exists an axiom $a_2 \in A \cap \Sigma^*$ such that $a_2 = sxv$.

We have $a_1 + a_2 = uxt + sxv \Rightarrow uxv + sxt$, which implies that $uxv = w \in L(R)$.

If w contains more than one plug, then $w = u\gamma v$ where $\gamma = xl = rx'$, $x, x' \in J$, $ux = L_J(w) \subseteq L_J(A)$ and $x'v = R_J(w) \subseteq R_J(A)$. Consequently, there exist axioms $a_1, a_2 \in A \cap \Sigma^*$ such that $a_1 = uxt$, $a_2 = sx'v$.

By Lemma 32.4, there exist $\alpha, \beta \in \Sigma^*$ such that $\alpha\gamma\beta$ or $\bullet\alpha\gamma\beta$ is in $L(R)$.

We can then recombine

$uxt + \alpha\gamma\beta + sx'v = uxt + \alpha xl\beta + sx'v \Rightarrow uxl\beta + \alpha xt + sx'v =$
$urx'\beta + \alpha xt + sx'v \Rightarrow urx'v + \alpha xt + sx'\beta = u\gamma v + \alpha xt + sx'\beta$

or, in the circular case,

$uxt + \bullet\alpha\gamma\beta + sx'v = uxt + \bullet\alpha xl\beta + sx'v \Rightarrow uxl\beta\alpha xt + sx'v =$
$urx'\beta\alpha xt + sx'v \Rightarrow urx'v + sx'\beta\alpha xt = u\gamma v + sx'\beta\alpha xt.$

In both cases, $u\gamma v = w \in L(R)$.

If $\bullet w \in \Sigma^\bullet$ is a circular word that contains at least one plug ($E_J(\bullet w) \neq \emptyset$) then $\bullet w = \bullet ux$ for some $x \in J$. The word xux satisfies the conditions of Lemma 32.4 therefore $\alpha xux\beta$ or $\bullet\alpha xux\beta$ is in $L(R)$.

Then we have either $\alpha xux\beta \Rightarrow \bullet ux + \alpha x\beta$ or $\bullet\alpha xux\beta \Rightarrow \bullet ux + \bullet\alpha x\beta$ which both imply that $\bullet ux = \bullet w \in L(R)$.

For the converse inclusion "$L(R) \subseteq X$" note that if $w \in L(R)$, $E_J(w) = \emptyset$, $L_J(w) = \emptyset$, $R_J(w) = \emptyset$, and $w \in A$ then by definition $w \in X$. Otherwise, if some words in $L(R)$ belong to X, the result of their recombinations have the necessary properties that ensure their belonging to X by Proposition 32.1. Therefore, $L(R) \subseteq X$. □

The theorem above leads to the conclusion of our paper after we show that the language X is regular, being accepted by a finite automaton.

Definition 32.7 *Given a finite automaton $\mathcal{A}$, the circular language accepted by $\mathcal{A}$, denoted by $L(\mathcal{A})^\bullet$, is defined as the set of all words $\bullet w$ such that $\mathcal{A}$ has a cycle labelled by w.*

The circular/linear language accepted by a finite automaton $\mathcal{A}$ is defined as $L(\mathcal{A}) \cup L(\mathcal{A})^\bullet$, where $L(\mathcal{A})$ is the linear language accepted by the automaton $\mathcal{A}$ defined in the usual way.

Definition 32.8 *A circular/linear language $L \subseteq \Sigma^* \cup \Sigma^\bullet$ is called regular if there exists a finite automaton $\mathcal{A}$ such that it accepts the circular and linear parts of L, i.e. that accepts $L \cap \Sigma^*$ and $L \cap \Sigma^\bullet$.*

Theorem 32.2 *Let $J \subseteq \Sigma^*$ be a set of plugs and let $A \subseteq \Sigma^* \cup \Sigma^\bullet$ be a finite axiom set. Then X defined as in Theorem 32.1 equals the circular/linear language accepted by a finite automaton $\mathcal{A}$ and is therefore regular.*

Proof. If the set J is not finite, then we should start by eliminating plugs that do not appear in any elementary cables of A. As the axiom set A is finite, the number of elementary cables is finite, and the set of (useful) plugs is finite as well. Consequently, we can assume, without loss of generality, that the set J is finite.

Let $\mathcal{A} = (S, \Sigma, \delta, s_0, s_f)$ be a finite automaton constructed as follows.

The set of states is

$$S = \{s_x |\ x \in J\} \cup \{s_0, s_f\}$$

and the transition relation δ is defined as follows:

(i) $\delta(s_x, u) = \{s_y |$ for each $e = xu = vy \in E_J(A)\}$,

(ii) For each $ux \in L_J(A)$ we have the transition $\delta(s_0, ux) = s_x$,

(iii) For each $xu \in R_J(A)$ we have $\delta(s_x, u) = s_f$.

From the above construction and Theorem 32.1 one can prove that $L(\mathcal{A}) = X$. □

Note that Definition 32.7 of the acceptance of a circular language by an finite automaton and Definition 32.8 of regularity of a circular/linear language overlap but do not coincide with existing definitions. We will therefore conclude with a comparison between various definitions and an argument in favour of our choice. We can define circular languages accepted by automata in two more ways.

Definition 32.9 *Given a finite automaton $\mathcal{A}$, the circular language accepted by $\mathcal{A}$, denoted by $L(\mathcal{A})_1^\bullet$, is defined as the set of all words $\bullet w$ such that $\mathcal{A}$ has a cycle labelled by w that contains at least one final state.*

Definition 32.10 *Given a finite automaton $\mathcal{A}$, the circular language accepted by $\mathcal{A}$, denoted by $L(\mathcal{A})_2^\bullet$ is defined as the set of all words $\bullet w$ such that $w = uv$ and $vu \in L(\mathcal{A})$.*

The circular languages accepted by finite automata using Definition 32.10 coincide with the regular circular languages as introduced by Head in [3], while acceptance of circular languages as in Definition 32.9 will be proven to coincide with the definition in [11]. The following result holds.

Proposition 32.2 *(i) The family of circular languages accepted by finite automata under Definition 32.7 is strictly included in the family of circular languages accepted by finite automata under Definition 32.9. (ii) The family of circular languages accepted by finite automata under Definition 32.9 is strictly included in the family of circular languages accepted by finite automata under Definition 32.10.*

Proof. (i) If $L \subseteq \Sigma^\bullet$ is a language accepted by an automaton A under Definition 32.7, then there exists an automaton B such that L is accepted by B under Definition 32.9. Indeed, one can construct an automaton B from A by making every state a final state. Then every cycle contains a final state.

The inclusion is strict as shown by the two state automaton with transitions $\delta(q,a) = p, \delta(p,a) = q, \delta(p,b) = p$ where q is the only final state. All words $ab^n a$ are accepted, which implies a loop labeled by bs only. This implies that in the sense of Definition 32.7 words $\bullet b^n$ for some n are accepted although they are not accepted by the automaton in the sense of Definition 32.9.

(ii) If $L \subseteq \Sigma^\bullet$ is a language accepted by an automaton B under Definition 32.9 then there exists an automaton C such that L is accepted by C under Definition 32.10. Indeed, we can construct an automaton C from B as follows. For every final state q of B we can construct a copy $B(q)$ of B where only the initial and final states have been changed: In $B(q)$ state q is the only initial and the only final state. Then we take the union of $B(q)$ for all final states q of B. The resulting automaton C accepts the same language in the sense of Definition 32.10 as the original B accepted in the sense of Definition 32.9.

Indeed, let q be an arbitrary final state of B. As q is the only final and initial state of $B(q)$ then the new machine $B(q)$ accepts in the sense of Definition 32.10 all circular words that label a loop that goes through state q. (If w is a label of such a loop, that starts and ends in state p, then $w = uv$ where u labels a path from p to q and v labels a path from q to p. Then vu labels a path that starts and ends in q, i.e., w is accepted by $B(q)$ i.e. of C in the sense of Definition 32.10. Conversely, if w is accepted by C in the sense of Definition 32.10 this means it is accepted by some $B(q)$ in the sense of Definition 32.10 then $w = uv$ where vu labels a path from q to q. That means vu labels a loop that goes through q, and then also $w = uv$ labels such a loop.)

The inclusion is strict as an automaton accepting languages using Definition 32.9 does not accept any finite nonempty languages because any loop can be repeated arbitrarily many times. Automata accepting languages using Definition 32.10 accept finite languages. □

As mentioned above, Definition 32.9 is equivalent to the definition in [11] (the circular language accepted by an automaton is the set of all words that label a loop containing at least one initial and one final state). Indeed, let A be an NFA that accepts a circular language L in the sense of [11], i.e., a word is accepted if it labels a cycle that contains both initial and final states. To accept the same language in the sense of Definition 32.9 we make two identical copies of A, say A and A', and we make λ-transitions (which we can eliminate later) between the copies as follows: For every initial state i we have a λ-transition from i to i', and for every final state f we make λ -transitions from f' to a new state f'', and from f'' to f. All states f'' are final states, and no other states are final. A cycle that contains a final state must go through some f'', which means it goes from A' to A. Consequently it must go from A to A' as well, i.e. the cycle contains some initial state i also, i.e. the word labels some cycle of the original machine that contains both i and f.

To summarize, Definition 32.7 of circular languages accepted by automata is strictly more restrictive than Definition 32.9, which is in turn strictly more restrictive that Definition 32.10. Definition 32.9 is equivalent to the definition in [11], and Definition 32.10 is equivalent to the definition of regular circular languages proposed in [3]. Note that, as shown in [11], the family of languages accepted by Definition 32.10, which are in addition closed under repetition (if w^n is in the language whenever w is in the language), equals the family of circular languages accepted by automata in the sense of Definition 32.9.

Our preference for Definition 32.7 was motivated by the fact that, under this definition, in Theorem 32.2 we can use the same automaton to accept both the linear and circular components of the language. This makes our definition more natural and Theorem 32.2 stronger that its counterpart following from [4]. (Theorem 5.2, Chapter 5, [4], implies that the result of combined splicing starting from a finite set is regular by showing that the linear and circular components are each regular – using Definition 32.10 – but possibly accepted by different automata.)

However, Theorem 32.2 and Proposition 32.2 show that the language of a context-free recombination system with a finite axiom set is regular under any of the existing definitions.

References

[1] Csuhaj-Varjú, E.; R. Freund; L. Kari & Gh. Păun (1996), DNA computing based on splicing: universality results, in L. Hunter & T. Klein, eds., *Proceedings of the 1st Pacific Symposium on Biocomputing*: 179–190. World Scientific, Singapore.

[2] Head, T. (1987), Formal language theory and DNA: an analysis of the generative capacity of specific recombinant behaviors, *Bulletin of Mathematical Biology*, 49: 737–759.

[3] Head, T. (1991), Splicing schemes and DNA, in G. Rozenberg & A. Salomaa, eds., *Lindenmayer Systems*: 371–383. Springer, Berlin.

[4] Head, T.; Gh. Păun & D. Pixton (1997), Language theory and molecular genetics, in G. Rozenberg & A. Salomaa, eds., *Handbook of Formal Languages*, II: 295–358. Springer, Berlin.

[5] Kari, L.; J. Kari & L. Landweber (1999), Reversible molecular computation in ciliates, in J. Karhumäki; H. Maurer; Gh. Păun & G. Rozenberg, eds., *Jewels are Forever*: 353-363. Springer, Berlin.

[6] Landweber, L.F. & L. Kari (1998), The evolution of cellular computing: nature's solution to a computational problem, in *Proceedings of the 4th DIMACS Meeting on DNA Based Computers*: 3–15, Philadephia. Also in *Biosystems*.

[7] Landweber, L.F. & L. Kari (1999), Universal molecular computation in ciliates, in L. Landweber & E. Winfree, eds., *Evolution as Computation*, Springer, Berlin.

[8] Meyer, E. & S. Duharcourt (1996), Epigenetic Programming of Developmental Genome Rearrangements in Ciliates, *Cell*, 87: 9–12.

[9] Păun, Gh. (1995), On the power of the splicing operation, *International Journal of Computer Mathematics*, 59: 27–35.

[10] Pixton, D. (1995), Linear and circular splicing systems, in *Proceedings of the First International Symposium on Intelligence in Neural and Biological Systems*: 181–188. IEEE Computer Society Press, Los Alamos, Ca.

[11] Pixton, D. (1996), Regularity of splicing languages, *Discrete Applied Mathematics*, 69.1-2: 99–122.

[12] Pixton, D., Splicing in abstract families of languages, ms. in preparation.

[13] Prescott, D.M. (1994), The DNA of ciliated protozoa, *Microbiological Reviews*, 58.2: 233–267.

[14] Salomaa, A. (1973), *Formal Languages*. Academic Press, New York.

[15] Siromoney, R.; K.G. Subramanian & V. Rajkumar Dare (1992), Circular DNA and splicing systems, in *Parallel Image Analysis*: 260–273. Springer, Berlin.

[16] Yokomori, T.; S. Kobayashi & C. Ferretti (1997), Circular Splicing Systems and DNA Computability, in *Proceedings of the IEEE International Conference on Evolutionary Computation'97*: 219-224.

Chapter 33

SIMPLIFIED SIMPLE H SYSTEMS

Kamala Krithivasan
Department of Computer Science and Engineering
Indian Institute of Technology
Chennai - 600036, Madras, India
kamala@iitm.ernet.in

Arvind Arasu
Department of Computer Science and Engineering
Indian Institute of Technology
Chennai - 600036, Madras, India

Abstract In this paper we define restricted versions of simple H systems of type (1,3) and (1,4) and show that the generative power is not decreased by this restriction. In the case of these systems the power is not increased by having a target alphabet whereas having permitting context increases the power.

1. INTRODUCTION

Splicing systems were defined in [3] to model the recombinant behavior of DNA. Currently there is lot of interest in such systems [8].

Simple H systems were defined in [6] by restricting the form of the rules. Four types of simple H systems were defined in [6]. (1,3), (2,4), (1,4) and (2,3). (1,3) and (2,4) essentially define the same system. It is known that simple H systems of all the 3 types generate only regular languages and the family of languages generated by them are not comparable.

In [2], a detailed study of simple H systems with target alphabet and permitting context is made. The generative power of these families are compared

*This research was partially supported by a project from Department of Science and Technology, Government of India.

C. Martin-Vide and V. Mitrana (eds.), Where Mathematics, Computer Science, Linguistics and Biology Meet, 377-386.

and presented in the form of a table. Simple H systems with target alphabet are called simple extended H systems. It is shown in [2] that $SEH_{(2,3)}$ is more powerful than $SEH_{(1,4)}$ and $SEH_{(1,3)}$. It is also shown in [2] that every context-free language can be generated by a $SEH_{(2,3)}$ system with permitting context.

In [5] it is shown that $SEH_{(2,3)}$ systems with permitting context are as powerful as $EH(FIN, p[1])$ systems.

In [2], some hierarchy questions are left open. In this paper we give solutions for some of them. We mainly consider (1,3) and (1,4) systems with target alphabet and permitting context.

This result confirms the fact proved in [2] that $SEH_{(2,3)}$ is more powerful than $SEH_{(1,4)}$ or $SEH_{(1,3)}$. The result proved here shows $SEH_{(1,4)}$ and $SEH_{(1,3)} \in REG$ (family of regular sets) whereas $SEH_{(2,3)}$ with permitting context includes all context-free languages.

In the next section we give basic definitions and results needed for the paper. In section 3, a restricted version of SEH is defined and some results regarding hierarchy proved. The paper concludes with a brief note on the implication of these results.

2. BASIC DEFINITIONS

2.1 SPLICING SYSTEMS

We present below the basic notations and definitions required for building the concept of simple test tube systems. For more information on splicing systems the reader is directed to some of the excellent material in this field [8, 4].

2.1.1 H systems and EH systems [4].

Definition 33.1 *A splicing rule (over an alphabet V) is a string $r = u_1\#u_2\$u_3\#u_4$, where $u_i \in V^*, 1 \leqslant i \leqslant 4$, and $\#, \$$ are special symbols not in V. For such a rule r and the strings $x, y, z \in V^*$, the splicing operation is defined as follows:*

$$\begin{aligned} (x, y) \vdash_r z \quad \textit{iff} \quad & x = x_1u_1u_2x_2, y = y_1u_3u_4y_2 \\ & z = x_1u_1u_4y_2 \\ & \textit{for some } x_1, x_2, y_1, y_2 \in V^* \end{aligned}$$

x, y are then said to be *spliced* at the *sites* u_1u_2, u_3u_4 respectively to obtain the string z. The strings x, y are called the *terms* of splicing. When understood from context, the index r is omitted from $\vdash_r$.

A *splicing scheme* (or an *H scheme*) is a pair $\sigma = (V, R)$ where V is an alphabet and R is a set of splicing rules (over V). For a language $L \subseteq V^*$, the following are defined:

$$\sigma(L) = \{w \in V^* | (x, y) \vdash_r w \text{ for } x, y \in L, r \in R\}$$

The following are also defined for the language L

$$\begin{aligned} \sigma^0(L) &= L \\ \sigma^{i+1}(L) &= \sigma^i(L) \cup \sigma(\sigma^i(L)) \\ \sigma^*(L) &= \bigcup_{i \geqslant 0} \sigma^i(L) \end{aligned}$$

Thus, $\sigma^*(L)$ is the smallest language containing L and closed under the splicing operation.

Definition 33.2 *An extended splicing system is a quadruple*

$$\gamma = (V, T, A, R)$$

where V is an alphabet, $T \subseteq V$ (the terminal alphabet), $A \subseteq V^*$ the set of axioms and $R \subseteq V^*\#V^*\$V^*\#V^*$ the set of rules. The pair $\sigma = (V, R)$ is called the underlying H schema of γ. The language generated by γ is defined as follows

$$L(\gamma) = \sigma^*(A) \cap T^*$$

An H system $\gamma = (V, T, A, R)$ is said to be of *type* (F_1, F_2) for two families of languages F_1, F_2 if $A \in F_1, R \in F_2$. $EH(F_1, F_2)$ is used to denote the family of languages generated by extended H systems of type (F_1, F_2). An H system $\gamma = (V, T, A, R)$ with $V = T$ is said to be *non-extended;* here the H system is denoted by $\gamma = (V, A, R)$. The family of languages generated by non-extended H systems of type (F_1, F_2) is denoted by $H(F_1, F_2)$. Obviously, $H(F_1, F_2) \subseteq EH(F_1, F_2)$.

The following are a few important results regarding H systems and EH systems [4].

Theorem 33.1 *(i) $H(FIN, FIN) \subseteq REG$.*
(ii) $EH(FIN, FIN) = REG$
(iii) $EH(FIN, REG) = RE$

Thus these EH systems are as powerful as TMs. However, it is not realistic to deal with an infinite number of rules even if they were a regular set of rules. For this purpose several variants were proposed. One such variant is the working of several of these H systems in unison in a parallel manner. Another variant is to use context information.

2.1.2 Simple H systems [6].

Definition 33.3 *A simple H system is a triple*

$$\Gamma = (V, A, M)$$

where V is an alphabet, $A \subseteq V^*$ is a finite set of axioms and $M \subseteq V$.

The elements of M are called *markers*. One can consider four ternary relations on the language V^*, corresponding to the splicing rules of the form

$$a\#\$a\#, \#a\$\#a, a\#\$\#a, \#a\$a\#$$

where a is an arbitrary element of M. These four rules are respectively called splicing rules of type $(1,3), (2,4), (1,4), (2,3)$.

Clearly, rules of type $(1,3)$ and $(2,4)$ define the same operation, for $x, y, z \in V^*$ and $a \in M$ we obtain

$$(x, y) \vdash^a_{(1,3)} z \quad \text{iff} \quad x = x_1 a x_2, y = y_1 a y_2, z = x_1 a y_2, \\ \text{for some } x_1, x_2, y_1, y_2 \in V^*.$$

For the other types, the splicing is performed as follows:

$$(x, y) \vdash^a_{(1,4)} z \quad \text{iff} \quad x = x_1 a x_2, y = y_1 a y_2, z = x_1 a a y_2, \\ \text{for some } x_1, x_2, y_1, y_2 \in V^*.$$

$$(x, y) \vdash^a_{(2,3)} z \quad \text{iff} \quad x = x_1 a x_2, y = y_1 a y_2, z = x_1 y_2, \\ \text{for some } x_1, x_2, y_1, y_2 \in V^*.$$

Then for $L \subseteq V^*$ and $(i,j) \in \{(1,3), (1,4), (2,3)\}$, the following are defined:

$$\sigma_{(i,j)}(L) = \{w \in V^* | (x, y) \vdash^a_{(i,j)} w \text{ for some } x, y \in L, a \in M\}$$

$$\begin{aligned} \sigma^0_{(i,j)}(L) &= L \\ \sigma^{k+1}_{(i,j)}(L) &= \sigma^k_{(i,j)}(L) \cup \sigma_{(i,j)}(\sigma^k_{(i,j)}(L)), k \geqslant 0 \\ \sigma^*_{(i,j)}(L) &= \bigcup_{k \geqslant 0} \sigma^k_{(i,j)}(L) \end{aligned}$$

Definition 33.4 *The language generated by $\Gamma = (V, A, M)$ in case (i,j) is defined as follows:*

$$L_{(i,j)}(\Gamma) = \sigma^*_{(i,j)}(A).$$

As in all the cases mentioned, both the axiom set and the rule set are finite; it follows that for any simple H system Γ, the languages $L_{(i,j)}(\Gamma)$ are regular [6].

It has been shown that every two of the three family of languages obtained in this way are incomparable [6].

A simple extended H system with permitting context is a 4-tuple $S = (V, T, A, R)$ where the rules are of the form (a, b, c). $(x, y) \vdash z$ if the splicing alphabet is a, b is present in x and c is present in y. If $T = V$, then the system is a simple H system with permitting context.

3. SIMPLIFIED SIMPLE H SYSTEMS

In this section we define a restricted simple H system and show that the power is not decreased in the (1,4) and (1,3) cases. We mainly consider (1,4) and (1,3) systems only in this section.

Definition 33.5 *A restricted $SEH_{(i,j)}$ ($RSEH_{(i,j)}$) is a $SEH_{(i,j)}$ system with the following restrictions on splicing:*

1. *Every splicing operation should use an axiom on the right side. In other words (w, u) can be spliced only if u is an axiom.*
2. *Each letter participates in at most one splicing operation on the left side i.e in the (1,3) case this means that if $\alpha a \gamma$ has been obtained by splicing at a, then in the following steps this a cannot be used for splicing; in the (1,4) case this means that if $\alpha \underline{a} a \gamma$ has been obtained by splicing the letter a, then the marked a cannot be used for splicing in the following steps.*
3. *Splicing with a is not permitted if there is a letter b to the right of a in the same string that has been used in splicing.*

Example 33.1 *: Let $S = (\{0,1\}, \{101\}, 1)$ be an $RSEH_{(1,4)}$ system. Then,*

- $(101, 101) \vdash 101101$ *is permitted.*
- $(101101, 101101) \not\vdash 1011101$ *is not permitted since it does not use an axiom (rule 1).*
- $(101101, 101) \vdash 101101$ *is not permitted because it violates rule 2.*
- $(101, 101101) \not\vdash 1011$ *is not permitted since there is no axiom on the right side (rule 1).*
- $(101101, 101) \not\vdash 1101$ *is not permitted since it violates the rule 3.*

Notice that any string in a RSEH system is generated incrementally from left to right.

Theorem 33.2 *The generative power of the $SEH_{(1,4)}$ and $SEH_{(1,3)}$ systems does not change by introducing the restricted splicing rules:*

$$RSH_{(1,4)} = SH_{(1,4)}$$

$$\begin{aligned} RSEH_{(1,4)} &= SEH_{(1,4)} \\ RSH_{(1,3)} &= SH_{(1,3)} \\ RSEH_{(1,3)} &= SEH_{(1,3)} \end{aligned}$$

Proof. We prove the result only for $SEH_{(1,4)}$ systems. The proof for the rest follows in a similar manner. Consider a splicing system $S = (V, T, A, R)$ both under restricted rules $RSEH_{(1,4)}$ and non restricted rules $SEH_{(1,4)}$. Let L denote the language generated by the SEH system and L_r the language generated by the $RSEH$ system.

$L_r \subseteq L$ follows directly from the definition of the restricted rules. In order to prove $L \subseteq L_r$ we can use an inductive argument on the number of steps, n, needed to derive a string w in L, i.e we want to prove that for any string w, if $w \in L$, then w belongs to L_r. Trivially, the induction hypothesis is true when $n = 1$. Let $w \in L$ be any string which is not an axiom. Let the last step of derivation be

$$(\alpha a\beta, \gamma a\delta) \vdash \alpha aa\delta = w.$$

w can be generated in L_r as follows. By induction hypothesis both $w_1 = \alpha a\beta$ and $w_2 = \gamma a\delta$ can be generated in L_r. First, follow the derivation of w_1 in L_r until the letter a that is used in the splicing operation above is generated for the first time. Then follow the derivation of w_2 in L_r after the letter a has been generated in w_2 for the first time. This can bc casily verified to generate the string w in L_r. Thus $L = L_r$. Hence the proof follows. □

Theorem 33.3 *Let $S = (V, T, A, R)$ be a $RSEH_{(1,4)}$ system. Let $L = L(S)$ be the language generated by it. Then $L = L(S')$, where $S' = (V', T, A', R')$ is a $RSEH_{(1,4)}$ system, and*

$$\begin{aligned} V' &= T \cup \{N\}, N \notin T \\ A' &\subseteq (NT^*N \cup NT^* \cup T^*N \cup T^*) \\ R' &\subseteq T \end{aligned}$$

In other words in S':

- *No nonterminal is used in splicing.*
- *S' uses just one nonterminal symbol.*
- *The nonterminal symbol does not occur in the middle of any axiom.*

Proof. S' can be constructed as follows:

$$V' = T \cup \{N\}$$

$$R' = R \cap T$$

If $\alpha_1 N_1 \alpha_2 N_2 \ldots \alpha_r N_r \alpha_{r+1} \in A, (\alpha_i \in T^*, 1 \leqslant i \leqslant r+1), (N_i \in V-T, 1 \leqslant i \leqslant r+1)$, then $\alpha_1 N, N\alpha_2 N, \ldots, N\alpha_r N, N\alpha_{r+1} \in A'$. If $\alpha \in T^*$, is an axiom, then $\alpha \in A'$.

First notice that no nonterminal is used for splicing in S. If a non terminal is used for splicing in S then it cannot be removed because of the rule 3 given above for restricted splicing. Whenever the axiom uN_ivawN_jy is used in S to splice at the site a, we can use the corresponding axiom $NvawN$ in S' to splice at the site a and vice versa. It is easy to see that any string generated by $L(S)$ can be generated by $L(S')$ as well and vice versa. Hence the proof follows. □

This result can be proved for $RSEH_{(1,3)}$ system also.

Since we have a finite set of axioms the language generated in these cases is only regular.

Theorem 33.4 *Let $S = (V, T, A, R)$ be an $RSEH_{(1,4)}$ system. Let $L = L(S)$. Then there exists an $RSH_{(1,4)}$ system S', $S' = (V, A', R)$ such that $L(S') = L(S)$.*

Proof. This proof makes use of the fact that

$$SEH_{(1,4)} = RSEH_{(1,4)} \subseteq REG$$

Since $L = L(S)$ is regular, there exists a finite state automaton to recognize L. Let N be the minimum number of states in a DFSA that recognizes L. Define $M = \max\{\mid w \mid \mid w \in A\}$. Define $S^{2M+N} = \{w \mid \mid w \mid \leqslant M + 2N, w \in L(S)\}$. Then consider $S' = (V, S^{2M+N}, R)$. Notice that S^{M+2N} is finite since $M + 2N$ is finite.

Case 1: $L(S') \subseteq L(S)$. This follows directly from the fact that $S^{2M+N} \subseteq L(S)$.

Case 2: $L(S) \subseteq L(S')$. This is proved by contradiction. Let $w \in L(S)$. Assume that $w \notin L(S')$. Further assume that w is the smallest such string. Then w can be written as $\delta\gamma\omega$ where, $(\mid \gamma \mid > M, \mid \delta \mid = N, \mid \delta \mid = N)$. Let D be the DFSA that recognizes $L(S)$ using just N states.

Since ω is of length N, the finite state automaton enters some state twice during processing ω. Therefore there exists some ω' such that

$$\delta\gamma\omega' \in L(S) \ \ and \ \ \mid \omega' \mid < \mid \omega \mid$$

Similarly there exists δ' such that

$$\delta'\gamma\omega \in L(S) \ \ and \ \ \mid \delta' \mid < \mid \delta \mid$$

By our initial assumption it follows that $\delta\gamma\omega'$ and $\delta'\gamma\omega$ both belong to $L(S')$. Moreover since $\gamma > M$ it contains a site i.e there exists a letter $a \in R$ such that $\gamma = xaay$. But now by splicing the strings as shown below we get the string w, which is a contradiction.

$$(\gamma xaay\delta', \gamma' xaay\delta) \vdash \gamma xaay\delta = w$$

Thus $w \in L(S')$. Hence the proof. □

This result also can be proved for $SEH_{(1,3)}$ systems.

Theorem 33.5 *There exists a $SH_{(1,4)}(p)$ language that is not a $SH_{(1,4)}$ language.*

Proof. Consider the $SH_{(1,4)}(p)$ system $S = (V, A, R)$.

$$V = \{a, b, c, d, e, f, x, y, 0, 1\}$$

$$A = \{a1001, b1001x, c10001, d10001y, e1, f1\}$$
$$R = \{(1, a, b), (1, c, d), (1, e, x), (1, f, y), (1, e, f,)\}$$

using the rule $(1, a, b)$, strings of the form below are generated.

$$a1^+0011^+ \ldots 001^+x$$

using the rule $(1, c, d)$ strings of the form below are generated.

$$c1^+00011^+ \ldots 0001^+y$$

using the rule $(1, e, x)$ strings of the form below are generated.

$$e11^+0011^+ \ldots 001^+x$$

using the rule $(1, f, y)$ strings of the form below are generated.

$$f11^+00011^+ \ldots 0001^+y$$

using the rule $(1, e, f)$ strings of the form below are generated.

$$e11^+0011^+0011^+ \ldots 0011^+00011^+00011^+ \ldots 0001^+y$$

The last class of strings starting with e and ending with y are interesting. Notice that all the zeros in these strings occur in groups of two or groups of three. In addition, all the groups of two zeros occur before any group containing three zeros occurs.

The language generated by this system is clearly not generated by any $SH_{(1,4)}$ system. If there exists a $SH_{(1,4)}$ system that generates this, then 1 has to be a rule. This implies we can generate strings of the form $e11^{+}00011^{+}001^{+}y$ which does not belong to the above language. □

Theorem 33.6 *There exists a $SH_{(1,3)}(p)$ language that is not a $SH_{(1,3)}$ language.*

Proof. Consider the $SH_{(1,3)}(p)$ system $S = (V, A, R)$.

$$V = \{a, b, c, d, e, f, x, y, 0, 1\}$$

$$A = \{a1001, b1001x, c10001, d10001y, e1, f1\}$$

$$R = \{(1, a, b), (1, c, d), (1, e, x), (1, f, y), (1, e, f,)\}$$

using the rule $(1, a, b)$, strings of the form below are generated.

$$a1001 \ldots 001x$$

using the rule $(1, c, d)$ strings of the form below are generated.

$$c10001 \ldots 0001y$$

using the rule $(1, e, x)$ strings of the form below are generated.

$$e1001 \ldots 001x$$

using the rule $(1, f, y)$ strings of the form below are generated.

$$f10001 \ldots 0001y$$

using the rule $(1, e, f)$ strings of the form below are generated.

$$e1001001 \ldots 00100010001 \ldots 0001y$$

The last class of strings starting with e and ending with y are interesting. Notice that all the zeros in these strings occur in groups of two or groups of three. In addition, all the groups of two zeros occur before any group containing three zeros occurs.

The language generated by this system is clearly not generated by any $SH_{(1,3)}$ system. If there exists a $SH_{(1,3)}$ system that generates this, then 1 has to be a rule. This implies we can generate strings of the form $e10001001y$ which does not belong to the above language. □

4. CONCLUSIONS

In this paper we have seen that restricting the way the splicing takes place in a simple H system in a particular manner does not affect the generative power in cases (1,3) and (1,4). Also having target alphabets (i.e nonterminals and terminals) does not increase the power of systems of type (1,3) or (1,4) whereas having permitting context increases the power.

We think such results cannot be proved for (2,3) type systems. These and results in [2, 5] show that (2,3) systems behave in a different manner from (1,3) or (1,4) systems. It will also be interesting to find whether $SEH_{(1,3)}$ and $SEH_{(1,4)}$ systems with permitting context generate nonregular sets.

References

[1] Chakaravarthy, V.T. & K. Krithivasan (1997), A note on extended H systems with permitting/forbidden context of radius one, *Bulletin of the European Association for Theoretical Computer Science*, 208–213.

[2] Chakaravarthy, V.T. & K. Krithivasan (1998), Some Results on Simple Extended H-systems, *Romanian Journal of Information Science and Technology*, I.3: 203–215.

[3] Head, T. (1987), Formal language theory and DNA: an analysis of the generative capacity of specific recombinant behaviors, *Bulletin of Mathematical Biology*, 49: 737–759.

[4] Head, T.; Gh. Păun & D. Pixton (1997), Language theory and molecular genetics: generative mechanisms suggested by DNA recombination, in G. Rozenberg & A. Salomaa, eds., *Handbook of Formal Languages*: 295–360. Springer, Berlin.

[5] Lakshminarayanan, S. & K. Krithivasan, On the generative power of simple H systems, submitted.

[6] Mateescu, A.; Gh. Păun; G. Rozenberg & A. Salomaa (1997), Simple Splicing Systems, *Discrete Applied Mathematics*.

[7] Păun, Gh. (1996), Computing by Splicing. How simple rules?, *Bulletin of the European Association for Theoretical Computer Science.*

[8] Păun, Gh.; G. Rozenberg & A. Salomaa (1998), *DNA Computing: New Computing Paradigms*, Springer, Berlin.

Chapter 34

ON SOME FORMS OF SPLICING

Vincenzo Manca
Department of Informatics
University of Pisa
Corso Italia 40, 56125 Pisa, Italy
mancav@di.unipi.it

Abstract Some representations and extensions of the combinatorial mechanism of splicing are introduced that help to analyze in a simple manner important concepts about splicing processes and about the generative power of H systems.

1. INTRODUCTION

Splicing is the basic combinatorial operation on which DNA Computing is based. It was introduced in [2] as a formal representation of DNA recombinant behavior: DNA strands are cut in specific *sites*, by enzymes, and strands whose sticky ends match are combined into new DNA molecules.

This mechanism suggested new generative systems in Formal Language Theory and introduced new perspectives in the combinatorial analysis of strings, languages, grammars, and automata. In fact, the biological trend, initiated by Kleene's finite state automata [4] and Lindermayer's L-systems [5], continued in a new direction, because biochemical interpretations were found for concepts and results in formal language theory [12, 2, 3]. The field covering these subjects is now referred as DNA Computing or, in wider perspectives, Molecular Computing, or Natural Computing. The benefits of this synergy between Mathematics, Computer Science and Biology are apparent and constitute an exceptional stimulus for a deep understanding of the combinatorial mechanism underlying splicing processes.

In [7] we introduced the notion of the *Derivation System* that allows us to analyze in a uniform way a great quantity of symbolic systems, and to determine their common structure based on two main aspects: the combinatorial mechanism of the rules (e.g. replacement, parallel replacement, insertion/deletion,

C. Martin-Vide and V. Mitrana (eds.), Where Mathematics, Computer Science, Linguistics and Biology Meet, 387-398.

splicing, ...), and the regulation strategy that specifies the ways rules can be applied. The study of different forms that a given combinatorial mechanism or a given regulation strategy can assume is important in suggesting new perspectives in the analysis and comparison of string manipulation systems.

In this paper we express the combinatorial form of splicing by four cooperating combinatorial rules: two cut rules (suffix and prefix deletion), one paste rule, and one (internal) deletion rule. This natural translation of splicing allows us to prove in a very simple manner the regularity of (nonextended) splicing with finite rules.

Then, we concentrate on *linear* splicing, a form of splicing studied in [8], and show a sort of geometrical representation of splicing where some notions of *redundancy* and *monotony* can be defined.

Finally, we show that a natural extension of splicing, called *distant splicing*, provides, in a very simple manner, the computational universality of splicing systems with finite axioms and finite rules. Other forms of splicing are only suggested as glimpses for future investigations.

2. PRELIMINARIES

Consider an alphabet V and two symbols $\#, \$$ not in V. A *splicing rule* over V is a string $r = u_1\#u_2\$u_3\#u_4$, where $u_1, u_2, u_3, u_4 \in V^*$. For such a rule r and for any $x_1, x_2, y_1, y_2 \in V^*$ we define the (ternary) splicing relation $\Longrightarrow_r$

$$x_1u_1u_2x_2\ ,\ \ y_1u_3u_4y_2 \Longrightarrow_r x_1u_1u_4y_2.$$

We say that the string $x_1u_1u_2x_2$ is the *up premise*, the string $y_1u_3u_4y_2$ is the *down premise*, and the string $x_1u_1u_4y_2$ is the *conclusion* of an *r-splicing step*. The pairs of strings (u_1, u_2), and (u_3, u_4) are the *up site* and the *down site* of the rule r, or also the *splicing points* (of the up and down premises) of r. A splicing point (u, v) is included in a string x if the string uv is a substring of x. If a string x includes a splicing point (u, v) of a rule r, then the prefix of x that includes u is the *r-head* of x and the suffix of x that includes v is the r-tail of x.

Therefore, when an r-splicing step is applied to two strings, they are cut in between the left and right components of the splicing points of r, and then the $r-$*head* of the up premise is concatenated with the $r-$*tail* of the down premise. The resulting string is the conclusion of the step.

An *H system* system [10] $\Gamma = (V, A, R)$ is given by: an alphabet V, a set A of strings over this alphabet, called axioms of the system, and a set R of splicing rules over this alphabet. The language $L(\Gamma)$ generated by Γ consists of the axioms and the strings that we can obtain starting from the axioms, by applying to them iteratively the splicing rules of Γ. If a terminal alphabet $T \subset V$ is considered, and $L(\Gamma)$ consists of the strings over T^* that can be generated from the axioms (by applying the splicing rules of Γ to them iteratively), then we obtain an *extended* H system $\Gamma = (V, T, A, R)$. We say an H system to be *finitary* when it has a finite number of splicing rules. H systems are usually classified by means of two classes of languages FL_1, FL_2: a H system is of type $H(FL_1, FL_2)$ when its axioms are a language in the class FL_1, and its rules (viewed as strings) are a language in the class FL_2; $EH(FL_1, FL_2)$ is the class of extended H systems of type $H(FL_1, FL_2)$. Let us identify a type of H systems with the classes of languages generated by them, and let FIN, REG, RE indicate the classes of finite, regular, and recursively enumerable languages respectively. It is known that: $H(FIN, FIN) \subset REG$, $H(REG, FIN) = REG$, $EH(FIN, FIN) = REG$, $EH(FIN, REG) = RE$. Comprehensive details can be found in [10].

3. CUT-AND-PASTE SPLICING

One of the most important mathematical properties of splicing, in its original formulation, was that the class of languages generated by finite splicing rules, from an initial set of strings, is a subclass of regular languages ($H(FIN, FIN) \subset REG$ and more generally $H(FIN, FIN) = REG$). Nevertheless, the proof of this result has a long story. It originates in [1] and was developed in [11], in terms of a complex inductive construction of a finite automaton. In [10] Pixton's proof is presented (referred as *Regularity preserving Lemma*). More general proofs, in terms of closure properties of abstract families of languages, are given in [3, 2]. In [8] a direct proof was obtained by using linear splicing (see the next section).

In this section we give another and easier direct proof of this lemma, as a natural consequence of a representation of splicing rules.

The language $L(\Gamma)$ generated by an H system Γ can be obtained in a different way. Replace every splicing rule $u_1\#u_2\$u_3\#u_4$ of Γ by the following four rules (two cut rules, a paste rule, a deletion rule) where $\#$ and $\bullet$ are two new nonterminal symbols.

$$xu_1u_2y \Longrightarrow xu_1\#$$

$$xu_3u_4y \Longrightarrow \#u_4y$$

$$xu_1\#, \#u_4y \Longrightarrow xu_1 \bullet u_4y$$

$$x \bullet y \Longrightarrow xy$$

Then, if we apply these rules iteratively, starting from the axioms of Γ, the strings so obtained, where only terminal symbols occur, constitute the language $L(\Gamma)$. We say that the system Γ' obtained with this formulation of splicing rules of Γ is the *cut-and-paste* representation of Γ and any pair $u_1 \bullet u_4$ is also called a *bullet pair*. The bullet is used in order to distinguish a substring u_1u_4 that occurs in the axioms from another one that was obtained after splicing; in fact, this difference is essential in the use we develop for this representation.

This representation of H systems allows us to prove in a natural way the regularity lemma for splicing languages. Let us consider the notion of *knot* for an H system Γ' in cut-and-paste format. A knot is a string of the following type (x, y, z any strings of terminal symbols):

$$xu_1 \bullet u_4yu_1 \bullet u_4z$$

that can be obtained from the axioms by applying the rules of the system, and where the string y does not include two occurrences of the same bullet pair.

The string $u_4yu_1\bullet$ is called the *cross* of the knot and the string $u_1 \bullet u_4$ the *tie* of the knot.

Lemma 34.1 *If* Γ *is a finitary H system with a finite set of axioms, then* $L(\Gamma)$ *is infinite iff its cut-and-paste representation can generate some knot.*

Proof. If a knot $xu_1 \bullet u_4yu_1 \bullet u_4z$ can be generated in the cut-and-paste representation Γ' of Γ, then $xu_1(u_4yu_1)^iu_4z$ can also be generated in Γ' for any natural number i. In fact, in order to generate the knot, we are sure that we have already generated the three strings $xu_1\#, \#u_4yu_1\#, \#u_4z$ in Γ'; therefore, if we apply the past rule to the first two strings, we get $xu_1 \bullet u_4yu_1\#$; if we paste this string again with $\#u_4yu_1\#$ and then with $\#u_4z$, we generate $xu_1\bullet u_4yu_1\bullet u_4yu_1\# = xu_1\bullet(u_4yu_1\bullet)^2u_4z$. In general, if we paste $\#u_4yu_1\#$ i times consecutively, starting from $xu_1\#$ and finally pasting $\#u_4z$ we generate $xu_1 \bullet (u_4yu_1\bullet)^iu_4z$. Now if we delete the bullets we get what we claim to generate.

Conversely, if no knot can be generated, then $L(\Gamma)$ is finite, in fact, no string can be generated where $u_1\bullet u_4$ i.e. the tie of some knot occurs twice; therefore, as Γ' only has finite axioms and rules, we have a finite number of possibilities in generating other strings which are different from the axioms by splicing.

□

In its weak form, Regularity Preserving Lemma can be expressed by the following theorem.

Theorem 34.1 *(Regularity Preserving Lemma)*
If Γ *is a finitary H system with a finite set of axioms, then* $L(\Gamma)$ *is a regular language.*

Proof. Consider all the strings that can be generated with no knots. There are a finite number. Consider all the possible knots $\{k_1, \dots k_n\}$ that can be generated in the cut-and-paste representation Γ' of Γ. They are also of a finite number. The language $L(\Gamma)$ is the union of two sets: the set of strings generated without knots and the set of strings represented by the regular expressions generated during the following *expansion* procedure applied to all the knots $\{k_1, \dots k_n\}$. Given a knot $k = xu_1 \bullet u_4yu_1 \bullet u_4z$, its expansion of level 1 is the regular expression $xu_1 \bullet (u_4yu_1\bullet)^* \bullet u_4z$. Now assume that in the expansion E of the knot k, possibly after the deletion of some bullets, we find the occurrence of a tie $u'_1 \bullet u'_4$ different from $u_1 \bullet u_4$ (a substring occurs in a regular expression $\alpha\beta^*\gamma$ if it occurs in α or β, or γ). In this case, if γ' is the cross of a knot k' with tie $u'_1 \bullet u'_4$, then we expand the expression E into E' again by replacing the bullet of $u'_1 \bullet u'_4$ by γ'^*. The expression E' is called an expansion of level 2 of k. Analogously we will proceed in getting the expansions of k of level greater than 2.

We claim that there are some numbers $m_1, \dots m_n$ such that after m_1 levels for $k_1, \dots m_n$ levels for k_n we are sure we generated all the possible different expansions of $k_1, \dots k_n$. In fact, consider again the knot k; if at some level, say in an expansion E'' of the third level, we meet a tie that is already expanded, say $u'_1 \bullet u'_4$, then it is easy to verify that E'' could be factorized as $E'' = xu'_1yu'_4zu'_1wu'_4t$ (for some strings x, y, z, w, t); this means that E'' should be either a knot with tie $u'_1 \bullet u'_4$, or an expansion, at some level, of such a knot.

In order to conclude the proof we need to consider the possibility that a tie is not included in an expansion $E = xy^*z$, but in some instance of it xy^iy^*z or xy^*y^iz, for some positive integer i. But the possible length of the ties is bound; therefore the value of i that covers all these possible cases is determined by the rules of our H system Γ.

In conclusion, apart the finite set of strings generated without knots, the language generated by Γ' is expressed by a finite set of regular expression; this means that it is regular; therefore $L(\Gamma)$ is regular. □

The argument of the previous theorem remains essentially the same if the axioms of Γ are a regular language; in that case we have the finite (regular) expansions of a finite set of regular expressions, but this is of course a regular set.

4. LINEAR SPLICING

In this section, we consider *linear* splicing and show that we can express several interesting concepts, within its terms.

Definition 34.1 *Given an H system $\Gamma = (V, A, R)$, a splicing derivation δ of Γ, of length n, is a sequence of n strings and n labels, where each string associated with a corresponding label (written before an arrow that points to the string). A label is, either a triple (rule, string, string), or a special label indicated by λ:*

$$(\lambda \to \delta(1),\ l(1) \to \delta(2),\ \dots,\ l(n-1) \to \delta(n))$$

where for $1 \leqslant i < n$:

- *if* $l(i) \neq \lambda$ *then* $l(i) \in R \times \{\delta(1), \ldots \delta(i-1)\} \times \{\delta(1), \ldots \delta(i-1)\}$;
- *if* $l(i) = \lambda$, *then* $\delta(i+1) \in A$;
- *if* $l(i) = (r_i, \beta(i), \gamma(i))$, *then* $\beta(i), \gamma(i) \Longrightarrow_{r_i} \delta(i+1)$;
- $\forall i\ i < n\ \exists j,\ n \geqslant j > i$ *such that* $\beta(j), \gamma(j) \Longrightarrow_{r_i} \delta(j+1)$ *and either* $\delta(i) = \beta(j)$ *or* $\delta(i) = \gamma(j)$;
- *if* $n \neq 1$, *then* $\delta(n) \notin A$.

At each step i, $\delta(i)$ is called the *current string* of that step. According to the definition, any current string, apart the final one, has to be the conclusion of some splicing step with premises which are current strings of steps preceding i (no useless strings can occur in a derivation δ).

We indicate by $\Delta(\Gamma)$ the set of *(splicing) derivations* of an H system Γ.

If the last element $\delta(n)$ of a derivation δ is the string α we say that δ derives α and we write

$$\delta \vdash_\Gamma \alpha.$$

Two derivations of $\Delta(\Gamma)$ are said to be *equivalent* if the same string derives from them.

A derivation $\delta \in \Delta(\Gamma)$ of length n is said to be *linear* when, for any $1 \leqslant i < n$, if the element $\delta(i+1)$ is not an axiom, then it is obtained by applying a rule of R with $\delta(i)$ as a premise.

Lemma 34.2 *(Linearity Lemma)*
Given a derivation $\delta \in \Delta(\Gamma)$ *there is always a linear derivation equivalent to it.*

Proof. By induction on the length of derivations. For derivations where no rules are applied the linearity is trivial. Assume that for derivations of length smaller or equal to n we have linear derivations equivalent to them. Let δ be a derivation of length $n+1$ the string $\delta(n+1)$ has two premises $\delta(i), \delta(j)$ with $i, j \leqslant n$.

By induction hypothesis there are two linear derivations δ', δ'' of length at most n that derive $\delta(i), \delta(j)$ respectively. Consider the concatenation of δ', δ'' where the steps of δ'' that are already in δ' are removed, then add to it the last splicing step of δ. This derivation is a linear derivation equivalent to δ. □

A linear derivation can be represented in the following way:

$$(\lambda \to \delta(1),\ l(1) \to \delta(2),\ \ldots,\ l(n-1) \to \delta(n))$$

where, if at step i, the label is different from λ, then it is $(r_i, \beta(i), p)$ with $\beta(i) \in \{\delta(1), \ldots \delta(i-1)\}$, and $p \in \{0, 1\}$, in such a way that:

- if $l(i) = (r_i, \beta(i), 0)$, then $(\beta(i), \delta(i)) \Longrightarrow_{r_i} \delta(i+1)$;
- if $l(i) = (r_i, \beta(i), 1)$, then $(\delta(i), \beta(i)) \Longrightarrow_{r_i} \delta(i+1)$.

All the derivations we consider in this section are (tacitly) assumed to be linear derivations.

The notion of ω-splicing allows us to consider infinite processes of splicing (ω stands for the set of natural numbers).

Given two (linear) derivations δ, δ' we say that δ is an *expansion* of δ', and we write $\delta < \delta'$, if δ' is obtained by interrupting δ at some step i; Performing some further steps that derive a string, say β, we then continue from β by applying the same labels of the steps that are after the step i in δ.

An ω-splicing is an infinite sequence of derivations:

$$\delta = (\delta^i \mid i \in \omega)$$

where, for every $i \in \omega$ δ^i is called a *component* of δ, and for every $i > 1$ there is a $j \leqslant i$ such that $\delta^j < \delta^i$.

The set $\Delta^\omega(\Gamma)$ is the set of ω-splicings δ such that all their components belong to $\Delta(\Gamma)$. Any ω-splicing δ determines a language $L(\delta)$ constituted by the strings derived by its components.

The notion of ω-splicing was the basis for proving the Regularity Preserving Lemma, via linear splicing, in [8]. It is easy to prove that, given a finitary H system Γ whose axioms belong to a class FL of formal languages closed with respect to the noniterated splicing (see [10]), if $\Delta^\omega(\Gamma) = \emptyset$, then $L(\Gamma)$ also belongs to FL. Therefore, when Γ has finite axioms, if $L(\Gamma)$ is infinite, then $\Delta^\omega(\Gamma) \neq \emptyset$. If we consider only *nonredundant* ω-splicings (see later on), then we can reverse this implication; that is, $L(\Gamma)$ turns out to be infinite iff the set of nonredundant ω-splicings of $\Delta^\omega(\Gamma)$ is not empty.

Let $\Gamma = (V, A, R)$ be an H system. A linear derivation δ is called an *axiomatic* derivation if at any splicing step of δ at least one of the two premises is an axiom. The derivation δ is said to be *right-down* or RD if at any splicing step of δ the down premise is an axiom; while δ is said to be *left-up* or LU if at any splicing step of δ the up premise is an axiom.

The terms *right-down* and *left-up* are due to a graphical representation of splicing where the up premise is a (rectangular) frame, and the down premise another (rectangular) frame put under it.

In this manner the r-head of the superior frame (up premise) and the r-tail of the inferior frame (down premise), connected by an arrow, provide a representation of the result of the r-splicing. If the result of this step is the up premise of another step, we go in the *right down* direction (this kind of '*staired*' representation was adopted in [9]); otherwise if it is the down premise of a further step we go in the *left up* direction.

The following is the graphical representation of an axiomatic RD derivation in three steps.

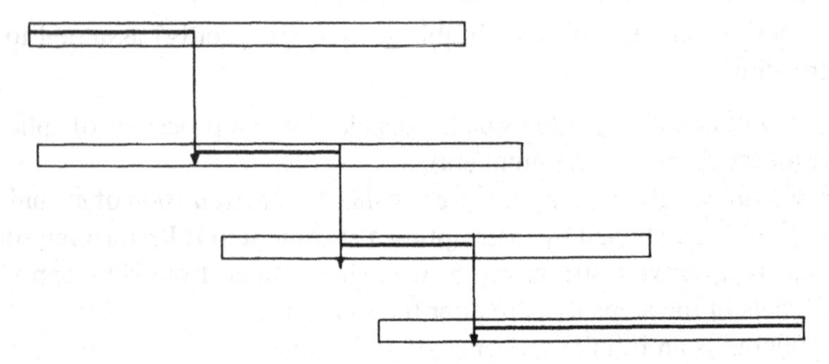

The graphical representation of an axiomatic LU derivation in three steps is the same, but with the arrows reversed.

Note that, in general, an axiomatic *LU* derivation cannot be transformed into an axiomatic *RD* derivation by changing the verse of the arrows in the 'staired' representation. For example, given the splicing rule $\gamma\delta\#\beta\$\gamma\#\delta$ with $\gamma \neq \delta$, you can go in the left up direction any number of times, but only one step in the right down direction (analogously, a rule $\alpha\#\beta\$\gamma\#\alpha\beta$ with $\gamma \neq \alpha$ shows that, in general, axiomatic *RD* derivations cannot be transformed into *LU* axiomatic derivations by reversing the arrows).

Any derivation can be factorized in terms of different *nested levels* of axiomatic derivations. We can illustrate this factorization by using the following diagram, where horizontal lines indicate sequence of steps and bullets represent start or end strings of these sequences. Strings α_2, α_3 are premises of a splicing step of conclusion α_4; and α_1, α_5 are premises of a splicing step of conclusion α_6.

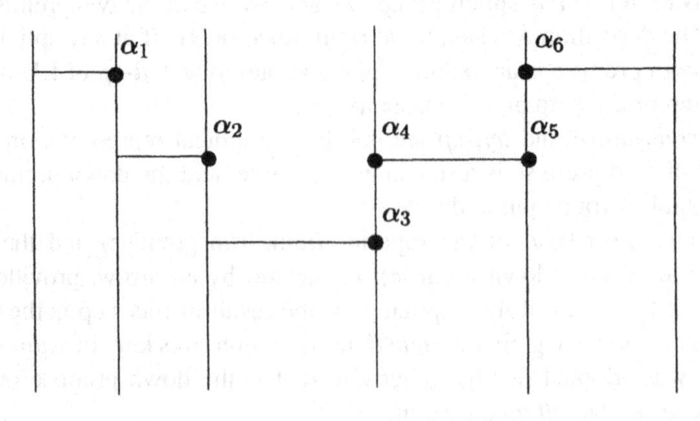

Given a derivation δ of length n, a *subderivation* of δ is a subsequence $\delta' = (l(j_1) \rightarrow \delta(j_1), ..., l(j_m) \rightarrow \delta(j_m))$, with $1 \leqslant j_1 \leqslant j_m \leqslant n$, which is also a derivation. A subderivation of δ is a *proper subderivation* if it is different from δ.

A derivation is *nonredundant* if it has no proper subderivation equivalent to it. For example, an H system with a splicing rule $u\#v\$u'\#v$ can generate derivations of any length that derive the same string. A less trivial situation of redundancy is illustrated by the following diagram, where horizontal lines model (parts of) strings involved in a splicing derivation.

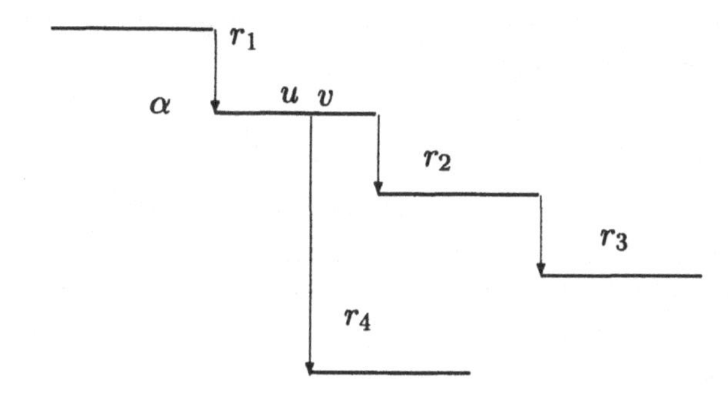

Consider the sequence of splicing steps r_1, r_2, r_3, r_4. Let α be the r_1-tail of the conclusion of the first splicing step. The up site (u, v) of the fourth step is included in α, then the sequence of rules r_1, r_2, r_3, r_4 is equivalent to the sequence r_1, r_4, where the two steps r_2, r_3 are avoided.

An ω-splicing is *nonredundant* iff all its components are nonredundant.

In a linear splicing derivation, each splicing step is constituted by two substeps: the current string is cut in at a splicing point: a part of this string is kept (the left part in the case of an RD step, or the right part in the case of an LU step) and a string is appended to this point (on the right in an RD step, and on the left in an LU step). If at a step $i > 2$, the splicing point is internal to a part of the current string that was appended in a step $j < i - 1$, then we say that for any k, $j < k < i$, the step i is *anti-directional* with respect to the step k (a splicing step is anti-directional if it is anti-directional with respect to some splicing step). For example, in the derivation (r_1, r_2, r_3, r_4), represented by the diagram above, the fourth splicing step is anti-directional with respect to the second step and to the third step. In other words, an anti-directional step removes, in the current string, the splicing point of a previous step.

A step is *monotone* if it is not anti-directional; a derivation is monotone, if all its splicing steps are monotone. An ω-splicing is monotone if its components are monotone.

In any string we have a finite number of splicing points; moreover, from the definition above, all the splicing points of the anti-directional steps of a given step are located in the same string. Therefore the following lemma holds.

Lemma 34.3 *(Finite Anti-directionality Lemma)*
In a nonredundant ω-splicing the steps that are anti-directional with respect to a given step i, if there are some of them, are of a finite number.

5. DISTANT SPLICING

A *Distant Splicing* rule r, indicated by $u_1\#\#u_2\$u_3\#\#u_4$, works in the following manner (x, x', y, y' any strings):

$$xu_1yu_2z,\quad x'u_3y'u_4z' \Longrightarrow_r xu_1yy'u_4z'.$$

DH systems are H systems where distant splicing replaces the usual splicing. $EDH(FL_1, FL_2)$ indicates the class of languages generated by extended DH systems with axioms belonging to the class FL_1 and rules (viewed as strings) to the class FL_2. The following theorem expresses the computational universality of extended DH systems.

Theorem 34.2 $EDH(FIN, FIN) = RE$

Proof. Consider the proof of lemma 7.16 in [10]. In this proof a type-0 grammar is simulated by an H systems with a regular set of rules. It is easy to verify that those rules are essentially distant splicing rules. □

If we formulate a distant splicing rule in a cut-and-paste form, we see that the only difference with the cut-and-paste representation of a usual splicing rule consists in the number of markers we need in the cutting process. In fact, we can introduce four nonterminal symbols, say a_1, a_2, a_3, a_4 in correspondence to the strings u_1, u_2, u_3, u_4 of a distant splicing rule and other two symbols, say b_1, b_4 in correspondence to the strings u_1, u_4 of the rule. The following set of seven rules simulates the application of the distant splicing rule $u_1\#\#u_2\$u_3\#\#u_4$.

$xu_1y \Longrightarrow xu_1a_1, a_1y$; $xu_2y \Longrightarrow xa_2$;
$xu_4y \Longrightarrow xa_4, a_4u_4y$; $xu_3y \Longrightarrow a_3y$;
$a_1xa_2, a_3ya_4 \Longrightarrow b_1xyb_4$; $xu_1a_1, b_1y \Longrightarrow xu_1y$; $yb_4, a_4u_4x \Longrightarrow yu_4x$.

In this representation it is apparent that distant splicing can be performed by agents (enzymes) with the same competence as agents performing usual splicing. They need: i) to recognize substrings, to cut in strings at the end of a recognized substring (or in general, at some internal point of it), and to mark the strings after cutting them in some way(with sticky ends, marks, features);

ii) to paste strings whose ends match in some way and to delete some internal symbols. (In order to be completely precise, 'cut-and-paste' means that some symbols can be appended to a string resulting from a cutting, and that some ending symbols can be deleted before pasting two strings).

This means that many agents with these abilitie, which cooperate according an intrinsic strategy, can extend the power of splicing rules significantly.

Distant splicing can be formulated in ways that show their resemblance to typical patterns in molecular biology (e.g., in the elimination of nonencoding DNA strands, or in the genetic determination of immunoglobulins with the recombination of genes V, J, D, C). The basic operations of these representations are pasting, contextual deletion (with markers) and suffix-prefix deletion (x, x', y, y', z, w any strings):

$x, y \Longrightarrow xy$ (pasting)
$xu_2yu_3z \Longrightarrow x\#z$ ((u_2, u_3) contextual marked deletion)
$xu_1y\#zu_4w \Longrightarrow xu_1yzu_4w$ ((u_1, u_4) contextual deletion)

or also:

$xu_2x', y'u_3y \Longrightarrow x\#y$ (prefix-suffix pasting)
$xu_1y\#zu_4w \Longrightarrow xu_1yzu_4w$ ((u_1, u_4) contextual deletion).

The cut-and-paste representation of distant splicing, and the subsequent remarks suggest several possible extensions of this combinatorial mechanism to us. For example, it could be interesting to consider conditions driving the cut and paste mechanisms that are more general than the recognizing of substrings. In [6], we suggest a splicing formulation of operations typical in the syntax of natural languages. In that case it would be very useful to consider cut and paste procedures determined by logical conditions concerning strings and classes of strings.

References

[1] Culik II, K. & T. Harju (1991), Splicing semigroups of dominoes and DNA, *Discrete Applied Mathematics*, 31: 261–277.

[2] Head, T. (1987), Formal language theory and DNA: an analysis of the generative capacity of specific recombinant behaviors, *Bulletin of Mathematical Biology*, 49: 737–759.

[3] Head, T.; Gh. Păun & D. Pixton (1997), Language theory and molecular genetics, in G. Rozenberg & A. Salomaa, eds., *Handbook of Formal Languages*, II: 295–360. Springer, Berlin.

[4] Kleene, S. C. (1956), Representation of events in nerve nets and finite automata, in *Automata Studies*, Princeton University Press, Princeton.

[5] Lindenmayer, A. (1968), Mathematical models for cellular interaction in development, I and II, *Journal of Theoretical Biology*, 18: 280–315.

[6] Manca, V. (1998), Logical Splicing in Natural Languages, in M. Kudlek, ed., *Proceedings of the MFCS'98 Satellite Workshop on Mathematical Linguistics*: 127–136, Universität Hamburg.

[7] Manca, V. (1998), Logical String Rewriting, *Theoretical Computer Science*, special issue devoted to the 23rd International Symposium on Mathematical Foundations of Computer Science, to appear.

[8] Manca, V. (2000), Splicing Normalization and Regularity, in C. Calude & Gh. Păun, eds., *Finite versus Infinite. Contributions to an Eternal Dilemma*: 199-215. Springer, London.

[9] Păun, Gh. (1996), On the splicing operation, *Discrete Applied Mathematics*, 70.1: 57–79.

[10] Păun, Gh.; G. Rozenberg & A. Salomaa (1998), *DNA Computing. New Computing Paradigms*, Springer, Berlin.

[11] Pixton, D. (1996), Regularity of splicing languages, *Discrete Applied Mathematics*, 69.1-2: 101–124.

[12] Rozenberg, G. & A. Salomaa, eds., *Handbook of Formal Languages*, Springer, Berlin.

Chapter 35

TIME-VARYING DISTRIBUTED H-SYSTEMS OF DEGREE 2 GENERATE ALL RECURSIVELY ENUMERABLE LANGUAGES

Maurice Margenstern
Metz Group of Foundational Computer Science
University Institute of Technology
Ile du Saulcy, 57045 Metz Cedex, France
margens@antares.iut.univ-metz.fr

Yurii Rogozhin
Institute of Mathematics and Informatics
Academy of Sciences of Moldova
Str. Academiei 5, MD-2028 Chisinau, Moldova
rogozhin@math.md

Abstract A time-varying distributed H system is a splicing system which has the following feature: at different moments one uses different sets of splicing rules. The number of these sets is called the degree of the system. The passing from one set of rules to another is specified in a cycle. It is known that any formal language can be generated by a time-varying distributed H-system of degree at least 4. We already proved that there are universal time-varying distributed H-systems of degree 2. In this paper we strengthen that result by showing, for any recursively enumerable language, how to construct a time-varying distributed H-system of degree 2 that generates that language exactly. We also indicate that such a construction is impossible for time-varying distributed H-systems of degree 1.

1. INTRODUCTION

Time-varying distributed H-systems were recently introduced in [7], [8, 6] as another theoretical model of biomolecular computing (DNA-computing), based on splicing operations. We refer the reader to [4, 5] for more details

C. Martín-Vide and V. Mitrana (eds.), Where Mathematics, Computer Science, Linguistics and Biology Meet, 399-407.

on the history of this model and its connections with previous models based on splicing computations. That new model introduces *components* (see the formal definition below) which cannot all be used at the same time but one after another, periodically.

This paper aims at giving an account of real biochemical reactions where the work of enzymes depends essentially on environment conditions. In particular, at any moment, only one subset of all available rules is in action. If the environment is periodically changed, then the active enzymes also change periodically.

In [8], it is proved that 7 different *components* are enough to generate any recursively enumerable language. In [6], the number of components was reduced to 4. Recently, see [3, 4], both authors proved that two components are enough to construct a *universal* time-varying distributed H-system, *i.e.* a time-varying distributed H-system, capable of simulating the computation of any Turing machine. Universality of computation and the ability of generate any recursively enumerable language are equivalent properties, but it is not necessarily true *a priory*, that the universality of some time-varying distributed H-systems with two components entails that there are time-varying distributed H-systems which generate all recursively enumerable languages, with only two components. Here we prove the latter result, namely that 2 different components are enough to generate any recursively enumerable language. It will be enough to show an algorithm which allows us to transform any recursively enumerable language $\mathcal{L}$ into a time-varying distributed H-system that generates $\mathcal{L}$. Taking $\mathcal{L}$ as a universal recursively enumerable language, the algorithm transforms $\mathcal{L}$ into a time-varying distributed H-system that generates any recursively enumerable language. The result was presented by both authors at FBDU'98, a satellite workshop of MFCS/CSL'98 Federal Conference, August 22-30, 1998, at Brno, Czech Republic.

The question of the status of time-varying distributed H-system of degree one is now solved: such systems generate only regular languages as was announced in a talk by the first author given to the seminar of theoretical computer science of the Mathematical Institute of the Academy of Sciences of Moldova, in August 1999, see [5] for a (simple) proof.

2. MAIN RESULT

We take definitions of *Head-splicing-system* and notations from [4, 5].

A *time-varying distributed H-system* (of degree n, $n \geqslant 1$), short $TVDH$-system, is a construct

$$\Gamma = (V, T, A, R_1, R_2, \ldots, R_n),$$

where V is an alphabet, $T \subseteq V$ is called the *terminal alphabet*, A is a finite subset of V^* called the *set of axioms*, and R_i are finite sets of splicing rules over V, $1 \leqslant i \leqslant n$. Sets R_i, $1 \leqslant i \leqslant n$, are called the *components* of the system.

At each moment $k = n \cdot j + i$, for $j \geqslant 0$, $1 \leqslant i \leqslant n$, only component R_i is used for splicing the currently available strings. Specifically, we define

$$\mathcal{L}_1 = A,$$
$$\mathcal{L}_{k+1} = \sigma_i(\mathcal{L}_k), \text{ for } i \equiv k(mod\, n),\ k \geqslant 1,$$

where $\sigma_i = \sigma_{(V,A,R_i)}$, $1 \leqslant i \leqslant n$, is the computation associated with H-system (V, A, R_i).

Therefore, from a step k to the next step, $k + 1$, one only passes the result of splicing the strings in $\mathcal{L}_k$ according to the rules in R_i for $i \equiv k(mod\ n)$; the strings in $\mathcal{L}_k$ which cannot enter a splicing are removed. That latter clause plays an important role in the sequel.

The language generated by Γ is defined by

$$\mathcal{L}(\Gamma) = (\bigcup_{k \geqslant 1} \mathcal{L}_k) \cap T^*.$$

By RE we denote the family of recursively enumerable languages, by VDH_n, $n \geqslant 1$, the family of languages generated by time-varying distributed H systems of degree at most n, and by VDH_* the family of all languages of this type.

Păun's latest result [8] shows that $VDH_n = VDH_* = RE$, $n \geqslant 7$. Later, it was proved in [6] that $VDH_n = RE$ for $n \geqslant 4$.

Independently of [6], we improved the former result by showing that $VDH_n = RE$ for $n \geqslant 2$.

Our goal is now to prove the following result :

Theorem 35.1 *Time-varying distributed H-systems of degree 2 generate all recursively enumerable sets, or in other words, $VDH_2 = RE$.*

3. SKETCHY PROOF OF THE RESULT

3.1 INFORMAL INDICATIONS

In our proof, we shall directly simulate the computation of a Turing machine. We assume that readers are familiar with that model and refer them to [5, 2] in order to remind them of the necessary background.

The part of the tape delimited by both ends of the configuration can be represented by a word in the alphabet of the symbols which can be read or written on the tape, called the *tape alphabet*. A slight modification of that word allows us to represent the configuration: let ω_ℓ be the part of that word representing the contents of the cells ranging from the left end of the configuration up to the left neighbour of the scanned cell. It is clear that ω_ℓ is empty if the scanned cell is the left end of the current configuration. We define ω_r symmetrically to be the part of the word representing the set of cells ranging from the right neighbour of the scanned up to the right end of the configuration. It is clear that ω_r is empty if the scanned cell is in the right end cell. If x denotes the scanned symbol and if the machine head is in state **a**, we can denote the current configuration by $\omega_\ell\,\mathbf{a}\,x\,\omega_r$. It will be enough that the states are denoted by letters which do not belong to the tape alphabet.

Assume that ω_ℓ is not empty. It can be rewritten as $\nu_\ell\, z$, where z is a tape symbol. Now, as an illustrative example, consider instruction $\boxed{\mathbf{a}\; x\; L\; y\; \mathbf{b}}$, say I, where notations are taken from [5] and mean the following: being under state **a**, the machine head reads x in the scanned cell, replaces x with y, moves one step to left and turns to state **b** for the next step of its computation. Then, the action of performing I can be rewritten in the following way, in terms of configurations:

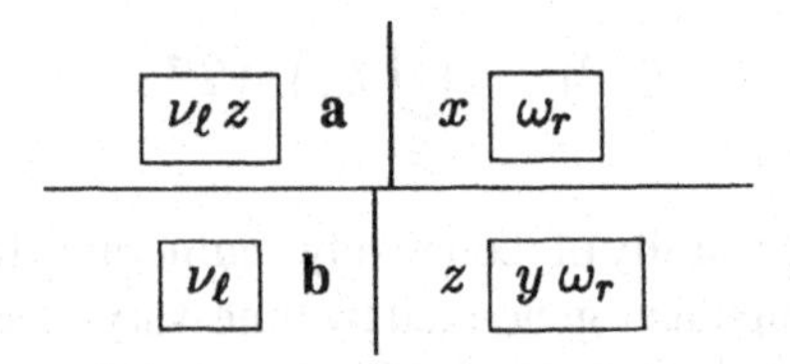

Figure 35.1 From the current configuration to the next one.

This looks rather like a splicing rule.

Indeed, what we do in effect to make the scheme just written enter the frame of splicing rules.

Using the previous figure, we see that we cannot go directly from the upper line to the lower one by a splicing rule because a first effect of the splicing rule is to split the current *molecule* into two parts. And so, we need a second rule for recombining the former right and left parts after a second splitting. This can be done simply by using an intermediary configuration:

$$\begin{array}{r|l} \boxed{\nu_\ell} & z\;\mathbf{a}\;x\;\boxed{\omega_r} \\ \hline L & \mathbf{b}\;z\;y \end{array} \quad \vdash \quad \begin{array}{l} \nu_\ell\;\mathbf{b}\;z\;y \\ L\;z\;\mathbf{a}\;x\;\omega_r \end{array}$$

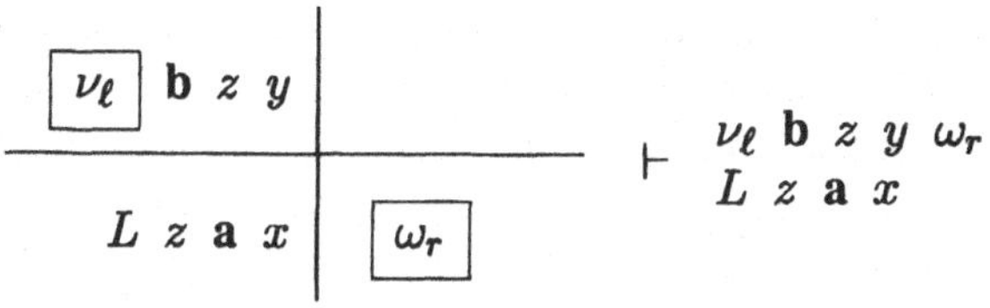

$\boxed{\nu_\ell}$ b z y	
L z a x	$\boxed{\omega_r}$

$\vdash$ ν_ℓ b z y ω_r
L z a x

Figure 35.2 Simulating a move to the left.

taking into account that $\omega_\ell = \nu_\ell z$ is assumed.

Notice the following particularity of those figures. The second rule applies to both words produced by the application of the first rule. This is not the case of the first rule where the lower word of the figure is not the representation of a configuration: L b z y. That auxiliary word is necessary for applying the rules. It must be brought in somehow. This is the reason for the axioms which will be introduced into the formal definitions and for special rules whose role will be to keep the required axioms ready for use. Special rules are also needed to produce auxiliary words without which the transformation could not be carried out.

3.2 FORMAL DEFINITIONS

As already indicated in our introduction, we shall consider recursively enumerable sets of natural numbers instead of recursively enumerable languages. This will make our proof simpler.

It is a well-known fact from theoretical computer science that for every recursively enumerable set $\mathcal{L}$ of natural numbers, there is a Turing machine $T_\mathcal{L}$ which generates $\mathcal{L}$ as follows. Let $f_\mathcal{L}$ be a total recursive function enumerating the elements of $\mathcal{L}$, *i.e.* $\mathcal{L} = \{f_\mathcal{L}(x)\ ;\ x \in \mathbb{N}\}$. How to construct a Turing machine $T_\mathcal{L}$ from the definition of $f_\mathcal{L}$ is well-known. For every natural number x, the machine would transform any initial configuration $q_1 01^x$ into the final configuration $q_0 01^{f_\mathcal{L}(x)}0 \ldots 0$.

Let us assume for a while that we succeeded in simulating machine $T_\mathcal{L}$. This is still not enough to prove the theorem. Indeed, starting from one word, say $q_1 01^x$ for some x, we would obtain an encoding of $f_\mathcal{L}(x)$ as $q_0 01^{f_\mathcal{L}(x)}$. But this is not what we need, because this gives us a single word.

One can easily see that starting from machine $T_\mathcal{L}$, there is a Turing machine $T'_\mathcal{L}$ which, starting from input word $q_1 01^x$ computes word $w = 01^{x+1} q_0 01^{f_\mathcal{L}(x)}0 \ldots 0$ as an ouptut. As we shall see, it will be possible for us to devise a time-varying distributed H-system Γ which, on arriving at word w will split it into two parts: $1^{f_\mathcal{L}(x)}$ if $f_\mathcal{L}(x) \neq 0$ and 0 if $f_\mathcal{L}(x) = 0$ as a *generated* word and $q_1 01^{x+1}$ as a new starting configuration. This will allow us to obtain $\mathcal{L}$ as $\mathcal{L}(\Gamma)$ exactly.

Definition of Γ. Consequently, we assume that $\{0,1\}$ is the tape alphabet of Turing machine $T'_{\mathcal{L}}$; let us number its states $q_0, q_1, \ldots, q_n$, where q_1 is the initial state and q_0 the final one. From now on, we shall use latin small letters to denote tape symbols, greek letters for words in the alphabet of the tape symbols, and bold latin small letters for states of the Turing machine.

We define $TVDH$-system Γ as follows:
Its terminal is $T = \{0,1\}$, the tape alphabet of $T'_{\mathcal{L}}$. Let us introduce variables for $q_0, q_1, \ldots, q_n$, namely **a** and **b** with the additional assumption that **b** ranges over all q_i's but **a** ranges over all of them except q_0. Moreover, we shall introduce two copies of the set of q_i's, namely $\{q'_0, q'_1, \ldots, q'_n\}$ and $\{q''_0, q''_1, \ldots, q''_n\}$. Accordingly, if **a** denotes one of the q_i's, say q_j, then $\mathbf{a}'$ denotes q'_j and $\mathbf{a}''$ denotes q''_j. We shall use the same convention with **b**.
Let also $x, y, z \in \{0,1\}$. Define alphabet V by setting
$V = \{0,1\} \cup \{X, Y, Y', \mathbf{b}, \mathbf{b}', \mathbf{b}'', t_1, t_2, t_3, t_4, t_5, R, L, S, Q, Z, \rightarrow, \leftarrow\}$,
with $\mathbf{b}, \mathbf{b}', \mathbf{b}''$ taking all their possible values.

$$1.1: \begin{array}{c|l} s & \mathbf{a}Y \\ \hline t_1 & \mathbf{a}'0Y \end{array}$$
$\forall s \in \{0,1,X\}$, $\forall \mathbf{a}$,

$$1.2: \begin{array}{c|l} s & \mathbf{a}\,x \\ \hline R & y\,\mathbf{b} \end{array}$$
$\forall s \in \{0,1,X\}$,
$\forall \mathbf{a}, x, y, \mathbf{b} \in \mathcal{I}_R$,

$$1.3: \begin{array}{c|l} s & \mathbf{a}\,x \\ \hline S & \mathbf{b}\,y \end{array}$$
$\forall s \in \{0,1,X\}$,
$\forall \mathbf{a}, x, y, \mathbf{b} \in \mathcal{I}_S$,

$$1.4: \begin{array}{c|l} s & z\,\mathbf{a}\,x \\ \hline L & \mathbf{b}\,z\,y \end{array}$$
$\forall s \in \{0,1,X\}$, $\forall z$,
$\forall \mathbf{a}, x, y, \mathbf{b} \in \mathcal{I}_L$,

$$1.5: \begin{array}{r|c} X\,\mathbf{a}\,x & s \\ \hline X\,\mathbf{b}''\,0\,y & t_3 \end{array}$$
$\forall s \in \{0,1,Y\}$,
$\forall \mathbf{a}, x, y, \mathbf{b} \in \mathcal{I}_L$,

$$1.6: \begin{array}{c|l} 1 & q_0\,0\,1 \\ \hline R & Q\!\rightarrow \end{array}$$

$$1.7: \begin{array}{c|l} 1 & q_0\,0\,0 \\ \hline & Q\!\leftarrow\!0 \end{array}$$

$$1.8: \begin{array}{c|l} s & \rightarrow\!1 \\ \hline t_3 & 1\!\rightarrow \end{array}$$
$\forall s \in \{1, Q\}$,

$$1.9: \begin{array}{c|l} & \rightarrow\!0 \\ \hline t_5 & \leftarrow \end{array}$$

$$1.10: \begin{array}{r|c} \leftarrow\!1 & \\ \hline t_4 1\!\leftarrow & \end{array}$$

$$1.11: \begin{array}{r|c} Z & 1 \\ \hline t_4\!\leftarrow & Z \end{array}$$

$$1.12: \begin{array}{c|l} & Q\!\leftarrow\!Y' \\ \hline & \leftarrow\!Y \end{array}$$

Diagram 1.

Axioms A are given by

$$A=\{Xq_1Y,\ t_1\mathbf{a}'0Y,\ t_2\mathbf{a}0Y,\ Z\mathbf{b},\ RyZ,\ S\mathbf{b}Z, L\mathbf{b}xZ,\ X\mathbf{b}''0yt_3,$$
$$X\mathbf{b}t_4,\ Z\!\rightarrow,\ RQZ,\ t_31Z,\ t_5\!\leftarrow,\ Zx,\ t_4\!\leftarrow\! Z,\ Y',\ \leftarrow\! Y,$$
$$Xq_1Z, Xq_10Z, Q\!\leftarrow Z\},$$

with $\mathbf{a}, \mathbf{a}', \mathbf{b}, \mathbf{b}'', \mathbf{x}, \mathbf{y}$ taking all their possible values.

$2.1:\ \begin{array}{c|c} s & \mathbf{a}'0Y \\ \hline t_2 & \mathbf{a}0Y \end{array}$
$\forall s \in \{0,1,X\}, \forall \mathbf{a},$

$2.2:\ \begin{array}{c|c} y\mathbf{b} & \\ \hline R\mathbf{a}x & \end{array}$
$\forall \mathbf{a}, x, y, \mathbf{b} \in \mathcal{I}_R,$

$2.3:\ \begin{array}{c|c} Z & \mathbf{b} \\ \hline Ry & Z \end{array}$
$\forall \mathbf{b}, y,$

$2.4:\ \begin{array}{c|c} s\mathbf{b}y & \\ \hline S\mathbf{a}x & \end{array}$
$\forall s \in \{0,1,X\},$
$\forall \mathbf{a}, x, y, \mathbf{b} \in \mathcal{I}_S,$

$2.5:\ \begin{array}{c|c} Z & y \\ \hline S\mathbf{b} & Z \end{array}$
$\forall \mathbf{b}, y$

$2.6:\ \begin{array}{c|c} s\mathbf{b}zy & \\ \hline Lz\mathbf{a}x & \end{array}$
$\forall s \in \{0,1,X\}, \forall z,$
$\forall \mathbf{a}, x, y, \mathbf{b} \in \mathcal{I}_S,$

$2.7:\ \begin{array}{c|c} Z & y \\ \hline L\mathbf{b}z & Z \end{array}$
$\forall \mathbf{b},\ \forall y, z,$

$2.8:\ \begin{array}{c|c} X\mathbf{b}'' & 0 \\ \hline X\mathbf{b} & t_4 \end{array}$
$\forall \mathbf{b},$

$2.9:\ \begin{array}{c|c} Q\!\rightarrow & \\ \hline Rq_00 & \end{array}$

$2.10:\ \begin{array}{c|c} Z & \rightarrow \\ \hline RQ & Z \end{array}$

$2.11:\ \begin{array}{c|c} Z & 0 \\ \hline Q\!\leftarrow & Z \end{array}$

$2.12:\ \begin{array}{c|c} 1\!\rightarrow & \\ \hline t_3\!\rightarrow\!1 & \end{array}$

$2.13:\ \begin{array}{c|c} Z & \rightarrow \\ \hline t_31 & Z \end{array}$

$2.14:\ \begin{array}{c|c} s & 1\!\leftarrow \\ \hline t_4 & \leftarrow\!1 \end{array}$
$\forall s \in \{0,1,Q\},$

$2.15:\ \begin{array}{c|c} Q\!\leftarrow & x \\ \hline & Y' \end{array}$
$\forall x,$

$2.16:\ \begin{array}{c|c} X0\!\leftarrow & 1 \\ \hline Xq_10 & Z \end{array}$

Diagram 2.

Notice that for any pair $\mathbf{a}$, x, there is at most one instruction in the program of $T'_{\mathcal{L}}$ of the form $\boxed{\mathbf{a}\,x\,M\,y\,\mathbf{b}}$, M, y, $\mathbf{b}$ being uniquely determined. Using $\mathcal{I}$ to denote the set of instructions of $T'_{\mathcal{L}}$, the expressions $\mathbf{a}$, $x \in \mathcal{I}$ as well as $\mathbf{a}$, $x, \mathbf{b}$, $y \in \mathcal{I}$ will be used to indicate an instruction by its constituting elements. In particular, we will use quantifiers in such expressions. Moreover, we shall specialize $\mathcal{I}$ by using $\mathcal{I}_R$ to denote the set of all instructions with a move to right, by $\mathcal{I}_L$ the set of all instructions with a move to left and by $\mathcal{I}_S$, the set of all stationary instructions, *i.e.* with $M = S$. The first component of rules, R_1 is given by Diagram 1.

The second component of rules, R_2 is given by Diagram 2.

Both components R_1 and R_2 also contain the following rules:

$$\begin{array}{c|c} \alpha & \\ \hline \alpha & \end{array}$$

for each axiom $\alpha \in A$, except Xq_1Y. As can be checked in the computation -see [5]- axioms must always be present so that it is possible to apply most of the rules.

It is not possible here to give details of the proof which boils down to checking the computation on key configurations. See [5] for details.

Now, in order to achieve the proof, it is sufficient to apply the algorithm just given to any universal Turing machine on $\{0, 1\}$, for instance the machine of [10].

4. CONCLUSION

The proof given in [5] shows clearly that our computation does not make any use of the parallelism which is contained in the definition of time-varying distributed H-systems as well as other kinds of H-systems.

Therefore, it seems reasonable to ask how many components are needed in order to obtain a *parallel* universal computation? How many of them are needed to generate any recursively enumerable set with a parallel computation? A partial answer is given in [6], as the proof of the result preserves the original parallel structure of time-varying distributed H-system: four components are enough to generate all recursively enumerable sets in a parallel way. This last number is near our result. And so the question is now: is it possible to obtain two components with a parallel computation?

Acknowledgments

The authors acknowledge the very helpful contribution of INTAS project 97-1259 to enhance their cooperation, which provided the best conditions for producing the present result.

References

[1] Kleene, S.C. (1952), *Introduction to Metamathematics*, Van Nostrand, New York.

[2] Margenstern, M. (1997), The laterality problem is completely solved, *Theoretical Informatics and Applications, RAIRO/ITA*, 31.2: 159–204.

[3] Margenstern, M. & Yu. Rogozhin (1998), A universal time-varying distributed H-system of degree 2, in *Preliminary Proceedings of the Fourth*

International Meeting on DNA Based Computers: 83–88, University of Pennsylvania, Philadelphia.

[4] Margenstern, M. & Yu. Rogozhin (1999), A universal time-varying distributed H-system of degree 2, *Biosystems*, 52: 73–80.

[5] Margenstern, M. & Yu. Rogozhin (1999), Generating all recursively enumerable languages with a time-varying distributed H-system of degree 2, Technical Report, Institut Universitaire de Technologie de Metz, ISBN 2-9511539-5-3.

[6] Păun, A. (1999), On Time-Varying H Systems, *Bulletin of the European Association for Theoretical Computer Science*, 67: 157–164.

[7] Păun, Gh. (1997), DNA computing: distributed splicing systems, in *Structures in Logic and Computer Science. A Selection of Essays in honor of A. Ehrenfeucht*: 353–370, Springer, Berlin.

[8] Păun, Gh. (1998), DNA Computing Based on Splicing: Universality Results, in *Proceedings of MCU'98*, I, ISBN 2-9511539-2-9: 67-91.

[9] Păun, Gh.; G. Rozenberg & A. Salomaa (1998), *DNA Computing: New Computing Paradigms*, Springer, Berlin.

[10] Rogozhin, Yu. (1996), Small universal Turing machines, *Theoretical Computer Science*, 168.2: 215–240.

Chapter 36

ON MEMBRANE COMPUTING BASED ON SPLICING

Andrei Păun
Department of Computer Science
University of Western Ontario
London, ON Canada N6A 5B7
apaun@csd.uwo.ca

Mihaela Păun
Department of Computer Science
University of Western Ontario
London, ON Canada N6A 5B7

Abstract This paper is a direct continuation of [11]. Characterizations of recursively enumerable languages are given, by means of splicing P systems, having splicing rules of small size (that is, involving short context strings). Also it is shown that with only two membranes we can generate all the recursively enumerable languages; this improves a result from [11], where three membranes are used.

1. INTRODUCTION

The P systems are a class of distributed parallel computing devices of biochemical inspiration (they can be considered as a possible branch of natural/molecular computing), introduced in [7]. The reader can find results in this area in [2,10,12] etc.

In short, a *membrane structure* is considered to be several cell-membranes which are hierarchically embedded in a main membrane, called the *skin* membrane. The membranes delimit *regions*, where we place *objects*. The objects evolve according to given *evolution rules*, which are associated with the regions. In the present paper we deal with the case when the objects are represented by strings and the evolution rules are splicing rules, as formalized in [3].

C. Martin-Vide and V. Mitrana (eds.), Where Mathematics, Computer Science, Linguistics and Biology Meet, 409-422.

Starting from an initial configuration (identified by the membrane structure, and the objects placed in its regions) and using the evolution rules in parallel, we get a *computation*. A computation identifies a language, the set of all terminal strings which leave the system. P systems of this form were investigated in [11], where several characterizations of recursively enumerable languages were obtained. Here we improve one of these results in what concerns the number of membranes and we also consider the size of the splicing rules – in the sense of [4], we bound the length of the contexts used in the splicing rules. We find that P systems with splicing rules of a rather small complexity suffice.

2. SPLICING P SYSTEMS

We will first recall the splicing operation introduced in [3] as a formal model of the DNA recombination under the influence of restriction enzymes and ligases.

A *splicing rule* (over an alphabet V) is a string $r = u_1\#u_2\$u_3\#u_4$, where $u_1, u_2, u_3, u_4 \in V^*$ and $\#, \$$ are two special symbols not in V. (V^* is the free monoid generated by the alphabet V under the operation of concatenation; the empty string is denoted by λ; the length of $x \in V^*$ is denoted by $|x|$.)

For $x, y, w, z \in V^*$ and r as above we write

$$(x, y) \vdash_r (w, z) \quad \text{iff} \quad x = x_1u_1u_2x_2,\ y = y_1u_3u_4y_2,$$
$$w = x_1u_1u_4y_2,\ z = y_1u_3u_2x_2,$$
$$\text{for some } x_1, x_2, y_1, y_2 \in V^*.$$

We say that we splice x, y at the *sites* u_1u_2, u_3u_4. These sites encode the patterns recognized by restriction enzymes able to cut the DNA sequences between u_1, u_2, respectively between u_3, u_4.

When r is understood, we write $\vdash$ instead of $\vdash_r$. For clarity, we usually indicate by a vertical bar the place of splicing: $(x_1u_1|u_2x_2, y_1u_3|u_4y_2) \vdash (x_1u_1u_4y_2, y_1u_3u_2x_2)$.

The *radius* of a splicing rule $u_1\#u_2\$u_3\#u_4$ is the length of the longest string u_1, u_2, u_3, u_4.

A pair $\sigma = (V, R)$, where V is an alphabet and R is a set of splicing rules over V, is called an *H scheme*. With respect to an H scheme $\sigma = (V, R)$ and a language $L \subseteq V^*$ we define

$$\sigma(L) = \{w \in V^* \mid (x, y) \vdash_r (w, z) \text{ or } (x, y) \vdash_r (z, w),$$
$$\text{for some } x, y \in L, r \in R, z \in V^*\},$$
$$\sigma^*(L) = \bigcup_{i \geqslant 0} \sigma^i(L), \text{ for}$$
$$\sigma^0(L) = L \text{ and } \sigma^{i+1}(L) = \sigma^i(L) \cup \sigma(\sigma^i(L)),\ i \geqslant 0.$$

An *extended H system* is a construct $\gamma = (V, T, A, R)$, where V is an alphabet, $T \subseteq V$, $A \subseteq V^*$, and $R \subseteq V^*\#V^*\$V^*\#V^*$. ($T$ is the *terminal* alphabet, A is the set of *axioms*, and R is the set of *splicing rules*.) When $T = V$, the system is said to be *non-extended*. The pair $\sigma = (V, R)$ is the *underlying H scheme* of γ.

We define the *diameter* of γ, [4], by $dia(\gamma) = (n_1, n_2, n_3, n_4)$, where

$$n_i = \max\{|u_i| \mid u_1\#u_2\$u_3\#u_4 \in R\},\ 1 \leqslant i \leqslant 4.$$

For any $L \subseteq V^*$ and $\gamma = (V, T, A, R)$ we define

$$\sigma(L) = \{w \mid (x, y) \vdash_r (w, z) \text{ or } (x, y) \vdash_r (z, w),\ \text{for } x, y \in L, r \in R\},$$
$$\sigma^*(L) = \bigcup_{i \geqslant 0} \sigma^i(L),\ \text{for}$$
$$\sigma^0(L) = L,\ \text{and } \sigma^{i+1}(L) = \sigma^i(L) \cup \sigma(\sigma^i(L)),\ i \geqslant 0.$$

Then, the language generated by γ is $L(\gamma) = \sigma^*(A) \cap T^*$. (We iterate the splicing operation according to rules in R, starting from strings in A, and we only keep the strings composed of terminal symbols.)

It is known that extended H systems with finite sets of axioms and of splicing rules characterize the regular languages, [1,13], while H systems with regular sets of rules characterize the recursively enumerable languages, [6].

Let us now pass to splicing P systems, the object of our investigation.

We identify a membrane structure with a string of correctly matching parentheses, placed in a unique pair of matching parentheses; each pair of matching parentheses corresponds to a membrane. Graphically, a membrane structure is represented by a Venn diagram.

A *P system* is a membrane structure with multisets of *objects* placed in its regions and provided with *evolution rules* for these objects. We define here only the splicing P systems, in the variant we investigate in this paper.

A *splicing P system* (of degree $m, m \geqslant 1$) is a construct

$$\Pi = (V, T, \mu, L_1, \ldots, L_m, R_1, \ldots, R_m),$$

where:

(i) V is an alphabet; its elements are called *objects*;

(ii) $T \subseteq V$ (the *output* alphabet);

(iii) μ is a membrane structure consisting of m membranes (labeled $1, 2, \ldots, m$);

(iv) $L_i, 1 \leqslant i \leqslant m$, are languages over V associated with the regions $1, 2, \ldots, m$ of μ;

(v) $R_i, 1 \leqslant i \leqslant m$, are finite sets of *evolution rules* associated with the regions $1, 2, \ldots, m$ of μ, given in the following form: $(r = u_1\#u_2\$u_3\#u_4;$

$tar_1, tar_2)$, where $r = u_1\#u_2\$u_3\#u_4$ is a usual splicing rule over V and $tar_1, tar_2 \in \{here, out\} \cup \{in_j \mid 1 \leqslant j \leqslant m\}$.

Note that, as usual in H systems, when a string is present in a region of our system, it is assumed to appear in an arbitrarily large number of copies (any number of copies of a DNA molecule can be obtained by amplification). Thus, we do not use multisets here, as in basic P systems.

Any m-tuple $(M_1, \ldots, M_m)$ of languages over V is called a *configuration* of Π. For two configurations $(M_1, \ldots, M_m), (M'_1, \ldots, M'_m)$ of Π we write $(M_1, \ldots, M_m) \Longrightarrow (M'_1, \ldots, M'_m)$ if we can pass from $(M_1, \ldots, M_m)$ to $(M'_1, \ldots, M'_m)$ by applying the splicing rules from each region of μ, in parallel, to all possible strings from the corresponding regions, and following the target indications associated with the rules. More specifically, if $x, y \in M_i$ and $(r = u_1\#u_2\$u_3\#u_4, tar_1, tar_2) \in R_i$ such that we can have $(x, y) \vdash_r (w, z)$, then w and z will go to the regions indicated by tar_1, tar_2, respectively. If $tar_j = here$, then the string remains in M_i; if $tar_j = out$, then the string is moved to the region immediately outside the membrane i (maybe, in this way the string leaves the system); if $tar_j = in_k$, then the string is moved to the region k, providing that this is immediately below; if not, then the rule cannot be applied. Note that the strings x, y are still available in region M_i, because we have supposed that they appear in an arbitrarily large number of copies (an arbitrarily large number of them were spliced, arbitrarily therefore many remain), but if a string w, z is sent out of region i, then no copy of it remains here.

A sequence of transitions between configurations of a given P system Π, starting from the initial configuration $(L_1, \ldots, L_m)$, is called a *computation* with respect to Π. The result of a computation consists of all strings over T which are sent out of the system at any time during the computation. We denote by $L(\Pi)$ the language of all strings of this type. We say that $L(\Pi)$ is *generated* by Π.

Note two important facts: if a string leaves the system but it is not terminal, then it is ignored; if a string remains in the system, even if it is terminal, then it does not contribute to the language $L(\Pi)$. It is also worth mentioning that we do not consider here halting computations. We leave the process to continue forever and we just observe it from outside and collect the terminal strings leaving it.

We denote by $SPL(tar, m, p)$ the family of languages $L(\Pi)$ generated by splicing P systems as above, of degree at most $m, m \geqslant 1$, and of depth at most $p, p \geqslant 1$. (The depth of a P system is equal to the height of the tree describing its membrane structure.) If all target indications tar_1, tar_2 in the evolution rules of a P system are of the form *here, out, in*, then we say that Π is of the *i/o type*; the strings produced by splicing, having the associated *in* indication, are moved into any lower region immediately below the region where the rule is used. The family of languages generated by P systems with this weaker target indication and of degree at most m and depth at most p is denoted by $SPL(i/o, m, p)$.

We define the *diameter* of a splicing P system $\Pi = (V, T, \mu, L_1, \ldots, L_m, R_1, \ldots, R_m)$, in a similar way to the case of extended H systems ([4]), by $dia(\Pi) = (n_1, n_2, n_3, n_4)$, where

$$n_i = \max\{|u_i| \mid u_1\#u_2\$u_3\#u_4 \in R_1 \cup \ldots \cup R_m\},\ 1 \leqslant i \leqslant 4.$$

We denote the family of languages generated by P systems with the weaker target indication and of degree at most m and depth at most p and with diameter (n_1, n_2, n_3, n_4) by $SPL(i/o, m, p, (n_1, n_2, n_3, n_4))$.

By RE we denote the family of recursively enumerable languages.

3. THE POWER OF SPLICING P SYSTEMS

We start by improving a result from [11], where it is proved that $RE = SPL(i/o, 3, 3)$. In the proof of this result, no attention is paid to the diameter of the used system.

Theorem 36.1 $SPL(i/o, 2, 2) = SPL(tar, 2, 2) = RE.$

Proof. Let $G = (N, T, S, P)$ be a type-0 Chomsky grammar and let B be a new symbol. Assume that $N \cup T \cup \{B\} = \{\alpha_1, \ldots, \alpha_n\}$ and that P contains m rules, $u_i \to v_i, 1 \leqslant i \leqslant m$. Consider also the rules $u_{m+j} \to v_{m+j}, 1 \leqslant j \leqslant n$, for $u_{m+j} = v_{m+j} = \alpha_j$.

We construct the splicing P system (of degree 2):

$$\begin{aligned}
\Pi &= (V, T, \mu, L_1, L_2, R_1, R_2),\\
V &= N \cup T \cup \{B, X, X', Y, Y', Z, Z'\} \cup\\
&\quad \{X_i \mid 0 \leqslant i \leqslant n+m\} \cup \{Y_i \mid 0 \leqslant i \leqslant n+m\},\\
\mu &= [_1[_2]_2]_1,\\
L_1 &= \{XSBY, Z', X'Z, ZY'\} \cup \{ZY_i \mid 0 \leqslant i \leqslant n+m\} \cup\\
&\quad \{X_i v_i Z \mid 1 \leqslant i \leqslant n+m\},\\
R_1 &= \{(X_i v_i\#Z\$X\#;\ in,\ out),\ (\#Y_i\$Z\#Y_{i-1};\ in,\ out) \mid\\
&\quad 1 \leqslant i \leqslant m+n\}\\
&\cup\ \{(\#Y_0\$Z\#Y';\ in,\ out),\ (X_0\#\$X'\#Z;\ out,\ here)\}\\
&\cup\ \{(\#BY\$Z'\#;\ here,\ out),\ (X\#\$\#Z';\ out,\ out)\}.\\
L_2 &= \{XZ, ZY\} \cup \{ZY_i \mid 1 \leqslant i \leqslant m+n\} \cup \{X_i Z \mid\\
&\quad 0 \leqslant i \leqslant m+n-1\},\\
R_2 &= \{(\#u_i Y\$Z\#Y_i;\ out,\ here),\ (X_i\#\$X_{i-1}\#Z;\ here,\ out) \mid
\end{aligned}$$

$$1 \leqslant i \leqslant m+n\}$$
$$\cup \ \{(X'\#\$X\#Z;\ here,\ here),\ (\#Y'\$Z\#Y;\ out,\ here)\}.$$

The idea of this proof is the "rotate-and-simulate" procedure, as used in many proofs in H systems theory (first in [6]). Here both the simulation of the rules in G and the circular permutation of strings are performed in Π in the same way: a suffix u of the current string is removed and the corresponding string, v, is added at the left end of the string. For $u \rightarrow v$ a rule in P, we simulate a derivation step in G. For $u = v$ a symbol in $N \cup T \cup \{B\}$, we have one symbol "rotation" of the current string.

At the end of a computation we want to have the word in the right permutation, so we have to mark the beginning of the word. For this purpose we use the new letter B which marks the beginning of the right sentential word.

Let us see in more detail how the system works.

The "main" axiom is $XSBY$; we will always process a string of the form Xw_1Bw_2Y (the axiom is of that form). We replace a suffix u_iY of this word with Y_i (in region 2) and the prefix X with X_jv_j (in region 1). Then we will repeatedly decrease the subscripts of Y_i and X_j by one. In the end we will replace Y_0 with Y and X_0 with X (this means that $i = j$; so we simulated the production $u_i \rightarrow v_i$). In this way we can simulate the productions from P and rotate the string.

At the end, in membrane 1 we cut the markers B and Y together (to be sure that we have the right permutation of the word) and finally we cut the marker X and send the string out. Thus, we get $L(G) \subseteq L(\Pi)$.

Now we will prove the converse inclusion. First we can observe that we send out strings from the first membrane, and that we send out strings with at least a marker Z, X, Y (or variants of them with subscripts or primed). The only exception is the following splicing rule: $(X\#\$\#Z';\ out,\ out)$. Of course, the first string produced by this splicing rule contains the markers X and Z', but the second string can be a terminal one.

Suppose that we start in the first membrane.

The rule $(\#Y_i\$Z\#Y_{i-1};\ in,\ out)$ can only be applied to two axioms ZY_i and ZY_{i-1} and we produce the same strings; the string ZY_{i-1} being sent to membrane 2. In membrane 2 we have these axioms but ZY_0 which cannot enter any splicing here.

The rule $(\#Y_0\$Z\#Y';\ in,\ out)$ also uses two axioms, namely ZY_0 and ZY', so we produce the same strings; the string ZY' will be sent to membrane 2 where it can be spliced using $(\#Y'\$Z\#Y;\ out,\ here)$ and then nothing can be produced by the word ZY which arrives in membrane 1.

The rule $(X_0\#\$X'\#Z;\ out,\ here)$ cannot be applied (we do not have X_0 now in the first membrane).

If we apply the rule $(\#BY\$Z'\#;\ here,\ out)$ then we still have the string XS in membrane 1. This string can enter splicing using the rule $(X_iv_i\#Z\$X\#;\ in,\ out)$ and the string X_iv_iS arrives in the second membrane where the subscript of X is decreased by one and then the string arrives in the first membrane. Here the string cannot enter any splicing.

Thus, we have to use first a rule $(X_iv_i\#Z\$X\#;\ in,\ out)$.

If we start in the second membrane, then the last two rules cannot be applied (we do not have X' or Y' here yet), while the outputs of the first two rules are the same strings that have entered the splicing, or strings which cannot enter any new splicing.

Consequently, we have to replace X by X_iv_i in membrane 1 and then to cut u_jY and replace it with Y_j in membrane 2. The string $X_iv_iwY_j$ gets in membrane 1, where the only possibility is of applying the rule $(\#Y_j\$Z\#Y_{j-1};\ in,\ out)$. The string $X_iv_iwY_{j-1}$ arrives in membrane 2, where the only possibility is of applying the rule $(X_i\#\$X_{i-1}\#Z;\ here,\ out)$, so the string $X_{i-1}v_iwY_{j-1}$ arrives in membrane 1. We iterate the process until at least one subscript of X or Y is 0.

If we got X_0, then we decreased the subscript of X in membrane 2 and sent the string $X_0v_iY_{j-i}$ in membrane 1. Here we have two possibilities: $j \neq i$ or $j = i$.

If $j \neq i$ then $j - i \neq 0$, so we can decrease the subscript of Y and send the string in membrane 2. But here the string $X_0v_iwY_{j-i-1}$ can enter no further splicings. Before decreasing the subscript of Y, in membrane 1 we can also apply the splicing rule $(X_0\#\$X'\#Z;\ out,\ here)$; the string $X'v_iwY_{j-i}$ is sent to membrane 1 and we continue as before. In this case, in membrane 2 we can replace X' by X using the rule $(X'\#\$X\#Z;\ here,\ here)$ and the string Xv_iY_{j-i-1} cannot enter any other splicings so it remains in membrane 2.

If $j = i$, then the only productions from region 1 that can be applied are $(\#Y_0\$Z\#Y';\ in,\ out)$ and $(X_0\#\$X'\#Z;\ out,\ here)$. If we apply the first one, then the string X_0v_iwY' is sent to membrane 2, here we can only apply the rule that replaces Y' with Y, so the string X_0v_iwY goes into membrane 1. This string will never lead to a terminal string because we cannot delete the left marker (we can replace X_0 with X' but the string remains in this membrane and that marker cannot be deleted).

If we start with the rule $(X_0\#\$X'\#Z;\ out,\ here)$, then we get the string $X'v_iwY_0$. The only possibility of continuing is to apply $(\#Y_0\$Z\#Y';\ in,\ out)$ and the string $X'v_iwY'$ goes into membrane 2. If we do not replace X' with X here, then again the string that gets into membrane 1 cannot lead to a terminal string. So first we replace X' with X (using the rule $X'\#\$X\#Z$) and then we replace Y' by Y by using the rule $(\#Y'\$Z\#Y;\ out,\ here)$. In this way we send the string Xv_iwY in membrane 1 and we can perform another step of rotating the word or simulating the rules from P.

Let us consider the case of Y_0, where we decreased the subscript of Y in membrane 1 and sent the string $X_{i-j+1}v_iwY_0$ in membrane 2. If the subscript of X is 0, then we cannot apply any splicing rule and the computation stops. Suppose now that the subscript of X is not 0: then the only splicing that we can apply is $(X_{i-j+1}\#\$X_{i-j}\#Z;\ here,\ out)$ so we send the string $X_{i-j}v_iwY_0$ in membrane 1. Suppose now that $i \neq j$ (we already dealt with the case when they are equal). The only splicing rule that we can apply is $(\#Y_0\$Z\#Y';\ in,\ out)$ so $X_{i-j}v_iwY'$ gets in membrane 2. At this moment we can apply the following two splicing rules: $(X_{i-j}\#\$X_{i-j-1}\#Z;\ here,\ out)$ and $(\#Y'\$Z\#Y;\ out,\ here)$. If we apply the first one, then the string $X_{i-j-1}v_iwY'$ gets into membrane 1 and does not lead to terminal words because we cannot delete the marker Y'. If we apply the second splicing rule, then the word $X_{i-j}v_iwY$ enters membrane 1 and we cannot apply splicing rules that send away this word, so it does not count for the language (we also cannot delete the X_k marker).

Therefore, the computations in Π correctly simulate rules in G or circularly permute the string. In the end we remove all markers from a string using $(\#BY\$Z'\#;\ here,\ out)$ and $(X\#\$\#Z';\ out,\ out)$.

In this way we get that $L(\Pi) \subseteq L(G)$. □

Now we will try to minimize the diameter of the splicing P systems of the i/o type. To be able to obtain the results, we had to consider systems with 3 components. The results are similar to those obtained for the extended H systems with permitting/forbidding contexts, [4,5]. The following auxiliary result is easy to prove.

Lemma 36.1 $SPL(i/o, m, p, (n_1, n_2, n_3, n_4)) = SPL(i/o, m, p, (n_3, n_4, n_1, n_2))$, *for all* $m, p \geqslant 1$ *and* $n_i \geqslant 0$, $1 \leqslant i \leqslant 4$.

Now, we give the anticipated results.

Lemma 36.2 $SPL(i/o, 3, 3, (0, 2, 1, 0)) = SPL(i/o, 3, 3, (1, 0, 0, 2)) = RE$

Proof. We will only prove that $SPL(i/o, 3, 3, (0, 2, 1, 0)) = RE$, the other equality follows from Lemma 36.1.

Let $G = (N, T, S, P)$ be a type-0 Chomsky grammar in the Kuroda normal form, and let B be a new symbol. Assume that $N \cup T \cup \{B\} = \{\alpha_1, \ldots, \alpha_n\}$ and that P contains m rules, $u_i \to v_i, 1 \leqslant i \leqslant m$. Consider also the rules $u_{m+j} \to v_{m+j}, 1 \leqslant j \leqslant n$, for $u_{m+j} = v_{m+j} = \alpha_j$.

We denote by P_1 the set of context-free rules considered above, and with P_2 the rest of the rules. One can see that the rules $u_{m+j} \to v_{m+j}, 1 \leqslant j \leqslant n$ are in P_1.

We construct the splicing P system (of degree 3):

$$\begin{aligned}
\Pi &= (V, T, \mu, L_1, L_2, L_3, R_1, R_2, R_3),\\
V &= N \cup T \cup \{B, X, X', Y, Z_X, Z_{X'}, Z_Y, Z_\lambda, Z'_\lambda\} \cup \{Y_i, Z_{Y_i} \mid \\
&\quad 0 \leqslant i \leqslant n+m\} \cup \{X_i, Z_{X_i}, Z_i, Y'_i, Z_{Y'_i} \mid 1 \leqslant i \leqslant m+n\},\\
\mu &= [_1[_2[_3]_3]_2]_1,\\
L_1 &= \{XSBY, Z_\lambda, Z'_\lambda\} \cup \{Z_{Y_i}Y_i \mid 0 \leqslant i \leqslant n+m\} \cup \{Z_{Y'_i}Y'_i \mid \\
&\quad 1 \leqslant i \leqslant n+m\},\\
R_1 &= \{(\#u_iY\$Z_{Y_i}\#; in, out), \mid u_i \to v_i \in P_1\}\\
&\cup \{(\#DY\$Z_{Y'_i}\#; here, out), (\#CY'_i\$Z_{Y_i}\#; in, out) \mid \\
&\quad u_i = CD \to v_i \in P_2\} \cup \{(\#Y_i\$Z_{Y_{i-1}}\#; in, out) \mid 1 \leqslant i \leqslant n+m\}\\
&\cup \{(\#BY\$Z_\lambda\#; here, out),\ (\#Z'_\lambda\$X\#; out, out)\},\\
L_2 &= \{XZ_X, X'Z_{X'}\} \cup \{X_iv_iZ_i \mid 1 \leqslant i \leqslant m+n\} \cup \{X_iZ_{X_i} \mid \\
&\quad 1 \leqslant i \leqslant m+n-1\},\\
R_2 &= \{(\#Z_i\$X\#; out, in) \mid 1 \leqslant i \leqslant n+m\}\\
&\cup \{(\#Z_{X_{i-1}}\$X_i\#; out, in) \mid 2 \leqslant i \leqslant n+m\}\\
&\cup \{(\#Z_{X'}\$X_1\#; in, in),\ (\#Z_X\$X'\#; out, in)\},\\
L_3 &= \{Z_YY\},\\
R_3 &= \{(\#Y_0\$Z_Y\#; out, out)\}.
\end{aligned}$$

This proof follows the proof of Theorem 36.1 from [11] closely, with special attention paid to the diameter of the splicing rules.

The sentential forms generated by G are simulated in Π in a circular permutation: Xw_1Bw_2Y, maybe with variants of X, Y, will be present in a region of Π if and only if w_2w_1 is a sentential form of G. Note that we can only remove the nonterminal symbol Y together with B from a string of the form $XwBY$. In this way, we ensure that the string is in the right permutation.

The simulation of rules in P and the rotation are done in the same way. Assume that some string Xwu_iY is present in region 1. We have now two cases: if $u_i \to v_i \in P_1$, then right away we simulate the rule $u_i \to v_i$ with a splicing rule of the form $(\#u_iY\$Z_{Y_i}, in, here)$. If $u_i = CD \to v_i \in P_2$, then the simulation requires two steps: first we replace DY with Y'_i and then we replace CY'_i with Y_i.

So in both cases we replace the suffix u_iY with Y_i, $1 \leqslant i \leqslant m+n$: initially we have here the string $XSBY$. We can perform

$$(Xw|u_iY, Z_{Y_i}|Y_i) \vdash (XwY_i, Z_{Y_i}u_iY) \text{ for } u_i \to v_i \in P_1,$$

or else,

$$(XwC|DY, Z_{Y'_i}|Y'_i) \vdash (XwCY'_i, Z_{Y'_i}DY)$$

and

$$(Xw|CY_i', Z_{Y_i}|Y_i) \vdash (XwY_i, Z_{Y_i}CY_i').$$

The string XwY_i is sent to region 2, the "by products" are sent out and do not enter the language generated by Π because they contain at least one nonterminal. In region 2 we can only perform a splicing of the form $(X|wY_i, X_jv_j|Z_j) \vdash (XZ_j, X_jv_jwY_i)$, for some $1 \leqslant j \leqslant n+m$. The string $X_jv_jwY_i$ is sent back to region 1, XZ_j is sent to membrane 1 and cannot enter other splicings. Now, in region 1 the only splicing which can be applied to the string $X_jv_jwY_i$ is $(X_jv_jw|Y_i, Z_{Y_{i-1}}|Y_{i-1}) \vdash (X_jv_jwY_{i-1}, Z_{Y_{i-1}}Y_i)$. The string $X_jv_jwY_{i-1}$ is sent to region 2, while $Z_{Y_{i-1}}Y_{i-1}$ is sent out and do not enter the generated language. In region 2 we now decrease by one the subscript of X_j, using the rule $(X_j|v_jwY_{i-1}, X_{j-1}|Z_{X_{j-1}}) \vdash (X_jZ, X_{j-1}v_jwY_{i-1})$. We iterate this process of decreasing the subscripts until either the subscript of X reaches 1 or the subscript of Y becomes 0.

If at some moment we reach X_1, hence in region 2 we have a string $X_1v_jwY_k$, then we perform $(X_1|v_jwY_k, X'|Z_{X'}) \vdash (X_1Z_{X'}, X'v_jwY_k)$ and $X'v_jwY_k$ is sent to membrane 3. If $k \neq 0$, then nothing can be done, the string is "lost". Otherwise, Y_0 is replaced with Y and the string $X'v_jwY$ is sent to region 2; X' is replaced here by X and the string Xv_jwY is sent to the skin membrane.

If at some moment in region 2 we get a string $X_kv_jwY_0$, for $k \geqslant 2$, this string cannot be processed in the skin membrane, hence it is "lost". Thus, we can only continue correctly when $i = j$, as we have passed from Xwu_iY to Xv_iwY; in this way we have either correctly simulated a rule from P or we have circularly permuted the string with one symbol. This is true because we cannot have "illegal" splicings: "by product" strings generated in membrane 1 are sent out, the ones produced in membrane 2 are sent to membrane 3 (none of them containing either Y_0 or Z_Y, thus they cannot enter splicings in membrane 3) and the "garbage" Z_YY_0 from membrane 3 is sent to membrane 2, where it cannot enter any splicing. Because of this, we know that when we have Z with a subscript in a word entering a splicing, then that word is a axiom (and one can see that there are no two axioms containing Z with the same subscript).

The process of simulating a rule or of rotating the string with one symbol can be iterated. Therefore, all derivations in G can be simulated in Π and, conversely, all correct computations in Π correspond to correct derivations in G. Because we collect only terminal strings which leave the system, we have the equality $L(G)) = L(\Pi)$. It is easy to see that the diameter of the P system is $(0, 2, 1, 0)$. □

In the proof of Lemma 36.2, the application of rules of grammar G was simulated in the equivalent P system at the right hand end of the strings. It is easy to see that we can perform this simulation also at the left hand end of the strings. The rotation can also be done in the reverse direction to that in the proof of Lemma 36.2: cut a symbol from the left end of the string and add it to the right hand end, repeatedly. A counterpart of the result in Lemma 36.2 can be obtained in this way.

Lemma 36.3 $SPL(i/o, 3, 3, (2, 0, 0, 1)) = SPL(i/o, 3, 3, (0, 1, 2, 0)) = RE.$

Proof. Again we will prove only one equality, $SPL(i/o, 3, 3, 2001) = RE$, the second one following from Lemma 36.1.

We simply repeat the construction in the proof of Lemma 36.2, and make the changes so that the strings are rotated in the converse order. We only give the construction of the system here; its correctness can be proved as in the previous proof. Let $G = (N, T, S, P)$ be a type-0 Chomsky grammar in the Kuroda normal form, and let B be a new symbol. Assume that $N \cup T \cup \{B\} = \{\alpha_1, \ldots, \alpha_n\}$ and that P contains m rules, $u_i \to v_i, 1 \leqslant i \leqslant m$. Consider also the rules $u_{m+j} \to v_{m+j}, 1 \leqslant j \leqslant n$, for $u_{m+j} = v_{m+j} = \alpha_j$. We denote by P_1 the set of context-free rules, and with P_2 the rest of the rules.

We construct the splicing P system :

$$
\begin{aligned}
\Pi &= (V, T, \mu, L_1, L_2, L_3, R_1, R_2, R_3),\\
V &= N \cup T \cup \{B, X, Y, Y', Z_X, Z_Y, Z_{Y'}, Z_\lambda, Z'_\lambda\} \cup \{X_i, Z_{X_i} \mid\\
&\quad 0 \leqslant i \leqslant n+m\} \cup \{Y_i, Z_{Y_i}, Z_i, X'_i, Z_{X'_i} \mid 1 \leqslant i \leqslant m+n\},\\
\mu &= [_1[_2[_3]_3]_2]_1,\\
L_1 &= \{XBSY, Z_\lambda, Z'_\lambda\} \cup \{X_i Z_{X_i} \mid 0 \leqslant i \leqslant n+m\} \cup \{X'_i Z_{X'_i} \mid\\
&\quad 1 \leqslant i \leqslant n+m\},\\
R_1 &= \{(Xu_i\#\$\#Z_{X_i}; out, in), \mid u_i \to v_i \in P_1\}\\
&\cup \{(XC\#\$\#Z_{X'_i}; out, here), (X'_i D\#\$\#Z_{X_i}; out, in) \mid\\
&\quad u_i = CD \to v_i \in P_2\}\\
&\cup \{(X_i\#\$\#Z_{X_{i-1}}; out, in) \mid 1 \leqslant i \leqslant n+m\}\\
&\cup \{(XB\#\$\#Z_\lambda; out, here), \ (Z'_\lambda\#\$\#Y; out, out)\},\\
L_2 &= \{Z_Y Y, Z_{Y'} Y'\} \cup \{Z_i v_i Y_i \mid 1 \leqslant i \leqslant m+n\} \cup \{Z_{Y_i} Y_i \mid\\
&\quad 1 \leqslant i \leqslant m+n-1\},\\
R_2 &= \{(Z_i\#\$\#Y; in, out) \mid 1 \leqslant i \leqslant n+m\}\\
&\cup \{(Z_{Y_{i-1}}\#\$\#Y_i; in, out) \mid 2 \leqslant i \leqslant n+m\}\\
&\cup \{(Z_{Y'}\#\$\#Y_1; in, in), \ (Z_Y\#\$\#Y'; in, out)\},\\
L_3 &= \{XZ_X\},\\
R_3 &= \{(X_0\#\$\#Z_X; out, out)\}.
\end{aligned}
$$

Again we obtain $L(\Pi) = RE$, and it is clear that the diameter of the system is $(2, 0, 0, 1)$. □

Lemma 36.4 $SPL(i/o, 3, 3, (1, 2, 0, 1)) = SPL(i/o, 3, 3, (0, 1, 1, 2)) = RE.$

Proof. Again we will only prove one equality, $SPL(i/o, 3, 3, 1201) = RE$, the second one following from Lemma 36.1. We will only give the construction of the splicing P system again, because the proof idea (and the construction) follows the previous proofs closely.

With the notations from the previous proofs we construct the following system:

$$
\begin{aligned}
\Pi &= (V, T, \mu, L_1, L_2, L_3, R_1, R_2, R_3),\\
V &= N \cup T \cup \{B, X, X', Y, Z_X, Z_{X'}, Z_Y, Z_\lambda, Z'_\lambda, Z_T\} \cup \{Y_i, Z_{Y_i} \mid\\
&\quad 0 \leqslant i \leqslant n+m\} \cup \{X_i, Z_{X_i}, Z_i, Y'_i, Z_{Y'_i} \mid 1 \leqslant i \leqslant m+n\},\\
\mu &= [_1[_2[_3]_3]_2]_1,\\
L_1 &= \{XSBY, Z_\lambda, Z'_\lambda, Z_T\} \cup \{Z_{Y_i}Y_i \mid 0 \leqslant i \leqslant n+m\} \cup \{Z_{Y'_i}Y'_i \mid\\
&\quad 1 \leqslant i \leqslant n+m\},\\
R_1 &= \{(\#u_iY\$\#Y_i; in, out), \mid u_i \to v_i \in P_1\}\\
&\cup \{(C\#DY\$\#Y'_i; here, out), (\#CY'_i\$\#Y_i; in, out) \mid\\
&\quad u_i = CD \to v_i \in P_2\}\\
&\cup \{(Z_{Y_{i-1}}\#Y_{i-1}\$\#Y_i; out, in) \mid 1 \leqslant i \leqslant n+m\}\\
&\cup \{(\#BY\$\#Z_T; here, out), (Z_\lambda\#\$\#Z_T; out, here),\\
&\quad (X\#\$\#Z'_\lambda; out, out)\},\\
L_2 &= \{XZ_X, X'Z_{X'}\} \cup \{X_iv_iZ_i \mid 1 \leqslant i \leqslant m+n\} \cup \{X_iZ_{X_i} \mid\\
&\quad 1 \leqslant i \leqslant m+n-1\},\\
R_2 &= \{(X\#\$\#Z_i; in, out) \mid 1 \leqslant i \leqslant n+m\}\\
&\cup \{(X_i\#\$\#Z_{X_{i-1}}; in, out) \mid 2 \leqslant i \leqslant n+m\}\\
&\cup \{(X_1\#\$\#Z_{X'}; in, in), (X'\#\$\#Z_X; in, out)\},\\
L_3 &= \{Z_YY\},\\
R_3 &= \{(Z_Y\#\$\#Y_0; out, out)\}.
\end{aligned}
$$

It is easy to see that the diameter of this system is $(1, 2, 0, 1)$, and that $L(\Pi) = L(G)$. □

In a similar fashion as we proceeded in the case of Lemma 36.2, we can now get a new result using the construct in Lemma 36.4, but rotating the strings in the converse order:

Lemma 36.5 $SPL(i/o, 3, 3, (1, 0, 2, 1)) = SPL(i/o, 3, 3, (2, 1, 1, 0)) = RE.$

Synthesizing these results, we obtain the following characterizations of RE.

Theorem 36.2 $RE = SPL(i/o, 3, 3, (n_1, n_2, n_3, n_4))$, *for all* (n_1, n_2, n_3, n_4) *componentwise greater than or equal to any of the following four-tuples:* $(0,2,1,0)$, $(1,0,0,2)$, $(2,0,0,1)$, $(0,1,2,0)$, $(1,2,0,1)$, $(0,1,1,2)$, $(2,1,1,0)$, $(1,0,2,1)$.

4. FINAL REMARKS

Several problems remain to be investigated further. For instance, in Theorem 36.1 we have $RE = SPL(i/o, 2, 2, (2,2,2,2))$. Can this result be improved, by using systems of a diameter smaller than $(2,2,2,2)$? Are the previous results optimal? What about the case of P systems with target indications of the form in_j? (The target feature is stronger than the in/out indication; can the previous result be improved in this case?)

Another research direction is to consider other classes of languages. For example, what is the diameter needed to generate regular (or context-free) languages?

References

[1] Culik II, K. & T. Harju (1991), Splicing semigroups of dominoes and DNA, *Discrete Applied Mathematics*, 31: 261–277.

[2] Dassow, J. & Gh. Păun (1999), On the power of membrane computing, *Journal of Universal Computer Science*, 5.2: 33–49 (www.iicm.edu/jucs).

[3] Head, T. (1987), Formal language theory and DNA: an analysis of the generative capacity of specific recombinant behaviors, *Bulletin of Mathematical Biology*, 49: 737–759.

[4] Păun, A. (1997), Controlled H systems of small radius, *Fundamenta Informaticae*, 31.2: 185–193.

[5] Păun, A. & M. Păun (1998), Controlled and distributed H systems of a small diameter, in Gh. Păun, ed., *Computing with Bio-Molecules: Theory and Experiments*, Springer, Singapore.

[6] Păun, Gh. (1996), Regular extended H systems are computationally universal, *Journal of Automata, Languages and Combinatorics*, 1.1: 27–36.

[7] Păun, Gh. (1998), Computing with membranes, *Journal of Computer and System Sciences*, to appear. (See also TUCS Research Reports, No. 208, November 1998, www.tucs.fi)

[8] Păun, Gh. (1999), Computing with membranes. An introduction, *Bulletin of the European Association for Theoretical Computer Science*, 67: 139–152.

[9] Păun, Gh.; G. Rozenberg & A. Salomaa (1998), *DNA Computing. New Computing Paradigms*, Springer, Berlin.

[10] Păun, Gh.; G. Rozenberg & A. Salomaa (1999), Membrane computing with external output, submitted. (See also TUCS Research Reports, No. 218, December 1998, www.tucs.fi)

[11] Păun, Gh. & T. Yokomori (1999), Membrane Computing Based on Splicing, in E. Winfree & D. Gifford, eds., *Preliminary Proceedings of the Fifth International Meeting on DNA Based Computers*: 213–227. MIT Press, Cambridge, Mass.

[12] Păun, Gh. & S. Yu (1999), On synchronization in P systems, *Fundamenta Informaticae*, 38.4: 397–410.

[13] Pixton, D. (1996), Regularity of splicing languages, *Discrete Applied Mathematics*, 69: 101–124.

Chapter 37

IS EVOLUTIONARY COMPUTATION USING DNA STRANDS FEASIBLE?

José Rodrigo
Department of Artificial Intelligence
Faculty of Informatics
Technical University of Madrid
Campus de Montegancedo, 28660 Boadilla del Monte, Madrid, Spain
jrodrigo@asterix.fi.upm.es

Juan Castellanos
Department of Artificial Intelligence
Faculty of Informatics
Technical University of Madrid
Campus de Montegancedo, 28660 Boadilla del Monte, Madrid, Spain
jcastellanos@fi.upm.es

Fernando Arroyo
Department of Computer Languages, Projects and Systems
Technical College of Informatics
Technical University of Madrid
Carretera de Valencia km. 7, 28031 Madrid, Spain
farroyo@eui.upm.es

Luis Fernando Mingo
Department of Computer Languages, Projects and Systems
Technical College of Informatics
Technical University of Madrid
Carretera de Valencia km. 7, 28031 Madrid, Spain
lfmingo@eui.upm.es

C. Martín-Vide and V. Mitrana (eds.), Where Mathematics, Computer Science, Linguistics and Biology Meet, 423-434.

Abstract Until now, there have not been many attempts at using DNA strands as the technological base for evolutionary computing. This paper tries to prove that such a computing paradigm can be achieved using DNA strands and also that it seems to be the most appropriate computing paradigm when computing with DNA strands. Classical genetic algorithm operations are translated into DNA strands and DNA operations in order to implement them. This new approach will solve the inconvenience of having great amounts of DNA strands if a $\mathcal{NP}$ problem must be solved.

1. INTRODUCTION

Evolutionary Computation is derived from a very simple and truthful observation: *The Human Being is the best machine in the universe capable of performing the most complex operations ever performed by a machine.* This observation, and the certainty that the Human Being has improved its abilities from the origin of its existence until now thanks to evolution, gives rise to the Evolutionary Computing Paradigm.

The Evolutionary Computing Paradigm is based on the concept of Evolution. With this paradigm, problems are solved applying evolution over a possible range of solutions for a given problem until a quasi-optimal solution is obtained. This way, the evolution paradigm may be translated into Computer Science to solve $\mathcal{NP}$ problems.

The evolutionary mechanism uses DNA strands [4] as its physical base. DNA strands store information. Evolution uses and combines them to develop better and more adapted individuals using biologically available operations over DNA strands.

This paper tries to show that Evolutionary Computation is feasible using DNA strands as the technological base. The way to do this could be to translate computer simulated evolutionary algorithms to DNA computing which seems to be the most appropriate medium to develop this kind of Computation.

Later sections will show the main stages that will permit such a translation process to be performed. Section 2 deals with the main algorithms used in Evolutionary Computation. The main features that evolutionary and DNA computations have in common are explained in section 2.1. Next, genetic algorithms are described in order to establish a first order relationship with DNA strands according to standard computer genetic operations [3], in section 3. These operations will be translated into DNA strand operations. Finally, an overall analysis of the ideas presented is offered.

2. EVOLUTIONARY COMPUTATION

Evolutionary Computation is based on individuals belonging to a population and their adaptation to a given environment. This adaptive process is produced by modifying the main characteristics of the members population's through successive generations, in which the best adapted members survive from one generation to the next. In other words, in a population some individuals will be worse adapted than others to the environment. These individuals will be more likely to be eliminated from the population. Meanwhile, other individuals in the population will change some characteristics and will improve the degree of their environment fitness. In Evolutionary Computation a problem is represented as a population of possible solutions, with a fitness degree for every one of them. Each solution is an individual of the problem population. The general process of a genetic algorithm consists of generating a random initial population of possible solutions to the problem. These initial individuals are probably bad solutions as they may have low fitness degree values and therefore they do not represent viable solutions to the problem. So it is necessary to let the population evolve through successive generations in order to obtain a set of better solutions. In this process, new individuals will appear with higher degrees of fitness. How can these new individuals appear? In each successive generation, individuals interchange internal information related to their characteristics in order to achieve better fitness and to improve the chances of survival in future generations. Individuals with worse fitness values might be eliminated from the population in future generations. The general process will continue creating new population generations until an adequate degree of fitness to the problem is reached. According to the previous description, we can define evolutionary algorithms as the sequence of steps shown in figure 37.1.

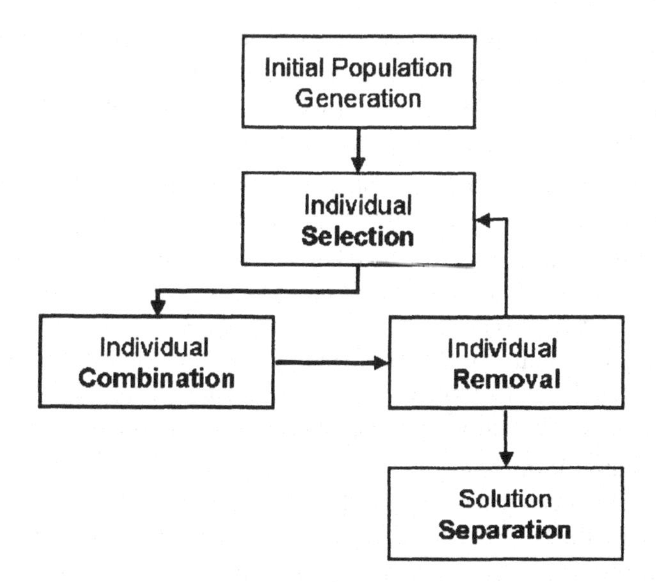

Figure 37.1 Steps involved in any Evolutive System.

Steps performed by Evolutionary Algorithms are very simple, therefore they could be used to solve any problem but in fact, they are not easy to apply. Evolutionary Algorithms are as difficult to apply as it is difficult to obtain the degree of fitness for individuals in the population. For some problems, it is quite difficult to obtain a fitness function which represents the appropriateness of a possible solution to the problem. For some problems it is quite easy, such as the maximization of a mathematical function, but this is not a real situation in many cases.

2.1 EVOLUTIONARY COMPUTATION AND DNA COMPUTING

So far, DNA Computing [6] has not been used together with the Evolutionary Paradigm, as this would require the amount of DNA strands to be drastically reduced to just using a few of them as an initial population of the Evolutionary Paradigm. It seems to be interesting and productive to use the Evolutionary Paradigm in DNA Computing since Biological Evolution has its origins in processing DNA strands.

When computing any evolutionary algorithm, some problems may arise, mainly due to the sequentiality of actual computers. The most important disadvantages are:

- Sequential Processing: individuals and their genotypes are treated sequentially, since actual digital computers cannot treat data in parallel on a massive scale.
- Fitness Evaluation: as a consequence of the previous disadvantage, the degree of fitness of individuals is sequentially evaluated. Another fact, related to fitness evaluation, consists of obtaining a good fitness function which gives an accurate degree of fitness for a given problem.
- Individual Selection: individuals are selected sequentially and they are chosen according to the individual's degree of fitness.

Previous disadvantages must be taken into account when trying to solve a problem with an evolutionary algorithm. Due the inner massive parallelism [9] that all DNA processes have, it would be a good idea to try to implement Evolutionary Computing techniques using DNA strands, that is, to try to solve problems using DNA Computation. The following ideas suggest solutions to the disadvantages Evolutionary Computation in this new approach.

- Sequential Processing: DNA Computing is well known for the massive parallelism that can be achieved using DNA strands. Using DNA strands, all individuals in the population could be treated as only one individual since one single operation could be applied to the whole population taking the same time as would be applied to one single individual.

- Fitness Evaluation: as a consquence of the previous fact, degree of fitness could be achieved in parallel. Another point is that an inner correction mechanism seems to exists, which evaluates a kind of degree of fitness over individuals with DNA strands, the worst individuals eliminating from the species.
- Individual Selection: individuals can be selected in parallel in order to recombine their information or to remove them. Moreover, natural selection is based on this idea and it is biologically implemented in DNA strands.

These are a few reasons why DNA strands seem to be well-suited as a technical base for Evolutionary Computing.

3. GENETIC ALGORITHMS

Genetic Algorithms are a kind of Evolutionary Algorithm. In Genetic Algorithms, individuals are treated sequentially in the population and they have proved to be a good Evolutionary Algorithm for simulation using DNA strands. The main process in a genetic algorithm could be specified using the following pseudo-code.

```
PROCEDURE    GeneticAlgorith;
BEGIN
     Generate initial population
     Evaluate fitness for each individual
     WHILE NOT stop DO
           Select population's individuals
           Cross selected individuals
           Mutate crossed individuals
           Evaluate fitness of generated individuals
           Introduce individuals on the population
           IF Population has converged THEN
                         stop:=TRUE;
           END;/*IF*/
     END;/*WHILE*/
END;/*PROCEDURE*/
```

It can be seen that the main steps of an Evolutionary Algorithm are performed in a previous code, so Genetic Algorithms fit into the Evolutionary paradigm.

The simulation of a Genetic Algorithm using DNA strands is set out in [7]. This simulation shows that Evolutionary Computation is feasible using DNA strands. Throughout this paper, the means of handling of the individuals in the population is inherited from Computer Science Genetic Algorithms and this makes the simulation impractical as a way of solving problems. Therefore, it seems more feasible to use raw Evolutionary Algorithms with DNA strands.

4. MAIN GENETIC OPERATIONS USING DNA STRANDS

Throughout this section, some genetic operation will be translated into DNA strand management. This approach will overcome some of previously stated disadvantages of Evolutionary Computation, though the realization of subsequent operations must be confirmed by biological experiments.

4.1 INDIVIDUAL ENCODING

The choice of codification is a key point for the correct evolution of the population towards the final solution. Individuals need to be coded so that it is possible to make combinations, duplications, copies, quick fitness evaluations and selections of specific individuals inside the population or mating pool without the need of a complete sequencing.

- Lipton encoding [5] is used to encode each individual as a sequence of ones and zeros, where the $ENC(b,i)$ function returns the codification of b (0 or 1) at the ith bit place. The codification returned from the ENC function is unique for each value of b and i. From now on, $ENC(b,i)$ will be represented by b_i.
- Between the DNA code bits at places i and $i+1$, a cutting or cleavage site for a restriction enzyme will be inserted.
- To allow a quick fitness evaluation, a field with the adaptation grade of the individual is included in the DNA strand. The fitness may be coded so that its length is proportional to the value it represents.
- A field indicating the numbering of that individual in the population will be included to locate it, and it will be inserted on both sides of the DNA strand with the peculiarity that on one side (sequence N_p') it is symmetrically complementary to a palindrome structure (sequence N_p) that will allow a later separation operation.

  Also one more cleavage site RE_p for a restriction enzyme will be included to separate its numbering.

Taking everything into account the final encoding of an individual has the following format (figure 37.2).

It must be considered that the individuals of the population may go to the mating pool from where they will be selected for later genetic operations, so some identification is needed in the usual codification to select a specific individual inside the mating pool. That identificator will be the sequence N_m. This

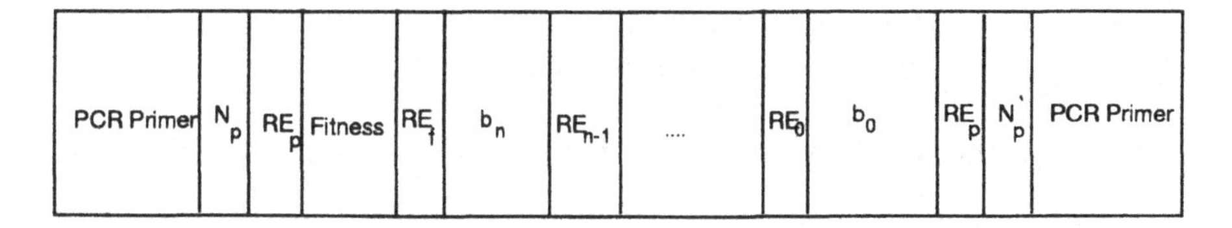

Figure 37.2 Final individual encoding.

value will be introduced in the fitness value field so that mating pool individuals have the same format as population ones with N_m values instead of the fitness field.

4.2 GENERATION OF THE INITIAL POPULATION

For a generic population with m individuals, in which each individual represents his genotype with n bits, $n - 1$ steps are needed to generate all the initial population. Also, in each step we would need to sequence $2m$ molecules. The steps of the initial population generation algorithm are explained in the following paragraph.

Half of the population is synthesized with a value 0 in the first bit place and with N having values for numbering half of the population. These sequences will have the following format: RE_0–0_0–RE_p–N'_p–PCR Primer, and the other half with a 1 in the first bit place with N having values for numbering the rest of the population. These sequences have the format: RE_0–1_0–RE_p–N'_p–PCR Primer.

The sequences RE_1–1_1–RE_0 and RE_1–0_1–RE_0 are also created. The previous chains are put together in a temporal test tube. The restriction enzyme corresponding to the site RE_0 is applied. The strands are cut and DNA ligase is applied to rejoin the fragments. Supposing that there are no mistakes in the operations of joining m molecules, the following sequences are obtained: RE_1–b_1–RE_0–b_0–RE_p–N'_p–PCR Primer with different values for b and N.

Until we arrive at the step $n-1$, in a generic step i, half of the individuals will be created with encoding format RE_i–0_i–RE_{i-1} and the other half with value 1: RE_i–1_i–RE_{i-1}. They will be joined in a temporal test tube with strands obtained in step $i - 1$. The restriction enzyme corresponds to site RE_{i-1} and DNA ligase will be applied to obtain the resulting strands for step i. Strands will have the following format: RE_i–b_i–RE_{i-1}–...–RE_0–b_0–RE_p–N'_p–PCR Primer.

When $i = n - 1$, RE_f should be used instead of RE_i.

Once we get the individuals with the last format the fitness function must be evaluated in order to append its value [2] and addthe number N_p. This number is encoded with a symmetrical sequence of N'_p. We need $\log_2 m$ steps of extraction and append to write the $\log_2 m$ bits of N_p.

4.3 INDIVIDUAL SELECTION FOR MATING POOL

Length separation is used to select individuals to fill the mating pool assuming that better adapted individuals are longer than worse adapted individuals. There are different methods for creating the mating pool.

4.3.1 Method of the best ones. The purpose is to take the best n individuals of the population. Length separation will be used so that the n best individuals are selected. For viewing individuals with their respective lengths we add a radioactive marker corresponding to each individual number, so that applying X-rays to the electropheresis gel all individuals sorted by fitness will be seen and the n longer will be taken.

4.3.2 Roulette method. This method will be carried out using a scale function. This function obtains the strands' fitness contribution—their length—to global fitness. Starting from scale function we obtain the inverse function that will return the individual identification through the degree of fitness contribution to global fitness in that particular population. Then, numbers will be generated randomly between one and hundred; and using the inverse function we get the individual identifications for these numbers. These individuals will be taken off the population separating the chains starting from the sequence corresponding to the individual number using magnetic beads.

4.3.3 Crowding method. This method consist of taking individuals randomly from the population; then, those individuals are introduced into a temporary test tube and the best of them will be taken to be introduced into the mating pool. The last step will be repeated until the mating pool is full

In order to take individuals randomly, a random number must be generated to obtain the individual from the population as described above. In order to select individuals randomly, instead of generating a random number, a small quantity of chains from the population tube could be taken and the best element is chosen separating it by length.

Once the individuals selected using one of the previously explained methods are introduced into the mating pool test tube, it is necessary to change the chain format to introduce individual numbers for the mating pool. This mating pool numbering will replace the fitness field in the population format and it will be introduced by applying site directed mutagenesis [1] over the mating pool tube individuals.

4.4 INDIVIDUAL CROSSING

For a crossing operation the mating pool individuals will be taken by pairs or the crowding method will be used for selecting an individual from the population who will be crossed with one from the mating pool. In order to obtain a pair of individuals from the mating pool, we use strand separation according to the subchain of the two mating pool numbers selected fro crossing. When the crowding method is used, two individuals with different format will be crossed, due to the different chain format between population and mating pool individuals. To solve this problem, it is enough to assign the individual of the population a mating pool individual number 0 never used by a mating pool individual.

If a one point crossing is to be made, a crossing point will be chosen randomly. It will be equivalent to choosing a restriction enzyme randomly among $RE_0 \ldots RE_{n-1}$. Then, the chosen restriction enzyme is applied and the chains are cut into two parts. Original individuals may be generated again; to avoid this, chains having palindrome structure will be separated from the rest of the resulting chains, so that chains with a palindrome structure are the chains for which the crossing has not been made, and these chains will have to be removed. The several points crossing can be made in several one point crossing steps.

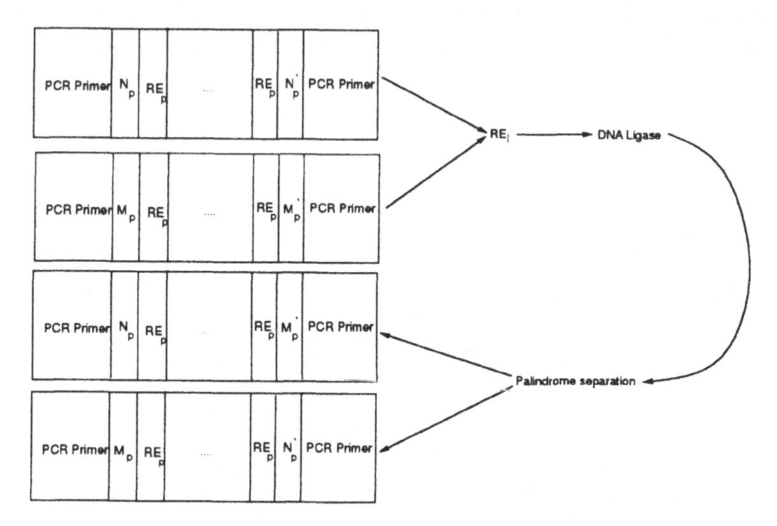

Figure 37.3 Individual crossing.

4.5 MUTATION

Once the individuals have been crossed, a mutation operation should be applied. This mutation can be achieved employing site directed mutagenesis

[8]. Previous to the substitution operation it is necessary to apply the probability of the mutation for each bit must be applied. To do so if the generic bit i is to be mutated the following pieces will be generated in two different temporary tubes: $RE_i - 0_i - RE_{i-1}$ and $RE_i - 1_i - RE_{i-1}$. One of the last short strands will be taken randomly, so that depending on the value of the bit i in the individual strand and on the value of the short strand chosen the mutation will be made or not. This mutation will be repeated for each bit.

4.6 INTRODUCING NEW INDIVIDUALS IN THE POPULATION

Recombined and mutated strands are evaluated and their fitness values are inserted. Once the final chains to be introduced into the new population have been obtained, the individuals from the last generation must be extracted. When those individuals have been selected, one by one they are taken from the mating pool by their identification number, the operation being carried out each individual as follows: the individual to be extracted and the individual from the mating pool are taken and introduced into separate temporary tubes and the restriction enzyme corresponding to RE_p is applied. Strands are separated by length to get the longer one from the tube containing the mating pool individual and the shorter one from the tube containing the individual to be extracted. They are introduced into a temporary tube and DNA ligase is applied (figure 37.4).

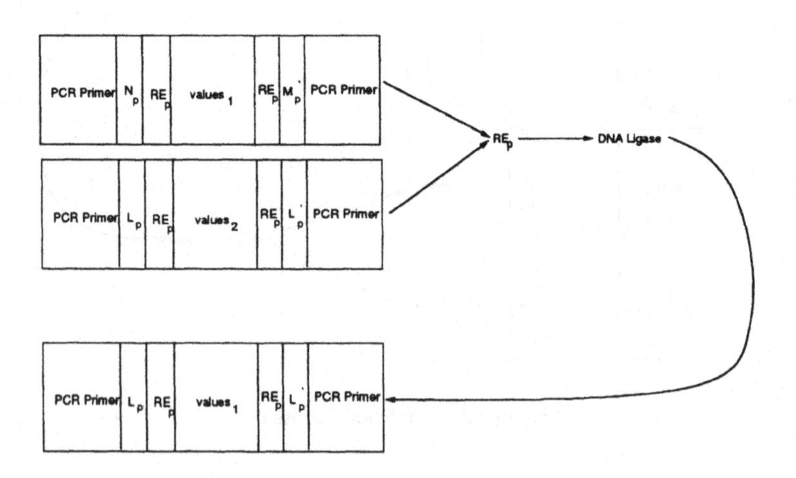

Figure 37.4 Individual replacement.

5. CONCLUSIONS

Evolutionary Computation seems to be one of the best Computation Paradigms for applying with DNA strands. Although DNA strands seem to offer the best technical base for Evolutionary Computing, there are a few problems and restrictions that have not been solved yet.

Getting a fitness function that will correctly evaluate individuals in the population is still difficult, as is the expensive handling of individuals in the population. It is better not to handle individuals in the population and let them evolve by their own.

A lot of work is being done to avoid these problems and make DNA Evolutionary Computing a real and practical way for solving problems. There is a very promising future in DNA Evolutionary Computing. Maybe in the future, DNA Evolutionary Computing can cover *in vivo* DNA so that individuals in the population could be living individuals dedicated to a specific task depending on their DNA encoding. They could be individuals who will perform extractions of decimals of a specified division and other individuals who only obtain the maximun value for a specified function. These individuals would require feeding and will die as normal living beings.

Maybe the use of DNA strands is not mandatory. With actual improvements in biomolecular technology, it may be possible in the future to develop a new molecule that will fit requirements in Evolutionary Computation and present massive parallelism.

References

[1] Beaver, D. (1996), Universality and Complexity of Molecular Computation, in *Proceedings of the 28th ACM Annual Symposium on the Theory of Computing (STOC)*.

[2] Boneh, D.; C. Dunworth & J. Sgall (1996), On the Computational Power of DNA, *Discrete Applied Mathematics*, 71.1-3: 79–94.

[3] Goldberg, D.E. (1989), *Genetic Algorithms in Search, Optimization and Machine Learning*, Addison-Wesley, Reading, Mass.

[4] Head, T. (1987), Formal language theory and DNA: an analysis of the generative capacity of specific recombinant behaviors, *Bulletin of Mathematical Biology*, 49: 737–759.

[5] Lipton, R.J (1995), DNA solution of hard combinatorial problems, *Science*, 268: 542–545.

[6] Păun, Gh. (1996), Five (plus two) universal DNA computing models based on the splicing operation, in *Proceedings of the Second Annual DNA Computing Workshop*, Princeton.

[7] Rodrigo, J.; A. Rodríguez-Patón; J. Castellanos & S. Leiva (1998), Molecular Computation for Genetic Algorithms, in *Rough Sets and Current Trends in Computing*, Warsaw.

[8] Sambrook, J.; E.F. Fritsch & T. Maniatis (1989), *Molecular Cloning: A Laboratory Manual*, 2nd ed. Cold Spring Harbor, New York.

[9] Yokomori, T.; S. Kobayashi & C. Ferretti (1995), On the power of circular splicing systems and DNA computability, Report CSIM 95-01, University of Electro-Communications, Chofu, Tokyo.

Chapter 38

SPLICING SYSTEMS USING MERGE AND SEPARATE OPERATIONS

Claudio Zandron
Department of Informatics, Systems Science and Communication
University of Milano-Bicocca
Via Bicocca degli Arcimboldi 8, 20126 Milano, Italy
zandron@dotto.usr.dsi.unimi.it

Giancarlo Mauri
Department of Informatics, Systems Science and Communication
University of Milano-Bicocca
Via Bicocca degli Arcimboldi 8, 20126 Milano, Italy
mauri@disco.unimib.it

Claudio Ferretti
Department of Informatics, Systems Science and Communication
University of Milano-Bicocca
Via Bicocca degli Arcimboldi 8, 20126 Milano, Italy
ferretti@disco.unimib.it

Paola Bonizzoni
Department of Informatics, Systems Science and Communication
University of Milano-Bicocca
Via Bicocca degli Arcimboldi 8, 20126 Milano, Italy

* Work supported by the Italian MURST Project "Innovative Computational Models".

C. Martin-Vide and V. Mitrana (eds.), Where Mathematics, Computer Science, Linguistics and Biology Meet, 435-446.

Abstract We study in detail how to control molecules in a set of splicing tubes, in such a way that we can generate any recursively enumerable language by allowing them to split and join (splice), but also by separating the molecules in a tube, according to the presence of a pattern in them, and by merging tubes when needed. We show that this new model of a distributed splicing system is computationally universal.

1. INTRODUCTION

In 1987 ([3]) Tom Head suggested studying interactions between DNA molecules in a formal way, considering the well-known biochemical reaction of splicing: molecules are first cut and later joined in a crossed way, according to the presence of specific patterns along the molecules themselves.

Recently, with the long term goal of producing computing devices based on biological molecules, a long series of studies considered this model again. Other models are also being considered, and a good early instance of the results which can be obtained practically by implementing these ideas can be found in [1]. More specifically, real splicing systems have been studied in [5].

The original theoretical model of splicing systems had a smaller computational power than that of a finite state automaton. This has motivated several new models based on these original ideas, aiming at obtaining more powerful computing molecular devices ([4]).

One direction has been that of working with many test tubes together, instead of using only one. In each tube molecules are free to perform splicing, but some specified molecules can be moved from a tube to another. Several scientific papers already studied this model, looking for computationally universal distributed splicing systems of this type, using fewer and fewer test tubes. Recent results are, for instance, in [4] and [6].

In this paper we describe a different distributed model based on splicing, where molecules can move according to these two ideas:

- the molecules of a test tube can be separated into two new test tubes, according to the presence or not of some pattern in the molecules themselves;
- two test tubes can be merged into one single new test tube.

We show, in a constructive way, that even this model is computationally universal. We observe that similar ideas were described even in [1].

2. NOTATION

Given a finite alphabet V, we denote with V^* the set of all (finite) strings over V, while $V^+ = V^* - \epsilon$ is the set of all strings in V^* but the empty string, ϵ.

The family of recursively enumerable languages and of finite languages are denoted by RE and FIN, respectively.

A *Head splicing system* (or H system) is a triple $H = (V, A, R)$, where V is the alphabet of H, $A \subseteq V^*$ is the set of *axioms*, and R is the set of *splicing rules*, with $R \subseteq V^*\#V^*\$V^*\#V^*$.

For $x, y, z, w \in V^*$ and $r = u_1\#u_2\$u_3\#u_4$ in R, we define

$$(x, y) \vdash_r (z, w) \quad \text{if and only if} \quad x = x_1u_1u_2x_2, y = y_1u_3u_4y_2, \text{ and}$$
$$z = x_1u_1u_4y_2, w = y_1u_3u_2x_2,$$
$$\text{for some } x_1, x_2, y_1, y_2 \in V^*.$$

For an H system $H = (V, A, R)$ and a language $L \subseteq V^*$, we write

$$\sigma(L) = \{z \in V^* | (x, y) \vdash_r (z, w) \text{ or } (x, y) \vdash_r (w, z),$$
$$\text{for some } x, y \in L, r \in R\},$$
$$\sigma^*(L) = \bigcup_{i \geqslant 0} \sigma^i(L)$$
$$\text{where } \sigma^0(L) = L,\ \sigma^{i+1}(L) = \sigma^i(L) \cup \sigma(\sigma^i(L)) \text{ for } i \geqslant 0$$

An H system is meant to operate starting from the set of strings A, and then generate new strings iterating the splicing step $\vdash_r$ on them and on the strings generated during this process. The language generated in this way is $\sigma^*(A)$.

In a Merge and Separate System we have a set of test tubes, each one containing a set of DNA strands and a set of restriction enzymes R. The restriction enzymes are the same for each tube. This system can do these operations:
- Create new strands using the restriction enzymes and the strands initially present in the tube.
- Merge two or more tubes, creating a new test tube with all the strands of the initial tubes.
- Separate a test tube to create two distinct test tubes, with different strands.

Formally, a *Merge and Separate System* (MS for short) is a construct $M = (V, T, R, A_1, \ldots, A_f)$ where V is an alphabet, T is the terminal alphabet, R is a set of splicing rules and each $A_i \subseteq V^*(1 \leqslant i \leqslant f)$ is a set of axioms. f is the number of tubes initially present in the system.

In each tube we create all the strings we are able to produce by applying the set of splicing rules R (as usually done in H systems).

Moreover, we can do two other operations:

MERGE: $M(T_i, T_j) = T_i \cup T_j$. Starting from two tubes, we produce a single tube containing all the strings of the two starting tubes. This operation can easily be extended to a finite number of tubes: by $M(T_1, T_2, ..., T_n)$ we indicate the Merge operation executed on n tubes.

SEPARATE: $+(T_i, s)$ and $-(T_i, s), s \in V^+$. Starting from a single tube, we produce two tubes. In the first one, $+(T_i, s)$, we put all the strings that contain the substring s and in the second one, -(T_i, s), we put all the strings not containing s as a substring.

We want to underline the following facts:
• The splicing rules R are the same in each tube, even for the tubes obtained after a Merge or after a Separate operation. Different tubes generate different strings only due to the strings present in those tubes.
• A separate operation can stop a reaction in a tube, by separating the strings involved in the application of a rule. When a reaction starts (due to the strings initially present in a tube or as a result of a Merge operation) we have to wait for a fixed amount of time, say $\hat{t}$, before we execute a Separate operation, to let new strings be created. We have to bear in mind that, even if we wait for that amount of time, some string might not be created, but the whole process is iterated many times and all the possible strings will eventually be created.

As computation proceeds, the number of test tubes will vary, depending on the operations we decide to undertake: if we do a Separate operation we will increase the number of test tubes by one, if we Merge two tubes we decrease the number of test tubes by one.

The language generated by a MS system is the set of all the strings from T^* present in any test tube of the system. The class of languages MS is the set of all the languages which can be generated by a MS system.

3. MAIN RESULT

We show now that such a model using only a finite number of rules and a finite number of axioms is able to generate the class of RE languages.

Theorem 38.1 $MS = RE$.

Proof. From the Turing-Church thesis we have $MS \subseteq RE$.

We have to show that $RE \subseteq MS$. Let us take a type 0 grammar $G = (T, N, S, P)$, where $T = \{t_1, ..., t_n\}$ is the set of terminal symbols, $N = \{n_1, ..., n_m\}$ is the set of nonterminal symbols, S is the starting symbol and P the set of productions.

We injectively map in $U = \{U_1, ..., U_k\}$ the set of symbols $T \cup N \cup \{B\}$ (B is a special symbol not in G). We have: $k = n + m + 1$.

The MS system is built as follows. The alphabet of the system contains terminal and nonterminal symbols of grammar G and two other sets of special symbols, not in G.

$$V = T \cup N \cup \{X, Y, B, Z_1, Z_2, Z_3, Z_H, Z_T, X_H, Y_T\} \cup \{\alpha_i | 1 \leqslant i \leqslant k\}.$$

The symbols in the set $\{\alpha_i | 1 \leqslant i \leqslant k\}$ are used to rotate the characters of the strings as explained below, while the set $\{X, Y, B, Z_1, Z_2, Z_3, Z_H, Z_T, X_H,$

$Y_T\}$ contains symbols used as brackets for strings and to recognize "work" strings, and the symbol B which marks the starting of the rotated string. Obviously, the terminal alphabet is the same in both systems.

R is the following union of different *types* of splicing rules:

1 (simulate productions of G):$\{\#uY\$Z_1\#vY|u \rightarrow v \in P\}\cup$
2 (prepare the string to rotate last symbol):$\{\#U_iY\$Z_1\#\alpha_iY_T|1 \leqslant i \leqslant k\}\cup$
3 (put a rotation symbol to the left):$\{X\#\$X_H\alpha_i\#Z_2|1 \leqslant i \leqslant k\}\cup$
4 (decode the rotation symbol) :$\{X_H\alpha_i\#\$XU_i\#Z_3|1 \leqslant i \leqslant k\}\cup$
5 (delete the rotation symbol from the right end):$\{\#\alpha_iY_T\$Z_3\#Y|1 \leqslant i \leqslant k\}\cup$
6 (delete the bracket from the left end of the string) :$\{XB\#\$\#Z_H\}\cup$
7 (delete the bracket from the right end of the string):$\{\#Y\$Z_T\#\}$

We start the computation with nine tubes; during the computations the number of tubes will vary due to the Merge and Separate Operations, as said before. During the computation we use these nine tubes as a sort of "main structure", as we will see in the following. We label the nine starting tubes $P_1, P_2, P_3, P_4, S_1, S_2, S_3, S_4, T_T$. P is for Primary, because these tubes will contain the strings from which we get the terminal ones, while S is for Secondary, because these tubes will contain the strings used to start the reaction and the "work" strings (i.e. string created during the computation which are, however, useless). T_T is the tube in which we collect the terminal strings.

The sets of starting axioms are:

$A_1 = \{XBSY\}$, $A_2 = \emptyset$, $A_3 = \emptyset$, $A_4 = \emptyset$
$A_5 = \{Z_1vY|u \rightarrow v \in P\} \cup \{Z_1\alpha_iY_T|1 \leqslant i \leqslant k\}$
$A_6 = \{X_H\alpha_iZ_2|1 \leqslant i \leqslant k\}$
$A_7 = \{XU_iZ_3|1 \leqslant i \leqslant k\} \cup \{Z_3Y\}$
$A_8 = \{Z_H, Z_T\}$
$A_9 = \emptyset$

We place the axioms in the tubes as follows: $A_1, ..., A_4$ in $P_1, \ldots, P_4$ respectively, and $A_5, \ldots, A_8$ in $S_1, \ldots, S_4$ respectively, and A_9 in T_T.

The computation proceeds as follows (Fig. 1). First we execute *in parallel* four Merge operations: every tube P_i is merged with the corresponding tube S_i $(1 \leqslant i \leqslant 4)$.

Then the reaction starts, so we wait for an amount of time $\hat{t}$ to let new strings be created.

Then we execute a series of Separate operations, following a fixed scheme, using the tubes obtained in this way to extract the strings we are interested in.

Finally, the tubes are Merged (and/or relabelled) to rebuild the "main structure": we get eight tubes (four labelled P_i and four S_i) and the terminal strings are collected in T_T.

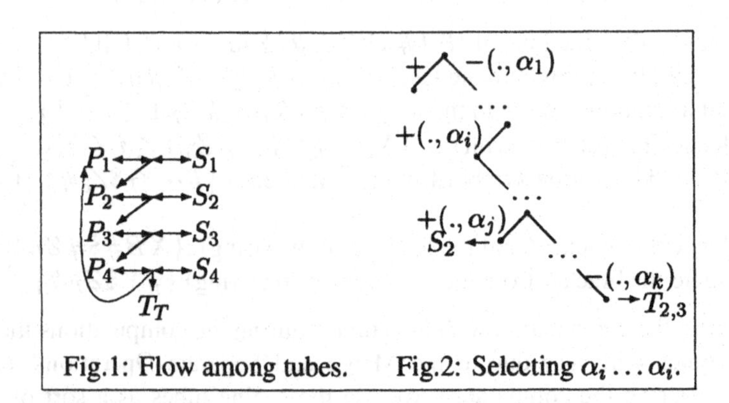

Fig.1: Flow among tubes. Fig.2: Selecting $\alpha_i \ldots \alpha_i$.

These four steps are iterated. A single sequence of these four steps will be denoted, from now on, as a Macro-step.

We start the first Macro-step by merging every tube P_i with the corresponding tube S_i. Let's denote by T_i the obtained tubes. We wait for time $\hat{t}$ to allow new strings to generate. The only tube able to create something new is T_1 ($T_2 = S_2$, $T_3 = S_3$, $T_4 = S_4$).

In T_1 we have the strings of the form Z_1vY and $Z_1\alpha_iY_T$ and the string $XBSY$. We can apply splicing rules of types 1 and 2 (we consider the general case of a string of the form XmY; for $XBSY$ we have $m = BS$)

i) $Xm_1uY, Z_1vY \vdash_1 Xm_1vY, Z_1uY$, where $u \to v \in P$ and $m_1u = m$, to simulate a production of grammar G on the right end of the string.

ii) $XwU_iY, Z_1\alpha_iY_T \vdash_2 Xw\alpha_iY_T, Z_1U_iY$, where $m = wU_i$, to prepare the string to rotate the last character (U_i, not Y which is used only as a bracket).

Moreover, we can apply rules like:

iii) $Z_1v_1U_iY, Z_1\alpha_iY_T \vdash_2 Z_1v_1\alpha_iY_T, Z_1U_iY$ (where $v = v_1U_i$)

iiii) $Z_1v_2uY, Z_1vY \vdash_1 Z_1v_2vY, Z_1uY$, (where $v = v_2u$)

After these operations, in tube T_1 we get strings of the form:

XmY (where $m \in U^+$),
Z_1vY (v is the right term of a production in G),
$Z_1\alpha_iY_T (1 \leqslant i \leqslant k)$, Z_1uY (u is the left term of a production in G),
Z_1v_2vY ($v_2 \in U^+$ and v is the right term of a production in G)
$Xw\alpha_iY_T, Z_1U_iY, Z_1m_1\alpha_iY_T$ (where $m_1 \in U^+$).

The strings of the form XmY, Z_1vY (v is the right term of a production in G) and $Z_1\alpha_iY_T$ can enter new splicings of type 1 and 2, creating new strings in which we simulate the production of G on the right end of the string and in which we prepare the right end character to rotate. After a series of these splicing operations, we can create nothing new.

The strings of the form Z_1uY can enter splicings of type 1 and 2. We have the following possibilities:

- $Z_1uY, Z_1vY \vdash_1 Z_1uY, Z_1vY$, hence creating nothing new
- $Z_1u_1U_iY, Z_1\alpha_iY_T \vdash_2 Z_1u_1\alpha_iY_T, Z_1U_iY$, where $u = u_1U_i$. The strings Z_1U_iY are already in T_1. The strings $Z_1u_1\alpha_iY_T$ cannot enter new splicing.

The strings of the form Z_1v_2vY can enter splicing of type 1 and 2. In both cases, we get strings of a form already present in T_1 The strings of the form $Xw\alpha_iY_T$ cannot enter new splicing in T_1.

The strings Z_1U_iY can enter splicing of type 1 and 2 but creating nothing new.

The strings of the form $Z_1m_1\alpha_iY_T$ ($m_1 \in U^*$) cannot enter new splicing in T_1.

Thus, after applying rules of type 1 and 2 a number of times, we are not able to create new strings.

After time $\hat{t}$, we start the phase of Separate operations.

We separate T_2, T_3 and T_4 to get the same tubes P_2, P_3, P_4, S_2, S_3 and S_4, which we initially started with (this can easily be done, because P_2, P_3 and P_4 were empty).

On T_1, we execute two Separate operations to select the strings ready to rotate the character in the right end (i.e. the strings of the form $Xw\alpha_iY_T$) by looking for the symbols Y_T and then X. We Merge these strings in P_2.

Moreover, to get a structure like the initial one, with nine tubes, we merge strings without Y_T, but with X in P_1: it contains the strings we can get from the strings of the form XmY and applying the rules of type 1 (that simulates the productions of grammar G on the right end of the strings). All the other strings go back to S_1. This contains the original axioms and other strings created during the computation (these strings are of no use; we have to make sure that they do not create terminal strings not in G).

Thus, after the first Macro-step, we get:

- P_1 containing the strings generated by applying the rules that simulate the rules in G to the right end of the strings.
- P_2 containing the strings of the form $Xw\alpha_iY_T$, ready to rotate the last character.
- P_3 and P_4 are empty.

• S_1 containing the strings of the forms $Z_1vY, Z_1\alpha_iY_T, Z_1uY, Z_1U_iY, Z_1m_1\alpha_iY_T, Z_1v_2vY$, i.e. the original axioms and the "work" strings.
• S_2, S_3 and S_4 containing the strings they contained when the computation started.

In other words, we simulate the rules of G on the right-end of the strings and we place the obtained strings in P_1. Moreover, we prepare some strings to rotate their last character. These strings are placed in P_2.

When all P_i and S_i have been rebuilt, we can start a new Macro-step. We Merge every tube P_i with the corresponding tube S_i, and we get T_i.
• In T_1 we get the same strings we get at the previous Macro-step (P_1 and S_1 have been rebuilt using strings in T_1). If in the previous step some strings have not been created (because the separate operations blocked the reaction), they could be created in this step.
• $T_3 = S_3$ and $T_4 = S_4$, as in the previous step.
• In T_2 we get the strings:

$$\{Xw\alpha_iY_T|1 \leqslant i \leqslant k\} \cup \{X_H\alpha_iZ_2|1 \leqslant i \leqslant k\}$$

We can apply only rules of type 3. We get:

$$Xw\alpha_iY_T, X_H\alpha_jZ_2 \vdash_3 X_H\alpha_jw\alpha_iY_T, XZ_2$$

After a series of such operations, no new string appears.
From T_2 and with Merge and Separate operations we rebuild tubes as we put: the strings with Z_2 in S_2, the strings without Z_2 or X_H in P_2, and the other strings in a new tube $T_{2,2}$.

The strings in $T_{2,2}$ are of the form $X_H\alpha_jw\alpha_iY_T(1 \leqslant i \leqslant k, 1 \leqslant j \leqslant k)$: the rotation characters on both ends are not necessarily the same.
In this set of strings, we have to select the subset containing only strings in which the rotation character on the left side is the same as that on the right side. To perform this operation, we are going to execute a series of Separate operations. Every string has exactly two α_i's. We consider the tubes produced after k separate operations, one for each α_i. The tubes generated after some $k-1$ negative separations, and a positive separation for α_i, will have necessarily $\alpha_i \dots \alpha_i$, and we put them in tube $T_{2,3}$. As soon as a tube is generated after any two positive separations, we know that it will have $\alpha_i \dots \alpha_j, i \neq j$, and we put it in S_2. This is visualized in Fig. 2.
In $T_{2,3}$ we only have strings in which the special characters of rotation (α_i) are the same on the left end and on the right end, i.e. it only contains strings of the form $\{X_H\alpha_iw\alpha_iY_T|1 \leqslant i \leqslant k\}$.

In $T_{2,3}$ we can create nothing new by applying splicing rules. We put the strings of $T_{2,3}$ in P_3.

After the second Macro-step, we get:

• P_1 containing the same strings it contained when the Macro-step started.

• P_2 containing the strings of the form $Xw\alpha_i Y_T$, ready to rotate the last character. These kinds of strings were already in P_2 when the Macro-step started.

• P_3 containing the strings of the form $\{X_H\alpha_i w\alpha_i Y_T|1 \leqslant i \leqslant k\}$, in which the special rotation symbol is the same on both right and left ends.

• P_4 is empty.

• S_1 containing the strings of the forms $Z_1vY, Z_1\alpha_i Y_T, Z_1uY, Z_1U_iY, Z_1m_1\alpha_i Y_T$, i.e. the original axioms and the "work" strings.

• S_2 containing the string of the form $X_H\alpha_i Z_2, XZ_2$ and the strings of the form $X_h\alpha_j w\alpha_i Y_T (1 \leqslant i,j \leqslant k, i \neq j)$, i.e. the original axioms and the "work" strings.

• S_3 and S_4 containing the strings they contained when the computation started.

In other words, we rotate the special symbol of the strings in P_2, by selecting the strings in which the special symbol is the same on both right and left end of the strings. We place these strings in P_3.

When all P_i and S_i have been rebuilt, we can start the third Macro-step. We Merge every tube P_i with the corresponding tube S_i, and we get T_i.

In T_1 and T_2 we get the same strings as the previous step. $T_4 = S_4$; after time $\hat{t}$, we rebuild P_1, S_1, P_2, S_2, P_4 and S_4 as in the previous steps. In T_3 we get the strings $\{X_H\alpha_i w\alpha_i Y_T|1 \leqslant i \leqslant k\} \cup \{XU_iZ_3|1 \leqslant i \leqslant k\} \cup \{Z_3Y\}$

By applying splicing rules of type 4 and 5, which are the only two types of rules we can apply to these strings, we get:

$$X_H\alpha_i w\alpha_i Y_T, XU_iZ_3 \vdash_4 XU_iw\alpha_i Y_T, X_H\alpha_i Z_3$$
$$XU_iw\alpha_i Y_T, Z_3Y \vdash_5 XU_iwY, Z_3\alpha_i Y_T$$

If we apply the rules of type 5 before the rules of type 4 we can create strings of the form $X_H\alpha_i wY$ too.

After a series of operations of these types, we get in T_3 the strings:

$\{XU_iw\alpha_i Y_T|1 \leqslant i \leqslant k\} \cup \{X_H\alpha_i wY|1 \leqslant i \leqslant k\} \cup \{XU_iwY|1 \leqslant i \leqslant k\} \cup \{Z_3\alpha_i Y_T|1 \leqslant i \leqslant k\} \cup \{X_H\alpha_i Z_3\}$,

in addition to the strings that were already present:

$\{X_H\alpha_i w\alpha_i Y_T|1 \leqslant i \leqslant k\} \cup \{XU_iZ_3|1 \leqslant i \leqslant k\} \cup \{Z_3Y\}$

After time $\hat{t}$, we start another series of Merge and Separate operations, easily building tubes P_3, P_4, S_3 such that after this third Macro-step we get:

- P_1 containing the same strings it contained when the Macro-step started.
- P_2 containing the same strings it contained when the Macro-step started.
- P_3 containing the strings of the form $\{X_H\alpha_i w\alpha_i Y_T | 1 \leqslant i \leqslant k\}$, the same strings it contained when the Macro-step started.
- P_4 containing the strings of the form $XU_i wY$; a character from the right end of the string has been rotated to the left end.
- S_1 containing the strings of the forms $Z_1 vY, Z_1\alpha_i Y_T, Z_1 uY, Z_1 U_i Y,$ $Z_1 m_1 \alpha_i Y_T$, i.e. the original axioms and the "work" strings.
- S_2 containing the strings of the form $X_H\alpha_i Z_2, XZ_2$ and the strings of the form $X_h\alpha_j w\alpha_i Y_T (1 \leqslant i, j \leqslant k, i \neq j)$, i.e. the original axioms and the "work" strings.
- S_3 containing the strings of the form $XU_i Z_3, Z_3 Y$ and the strings of the form $XU_i wY, X_H\alpha_i Z_3, Z_3\alpha_i Y_T, X_H\alpha_i wY$, i.e. the original axioms and the "work" strings.
- S_4 containing the strings it contained when the computation has started.

In other words, we delete the special rotation symbol from the right end of the strings and we decode the special symbol on the left end with the corresponding symbol in U. The strings obtained are placed in P_4.

When all P_i and S_i have been rebuilt, we can start the fourth Macro-step. We Merge every tube P_i with the corresponding tube S_i, and we get T_i. We let the tubes react for a time $\hat{t}$.

In T_1, T_2 and T_3 we get the same strings as the previous step. In T_4 we get the strings $\{XU_i wY | 1 \leqslant i \leqslant k\} \cup \{Z_H\} \cup \{Z_T\}$.

In T_4, we can apply the splicing rules of type 7. Moreover, if there are strings of the form $XBwY$ we can apply splicing rules of type 6. We get the following strings:

$$XBwY, Z_H \vdash_6 wY, XBZ_H$$
$$wY, Z_T \vdash_7 w, Z_T Y$$

If we apply rules of type 7 on strings of the form $XU_i wY$ we also get $XU_i w$.

After a series of operations of these types, we can create nothing new by applying splicing rules. In T_4, we have the strings:

$$\{XU_i wY | 1 \leqslant i \leqslant k\} \cup \{Z_H, Z_T\} \cup \{XBZ_H, Z_T Y\} \cup \{wY\} \cup$$
$$\cup \{XU_i w | 1 \leqslant i \leqslant k\} \cup \{w | w \in (T \cup N)^*\}$$

Now we rebuild S_4: with two Separate operations, one on Z_H and one on Z_T, and one merge we get its strings $Z_H, Z_T, XBZ_H, Z_T Y$.

From the other resulting strings we can separate and merge again the strings containing exactly one of the two symbols X and Y (XU_iw, wY) in S_4.

We are left with a tube with strings containing X and Y (XU_iwY), and one with strings not containing Y or X (w). If we execute a series of Separate operations on the first tube, each one based on a nonterminal symbol, we can extract all the strings which contain nonterminal symbols from this tube. The remaining strings are terminal ones, so we put them in T_T. All the other tubes just created can be merged with S_4, because the strings in those tubes are "work" strings. We get a new empty tube and we label it P_4 (we can obtain it with a Separate operation on one of the other tubes).

Now consider what happens if we put the strings from the other remaining tube into P_1: in P_1 we put new strings, which were not present before. In fact, the right end symbol of these strings has been rotated and it is now placed on the left end. Thus, in P_1 these strings can be prepared to rotate another character. The strings obtained are merged in P_2 where we start the rotating phase with the special symbols α_i and then we merge the strings in P_3. Here, we select the strings with the same special rotating symbol on both ends, we decode the special rotating symbols on the left end of the strings and we delete the one on the right end. We Merge the strings obtained in P_4, where we select the fully rotated terminal strings, and then we can send the strings with another rotated character back to P_1.

This means that this system is able to simulate the production of grammar G on the right end of the strings and to rotate the characters of the strings to keep a production, that is in the middle of the string, to the right end.

The symbol B indicates the correct place where a string starts; so we can recognize a fully rotated string, and we can control if it is terminal. In the last case, we collect it in T_T, which contains precisely the terminal strings generated by grammar G. □

4. OBSERVATIONS

The above universal MS system starts with exactly nine test tubes, and during the computation the number of tubes can vary. As has been suggested, using well-known splicing techniques we could start with a single test tube. The molecules of each of the families of starting axioms could be marked in four different ways, and put together in the single starting tube. Later, when each family has to be used, respective strands could be selected and then have the marker removed, so as to rebuild one of our starting test tubes.

The language of a system is eventually generated after many Macro-steps, but the process is autonomous: every step always repeats the same operations in parallel, eventually giving all the strings sufficient time and opportunity to be processed correctly.

Other observations could be made with respect to the largest number of test tubes needed at any time during a computation. In our simulation of a grammar having $O(k)$ symbols, a quick analysis shows that we need to perform $O(k^2)$ merge or separate operations, and we use $O(2k)$ test tubes at most. This can be done if we sort all the operations in each of the groups of operations described above in a smart way. But since these bounds depend on k, it would be interesting to look for a system operating with constant bounds with respect to the number of symbols.

References

[1] Adleman, L.M. (1996), On constructing a molecular computer, in R.J. Lipton & E.B. Baum, eds., *DNA Based Computers*, Proceedings of a DIMACS Workshop: 1–22. American Mathematical Society, Princeton.

[2] Head, T. (1987), Formal language theory and DNA: an analysis of the generative capacity of specific recombinant behaviors, *Bulletin of Mathematical Biology*, 49: 737–759.

[3] Laun, E. & K.J. Reddy (1997), Wet Splicing Systems, in *Proceedings of the 3rd DIMACS Workshop on DNA Based Computers*: 115–126.

[4] Păun, G. (1997), DNA computing: distributed splicing systems, in *Structures in Logic and Computer Science. A Selection of Essays in honor of A. Ehrenfeucht*, Springer, Berlin.

[5] Păun, Gh.; G. Rozenberg & A. Salomaa (1998), *DNA Computing: New Computing Paradigms*, Springer, Berlin.

[6] Priese, L.; Yu. Rogozhin & M. Margenstern (1988), Finite H-systems with 3 test tubes are not predictable, in *Pacific Symposium on Biocomputing*, Hawaii. World Scientific, London.